The 50th Anniversary of Gröbner Bases

ADVANCED STUDIES IN PURE MATHEMATICS 77

Chief Editor of the Series: Ken'ichi Ohshika (Osaka University)

The 50th Anniversary of Gröbner Bases

Edited by

Takayuki Hibi (Osaka University)

Mathematical Society of Japan

This book was typeset by $\mathcal{A}\mathcal{M}\mathcal{S}$-TEX and $\mathcal{A}\mathcal{M}\mathcal{S}$-LATEX, the TEX macro systems of the American Mathematical Society, together with the style files aspm.sty *and* aspmfm.sty *for $\mathcal{A}\mathcal{M}\mathcal{S}$-TEX written by Dr. Chiaki Tsukamoto and* aspmproc.sty *for $\mathcal{A}\mathcal{M}\mathcal{S}$-LATEX written by Dr. Akihiro Munemasa and* aspm.cls *for $\mathcal{A}\mathcal{M}\mathcal{S}$-LATEX provided by Livretech Co., Ltd.*

TEX is a trademark of the American Mathematical Society.

Edited by the Mathematical Society of Japan.

Published by the Mathematical Society of Japan.

Distributed by the Mathematical Society of Japan, the American Mathematical Society and the World Scientific Publishing Co., Ltd.
Distributed exclusively in North America by the American Mathematical Society.
Partially supported by Grant-in-Aid for Publication of Scientific Research Results, Japan Society for the Promotion of Science, Grant Number 17HP1002.

Advanced Studies in Pure Mathematics 77
ISBN 978-4-86497-052-5

PRINTED IN JAPAN
by Livretech Co., Ltd.

2010 Mathematics Subject Classification.
Primary 13P10.
Secondary 05E40, 14F10, 14M25, 33F10, 46F10, 52B20, 62H05, 62E15.

Foreword

When Takayuki Hibi invited me to the conference on Gröbner bases with the subtitle "The 50th Anniversary of Gröbner Bases," I felt deeply pleased and grateful that the subject of my PhD thesis back in 1965 was receiving and is receiving so much attention.

In fact, the beginning in 1965 was quite hard and I had absolutely no feeling for what the impact of my work could be. In 1964, I was a working student of mathematics with a full-time job as a programmer at the computing center of the University of Innsbruck. I had thoroughly studied Van der Waerden's and Gröbner's book on commutative algebra (polynomial ideal theory) and I was determined (in spring 1964) to take the first problem Gröbner would present in his research seminar as my PhD thesis problem in order to finish my study as quickly as possible. By this, I hoped that my strenuous life as both a student and a programmer would soon be over. Luckily (in retrospect), the first problem Gröbner presented was the problem of constructing, by an algorithm, a linearly independent basis for the residue ring of a polynomial ring modulo a polynomial ideal.

Unfortunately (in my subjective feeling at that time), the problem turned out to be quite tough although its specification was so easy to understand. Thus, I still suffered through quite a long time of working, during the day, as a programmer and working, in my "free time" but often also while waiting for computer output, on Gröbner's problem. The problem was basically always on my mind and I alternated between looking to examples, trying to invent and prove some lemmata and experimenting with the computer. I will always remember the big relief I felt when one day — I was riding on my bike to the university — I suddenly "saw" that the clue to the solution of the problem would be the consideration of what I then called the "S-polynomials." From there it was a comparatively short step to come up with a proof of the S-polynomial theorem and, in parallel, with the S-polynomial algorithm for computing the special kind of ideal bases which I later called "Gröbner" bases. And, it was then clear that, as soon as one has a Gröbner basis for an ideal, the original problem of Gröbner and also a couple of other fundamental problems about polynomial ideals could be solved by algorithms.

In retrospect, what appeared for me to be a big tension between pure mathematics, theorems, structure, and proving on the one side and algorithms, data structures, computer hardware intricacies, programming languages and efficiency on the other hand, spanned the axes for my entire professional life.

My work did not get any attention in the first years after submitting my PhD thesis in the last days of 1965 (not even after appearance of my 1970 aequationes paper) but, starting from 1976, many people both in pure mathematics and applications areas became heavily interested. When I observed this, I started to understand that Gröbner, by giving me this tough problem, had in fact provided a big intellectual gift to me and I decided to attach his name to the type of bases I had invented in my thesis. While, in 1964, even after having studied the books of Van der Waerden and Gröbner, I did not really understand the importance of the problem (but apparently had the intellectual freshness to solve it) Gröbner surely had been fully aware of the crucial relevance of the problem. He had an enormously broad view and mastery of mathematics always keeping coherence between analysis, algebra and geometry.

Over the years, I tremendously enjoyed my life between the world of abstract mathematics and algorithms. I am grateful for every moment and I am deeply impressed by all the wonderful contributions so many people have made to the theory, algorithmics and applications of Gröbner bases. In the future, we will move to higher and higher levels of algorithmic mathematics, in particular by moving from the object level of mathematics to the meta-level.

By Gödel, there is no upper bound to iterating these moves to higher levels and it is very important to understand and make it clear that, whatever catch words (like "artificial intelligence," "machine learning," "deep reasoning," etc.) people outside mathematics like to invent for these moves, the intellectual skill and culture that keeps this upward movement going, is just the ever-young power of human mathematical thinking. Thus, we can look forward with enthusiasm and optimism to the next layers and waves of algorithmic mathematics with more and more challenges for the brightest minds of the next generations of mathematicians.

Let me conclude with warm thanks to Takayuki Hibi for organizing the conference and guiding us in the process of preparing our papers for the proceedings.

Linz, April 2017
Bruno Buchberger

Preface

Tracking back through the history, the vague shade of Gröbner bases already appeared in the famous paper[1] by F. S. Macaulay in 1927. The problem which Macaulay studied was to find a combinatorial characterization of Hilbert functions of residue class rings of the polynomial ring modulo homogeneous ideals. Macaulay succeeded in discovering the fundamental fact that the Hilbert function of the residue class ring of an arbitrary homogeneous ideal coincides with that of a certain monomial ideal. Macaulay's work had stimulated the algebraic study on enumerative combinatorics and promoted the birth of the historic area called "Commutative Algebra and Combinatorics," which originated in the work by Richard Stanley in 1975 on the upper bound conjecture for spheres by using the algebraic theory of Cohen–Macaulay rings.

The modern definition of Gröbner bases was independently introduced by Heisuke Hironaka in 1964 and Bruno Buchberger in 1965. Hironaka caught the idea of standard bases in the process of solving the outstanding problem, the resolution of singularities of algebraic varieties. On the other hand, Buchberger created the notion of Gröbner bases in his dissertation whose research topic had been given by his advisor Wolfgang Gröbner. Hironaka's standard bases work in the local ring, while Buchberger's Gröbner bases work in the polynomial ring. Apart from discussions of difference between the idea of standard bases and that of Gröbner bases, it turned out that Buchberger algorithm had opened the fascinating research area called "Computer Algebra."

After the pioneering work of Hironaka and Buchberger, however, for about twenty years, Gröbner bases had been out of the limelight. A turning point arose in the middle of 1980s, when David Bayer and Michael Stillman developed the computer software **Macaulay**, which has a great influence on computational aspects of commutative algebra and algebraic geometry. Since the theory of Gröbner bases was indispensable for developing **Macaulay**, Gröbner bases became common knowledge for researchers on commutative algebra and algebraic geometry.

The entry of Gröbner bases into the world of applied mathematics was achieved by Conti and Traverso in 1991, who proposed an algorithm to solve problems on integer programming by means of Gröbner bases of toric ideals. Toric ideals spread rapidly in the middle of 1990s. As one of the effective techniques to compute the dimension of the solution

---

[1] F. S. Macaulay, Some properties of enumeration in the theory of modular systems, *Proc. London Math. Soc.* **26** (1927), 531–555.

space of a hypergeometric equation, in 1994, Gel'fand, Kapranov and Zelevinsky introduced the notion of regular triangulations. It was shown by Sturmfels that the Stanley–Reisner ideal of a regular triangulation is just the radical of an initial ideal of a toric ideal. As a result, Gröbner bases of toric ideals turned out to be a brilliant bridge between the theory of monomial ideals and that of regular triangulations of convex polytopes. Since then, the algebraic theory of regular triangulations of convex polytopes has developed dramatically.

The study of Gröbner bases in the ring of differential operators started gradually in 1980s. A breakthrough was done by Oaku, who created new and effective algorithms on D-modules based on Buchberger algorithm. Since regular triangulations originated in the study of hypergeometric equations, the algebraic development of toric ideals naturally had a great influence on the study of hypergeometric equations.

An epoch-making application of Gröbner bases to statistics originated in the work by Diaconis and Sturmfels in 1998. When Markov chain Monte Carlo method is achieved in the examination of a statistical model, to find a Markov basis is required. It was shown by Diaconis and Sturmfels that a Markov basis corresponds to a system of generators of the toric ideal arising from a statistical model. Later, the new and exciting research area called "Algebraic Statistics" was born and has been developing rapidly. Algebraic statistics supplies commutative algebra with new problems and, conversely, toric ideals studied in commutative algebra supply algebraic statistics with new statistical models.

Taking the above background into consideration, in order to promote the international exchange of the study on Gröbner bases, the 8th Mathematical Society of Japan Seasonal Institute (MSJ-SI) entitled "Current Trends on Gröbner Bases" with the subtitle "The 50th Anniversary of Gröbner Bases" was held at Hotel Nikko Osaka, Osaka, Japan, during 1 – 10 July 2015. The organizing committee consisted of Satoshi Aoki (Kagoshima University), Takayuki Hibi (Osaka University, Chair), Masayuki Noro (Rikkyo University), Hidefumi Ohsugi (Kwansei Gakuin University), Nobuki Takayama (Kobe University) and Akimichi Takemura (The University of Tokyo). The present volume plays a role as the Proceedings of the conference.

The overall schedule of the 8th MSJ-SI 2015 is worth keeping a record. The first half, 1 – 4 July 2015, was a school for graduate students and the latter half, 6 – 10 July 2017, was a conference for researchers. There were 93 participants and among them 47 participants came from abroad. In the break of Sunday 5 July, a day trip to Mount Koya (Kōyasan), which is the center of Shingon Buddhism, an important Buddhist sect which was introduced to Japan in 805 by Kobo Daishi, also known as Kukai, was organized. There was a welcome party in the evening of Monday 6 July as well as an official banquet in the evening of Thursday 9 July. A half day trip to each of Kyoto, Nara and Kobe was arranged in the afternoon of Wednesday 8 July.

In the school, the following four series of expository lectures were presented: "Algebraic Problems in Structural Equation Modeling" by Mathias Drton; "Algorithms for D-modules, Integration, and Generalized Functions" by Toshinori Oaku; "Computing in Algebraic Geometry — Sheaf Cohomology and Its Applications in Geometry and Physics" by Michael Stillman; "Identifiability Problems in Statistics and Biology" by Seth Sullivant. Each of the series of expository lectures consisted of three 60-minute lectures together with one 60-minute exercise class.

In the conference, the attempt was made to organically disperse the invited talks into the three different aspects of Gröbner bases; theory (commutative algebra and algebraic geometry), application (algebraic statistics) and computation (D-modules and computational mathematics). The invited speakers were as follows: Bernd Sturmfels, Michael Stillman, Hyungju Park, Thomas Kahle, Toshinori Oaku, Mathias Drton, Seth Sullivant, Ezra Miller, Bruno Buchberger, Jürgen Herzog, Anna M. Bigatti, Anton Leykin, Dongming Wang, Viktor Levandovskyy, Mateusz Michałek, Anders N. Jensen, Caroline Uhler, Sonja Petrović, Raimundas Vidunas, Eva Riccomagno, Constantin Siriteanu and Henry Wynn. In addition, a session of contributed talks was organized from the afternoon through the night of Tuesday 7 July, in which twenty one 20-minute talks were given. At the end of Buchberger's talk in the morning of Wednesday 8 July, he made a highly amusing remark about the birth of the modern theory of Gröbner bases. The atmosphere of the conference had a cooperative spirit between the different areas.

Our sincere thanks must be given to the Mathematical Society of Japan for various support during the preparation of the 8th MSJ-SI 2015. The main source of funding for the 8th MSJ-SI 2015 is the JSPS (Japan Society for the Promotion of Science) Grants-in-Aid for Scientific Research (S) entitled "The Birth of Modern Trends on Commutative Algebra and Convex Polytopes with Statistical and Computational Strategies" (JP 26220701). Finally, the editor would like to thank Bruno Buchberger for contributing a foreword to the present volume.

Osaka, April 2017
Takayuki Hibi

All papers in this volume have been refereed and are in final form. No version of any of them will be submitted for publication elsewhere.

CONTENTS

Advanced Studies in Pure Mathematics 77, 2018
The 50th Anniversary of Gröbner Bases
pp. 1–24

Gröbner bases for everyone with CoCoA-5 and CoCoALib

John Abbott and Anna Maria Bigatti

Abstract.

We present a survey on the developments related to Gröbner bases, and show explicit examples in CoCoA.

The CoCoA project dates back to 1987: its aim was to create a "mathematician"-friendly computational laboratory for studying Commutative Algebra, most especially Gröbner bases. Always maintaining this "friendly" tradition, the project has grown and evolved, and the software has been completely rewritten.

CoCoA offers Gröbner bases for all levels of interest: from the basic, explicit call in the interactive system CoCoA-5 [5], to problem-specific optimized implementations, to the computer–computer communication with the open source C++ software library, CoCoALib [4], or the prototype OpenMath-based server.

The openness and clean design of CoCoALib and CoCoA-5 are intended to offer different levels of usage, and to encourage external contributions.

§1. Introduction

The CoCoA project traces its origins back to 1987 under the lead of Prof. L. Robbiano: the aim was to create a software laboratory for studying Commutative Algebra and especially Gröbner bases, which is welcoming even to mathematicians who are wary of new-fangled computers.

Received September 1, 2016.

2010 *Mathematics Subject Classification.* 13-04, 13P10, 68W30, 13P15, 14Q10, 05E40, 68W25.

Key words and phrases. Gröbner bases, elimination, term orderings, modular methods, software library.

(J. Abbott) J. Abbott is an INdAM-COFUND Marie Curie Fellow

(A. M. Bigatti) This work was partly supported by the "National Group for Algebraic and Geometric Structures, and their Applications" (GNSAGA – INdAM)

Since then the realm of applicability of Gröbner bases has continually expanded, so researchers interested in using them now come from a broad palette of subject areas ranging from the theoretical to quite practical topics. So there are still "pure" mathematicians as at the outset, but now also "programming" mathematicians, and statisticians, computer scientists, and so on. Another factor crucial in making Gröbner bases relevant to practical problems is the interim progress in computer hardware and software techniques.

The CoCoA project has evolved considerably from its original form, and the software has been rewritten: it now comes in the form of the very flexible software combination CoCoA-5/CoCoALib, while maintaining its tradition of being *user-friendly* so it offers Gröbner bases for all levels of interest and programming ability — *a Gröbner basis for everyone*! This means that the "CoCoA experience" covers a wide range: from the basic, explicit call in the interactive system CoCoA-5 [5] (see Section 2), to functions which use Gröbner bases implicitly (see Sections 3 and 4), to problem-specific optimized implementations (see Sections 5, 6 and 7), to the computer–computer communication with the open source C++ software library, CoCoALib [4], or with the prototype OpenMath-based server (see Section 8).

The importance that Gröbner bases have acquired derives from the fact that they enable or facilitate so many other computational mathematical results. A natural consequence is that a Gröbner basis is *almost never the final answer* that is sought, but just a stepping stone on the way to the goal, *e.g.* a Hilbert series or a primary decomposition. In this paper we concentrate on those computations in CoCoA which are directly related to Gröbner bases, illustrating the wide range of applications which have evolved over the last 50 years, and providing (explicitly or implicitly) Gröbner bases for everyone.

1.1. What is new in CoCoA-5? And what is not?

CoCoA-4 was widely appreciated for its ease of use, and the naturalness of its interactive language. However, it did have limitations, and several "grey areas". We designed the new CoCoA-5 language to strike a balance between backward-compatibility (to avoid alienating existing CoCoA-4 users) and greater expressibility with a richer and more solid mathematical basis (eliminating those "grey areas").

So, what's not new? Superficially the new CoCoA-5 language and system closely resemble CoCoA-4 because we kept it largely backward compatible. At the same time CoCoA-5 improves upon the underlying mathematical structure and robustness of the old system. We are very aware that a number of CoCoA users are mathematicians with only limited programming experience, for whom learning CoCoA was a "big investment", and who are reluctant to make another such investment — that is why we wanted to make the passage to CoCoA-5 as painless as possible.

So, if almost nothing has changed, *what is new*? The clearly defined semantics of the CoCoA-5 new language make it both more robust and more flexible; it provides greater expressibility and a more solid mathematical basis. In particular, it offers full flexibility for the field of coefficients: *e.g.* $\mathbb{Z}/\langle p \rangle$ with large p, fraction fields and algebraic extensions (see Section 2.2), and even heuristically verified floating point arithmetics with rational reconstruction (see Section 7.1).

However, under the surface, the change is radical since its mathematical core, CoCoALib, has been rewritten from scratch, to be faster, cleaner and more powerful than the old system, and also to be used as a C++ library.

1.2. How Do CoCoALib and CoCoA-5 differ?

The glib answer is: *As little as possible!*

One important idea behind the designs of CoCoALib and CoCoA-5 is that of making it easy to take a prototype implementation in CoCoA-5 and translate it into C++ using CoCoALib. We intend to facilitate the "translation" step as much as possible.

CoCoALib [4, 1], the C++ library, contains practically all the mathematical knowledge and ability whereas CoCoA-5 [5] offers convenient, interactive access to CoCoALib's capabilities. Most functions are accessible from both, and have identical names and behaviour (see Section 8).

More precisely, CoCoA-5 is an interactive, interpreted environment which makes it better suited to "rapid prototyping" than the relatively rigid, statically typed regime of C++. To keep it simple to learn, CoCoA-5 has only a few data types: for instance, a power-product in CoCoA-5 is represented as a monic polynomial with a single term, *i.e.* it is a ring element (of a polynomial ring). In contrast, in CoCoALib there is a dedicated class, `PPMonoidElem`, which directly represents each power-product, and allows efficient operations on the values (*e.g.* without the overhead of the superfluous coefficients present in the simplistic approach of CoCoA-5).

Programming with CoCoALib does tend to be more onerous than with CoCoA-5, largely because of C++'s demanding, rigid rules. However, the reward is greater flexibility and typically faster computation (sometimes much faster). Also, of course, those who want to use CoCoALib's abilities in their own C++ program necessarily have to use

CoCoALib. This is why our goal is that everything which can be computed in CoCoA-5 should be just as readily computable with CoCoALib. Currently a few CoCoA-5 functions are still implemented in CoCoA-5 packages, but these are being steadily translated into C++. (We don't say which ones, because the list is constantly shrinking)

§2. Gröbner Bases with Ease

The simplest cases of Gröbner bases are for ideals in $\mathbb{Q}[x_1, \ldots, x_n]$ or $\mathbb{Z}/\langle p\rangle[x_1, \ldots, x_n]$ with p a prime. These are also the easiest cases to give to CoCoA. Here is an example:

```
/**/ use QQ[x,y,z];    // or ZZ/(2)[x,y,z];
/**/ I := ideal(x^3+3,  y-x^2,  z-x-y);
/**/ GBasis(I);
[x +y -z,  y^2 -3*y +3*z,  y*z -3*y +3*z +3,  z^2 -4*y +3*z +6]
```

An essential ingredient in the definition of a Gröbner basis is the term-ordering: *i.e.* a total ordering on the power-products which respects multiplication, and where 1 is the smallest power-product.

In the example above the term-ordering was not explicitly indicated, so CoCoA assumes `StdDegRevLex` (with the common convention that the indeterminates generating the polynomial ring were given in decreasing order). In many computer algebra systems this is the default term-ordering because it *generally* gives the best performance and most compact answer. Another well-known family of orderings is `lex` (short for "lexicographic"). A `lex` Gröbner basis of a zero dimensional ideal in normal position has a particular *shape* which is theoretically useful for solving polynomial systems (see, for example, the Kreuzer–Robbiano book [20], Sec. 3.7). However its practical usefulness is limited by the fact that `lex` bases tend to be particularly big and ugly, and are frequently rather costly to compute.

There are various other gradings and orderings which are useful for studying specific problems: for instance, an important family are the *elimination orderings* which are used implicitly in Section 5. CoCoA also offers a fully general, matrix-based implementation of term-orderings (see Sections 3 and 5).

In CoCoA the term-ordering is specified at the same time as the polynomial ring; Gröbner bases of all ideals in that polynomial ring will automatically be computed with respect to that ordering. Thus, in CoCoA the term-ordering is an intrinsic property of each polynomial ring. This means that $\mathbb{Q}[x, y, z]$ with `lex` is regarded as a different ring from $\mathbb{Q}[x, y, z]$ with **StdDegRevLex**. Here is an example of computing a **lex** Gröbner basis.

```
/**/ use QQ[x,y,z], lex; // specify ordering together with ring
/**/ I := ideal(x^3 +3,  y-x^2,  z-x-y);
/**/ ReducedGBasis(I);    // basis is wrt lex ordering
[x +(1/4)*z^2 +(-1/4)*z +3/2,  y +(-1/4)*z^2 +(-3/4)*z -3/2,
   z^3 +9*z -6]
```

In the last example above we used the command **ReducedGBasis** which computes a reduced Gröbner basis: namely a "cleaned up" basis with only non-redundant, monic, fully reduced elements — it is unique (up to the order of its elements).

2.1. Verbosity and interruption

Sometimes it is handy to know what is happening inside a running function. For example, a Gröbner basis computation may be taking a long time, and we would like to know whether it is likely far from finishing, and if so, interrupt it.

A new feature in CoCoA-5.2.0 is the ability to set the verbosity level; there is also a companion function which tells you the current level.

```
/**/ SetVerbosityLevel(100);
/**/ VerbosityLevel();
100
```

This is a global setting, and higher verbosity levels trigger the printing of increasing amounts of internal "progress information" in several functions (both in CoCoA-5 and in CoCoALib).

For instance, the lowest level giving information on the progress of Gröbner bases is 100; every time a new polynomial is found, a line like this is printed:

```
myDoGBasis[1]: New poly in GB: len(GB) = 10 len(pairs) = 6
```

By setting the verbosity level before starting some hard Gröbner basis computations, one may see that the number of pairs yet to be processed is unfeasibly high. The user may then choose to interrupt the computation by typing **Ctrl-C**: the computation will be interrupted as soon as the reduction of the current S-polynomial terminates.

This interruption cancels the incomplete Gröbner basis computation, and returns the computer to the state it was in just before the Gröbner basis computation was begun (thanks to the clean, exception-safe design of CoCoALib).

Besides **GBasis**, verbose information can be produced by numerous functions: see, for instance, Sections 3 and 4. Indeed, the number of functions (both in CoCoALib and CoCoA-5) which respond to the verbosity setting is steadily increasing — details are in the documentation (type "`?verbose`"). Similarly the number of interruptible CoCoALib functions is gradually increasing; in any case, all interpreted CoCoA-5 functions can be interrupted.

2.2. More rings of coefficients

The easy examples above show the definition of polynomial rings with rational coefficients, but the choice of coefficients in CoCoA is quite wide. For example, coefficients in a finite field $\mathbb{Z}/\langle p\rangle$:

```
/**/ use ZZ/(10^29 + 319)[x];
/**/ ReducedGBasis(ideal(3*x-1));
[x -33333333333333333333333333440]
```

Or coefficients in algebraic extension fields:

```
/**/ use R ::= QQ[i];
/**/ K := R/ideal(i^2 +1);
/**/ use K[x,y,z];
/**/ I := ideal(i*x^3 -z, x^2*y^3 -i*y*z^2);
/**/ ReducedGBasis(I);
[x^3 +(i)*z,  x^2*y^3 +(-i)*y*z^2,  y^3*z +x*y*z^2]

/**/ use R ::= QQ[sqrt2, sqrt3];
/**/ K := R/ideal(sqrt2^2 -2, sqrt3^2 -3);
/**/ IsField(K);
true
/**/ use K[x,y,z];
/**/ I := ideal(sqrt3*x^2 -y, x*y -sqrt2*z);
/**/ ReducedGBasis(I);
[x*y +(-sqrt2)*z,  x^2 +((-1/3)*sqrt3)*y,  y^2 +(-sqrt2*sqrt3)*x*z]
```

Or coefficients in a fraction field:

```
/**/ use QQab ::= QQ[a,b];
/**/ K := NewFractionField(QQab);
/**/ use K[x,y,z];
/**/ I := ideal(x^3 -a*z, x^2*y^3 -b*y*z^2);
/**/ ReducedGBasis(I);
[x^3 -a*z,  x^2*y^3 -b*y*z^2,  y^3*z +(-b/a)*x*y*z^2]
```

One should note that in this last example K is actually the field $\mathbb{Q}(a,b)$ with no specialization of $a, b \in \mathbb{Q}$. So the Gröbner basis produced represents the *generic* case, meaning that every algebraic expression in a, b which is not identically zero is considered to be non-zero. The problem of considering all possible specializations of the parameters is known as *comprehensive Gröbner basis*, and is not (yet) implemented in CoCoA.

Another family of computationally interesting rings in CoCoA is given by `NewRingTwinFloat(BitPrec)`. These will be presented in detail in Sections 4 and 7.1.

§3. Universal Gröbner bases and Gröbner fans

There is a notion of **universal Gröbner basis** which is a Gröbner basis for every term-ordering. The CoCoA function "`UniversalGBasis`" will compute one such basis; this function is based on the computation of the Gröbner fan (a richer structure, described below) which gives all possible reduced Gröbner bases: we can take the union of all of them to produce the universal basis.

The following example shows that the maximal minors of a 3×4 matrix of indeterminates form a universal Gröbner basis of the ideal they generate:

```
/**/ use R ::= QQ[a,b,c,d,e,f,g,h,i,j,k,l];
/**/ I := ideal(minors(mat([[a,b,c,d],[e,f,g,h],[i,j,k,l]]),3));
/**/ indent(UniversalGBasis(I));
[
  d*g*j -c*h*j -d*f*k +b*h*k +c*f*l -b*g*l,
  d*g*i -c*h*i -d*e*k +a*h*k +c*e*l -a*g*l,
  d*f*i -b*h*i -d*e*j +a*h*j +b*e*l -a*f*l,
  c*f*i -b*g*i -c*e*j +a*g*j +b*e*k -a*f*k
]
/**/ EqSet(-1*gens(I), ReducedGBasis(I));
true
```

The **Gröbner fan** of an ideal was defined by Mora and Robbiano in 1988 ([22]): it is a (finite) fan of polyhedral cones indexing the reduced Gröbner bases of the ideal. This has been implemented by Jensen in his software *Gfan* ([19]) which he has recently linked into CoCoA; we note that CoCoA's fully general approach to representing term-orderings was essential in making this integration possible.

The Gröbner fan is useful because several well-known theoretical applications of Gröbner bases rely on the existence of a Gröbner basis of an ideal with prescribed properties, such as having a certain cardinality, or comprising polynomials of a specified degree, or all squarefree. For example, if an ideal $I \in K[x_1, \ldots, x_n]$ has a Gröbner basis for *some* term-ordering comprising just quadrics, then the algebra $K[x_1, \ldots, x_n]/I$ is Koszul.

The function `GroebnerFanIdeals(I)` computes all reduced Gröbner bases of the ideal I. We have chosen to express the result as a list of ideals, each generated by one of the various possible reduced Gröbner bases: the ideals are all "the same" but belong to *different* polynomial rings (remember that the term-ordering is an intrinsic property of the polynomial ring). An advantage of this approach is that further computation with any of these ideals automatically takes place in the corresponding polynomial ring equipped with an appropriate term-ordering. Furthermore, each of these ideals already knows its own reduced Gröbner basis, whose value is thus immediately available (*i.e.* without any computation).

The following ideal, Example 3.9 from Sturmfels's book [23], has 360 distinct reduced Gröbner bases:

```
/**/ use R ::= QQ[a,b,c];
/**/ I := ideal(a^5+b^3+c^2-1, a^2+b^2+c-1, a^6+b^5+c^3-1);
/**/ L := GroebnerFanIdeals(I);
/**/ len(L);
360
```

Since the computation easily becomes very cumbersome, it is interesting to see how it is progressing; for example, after setting the verbosity level to 10 (see Section 2.1), a * is printed every time a new Gröbner basis is added to the list (for more information see the manual by typing "`?GroebnerFan`"):

```
/**/ use QQ[x,y,z];
/**/ I := ideal(x^3 +x*y -z, x^2 -y*z);
/**/ SetVerbosityLevel(10);
/**/ GF := GroebnerFanIdeals(I);
********
/**/ indent(GF);
[ ideal(x^2 -y*z,  x*y*z +x*y -z,  y^2*z^2 +y^2*z -x*z),
  ideal(x^2 -y*z,  x*z -y^2*z^2 -y^2*z,  y^3*z^2 +x*y +y^3*z -z),
  ideal(x^2 -y*z,  x*z -y^2*z^2 -y^2*z,  x*y +y^3*z^2 +y^3*z -z,
    y^3*z^3 +2*y^3*z^2 +y^3*z -z^2),
  ideal(y*z -x^2,  x^3 +x*y -z),
  ideal(x*y +x^3 -z,  y*z -x^2,  x^3*z -z^2 +x^3),
  ideal(x*y -z +x^3,  y*z -x^2,  z^2 -x^3*z -x^3),
  ideal(z -x*y -x^3,  x*y^2 +x^3*y -x^2),
  ideal(z -x^3 -x*y,  x^3*y +x*y^2 -x^2) ]
```

Storing all the possible different (reduced) Gröbner bases is practicable only for small examples; larger ideals may have thousands or even millions of different Gröbner bases. Often we are interested only in those bases satisfying a certain property. So CoCoA offers the function **CallOnGroebnerFanIdeals** which calls a given function on each of the Gröbner fan ideals successively without storing them all in a big list (which may not even fit in the computer's memory!). Using this CoCoA function needs a little technical ability, but makes it possible to tackle larger computations.

In the following example we see explicitly that CoCoA represents some term-orderings via matrices of integers. Indeed, each such matrix is the only information necessary to be able to recalculate the corresponding reduced Gröbner basis (in this example, those having 3 elements). See Section 5 for an example of how to ask CoCoA to compute a Gröbner basis with a term-ordering given by a matrix.

```
define PrintIfGBHasLen3(I)
 if len(GBasis(I))=3 then
   println OrdMat(RingOf(I));
   indent(ReducedGBasis(I));
 endif;
enddefine;


/**/ use R ::= QQ[a,b,c];
/**/ I := ideal(a^5+b^3+c^2-1, b^2+a^2+c-1, c^3+a^6+b^5-1);
/**/ CallOnGroebnerFanIdeals(I, PrintIfGBHasLen3);
matrix(ZZ,
 [[3, 7, 7],
  [3, 6, 8],
  [0, 0, -1]])
[b^2+c+a^2-1,
 a^5+c^2-b*c-a^2*b+b-1,
 c^3+b*c^2+2*a^2*b*c+a^4*b-a*c^2+a*b*c+a^3*b-2*b*c-2*a^2*b-a*b+b+a-1]
matrix(ZZ,
 [[6, 7, 14],
  [6, 5, 15],
  [0, 0, -1]])
[c+b^2+a^2-1,
 -b^6-3*a^2*b^4-3*a^4*b^2+b^5+3*b^4+6*a^2*b^2+3*a^4-3*b^2-3*a^2,
 a^5+b^4+2*a^2*b^2+a^4+b^3-2*b^2-2*a^2]
```

§4. Leading Term Ideals and "gin"

Let $P = K[x_1, \ldots, x_n]$ be a polynomial ring over a field K, and let σ be a term-ordering on the power-products in P. Let I be an ideal in P then we define its **leading term ideal** with respect to σ, written $\mathrm{LT}_\sigma(I)$, to be the ideal generated by the leading power-products of all non-zero polynomials in I; some authors use the name "initial ideal" for this notion. A generating set for $\mathrm{LT}_\sigma(I)$ may easily be obtained: we compute a reduced σ-Gröbner basis for I then collect the σ-leading terms of the elements of the basis. Remarkably $\mathrm{LT}_\sigma(I)$ captures some interesting "combinatorial" information about the original polynomial ideal I: for instance, its Hilbert series. Hence in CoCoA calling `LT(I)` or `HilbertSeries(P/I)` actually contains a "hidden" call to `GBasis(I)`.

```
/**/ use P ::= QQ[x,y,z];
/**/ I := ideal(y^20 -x^4*z^16,  x^12*z^3 -y^13*z^2);
/**/ LT(I);
ideal(y^13*z^2,  y^20,  x^12*y^7*z^3,  x^24*z^4)

/**/ HilbertSeries(P/I);
(1 + 2*t + 3*t^2 + 4*t^3 + 5*t^4 + 6*t^5 + 7*t^6 + 8*t^7 + 9*t^8 + 10*t^9
+ 11*t^10 + 12*t^11 + 13*t^12 + 14*t^13 + 15*t^14 + 15*t^15 + 15*t^16
+ 15*t^17 + 15*t^18 + 15*t^19 + 14*t^20 + 13*t^21 + 12*t^22 + 11*t^23
+ 10*t^24 + 9*t^25 + 8*t^26 + 7*t^27 + 6*t^28 + 5*t^29 + 4*t^30 + 3*t^31
+ 2*t^32 + t^33) / (1-t)
```

A more sophisticated tool in Commutative Algebra is the **generic initial ideal** of a polynomial ideal I. This is useful because it encodes more geometrical properties of I into a monomial ideal. It is defined as $\mathrm{gin}_\sigma(I) = \mathrm{LT}_\sigma(\gamma(I))$ where γ is a generic change of coordinates, *i.e.* $\gamma(x_j) = \sum_{i=1}^n a_{ij}\, x_i$ in $K(a_{ij})[x_1, \ldots, x_n]$. And here we have to admit that the acronym *gin* sounds nicer than *gLT*!

The definition of *gin* suggests an obvious algorithm for computing it (see Section 2.2 for an example with *generic* coefficients). However, even knowing that it is enough to consider a *triangular* change of coordinates $\gamma(x_j) = \sum_{i=1}^j a_{ij}\, x_i$, it quickly becomes apparent that the coefficients in $K(a_{ij})$ grow to unwieldy sizes except for the very simplest cases; so the obvious approach is utterly hopeless. Instead we can pick an explicit, random change of coordinates $\tilde{\gamma}$, and then compute $\mathrm{LT}_\sigma(\tilde{\gamma}(I))$; the coordinate changes for which $\mathrm{gin}_\sigma(I) = \mathrm{LT}_\sigma(\tilde{\gamma}(I))$ form a non-empty Zariski-open set. This approach can be used when K is infinite: if the coefficients for the random change of coordinates are chosen from a large set then $\mathrm{LT}_\sigma(\tilde{\gamma}(I))$ will indeed be $\mathrm{gin}_\sigma(I)$ with high probability. This is what CoCoA does.

While choosing random changes of coordinates with large coefficients increases the probability of getting the correct result, it also tends to produce large coefficients in the transformed polynomials. In the example below the original polynomials have very small coefficients, but there is a coefficient with almost 50 digits in the transformed polynomials:

```
/**/ use P ::= QQ[x,y,z];
/**/ I := ideal(y^20 -x^5*z^6,  x^2*z^3 -y*z^2);
/**/ L := [sum([ random(-500,500)*indet(P,j) | j in 1..3])
                                                 | i in 1..3];
/**/ L;
[-414*x +341*y -141*z, -318*x +389*y +178*z, -498*x +498*y +28*z]
/**/ gamma := PolyAlgebraHom(P, P, L);
/**/ GI := ideal(apply(gamma, gens(I)));  GI;
ideal(11182589936405515962964647295818826649092311062937 6*x^20 +...,
     -21168433004832*x^5 +98376968843712*x^4*y - ...)
```

With coefficients like that, computing the Gröbner basis of the transformed ideal over the rationals would be quite expensive! Thus, one needs to strike a balance between picking coefficients from a wide range, so the transformation is "generic enough", but not so wide that there is excessive growth in the coefficients of transformed ideal generators.

To avoid the costs of computing with large coefficients, the implementation for computing gin in CoCoA uses a special representation for rational coefficients, namely *twin-floats* (see Section 7.1). The Gröbner basis of the twin-float transformed ideal will have only approximate twin-float coefficients, but this does not matter because we need only the leading power-products of the polynomials in the basis.

Twin-float numbers have fixed-precision (so do not grow in size the way rational numbers do), and employ heuristics to verify the correctness of results. This allows the implementation to make random coefficient choices from a wide range (in fact, integers between -10^6 and 10^6) without paying the price for calculating with transformed polynomials having cumbersome rational coefficients. If the initially chosen precision for the twin-floats is too low, this will be signalled; and the computation will be automatically restarted with a higher precision.

CoCoA's function `gin` does all this behind the scenes. Moreover, it verifies the result by trying a second random change of coordinates, and checking it gets the same leading term ideal. If the results differ, CoCoA repeatedly tries further random changes of coordinates until it gets the same answer twice in succession — though we have never seen the verification fail when picking random coefficients in the range

$(-10^6, 10^6)$. The internal workings can be seen via printed messages with the appropriate verbosity level (see Section 2.1).

```
/**/ SetVerbosityLevel(50);
/**/ J := gin(I);
RandIdeal: change coord = [
  -7426*x,
  695955*x +168758*y,
  -239080*x +304634*y +480790*z
]
TryPrecisions: -- trying with FloatPrecision 64
TryPrecisions: -- trying with FloatPrecision 128
RandIdeal: change coord = [
  -499447*x,
  -732749*x -840921*y,
  -466314*x -691911*y +554086*z
]
TryPrecisions: -- trying with FloatPrecision 128

/**/ J;
ideal(x^5,  x^4*y^16,  x^3*y^18,  x^2*y^20,  x*y^22,  y^24)
```

Since the gin ideal with respect to the ordering `StdDegRevLex` has many interesting properties, CoCoA offers the function `rgin` which computes it, independently of the term ordering inherent in the polynomial ring.

§5. Elimination and related functions

Elimination means: given an ideal $I \in K[t_1, \ldots, t_s,\ x_1, \ldots, x_n]$, find a set of generators of the ideal $I \cap K[x_1, \ldots, x_n]$ where the indeterminates $\{t_1, \ldots, t_s\}$ have been "eliminated". Elimination is a central topic in Computational Commutative Algebra (see for example the text book by Kreuzer and Robbiano [20], Sec. 3.4) and its applications are countless.

Given its usefulness, elimination is an operation offered in almost all Computer Algebra Systems. In general, such elimination functions internally compute a Gröbner basis with respect to an **elimination ordering** for the subset of indeterminates to be eliminated: with such an ordering the subset of polynomials in the Gröbner basis whose leading terms are not divisible by any of the t_j are exactly the generators we seek for the ideal $I \cap K[x_1, \ldots, x_n]$.

In the example below we see the process we described and compare it with the actual output of CoCoA's own function `elim`. Note that in both cases the generators are not minimal, but they are indeed a Gröbner basis of the elimination ideal (wrt the restriction of the elimination term-ordering used).

```
/**/ M := ElimMat([1], 4);  M;
matrix(ZZ,
 [[1, 0, 0, 0],
  [1, 1, 1, 1],
  [0, 0, 0, -1],
  [0, 0, -1, 0]])

/**/ P := NewPolyRing(QQ, "t, x,y,z", M, 0);  // 0: no grading
/**/ use P;
/**/ I := ideal(x-t, y-t^2, z-t^3);
/**/ GBasis(I);
[t -x,  x^2 -y,  x*y -z,  y^2 -x*z]
/**/ elim([t], I);
ideal(x^2 -y,  x*y -z,  y^2 -x*z)
/**/ MinSubsetOfGens(ideal(x^2 -y,  x*y -z,  y^2 -x*z));
[x^2 -y,  x*y -z]
```

The simple example above shows a particular application of `elim`: finding the **presentation of an algebra** $K[f_1, ..., f_n] \simeq K[x_1, ..., x_n]/I$. More precisely, let $f_1, ..., f_n \in K[t_1, \dots, t_s]$, where $\{t_1, \dots, t_s\}$ is another set of indeterminates (viewed as parameters) and consider the K-algebra homomorphism

$$\phi : K[x_1, \dots, x_n] \longrightarrow K[t_1, \dots, t_s] \text{ given by } x_i \mapsto f_i \quad \text{for } i = 1, \dots, n$$

Its kernel is a prime ideal; the general problem of **implicitization** (for a polynomial parametrization) is to find a set of generators for this ideal.

The Gröbner basis elimination technique consists of defining the ideal $J = \langle x_1 - f_1, \dots, x_n - f_n\rangle$ in the ring $K[t_1, \dots, t_s, x_1, \dots, x_n]$ and *eliminating* all the parameters t_i, as we saw in the example.

Unfortunately this extraordinarily elegant tool often turns out to be quite inefficient, resulting in long and costly computations. In the next two subsections we see that knowing how to exploit special properties of a given class of examples can make a huge difference.

5.1. Toric

If the algebra we want to present is generated by power-products then the elimination can be computed by the CoCoA function `toric`; toric ideals are prime and generated by binomials.

We consider $\mathbb{Q}[t, t^2, t^3]$ again, and also $\mathbb{Q}[t^3, t^4, t^5]$:

```
/**/ use QQ[x,y,z];
/**/ toric(RowMat([1,2,3])); // just the list of exponents
ideal(-x^2 +y, x^3 -z)

/**/ use QQ[x,y,z];
/**/ toric(RowMat([3,4,5]));
ideal(y^2 -x*z,  x^3 -y*z,  x^2*y -z^2)
```

With a very slightly more challenging example we can clearly measure the advantage in using the specialized function "`toric`" over the general function "`elim`":

```
/**/ use R ::= ZZ/(2)[x[1..6], s,t,u,v];
/**/ L := [s*u^20, s*u^30, s*t^20*v, t*v^20, s*t*u*v, s*t^2*u];
/**/ ExpL := mat([[ 1,   1,   1,   0,   1,   1],
                  [ 0,   0,  20,   1,   1,   2],
                  [20,  30,   0,   0,   1,   1],
                  [ 0,   0,   1,  20,   1,   0]]);

/**/ I := ideal([x[i] - L[i] | i in 1..6]);
/**/ t0 := CpuTime();  IE := elim([s,t,u,v], I);  TimeFrom(t0);
9.274
/**/ t0 := CpuTime();  IT := toric(ExpL);  TimeFrom(t0);
0.032
```

The CoCoA function `toric` employs a non-deterministic algorithm: so the actual set of ideal generators produced might vary.

For further details on the algorithms implemented in CoCoA see Bigatti, La Scala, Robbiano [14]. That article describes three different algorithms; the default one in CoCoA is EATI (*Elimination Algorithm for Toric Ideals*).

For more details on the specific function `toric` type `?toric` into CoCoA (or read the PDF manual, or the html manual on the web-site).

5.2. Implicitization of hypersurfaces

As mentioned earlier elimination provides a general solution to the implicitization problem, but this solution is more elegant than practical. We can do rather better in the special case of implicitization of a hypersurface. One immediate feature is that the result is just a single polynomial since the eliminated ideal must be principal.

It is well-known that Buchberger's algorithm usually works better with homogeneous ideals (though there are sporadic exceptions). Yet the very construction of the eliminating ideal $J = \langle x_1 - f_1, \ldots, x_n - f_n \rangle$ looks intrinsically non-homogeneous.

But with a little, well-guided effort we can transform the problem into the calculation of a Gröbner basis of a homogeneous ideal: we take a new indeterminate (say h) and use it to homogenize each f_j to produce F_j. Now we can work with the ideal $J' = \langle x_1 - F_1, \dots, x_n - F_n \rangle$ made homogeneous by giving weights to the x_i indeterminates: we set $\deg(x_i) = \deg(f_i)$ for each i.

Since we are in the special case of a hypersurface, it can be shown that the (non-zero) polynomial of lowest degree in $J' \cap K[x_1, \dots, x_n, h]$ is unique up to scalar multiples; its dehomogenization is then the polynomial we seek! We get two advantages from the homogeneous ideal J': we gain efficiency by using Buchberger's algorithm degree-by-degree, and we can stop as soon as the first basis polynomial is found — most probably there will still be many pairs to process. See Abbott, Bigatti, Robbiano [7] for all details and proofs, and also how to "correctly homogenize" parametrizations defined by rational functions.

```
/**/ use P ::= QQ[s,t, x,y,z];
/**/ elim([s,t],  ideal(x-s^2, y-s*t, z-t^2) );
ideal(y^2 -x*z)

/**/ use R ::= QQ[s,t];
/**/ P ::= QQ[x,y,z];
/**/ ImplicitHypersurface(P, [s^2, s*t, t^2], "ElimTH");
ideal(y^2 -x*z)
```

In the same paper we describe another algorithm which uses a completely different technique, a variant of the Buchberger-Möller algorithm (see Section 6), based on linear algebra. It is well-suited to low degree hypersurfaces.

```
/**/ ImplicitHypersurface(P, [s^2, s*t, t^2], "Direct");
ideal(y^2 -x*z)
```

For the case of rational coefficients, we use a modular approach in both algorithms: we compute the result modulo several primes, combine these using Chinese Remaindering, and finally reconstruct the rational coefficients of the answer using the fault-tolerant rational reconstruction described in Section 7.2.

5.3. MinPoly

Another popular application of elimination is for finding univariate polynomials in an ideal. If $I \subset P = K[x_1, \dots, x_n]$ is a 0-dimensional ideal, then we know that $I \cap K[x_i]$ is a principal ideal generated by some univariate polynomial $g_i(x_i) \neq 0$ which can be obtained by eliminating all x_j with $j \neq i$. These polynomials are used in several operations, such as computing the radical of a zero-dimensional ideal, or solving polynomial systems (see [21] and [8]).

This idea generalizes in a natural way to the following problem. Let $I \subset P$ be a 0-dimensional ideal, and let f be any polynomial in P, find $\mu_{f,I}(z) \in K[z]$, the minimal polynomial of $\bar{f} \in P/I$, or equivalently, the univariate monic polynomial of minimum degree whose evaluation at f yields an element of I. The corresponding algorithms have been recently implemented in CoCoA:

```
/**/ use P ::= QQ[x,y,z];
/**/ L := [ x^2-z^2, (y-3)*(y+2)*(y^3-2), z^3-1];
/**/ I := ideal(L);
/**/ IsZeroDim(I);
true

/**/ MinPolyQuot(x, I, x); -- 3rd arg is the indet for the answer
x^6 -1

/**/ f := x -2*y +3*z;
/**/ t0 := CpuTime();  MP := MinPolyQuot(f, I, x);  TimeFrom(t0);
0.036
/**/ MP;
x^30 +12*x^29 -84*x^28 -1544*x^27 +384*x^26 +62688*x^25 +119168*x^24 -629760*x^23 -4664832*x^22
-33803264*x^21 +107753472*x^20 +1318662144*x^19 -3480064000*x^18 -20059865088*x^17
+151993466880*x^16 -50058002432*x^15 -1931977162752*x^14 +9312278544384*x^13 +1002303913984*x^12
-113944836440064*x^11 +553708192530432*x^10 +720752546414592*x^9 -6749908862238720*x^8
+4995175176732672*x^7 +33972228030726144*x^6 +22154393721765888*x^5 -21399162914340864*x^4
-112685231584051200*x^3 +3245139849904128*x^2 -3103199770705920*x -16498446852685824
```

Needless to say, even in this small example, the standard elimination approach is considerably slower:

```
/**/ use Paux ::= QQ[x,y,z, aux];
/**/ phi := PolyAlgebraHom(P, Paux, [x,y,z]);
/**/ J := ideal(apply(phi,L)) + ideal(aux - phi(f));
/**/ t0 := CpuTime();  JE := elim([x,y,z], J);  TimeFrom(t0);
1.850
```

As we did for hypersurface implicitization (in Section 5.2), when computing with rational coefficients we use a modular approach and the fault-tolerant rational reconstruction described in Section 7.2.

The good computational speed of `MinPolyQuot` is the key point for a new algorithm for computing the primary decomposition of zero-dimensional ideals. See [8] for details on the algorithms for `MinPolyQuot` and some interesting applications.

```
/**/ PD := PrimaryDecomposition0(I);
/**/ indent([IdealOfGBasis(Qi) | Qi in PD]);
[
  ideal(y +2,  x +1,  z -1),
  ideal(y -3,  x +1,  z -1),
  ideal(x +1,  z -1,  y^3 -2),
  ideal(y +2,  x -1,  z -1),
  ideal(y -3,  x -1,  z -1),
  ideal(x -1,  z -1,  y^3 -2),
  ideal(y +2,  z^2 +z +1,  x +z),
  ideal(y -3,  z^2 +z +1,  x +z),
  ideal(z^2 -x +1,  y^3 -2,  x +z),
  ideal(y +2,  z^2 +z +1,  x -z),
  ideal(y -3,  z^2 +z +1,  x -z),
  ideal(z^2 +x +1,  y^3 -2,  x -z)
]
```

§6. Ideals of Points, 0-Dimensional Schemes

Let X be a non-empty, finite set of points in K^n, then the set of all polynomials in $K[x_1, \ldots, x_n]$ which vanish at all points in X is an ideal, I_X. One reason this ideal is interesting is because it captures the "ambiguity" present in a polynomial function which has been interpolated from its values at the points of X. How best to compute a set of generators for I_X, or a Gröbner basis, knowing just the points X?

If X contains a single point $(a_1, \ldots, a_n)$ then we can write down immediately a Gröbner basis, namely $[x_1 - a_1, \ldots, x_n - a_n]$. If X contains several points we could just intersect the ideals for each single point, and these intersections may be determined via Gröbner basis computations; while fully effective and mathematically elegant this approach is computationally disappointing.

A far more efficient method is the Buchberger-Möller algorithm [15]. Somewhat astonishingly it uses just simple linear algebra to determine the Gröbner basis. In [6] there is a detailed complexity analysis of the original algorithm, and also an extension to the projective case. It was later further generalized to zero-dimensional schemes [12], where it turned out that it also incorporates the well-known FGLM algorithm for "changing term-ordering" of a Gröbner basis (see [17]).

Much as we have seen in the previous sections for computing with rational coefficients, the Buchberger-Möller algorithm also benefits from a modular approach, and naturally the CoCoA implementation uses this technique.

```
/**/ P ::= QQ[x,y];
/**/ points := mat([[10, 0], [-10, 0], [0, 10], [0, -10],
                    [7, 7], [-7, -7], [7, -7], [-7, 7]]);
/**/ indent(IdealOfPoints(P, points));
```

```
ideal(
  x^2*y +(49/51)*y^3 +(-4900/51)*y,
  x^3 +(51/49)*x*y^2 -100*x,
  y^4 +(-2499/2)*x^2 +(-2699/2)*y^2 +124950,
  x*y^3 -49*x*y  )
```

The use of simple linear algebra in the Buchberger-Möller algorithm makes it a good candidate for identifying "almost-vanishing" polynomials for sets of *approximate points*: for instance, the points in the example above "almost lie on" a circle of radius 9.95 centred on the origin, though we cannot tell this from the *exact* Gröbner basis.

In fact, the notion of Gröbner basis does not generalize well to an "approximate context" because the algebraic structure of a Gröbner basis is determined by Zariski-closed conditions (*i.e.* the structure is valid when certain polynomials vanish); instead, the notion of a *Border Basis* is better suited since the validity of its structure depends on a Zariski-open condition (*i.e.* provided a certain polynomial *does not* vanish). So long as the approximate points are not too few nor too imprecise the NBM (*Numerical Buchberger-Möller*) algorithm can compute at least a partial Border Basis, and this should identify any "approximate polynomial conditions" which the points almost satisfy (see Abbott, Fassino, Torrente [11] and Fassino [16]). We can ask CoCoA to allow a certain approximation on the coordinates of the points:

```
/**/ epsilon := [0.1, 0.1]; // coord approximation 0.1
/**/ AP01 := ApproxPointsNBM(P, mat(points), mat([epsilon]));
/**/ indent(AP01.AlmostVanishing);
[
  x^2 +(4999/5001)*y^2 -165000/1667, // almost a circle
  x*y^3 -49*x*y,
  y^5 -149*y^3 +4900*y
]

/**/ epsilon := [0.01, 0.01]; // approximation 0.01 for each coord
/**/ AP001 := ApproxPointsNBM(P, mat(points), mat([epsilon]));
/**/ indent(AP001.AlmostVanishing);  // not "epsilon-near" a conic
[
  x^2*y +(49/51)*y^3 +(-4900/51)*y,
  x^3 +(51/49)*x*y^2 -100*x,
  y^4 +(-2499/2)*x^2 +(-2699/2)*y^2 +124950,
  x*y^3 -49*x*y
]
```

§7. Gröbner bases and rational coefficients

It is well known that computations with coefficients in $\mathbb{Q}$ can often be very costly in terms of both time and space. For Gröbner bases over $\mathbb{Q}$ we are free to multiply the polynomials by any non-zero rational; so we can clear denominators and remove integer content. Avoiding rational arithmetic this way does yield some benefit, but is not wholly satisfactory.

Sometimes the Gröbner basis has complicated coefficients (*i.e.* we mean *big numerators and denominators*), but more often the coefficients in the answer are reasonably sized, while the computation to obtain them involved far more complicated coefficients: this problem is known as *intermediate coefficient swell.*

The phenomenon of coefficient swell is endemic in computer algebra, and many techniques have been investigated to tackle this problem. We illustrate two techniques used in CoCoA.

7.1. TwinFloat

CoCoA offers floating-point arithmetic with a heuristic verification of correctness: the aim is to offer a good compromise between the speed of floating-point computation and the reliability of exact rational arithmetic — for a fuller description see the article [2]. Normally a twin-float computation will produce either a good approximation to the correct result or an indication of failure; strictly, there is a very small chance of getting a wrong result, but this never happens in practice.

To perform a computation with twin-floats the user must first specify the required precision; CoCoA will then perform the computation checking heuristically that the result of every twin-float operation has at least that precision. If the check fails then CoCoA signals an "insufficient precision" error; the user may then restart the computation specifying a higher precision. Although twin-float values are, by definition, approximate, all input values are assumed to be exact (so they can be converted to a twin-float of any precision).

It is also possible to convert a twin-float value to an exact rational number. Like all other twin-float operations, this conversion may fail because of "insufficient precision". Printing out a twin-float value automatically attempts conversion to a rational as rationals are easier to read and comprehend in the context of exact computations.

```
/**/ RR16 := NewRingTwinFloat(16);
/**/ use RR16_X ::= RR16[x,y,z];
/**/ f := 12345678*x+1/456789;
/**/ f;            // both coeffs are printed as rationals
```

```
12345678*x +1/456789
/**/ f * 10^3;    // first coeff is printed in "floating-point"
0.12345678*10^11*x +1000/456789
/**/ f * 10^5;    // both coeffs are printed in "floating-point"
0.12345678*10^13*x +0.2189194573
```

Twin-floats include a (heuristically verified) test for zero; this means it is possible to compute Gröbner bases with twin-float coefficients. One reason for wanting to do this is that often the computation of a Gröbner basis over the rationals involves "complicated fractions" (*i.e.* whose numerator and denominator have many digits), and arithmetic with such complicated values can quickly become very costly. In contrast, with twin-floats the arithmetic has fixed cost (dependent on the precision chosen, of course). These characteristics are exploited in CoCoA for the computation of the gin, described in Section 4.

7.2. (Fault-tolerant) Rational reconstruction

A widely used technique for avoiding intermediate coefficient swell is to perform the computation modulo one or more prime numbers, and then lift/reconstruct the final result over $\mathbb{Q}$. We call this the **modular approach**. There are two general classes of method: **Hensel Lifting** and **Chinese Remaindering**, the first is not universally applicable but does work well for polynomial gcd and factorization, while the second is widely applicable and works well in most other contexts.

The modular approach has been successfully used in numerous contexts, here are a few examples: polynomial factorization [24], determinant of integer matrices [10], ideals of points (see Section 6), implicitization (see Section 5.2), and minimal polynomial (see Section 5.3).

In any specific application there are two important aspects which must be addressed before a modular approach can be adopted, and there is no universal technique for addressing these issues:

- knowing how many different primes to consider to guarantee the result (*i.e.* find a realistic bound for the size of coefficients in the answer);
- handling *bad primes*: namely those whose related computation follows a different route, yielding an answer with the wrong "shape" (*i.e.* which is not the modular reduction of the correct, non-modular result).

In the context of Gröbner bases we do not have good, general solutions to either of these issues. One of the first successes in applying modular techniques to certain instances of Gröbner basis computation appeared in [13]. Finding good ways to employ a modular approach for general Gröbner bases is still an active area. CoCoA does not currently use a modular approach for general Gröbner basis computations.

A vital complement to the modular computation is the reconstruction of the final, rational answer from the modular images. CoCoA offers functions for

- combining two residue-modulus pairs into a single "combined" residue-modulus pair (using the Chinese Remainder Theorem);
- determining a "simple" rational number corresponding to a residue-modulus pair; this is called **rational reconstruction**.

Correct reconstruction can still be achieved even in the presence of a few "faulty residues" (see [3]); this fault-tolerance was exploited in the functions for hypersurface implicitization (see Section 5.2) and minimal polynomials (see Section 5.3).

Here we see how two modular images can be combined in CoCoA (using "`CRTPoly`"), and then the correct rational result is reconstructed from the combined residue-modulus pair (using "`RatReconstructPoly`").

```
/**/ P1 ::= ZZ/(123457)[x];
/**/ P2 ::= ZZ/(234571)[x];
/**/ RingElem(P1, "3*x/11-1/5");   // modular image in P1
22447*x -49383
/**/ RingElem(P2, "3*x/11-1/5");   // modular image in P2
-42649*x +46914

/**/ use P ::= QQ[x];
/**/ combined := CRTPoly(22447*x -49383, 123457,
                         -42649*x +46914, 234571);
/**/ combined;  // in P
record[modulus:=28959431947, residue:=13163378158*x-11583772779]
/**/ RatReconstructPoly(combined.residue, combined.modulus);
(3/11)*x -1/5
```

§8. Gröbner bases in C++ with CoCoALib

As mentioned in Section 1.2, our aim is to make computation using CoCoALib as easy as using CoCoA-5. To illustrate this, here is the first example from Section 2 but in C++:

```
ring P = NewPolyRing(RingQQ(), symbols("x,y,z"));
ideal I = ideal(RingElem(P, "x^3+3"),
                RingElem(P, "y-x^2"),
                RingElem(P, "z-x-y") );
cout << GBasis(I);
```

In comparison to CoCoA-5, this C++ code is more cumbersome and involved, though we maintain that it is still reasonably comprehensible (once you know that `cout <<` is the C++ command for printing).

We have designed CoCoA-5 and CoCoALib together with the aim of making it easy to develop a prototype implementation in CoCoA-5, and then convert the code into C++. To facilitate this conversion we have used the same function names in both CoCoA-5 and CoCoALib, whenever possible, and we have preferred using traditional "functional" syntax in CoCoALib over object oriented "method dispatch" syntax (*e.g.* `GBasis(I)` rather than `I.GBasis()`). This means that most of the CoCoA-5 examples given here require only minor changes to become equivalent C++ code for use with CoCoALib.

To maintain the "friendly" tradition of CoCoA software for mathematicians, and to extend it to "mathematical programmers", our design of CoCoALib follows these aims:

- Designed to be **easy and natural to use**
- Motto: **"No nasty surprises"** (*e.g.* avoid ambiguities)
- Execution speed is **good**
- Well-documented, including **many example programs**
- **Free and open source** C++ code (GPL3 licence)
- Source code is **clean and portable** (currently C++03)
- Design respects the underlying **mathematical** structures (using C++ inheritance, no templates)
- **Robust** exception-safe, thread-safe

§9. Conclusion

The CoCoA software aims to make it easy for everyone to use Gröbner bases, whether directly or indirectly through some other function. The CoCoA-5 system is designed to be welcoming to those with little computer programming experience, while the CoCoALib library aims to make it easy for experienced programmers to use Gröbner bases in their own programs.

We hope this helps everyone to have their Gröbner basis!

References

[1] J. Abbott, *The design of CoCoALib*, International Congress on Mathematical Software, 2006, pp. 205–215.

[2] J. Abbott, *Twin-Float Arithmetic*, Journal of Symbolic Computation, **47**, 2012, pp. 536–551.

[3] J. Abbott, *Fault-tolerant modular reconstruction of rational numbers*, Journal of Symbolic Computation **80**, 2017, pp. 707–718.

[4] J. Abbott and A.M. Bigatti, *CoCoALib: a C++ library for doing Computations in Commutative Algebra.* Available at `http://cocoa.dima.unige.it/cocoalib`

[5] J. Abbott, A.M. Bigatti, G. Lagorio, *CoCoA-5: a system for doing Computations in Commutative Algebra.* Available at `http://cocoa.dima.unige.it/`

[6] J. Abbott, A.M. Bigatti, M. Kreuzer, L. Robbiano, *Computing Ideals of Points*, Journal of Symbolic Computation **30**, 2000, pp. 341–356.

[7] J. Abbott, A.M. Bigatti, L. Robbiano, *Implicitization of Hypersurfaces*, Journal of Symbolic Computation **81**, 2017, pp. 20–40.

[8] J. Abbott, A.M. Bigatti, E. Palezzato, L. Robbiano, *Computing and Using Minimal Polynomials*, submitted, arXiv preprint `arXiv:1702.07262`.

[9] J. Abbott, A.M. Bigatti, C. Söger, *Integration of libnormaliz in CoCoALib and CoCoA 5* Proceedings of ICMS 2014, LNCS Springer Verlag, 2014.

[10] J. Abbott, M. Bronstein, T. Mulders, *Fast deterministic computation of determinants of dense matrices*, ISSAC '99 Proceedings, ACM New York, 1999, pp. 197–204.

[11] J. Abbott, C. Fassino, M.L. Torrente, *Stable border bases for ideals of points*, Journal of symbolic computation **43** (12), 2008, pp. 883–894.

[12] J. Abbott, M. Kreuzer, L. Robbiano, *Computing zero-dimensional schemes*, Journal of Symbolic Computation, **39**, 2005, pp. 31–49.

[13] E.A. Arnold *Modular Algorithms for Computing Groebner Bases*, Journal of Symbolic Computation, **35**, pp. 403–419 (2003).

[14] A.M. Bigatti, R. La Scala, L. Robbiano, *Computing Toric Ideals*, Journal of Symbolic Computation, **27**, pp. 351–365 (1999).

[15] B. Buchberger, H.M. Möller, *The construction of multivariate polynomials with preassigned zeros*, Proceedings of the European Computer Algebra Conference (EUROCAM '82), **144**, Lecture Notes in Comp.Sci., pp. 24–31 (1982).

[16] C. Fassino, *Almost vanishing polynomials for sets of limited precision points*, Journal of symbolic computation **45** (1), pp. 19–37 (2010).

[17] J.C. Faugère, P. Gianni, D. Lazard, T. Mora. *Efficient Computation of Zero-dimensional Gröbner Bases by Change of Ordering.* Journal of Symbolic Computation, **16**(4), 329-344 (1993).

[18] M. Galassi et al, *GNU Scientific Library Reference Manual* (*3rd Ed.*), ISBN 0954612078. *GSL.* Available at `http://www.gnu.org/software/gsl`

[19] Anders N. Jensen, *Gfan, a software system for Gröbner fans and tropical varieties*. Available at
`http://home.imf.au.dk/jensen/software/gfan/gfan.html`

[20] M. Kreuzer, L. Robbiano, *Computational Commutative Algebra 1*, Springer Verlag, 2000.

[21] M. Kreuzer, L. Robbiano, *Computational Linear and Commutative Algebra*, Springer, Heidelberg 2016.

[22] Teo Mora and Lorenzo Robbiano, *The Gröbner fan of an ideal*, J. Symb. Comput., **6** (2/3), pp. 183–208 (1988).

[23] B. Sturmfels, *Gröbner bases and convex polytopes*, American Mathematical Soc. (1996).

[24] H Zassenhaus, *On Hensel Factorization, I*, J. Number Theory vol. 1, pp. 291–311 (1969).

(John Abbott) *Institut für Mathematik, Universität Kassel*
E-mail address: `abbott@dima.unige.it`

(Anna Maria Bigatti) *Dipartimento di Matematica, Universitàdegli Studi di Genova*
E-mail address: `bigatti@dima.unige.it`

Advanced Studies in Pure Mathematics 77, 2018
The 50th Anniversary of Gröbner Bases
pp. 25–33

Gröbner bases computation by triangularizing Macaulay matrices

Bruno Buchberger

Abstract.

In my PhD thesis 1965 and the subsequent publication 1970 in aequationes mathematicae, I introduced the notion of Gröbner bases and proved a characterization theorem for Gröbner bases on which an algorithm for constructing Gröbner bases can be based. The main idea for the theorem and the algorithm was the notion of "S-polynomials". Most of the subsequent work on the algorithmic theory of Gröbner bases, including the implementation of the Gröbner bases technology in mathematical software systems like Mathematica, Maple, etc. was based on this approach.

In the early eighties, I proposed a completely different strategy for computing Gröbner bases proceeding by the following three steps:
1. Produce the set of all multiples $u.f$ of the polynomials f in the initial basis F by all power products u (which we call "generalized Sylvester matrix" or "Macaulay matrix" of F).

2. Triangularize this matrix.

3. Take the "contour" in the diagonal of the matrix, i.e. the set of all those polynomials in the diagonal whose leading power product is not a multiple of the leading power product of any other polynomial in the diagonal.

It is easy to prove that the above procedure yields a Gröbner basis if one starts with the infinite matrix of *all* shifts $u.f$. However, of course, this is not an algorithm. So I posed the question whether one can give an a-priori bound on D so that, if one puts all shifts $u.f$ with degree of u smaller than D to the initial matrix, the matrix resulting by triangularization and contour formation will be a Gröbner basis for F. Over the years, several people tried to find such a bound but only recently (2014) my PhD student Manuela Wiesinger-Widi was able to establish such a bound combining known bounds by G. Hermann and T. W. Dubé for the ideal membership and the Gröbner bases degree problem, respectively.

Received November 6, 2016.
2010 *Mathematics Subject Classification.* 33F10, 13P10.
Key words and phrases. Gröbner basis, S-polynomial, Macaulay matrix.

§1. Computing Gröbner Bases by S-Polynomials and the Problem of Constructing a Corresponding Macaulay Matrix

My PhD thesis[1, 2] contains the following results (for sets of multivariate polynomials F):

- Introduction of the notion of Gröbner bases: F is Gröbner basis iff reduction w.r.t. F is unique.
- Characterization Theorem: F is Gröbner basis iff all S-polynomials of F reduce to 0 w.r.t. F.
  (S-polynomial of f and $g := u.\ f - v.\ g$, with certain power products u, v.)
  (Note: The Characterization Theorem is an algorithm for deciding whether a given F is a Gröbner basis!)
- Algorithm for constructing Gröbner bases: Iterate the formation of S-polys and add non-zero remainders.
- Correctness of the algorithm: by the Characterization Theorem.
- Termination of the algorithm: by (a re-invention and new proof) of Dixon's Lemma.
- Application: Algorithm for constructing a linearly independent basis for the residue class ring modulo the ideal generated by F.
- Application: Algorithm for constructing the multiplication table for the associative algebra formed by this linearly independent basis.
- Application: Algorithm for deciding the solvability and zero-dimensionality of systems F of algebraic equation and computing all roots with multiplicities (in the zero-dimensional case).
- Application: Computation of the Hilbert function for Ideal(F).
- Complete implementation of the algorithm on the ZUSE Z23 computer and example computations.
- A first complexity analysis of the algorithm for the bivariate case.

Later, in [3], I introduced the technique of "criteria" (in particular, the chain criterion) for eliminating the consideration of (many) unnecessary S-polynomials in the course of computing Gröbner bases. In many cases, this makes the algorithm much faster. In [5], I proposed the technique "first auto-reduce, only then consider S-polynomials". (In the terminology of the current paper, "auto-reduction" basically is triangularization of the coefficient matrix of polynomials that have been computed up to a certain stage of the algorithm.) Many times, after auto-reduction, no or only a few S-polynomials have to be considered. However, only heuristics are available on when this happens.

There is a huge literature on the theory of Gröbner bases and their computation using the S-polynomials approach, see [6]. The use of S-polynomials was the prevailing approach to algorithmic Gröbner bases theory. However, there were exceptions:

- Gröbner's original 1954 idea [11] for obtaining a multiplication table for the associative algebra modulo Ideal(F). (Termination was a question and this lead to [1].)
- Mayr's approach [14] for obtaining an exponential space upper bound for Gröbner bases computation.
- Faugére's et al. approach (F4, F5) [8, 9]: Termination of F4 was still based on S-polynomials.
- Grigoriev's approach [10] for bounds for the number of vectors in Macaulay-like matrices.

§2. Gröbner Bases and Resultants

I have often been asked about the relation of Gröbner bases with resultants and related notions (Sylvester matrices, Macaulay matrices) and some people thought that the effect of repeated resultants on bivariate polynomials (eliminating variables one by one) is the same as a Gröbner basis computation. However, definitely, this is not the case.

A more appropriate view is the following:

- In the univariate case, an alternative for computing the GCD of a set F of two polynomials by Euclid's algorithm is the triangulariziation of the Sylvester matrix of F and solvability can also be decided by calculating the determinant of the Sylvester matrix (the resultant). By Habicht's theory [12, 15] one can in fact understand Euclid's algorithm as a particular way of triangularizing the Sylvester matrix.
- In the linear multivariate case, an alternative for variable elimination by Gauss' algorithm is the triangularization of the extended coefficient matrix. Solvability can again be decided by calculating the determinant of the coefficient matrix. In this case, Gauss' algorithm on the linear polynomials and triangularization of the coefficient matrix is in fact basically identical.
- In the non-linear multivariate case, we can compute a Gröbner basis of the initial polynomials by my S-polynomial algorithm (which specializes to Euclid and Gauss in the univariate and linear multivariate cases, respectively). Now the question is,

whether there is a corresponding matrix whose triangularization yields a Gröbner basis and whether my S-polynomial algorithm can be considered as just a special way of triangularizing this matrix.

In her PhD thesis, my student M. Wiesinger-Widi [17], solved the first part of the question thoroughly, i.e. we now can construct a matrix from the coefficients of a given set F of polynomials (a "generalized Sylvester matrix" or "Macaulay matrix") whose triangularization contains a Gröbner basis for F. It is not yet clear, though, whether the computation of Gröbner bases by S-polynomials can be considered as just one special way of triangularizing this matrix, i.e. whether a kind of "generalized Habicht's theory" could be established.

§3. An Inconstructive Method for Computing Gröbner Bases by Triangularizing a Macaulay Matrix

In [4], I proved that the following steps yield a Gröbner basis for any polynomial set F. (Note: we use "Sylvester" interchangeably with "Generalized Sylvester" and "Macaulay"):

S := Sylvester(F) := set of all multiples $u.f$ of the polynomials f in F with all power products u.

Consider the elements in Sylvester(F) as rows of an (infinite) matrix with the columns numbered by the power products and ordered according to the admissible ordering of power products w.r.t. to which one wants to find the Gröbner basis for F.

T := Triangularized(S). (In fact this is nothing else but a special kind of auto-reduction in the terminology of [5].)

C := Contour(T) := the set of those polynomials in T whose leading power products are not multiple of the leading power product of any other polynomial in T.

Then C is a finite Gröbner basis of the original set F. (Finiteness can be proved, again, by applying Dixon's lemma.)

Proof Sketch:

VectorSpace(Sylvester(F)) = Ideal(F).

VectorSpace(Sylvester(F)) = VectorSpace(Triangularized (Sylvester(F)).

The leading power product of any f in Ideal(F) must occur in Triangularized(Sylvester(F)) and, hence can be reduced by a polynomial in Contour(Triangularized(Sylvester(F))).

In fact, I had this result much earlier but I did not think it was worth publishing because it is only a "method", not an algorithm.

If we knew a finite a priori bound on the degrees of the multiplies $u.f$ that have to go into Sylvester(F) in order to guarantee that Contour(Triangularized (Sylvester(F))) is a Gröbner basis for F, then the above method would be an algorithm. The upper bound should be expressed in terms of

n (number of polys in F),
r (number of polynomials in F),
and d (maximum degree of polynomials in F).

Over the years, I proposed this problem of finding such an upper bound a couple of times to my PhD students. However, only recently (2014), Manuela Wiesinger-Widi, solved it by combining Hermann's and Dube's bounds in a clever way, see the theorems in the next section.

§4. Turning the Inconstructive Method into an Algorithm

Theorem 4.1. (***Manuela Wiesinger-Widi 2014***): *In the above procedure, it suffices to take the power products u with degree less or equal to*

$$2\left(\frac{d^2}{2}+d\right)^{2^{n-1}}+\sum_{j=0}^{n-1}(r\ d)^{2^j}.$$

In fact, the resulting Gröbner basis will be head-reduced.

Also, by the same approach, the following theorem can be proved.

Theorem 4.2. (***Manuela Wiesinger-Widi 2014***): *If, in the above procedure, one considers the matrix of multiples u.f with power products u whose degree is less or equal*

$$\sum_{j=0}^{n-1}(r\ d)^{2^j}$$

then 1 is in Ideal(F) if and only if the above procedure yields a matrix containing a polynomial with leading power product 1.

Note: The above **(generalized) Sylvester matrix** is the appropriate analogue to the univariate Sylvester matrix in the general, multivariate, case in the sense I explained in the introduction. The theorems were proved in 2014, the thesis appeared in 2015.

Proof of the Wiesinger-Widi theorems: The proof uses Hermann's and Dube's bound:

Hermann's Bound [13]**:** If g is in Ideal(F) then there exist $q_1, \ldots, q_r$ such that

$$g = \sum_{i=1}^{r} (q_i \, . \, F_i)$$

and, for all i,

$$\text{degree}\,(q_i) \leq \text{degree}(g) + \sum_{j=0}^{n-1} (r \; d)^{2^j}.$$

Dubé's Bound [7]: If G is the reduced Gröbner basis of F then, for all g in G,

$$\text{degree}(g) \leq 2\left(\frac{d^2}{2} + d\right)^{2^{n-1}}.$$

The proof of Wiesinger-Widi's theorems now proceed as follows:

Lemma: If G is a finite Gröbner basis for F and the finite (truncated Sylvester) matrix S that contains all the multiples $u.f$ with f in F and the power product u occuring in one of the q_i of the presentations

$$g = \sum_{i=1}^{r} (q_i \, . \, F_i)$$

of the polynomials g in G, and T = Triangularized(S), then Contour(T) is also a (head-reduced) Gröbner basis of F.

Proof of Lemma: $G \subseteq$ VectorSpace(S) = Vectorspace(Triangularized(S)). Every leading power product of a polynomial g in G must occur among the leading power products of T since T is triangularized. By a well-known property of Gröbner bases, all polynomials in G that are not on Contour(T) can be canceled.

Proof of Theorem 4.1:

By Dubé, we know that there exists a Gröbner basis G for F with

$$\text{degree}(g) \leq 2\left(\frac{d^2}{2} + d\right)^{2^{n-1}},$$

for all g in G.

By Hermann, each of these g has a presentation

$$g = \sum_{i=1}^{r} (q_i \, . \, F_i)$$

such that, for all i,

$$\text{degree}\,(q_i) \leq \text{degree}\,(g) + \sum_{j=0}^{n-1} (r\ d)^{2^j}.$$

Hence, if we take all the multiples $u.f$ described in the Theorem into the initial (truncated) Sylvester matrix, by the above Lemma, the contour of the triangularized matrix is a (head-reduced) Gröbner basis.

Proof of Theorem 4.2: Similar. Note that $\{1\}$ is a Gröbner basis. Hence degree(g) in the previous proof becomes zero.

§5. Remarks on Applications and Future Research

The method is not "practical" for computing a Gröbner basis of F:

- The polys in the S-poly algorithm for Gröbner bases, typically, stay way below the above upper degree bounds!
- The S-poly algorithm for Gröbner bases, typically, only produces very few of the rows in Sylvester(F).

(Example: The Gröbner basis computation of

$$\{-x + xy^2,\ x^2y - x\}$$

w.r.t. a total degree ordering, by S-polynomials, does not exceed degree four whereas the above bound, for this case, would request us to first set up a matrix with polynomials of up to degree 155. In [17], in fact, the author gives much lower bounds for the special case of binomials. However, still, concrete computations using S-polynomials typically stay significantly below these bounds.)

In other words, my 1965 Gröbner bases algorithm based on S-polynomials can be considered as an efficient way of avoiding to work with big Macaulay matrices. Analogy in case n=2: Euclid's algorithm can be considered as an efficient way of avoiding to work with big Sylvester matrices.

Anyway, the above results can be seen as a theoretical frame for the Gröbner 1954 approach and more recent algorithms for constructing Gröbner bases ([8, 9, 10]).

Also, the above theorem and algorithm suggests to extend Habicht's 1948 theory of subresultants (introduced for the univariate case) to the general case of Gröbner bases. Habicht gives a priori estimates on the coefficients that may appear in GCD computations, see also [15]. A general Habicht theory could also have applications for the numeric computation of Gröbner bases: If we knew exactly, which subdeterminants of the generalized Sylvester matrices may occur in the computation of Gröbner bases, we could derive lower bounds for the size of possible coefficients in the computation and distinguish between "very small" and "zero" in numerical computation. In addition, Habicht's theory gives very subtle predictive information on common factors in the coefficients of polynomials occuring in GCD computation. Similar results can be expected from a detailed study of the generalized Sylvester matrices. This study is not yet undertaken but I think it would be worthwhile. These investigations are quite demanding in terms of proof technology and, for me, were an important motivation for initiating the Theorema project on automated theorem proving in 1995. A recent PhD thesis of another student of mine [16] in the frame of the Theorema project gives some impression about how formalization and automated theorem proving could help in such investigations.

References

[1] B. Buchberger, An Algorithm for Finding the Basis Elements in the Residue Class Ring Modulo a Zero Dimensional Polynomial Ideal (German). PhD thesis, Mathematical Institute, University of Innsbruck, Austria, 1965. (English Translation in Journal of Symbolic Computation, Special Issue on Logic, Mathematics, and Computer Science, **41**, 2006, pp. 475–511.)

[2] B. Buchberger, An Algorithmic Criterion for the Solvability of Algebraic Systems of Equations (German). Aequationes Mathematicae, **3**, 1970, pp. 374–383.
(English translation in B.Buchberger, F.Winkler eds.: Gröbner Bases and Applications, London Math, Society Lecture Note Series, **251**, Cambridge University Press, 1998, pp. 535–545.)

[3] B. Buchberger, A Criterion for Detecting Unnecessary Reductions in the Construction of Gröbner Bases, In: Proceedings of EUROSAM' 79, Springer LNCS, **72**, 1979, pp. 3–21.

[4] B. Buchberger, Miscellaneous Results on Gröbner-Bases for Polynomial Ideals II, Technical Report 83-1, Department of Computer And Information Sciences, University of Delaware, 1983.

[5] B. Buchberger, Gröbner Bases: An Algorithmic Method in Polynomial Ideal Theory, Chapter 6, In: Multidimensional Systems Theory - Progress, Directions and Open Problems in Multidimensional Systems Theory, (eds. N.K. Bose), Reidel Publishing Company, Dodrecht - Boston - Lancaster, 1985, pp. 184–232.

[6] B. Buchberger, M. Rady, H. Rahkooy, T. Vo Ngoc, M. Wiesinger-Widi, W. Windsteiger, and A. Zapletal, Gröbner Bases Bibliography, http://www.risc.jku.at/Groebner-Bases-Bibliography/.

[7] T.W. Dubé, The Structure of Polynomial Ideals and Gröbner Bases, SIAM Journal on Computing, **19/4**, 1990, pp. 750–773.

[8] J.C. Faugére, A New Efficient Algorithm for Computing Groebner Bases (F4), Journal of Pure and Applied Algebra, **39**, 1999, pp. 61–88.

[9] J.C. Faugére, A New Efficient Algorithm for Computing Groebner Bases Without Reductions to Zero (F5), Proceedings of ISSAC 2002, pp. 75–83.

[10] D. Grigoriev, Bounds on Numbers of Vectors of Multiplicities for Polynomials which are Easy to Compute, Proc. ACM Intern. Conf. Symbolic and Algebraic Computations 2000, Scotland, 2000, pp. 137–145.

[11] W. Gröbner, Über die Eliminationstheorie (On Elimination Theory), Monatshefte für Mathematik, **55**, 1950, pp. 71–78.
(English translation by M. Abramson in ACM SIGSAM Bulletin, **32/2**, 1998, pp. 40–46.)

[12] W. Habicht, Eine Verallgemeinerung des Sturmschen Wurzelzählverfahrens (A Generalization of Sturm's Method for Root Counting), Comm. Math. Helvetici, **21**, 1948, pp. 99–116.

[13] G. Hermann, The Question of Finitely Many Steps in Polynomial Ideal Theory (German), Mathematische Annalen, **95**, 1926, pp. 736–788.
(English translation in ACM SIGSAM Bull., **39/3**, 1998, pp. 8–30.)

[14] K. Kühnle, and E.W. Mayr, Exponential Space Computation of Gröbner Bases, Proceedings of ISSAC'96, Zürich, ACM, pp. 63–71.

[15] R. Loos, Generalized Polynomial Remainder Sequences, In: Computer Algebra - Symbolic and Algebraic Computation, (eds. B. Buchberger, G.E. Collins, and R. Loos), Springer Verlag, Wien, 1982, pp. 115–137.

[16] A. Maletzky, Computer-Assisted Exploration of Gröbner Bases Theory in Theorema, PhD Thesis, Johannes Kepler University, Research Institute for Symbolic Computation, Linz / Hagenberg, Austria, 2016.

[17] M. Wiesinger-Widi, Gröbner Bases and Generalized Sylvester Matrices, Ph.D. Thesis, Johannes Kepler University, Research Institute for Symbolic Computation, Linz / Hagenberg, Austria, 2015.

Research Institute for Symbolic Computation (RISC),
Johannes Kepler University, Linz, Austria
E-mail address: `bruno.buchberger@risc.jku.at`

Advanced Studies in Pure Mathematics 77, 2018
The 50th Anniversary of Gröbner Bases
pp. 35–86

Algebraic problems in structural equation modeling

Mathias Drton

Abstract.

The paper gives an overview of recent advances in structural equation modeling. A structural equation model is a multivariate statistical model that is determined by a mixed graph, also known as a path diagram. Our focus is on the covariance matrices of linear structural equation models. In the linear case, each covariance is a rational function of parameters that are associated to the edges and nodes of the graph. We statistically motivate algebraic problems concerning the rational map that parametrizes the covariance matrix. We review combinatorial tools such as the trek rule, projection to ancestral sets, and a graph decomposition due to Jin Tian. Building on these tools, we discuss advances in parameter identification, i.e., the study of (generic) injectivity of the parametrization, and explain recent results on determinantal relations among the covariances. The paper is based on lectures given at the 8th Mathematical Society of Japan Seasonal Institute.

CONTENTS

Received September 28, 2016.
2010 *Mathematics Subject Classification.* 62H05, 13P25, 05C90.
Key words and phrases. algebraic statistics, covariance matrix, Gaussian distribution, graphical model, Gröbner basis, structural equation model.

Part I. Structural Equation Models

§1. Motivation

The following example serves well to introduce the statistical models we will consider. It features the simplest instance of what is known as an instrumental variable model. An empirical study that shows this type of model 'in action' can be found in [29].

Example 1.1. Does a mother's smoking during pregnancy harm the baby? To answer this question researchers conduct a study in which they record, for a sample of pregnancies, the baby's birth weight and the average number of cigarettes the mom smoked per day during the first trimester. The researchers observe a significant negative correlation between the *birth weight* and *smoking* and are tempted to conclude that smoking has a negative effect on the baby's health, with an increase in the number of cigarettes smoked leading to lower birth weight.

The cigarette companies are not surprised by this finding. They argue, however, that smoking does not harm baby. Instead, heavier smoking merely reflects underlying factors that are the true causes of low birth weight. Such confounding variables could, for instance, be of socio-economic nature. In the context of the smoking-lung cancer debate, prominent Statistician Ronald Fisher liked to argue that correlations can be attributed to unobserved variables of genetic nature [59].

Familiar with this type of counter-argument, the researchers cleverly recorded a third variable: The *tax rate* on tobacco products in the local jurisdictions of the mothers in the sample. It is not unreasonable to assume that the tax rate does not have a direct effect on the baby's health. If there is then variation in the tax rate and higher taxes have an effect on the amount of smoking, then the effect that smoking has on birth weight can be estimated in a model that allows for the presence of unobserved confounders, as we will see shortly.

The above narrative suggests a number of cause-effect relations, as well as the absence thereof. Qualitatively these are summarized in the

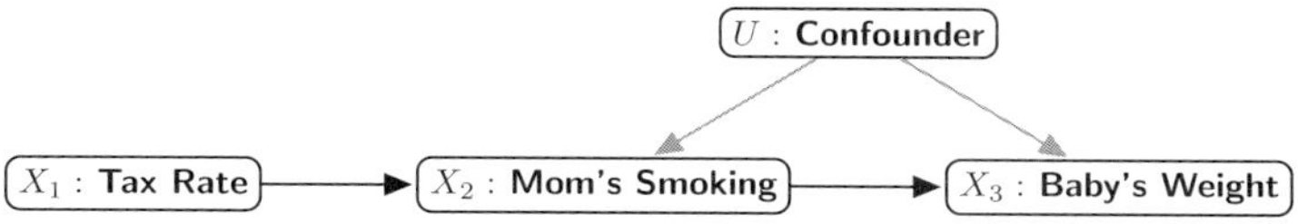

Fig. 1.1. Directed graph for an instrumental variable model.

graph in Figure 1.1. The variables in play are the nodes of the graph and cause-effect relations are indicated as directed edges. Variable U represents an unobserved confounder; we draw its edges in gray.

Structural equation models turn the qualitative descriptions of causes and effects into quantified functional relationships. In this article, the functional relationships will always be *linear*. The linear structural equation model for the present example is based on the following system of *structural equations*:

$$X_1 = \lambda_{01} + \varepsilon_1, \tag{1.1}$$

$$X_2 = \lambda_{02} + \lambda_{12}X_1 + \lambda_{u2}U + \varepsilon_2, \tag{1.2}$$

$$X_3 = \lambda_{03} + \lambda_{23}X_2 + \lambda_{u3}U + \varepsilon_3, \tag{1.3}$$

$$U = \lambda_{0u} + \varepsilon_u. \tag{1.4}$$

Here, the error terms ε_1, ε_2, ε_3, ε_u are independent random variables with zero mean. The eight coefficients λ_{01}, λ_{02}, λ_{03}, λ_{0u}, λ_{12}, λ_{23}, λ_{u2}, and λ_{u3} are unknown parameters. Equation (1.1) indicates that variable X_1, the tax rate, has expectation λ_{01}, from which it deviates according to the distribution assumed for ε_1. The analogous statement for the unobserved confounder U is made in (1.4). In (1.2), the amount of smoking, denoted X_2, is modeled to be a linear function of the tax rate and independent noise. Similarly, (1.3) introduces birth weight, denoted X_3, as a noisy linear function of smoking.

The quantity of primary interest is the coefficient λ_{23} that quantifies the relationship between smoking and birth weight. Using data, we can estimate the joint distribution and, in particular, the covariance matrix of the three observed variables X_1, X_2 and X_3. Because the error terms are independent, the covariance between X_3 and X_1 is

$$\mathrm{Cov}[X_1, X_3] = \lambda_{23}\,\mathrm{Cov}[X_1, X_2]. \tag{1.5}$$

Hence, as long as $\mathrm{Cov}[X_1, X_2] \neq 0$, statistical inference about λ_{23} may be based on the ratio of the two covariances in (1.5).

In some applications of structural equation models latent (that is, unobserved) variables are of direct interest. For instance, concepts such as intelligence or depression in psychology are of this nature and measured only indirectly through other variables such as exam results or answers in questionnaires. While problems with explicit latent variables are ubiquitous [5], we will focus on models in which the effects of latent variables are summarized and represented merely in terms of correlations among the error terms in structural equations. This representation of dependence induced by latent variables is discussed in detail in [33, 44, 49, 50, 72].

Example 1.2. We take up the instrumental variable model from Example 1.1. The effects of the confounding variable U can be summarized by absorbing U into the error terms in (1.2) and (1.3). Define

$$\tilde{\varepsilon}_2 = \lambda_{u2}U + \varepsilon_2, \qquad \tilde{\varepsilon}_3 = \lambda_{u3}U + \varepsilon_3. \tag{1.6}$$

Retaining only the equations for the observed variables X_1, X_2, and X_3, we are left with the equation system:

$$X_1 = \lambda_{01} + \varepsilon_1, \tag{1.7}$$

$$X_2 = \lambda_{02} + \lambda_{12}X_1 + \tilde{\varepsilon}_2, \tag{1.8}$$

$$X_3 = \lambda_{03} + \lambda_{23}X_2 + \tilde{\varepsilon}_3. \tag{1.9}$$

However, and this is the significance of the unobserved variable U, the new error terms may be correlated because

$$\begin{aligned} \omega_{23} &:= \operatorname{Cov}[\tilde{\varepsilon}_2, \tilde{\varepsilon}_3] \\ &= \operatorname{Cov}[\lambda_{u2}U + \varepsilon_2, \lambda_{u3}U + \varepsilon_3] = \lambda_{u2}\lambda_{u3}\operatorname{Var}[U] \neq 0. \end{aligned} \tag{1.10}$$

In the sequel, we will focus on models that are given by equations such as (1.7)-(1.9), with one equation for each observed variable but error terms that may be correlated. Graphically, such models may be represented by a *mixed graph* that features directed edges to encode which variables appear in each structural equation and bidirected edges that indicate possibly nonzero correlations between error terms. The mixed graph for the model given by (1.7)-(1.9) is depicted in Figure 1.2, which shows unknown parameters as edge weights. At the nodes, we show the variances of the error terms, namely, $\omega_{11} = \operatorname{Var}[\varepsilon_1]$, $\omega_{22} = \operatorname{Var}[\tilde{\varepsilon}_2]$, and $\omega_{33} = \operatorname{Var}[\tilde{\varepsilon}_3]$. In the statistical literature, the mixed graph for a structural equation model is also known as a *path diagram*.

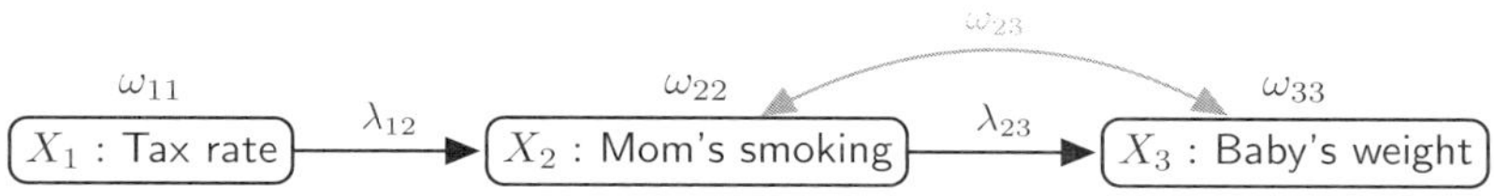

Fig. 1.2. Mixed graph for an instrumental variable model.

The ratio $\operatorname{Cov}[X_2, X_3]/\operatorname{Var}[X_2]$ is the regression coefficient when predicting X_3 from X_2. We have

$$\frac{\operatorname{Cov}[X_2, X_3]}{\operatorname{Var}[X_2]} = \lambda_{23} + \frac{\omega_{23}}{\operatorname{Var}[X_2]}. \tag{1.11}$$

Hence, linear regression predicting X_3 from X_2 only estimates the coefficient of interest if $\omega_{23} = 0$, as is the case when X_2 and X_3 do not depend on the latent variable U. When $\omega_{23} \neq 0$, the relation from (1.5), which involves all three variables, is needed to recover λ_{23}.

The remainder of this paper is organized as follows. Section 2 introduces linear structural equation models in full generality. We then formulate questions of statistical interest and the algebraic problems they correspond to (Section 3). Next, we examine the interplay between covariance matrices and mixed graphs. We treat the so-called trek rule (Section 4) and review useful results on subgraphs and graph decomposition (Sections 5 and 6). In Sections 7 and 8, we dive deeper into parameter identifiability, which in this paper means the question of whether a coefficient of interest can be recovered from the covariance matrix of the observed variables. Finally, we discuss relations among the entries of the covariance matrix (Sections 9-12).

§2. Linear Structural Equation Models

Let $\varepsilon = (\varepsilon_i : i \in V)$ be a random vector indexed by a finite set V. Define a new random vector $X = (X_i : i \in V)$ as the solution to the system of so-called structural equations

$$X = \Lambda^T X + \varepsilon. \tag{2.1}$$

Here, $\Lambda = (\lambda_{ij}) \in \mathbb{R}^{V \times V}$ is a matrix of unknown parameters. Suppose ε has covariance matrix $\Omega = (\omega_{ij}) = \operatorname{Var}[\varepsilon]$ with Ω being a positive definite matrix whose entries are again unknown parameters. Write I for the identity matrix. If $I - \Lambda$ is invertible, then the linear system in (2.1) is solved uniquely by $X = (I - \Lambda)^{-T}\varepsilon$, which has covariance matrix

$$\text{Var}[X] \;=\; (I-\Lambda)^{-T}\Omega(I-\Lambda)^{-1} \;=:\; \phi(\Lambda,\Omega). \tag{2.2}$$

Interesting settings are obtained by restricting the support of Λ and Ω, as is the case in our motivating example.

Example 2.1. Consider the setup from Example 1.2. If the equation system from (1.7)-(1.9) is written in vector form as in (2.1), then the coefficient matrix is

$$\Lambda = \begin{pmatrix} 0 & \lambda_{12} & 0 \\ 0 & 0 & \lambda_{23} \\ 0 & 0 & 0 \end{pmatrix}. \tag{2.3}$$

The error covariance matrix is

$$\Omega = \text{Var}[\varepsilon] = \begin{pmatrix} \omega_{11} & 0 & 0 \\ 0 & \omega_{22} & \omega_{23} \\ 0 & \omega_{23} & \omega_{33} \end{pmatrix}. \tag{2.4}$$

From (2.2), the covariance matrix of $X = (X_1, X_2, X_3)$ is found to be

$$\text{Var}[X] \;=\; \begin{pmatrix} \omega_{11} & \omega_{11}\lambda_{12} & \omega_{11}\lambda_{12}\lambda_{23} \\ \omega_{11}\lambda_{12} & \omega_{22}+\omega_{11}\lambda_{12}^2 & \omega_{23}+\lambda_{23}\sigma_{22} \\ \omega_{11}\lambda_{12}\lambda_{23} & \omega_{23}+\lambda_{23}\sigma_{22} & \omega_{33}+2\omega_{23}\lambda_{23}+\lambda_{23}^2\sigma_{22} \end{pmatrix}, \tag{2.5}$$

where σ_{22} denotes the $(2,2)$ entry of $\text{Var}[X]$. The relation from (1.5) can be confirmed from the $(1,2)$ and $(2,3)$ entry of $\text{Var}[X]$.

Restrictions on the support of a matrix naturally correspond to a graph. Specifically, we adopt mixed graphs because we are dealing with two matrices, Λ and Ω, whose rows and columns are indexed by the same set V. In structural equation modeling, this point of view originated in the work of Sewall Wright [73, 74].

A mixed graph with vertex set V is a triple $G = (V, D, B)$ where D and B are two sets of edges. The set $D \subset V \times V$ contains ordered pairs (i,j), which we also denote by $i \to j$ to visualize that such a pair encodes a directed edge pointing from i to j. Then i is the tail and j is the head of the edge. The elements of B are unordered pairs $\{i,j\}$ and encode bidirected edges that we also denote by $i \leftrightarrow j$. These edges have no orientation, so $i \leftrightarrow j \in B$ if and only if $j \leftrightarrow i \in B$. It will be convenient to call both endpoints i and j heads of $i \leftrightarrow j$. In our context, neither the bidirected part (V, B) nor the directed part (V, D) contain loops, so $i \to i \notin D$ and $i \leftrightarrow i \notin B$ for all $i \in V$. The mixed graph G is acyclic if (V, D) does not have any directed cycles $i \to \ldots \to i$.

Let $\mathbb{R}^D$ be the set of real $V \times V$-matrices $\Lambda = (\lambda_{ij})$ with support in D, that is,

$$\mathbb{R}^D = \{ \Lambda \in \mathbb{R}^{V \times V} : \lambda_{ij} = 0 \text{ if } i \to j \notin D \}. \tag{2.6}$$

Define $\mathbb{R}^D_{\text{reg}}$ to be the subset of matrices $\Lambda \in \mathbb{R}^D$ for which $I - \Lambda$ is invertible. If G is acyclic, then there is a permutation of V that makes $I - \Lambda$ unit upper triangular such that $\det(I - \Lambda) = 1$ for all $\Lambda \in \mathbb{R}^D$ and, thus, $\mathbb{R}^D = \mathbb{R}^D_{\text{reg}}$. Similarly, let PD_V be the cone of positive definite symmetric $V \times V$-matrices $\Omega = (\omega_{ij})$, and define $PD(B)$ to be the subcone of matrices supported over B, that is,

$$PD(B) = \{ \Omega \in PD_V : \omega_{ij} = 0 \text{ if } i \neq j \text{ and } i \leftrightarrow j \notin B \}. \tag{2.7}$$

Taking the error vector ε to be Gaussian (or in other words, to follow a multivariate normal distribution), we arrive at the following definition of a statistical model for the random vector X that solves (2.1). Readers looking for background such as the fact that linear transformations of a Gaussian random vector are Gaussian may consult a textbook on multivariate statistics, e.g., [2].

Definition 2.2. *The* linear structural equation model *given by a mixed graph $G = (V, D, B)$ is the family of all multivariate normal distributions on $\mathbb{R}^V$ with covariance matrix in the set*

$$\mathcal{M}_G = \{(I - \Lambda)^{-T}\Omega(I - \Lambda)^{-1} : \Lambda \in \mathbb{R}^D_{\text{reg}}, \Omega \in PD(B)\}.$$

The covariance parametrization *of the model is the map*

$$\phi_G : \mathbb{R}^D \times PD(B) \to PD_V, \qquad (\Lambda, \Omega) \mapsto (I - \Lambda)^{-T}\Omega(I - \Lambda)^{-1}.$$

The fiber *of a pair $(\Lambda, \Omega) \in \mathbb{R}^D_{\text{reg}} \times PD(B)$ is the preimage*

$$\mathcal{F}_G(\Lambda, \Omega) = \{(\Lambda', \Omega') \in \mathbb{R}^D_{\text{reg}} \times PD(B) : \phi_G(\Lambda', \Omega') = \phi_G(\Lambda, \Omega)\}.$$

As defined, a linear structural equation model does not impose any restrictions on the mean vector of the normal distributions. Consequently, the mean vector plays no role in our discussion. For instance, in maximum likelihood estimation we may assume without loss of generality that the mean vector is zero. Other questions we consider will directly concern covariances. Therefore, we may safely identify a linear structural equation model with its set of covariance matrices $\mathcal{M}_G$. On occasion, we will simply refer to $\mathcal{M}_G$ as the model.

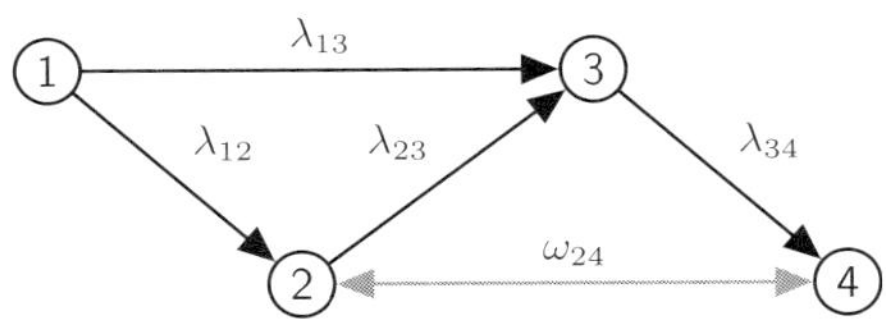

Fig. 2.1. An acyclic mixed graph known as the Verma graph.

Leaving statistics out of the picture, our interest is in the maps ϕ_G, their fibers $\mathcal{F}_G$ and their images $\mathcal{M}_G$. Algebra comes into play naturally.

Proposition 2.3. *For any mixed graph G, the map ϕ_G is a rational map whose image $\mathcal{M}_G$ and fibers $\mathcal{F}_G(\Lambda, \Omega)$ are semi-algebraic sets. The map ϕ_G is a polynomial map if and only if G is acyclic.*

Proof. That ϕ_G is rational follows from Cramer's rule for matrix inversion. The domain of ϕ_G is a semi-algebraic set and, thus, the fibers $\mathcal{F}_G(\Lambda, \Omega)$ are semi-algebraic as well. The Tarski-Seidenberg theorem implies that $\mathcal{M}_G$ is semi-algebraic. If $G = (V, D, B)$ is acyclic, then $\det(I - \Lambda) = 1$ for all $\Lambda \in \mathbb{R}^D$. Consequently, the entries of $(I - \Lambda)^{-1}$ are polynomial in Λ. If G is not acyclic, then $\det(I - \Lambda)$ is a non-constant polynomial. By the Leibniz formula, its terms correspond to collections of disjoint directed cycles in the graph; compare Theorem 1 in [41]. Q.E.D.

Example 2.4. The mixed graph $G = (V, D, B)$ in Figure 2.1 encodes the structural equations

$$\begin{aligned}
X_1 &= \lambda_{01} && + \varepsilon_1, \\
X_2 &= \lambda_{02} + \lambda_{12} X_1 && + \varepsilon_2 \\
X_3 &= \lambda_{03} + \lambda_{13} X_1 + \lambda_{23} X_2 && + \varepsilon_3 \\
X_4 &= \lambda_{04} + \lambda_{34} X_3 && + \varepsilon_4.
\end{aligned}$$

Only the errors ε_2 and ε_4 may be dependent and the error covariance matrix is

$$\Omega = \begin{pmatrix} \omega_{11} & 0 & 0 & 0 \\ 0 & \omega_{22} & 0 & \omega_{24} \\ 0 & 0 & \omega_{33} & 0 \\ 0 & \omega_{24} & 0 & \omega_{44} \end{pmatrix}.$$

Subtracting the coefficient matrix from the identity gives

$$I - \Lambda = \begin{pmatrix} 1 & -\lambda_{12} & -\lambda_{13} & 0 \\ 0 & 1 & -\lambda_{23} & 0 \\ 0 & 0 & 1 & -\lambda_{34} \\ 0 & 0 & 0 & 1 \end{pmatrix}$$

with $\det(I-\Lambda)=1$ and inverse

$$(I-\Lambda)^{-1} = \begin{pmatrix} 1 & \lambda_{12} & \lambda_{13}+\lambda_{12}\lambda_{23} & \lambda_{13}\lambda_{34}+\lambda_{12}\lambda_{23}\lambda_{34} \\ 0 & 1 & \lambda_{23} & \lambda_{23}\lambda_{34} \\ 0 & 0 & 1 & \lambda_{34} \\ 0 & 0 & 0 & 1 \end{pmatrix}.$$

To illustrate the form of the map ϕ_G, we display the coordinate function

$$\text{(2.8)} \quad \phi_G(\Lambda,\Omega)_{24} = \lambda_{12}\lambda_{13}\lambda_{34}\omega_{11} + \lambda_{12}^2\lambda_{23}\lambda_{34}\omega_{11} + \lambda_{23}\lambda_{34}\omega_{22} + \omega_{24}.$$

§3. Questions of Interest

Structural equation models are used to empirically estimate, test and possibly discover cause-effect relationships among a set of variables. In estimation and testing, the underlying graph is given. In discovery, we seek to infer the underlying graph, or in other words, perform model selection. This section gives a broad overview of algebraic problems that arise in the context of these statistical tasks. Only some of the problems are treated in the remainder of the paper, which focuses on parameter identifiability and polynomial relations between covariances.

3.1. Parameter identification

When specifying a model via a mixed graph $G=(V,D,B)$, a first question is whether the effects of interest are *identifiable*, that is, whether they are determined by the joint distribution of the observed variables. The importance of the question is clear: The joint distribution is what can be estimated from data. For linear and Gaussian models, the problem is equivalent to deciding whether the coefficients λ_{ij} in the structural equations can be recovered from the covariance matrix of the variables.

Different notions of parameter identifiability translate into related but different algebraic problems. The most stringent identifiability property is for a model to have all of its coefficients λ_{ij}, $i\to j\in D$, identifiable. In this case, we seek to answer the following:

Question 3.1. *Is the map ϕ_G is injective?*

Injectivity of ϕ_G can be decided efficiently, as we will discuss in Section 7. However, injectivity can be too strong of a requirement because all fibers are required to be singletons with $\mathcal{F}_G(\Lambda,\Omega)=\{(\Lambda,\Omega)\}$. Indeed, some interesting examples have fibers that are not singletons.

Example 3.2. The map ϕ_G fails to be injective when G is the graph for the instrumental variable model from Example 1.2. The relation from (1.5) shows that $\mathcal{F}_G(\Lambda,\Omega)=\{(\Lambda,\Omega)\}$ if $\lambda_{12}\neq 0$. If $\lambda_{12}=0$, however, then the fiber is infinite. Hence, all model parameters are identifiable as long as $\lambda_{12} \neq 0$. In the context of Example 1.2, this requires making an argument that higher tax rates impact the amount of smoking.

In the example just given, $\mathcal{F}_G(\Lambda, \Omega) = \{(\Lambda, \Omega)\}$ for generic choices of $(\Lambda, \Omega) \in \mathbb{R}^D_{\text{reg}} \times PD(B)$. In this case, we call ϕ_G *generically injective.* We are led to:

Question 3.3. *Is the map ϕ_G is generically injective?*

Generic injectivity turns out to be more difficult to decide. The problem's computational complexity has not yet been determined. In Section 8, we review methods to decide whether ϕ_G is generically injective as well as methods to decide when the fibers are generically infinite.

When ϕ_G is not generically injective, its generic fibers may be discrete sets. This property is known as *local identifiability* in the statistical literature. We will instead speak of ϕ_G being *generically finite-to-one* to highlight that in our case a discrete fiber is in fact finite because ϕ_G is rational. By the inverse function theorem, the question of whether ϕ_G is generically finite-to-one is the same as:

Question 3.4. *Does the Jacobian of ϕ_G have full column rank?*

The fiber $\mathcal{F}_G(\Lambda, \Omega)$ is defined by the equation system $\phi_G(\Lambda', \Omega') = \phi_G(\Lambda, \Omega)$. These equation systems have a generic number of complex solutions (i.e., the free entries of Λ and Ω are allowed to be complex numbers).

Definition 3.5. *The map ϕ_G is algebraically k-to-one if the equation systems defining its fibers generically have k complex solutions. We call the number k the algebraic degree of identifiability of G.*

The degree of identifiability may be determined by Gröbner basis methods (see Section 8). It is finite if and only if ϕ_G is generically finite-to-one. The main theorem in Section 7.6 shows that if ϕ_G is injective then its inverse is rational, which is the same as G having degree of identifiability one. Currently, there are no combinatorial results for when the degree is finite but larger than one. Example 8(b) in [34] has degree 3 but fibers whose cardinality over the reals is either one or three. To the author's knowledge, no example has been discovered in which ϕ_G is generically injective over the reals but algebraically k-to-one for $k \geq 2$.

An important question that we will not address in detail is the identifiability of only a single parameter λ_{ij} for a designated edge of interest $i \to j \in D$. This amounts to checking whether in every fiber the coefficient for the edge has only a single value. In other words, it must hold that $\lambda'_{ij} = \lambda''_{ij}$ whenever (Λ', Ω') and (Λ'', Ω'') are in the same fiber. In Example 1.2, the fiber of a pair (Λ, Ω) with $\lambda_{12} = 0$ is infinite but all pairs (Λ', Ω') such a fiber have $\lambda'_{12} = 0$. Two well-known graphical methods for identifying a single edge coefficient are the back-door and the front-door criterion [49]; see also [11, 36].

3.2. Model dimension

Statistical tests may be used to assess whether a model is compatible with empirical data. At an intuitive level, such tests are based on computing a distance between data and model and comparing this distance to typical distances that are obtained when data are generated from a distribution in the model. For linear Gaussian models, a test can be thought off as assessing the distance between the empirical (or sample) covariance matrix and the model $\mathcal{M}_G$. Recall that we have defined $\mathcal{M}_G$ as the set of covariance matrices.

The challenging part of designing a statistical test is to quantify, in a probabilistic manner, what typical distances between data and model are. Many procedures rely on asymptotic approximations that are obtained by letting the number of data points grow to infinity. Under regularity conditions, limiting distribution theory leads to consideration of so-called chi-square distributions, which are indexed by an integer parameter. In our context, when testing the model given by the mixed graph $G = (V, D, B)$, the chi-square parameter is set equal to the codimension of $\mathcal{M}_G$, where we think of $\mathcal{M}_G$ as embedded in the space of symmetric matrices. This gives concrete statistical motivation for:

Question 3.6. *What is the dimension of* $\mathcal{M}_G$*?*

The model $\mathcal{M}_G$ is parametrized by the coefficients and covariances associated with the edges in D and B as well as the variances associated with the nodes in V. Being a subset of the space of $V \times V$ symmetric matrices, $\mathcal{M}_G$ has expected dimension

$$\min\left\{|V| + |D| + |B|, \frac{|V|(|V|+1)}{2}\right\}.$$

The term $|D| + |V| + |B|$ counts the nodes and edges of G. Since $\mathcal{M}_G$ is the image of ϕ_G, its actual dimension is equal to the rank of the Jacobian of ϕ_G. A review of the connection between dimension and Jacobian in a statistical context is given in [38]. Question 3.6 is tied to parameter identifiability, most directly to Question 3.4. If ϕ_G is generically finite-to-one, then $\mathcal{M}_G$ has the expected dimension $|V| + |D| + |B|$.

3.3. Covariance equivalence

Different graphs may induce the same statistical model. For example, take $V = \{1, 2\}$, and let G_1 be the graph with the single edge $1 \to 2$. Let G_2 and G_3 be the graphs with single edge $1 \leftarrow 2$ and $1 \leftrightarrow 2$, respectively. Then $\mathcal{M}_{G_1} = \mathcal{M}_{G_2} = \mathcal{M}_{G_3}$ as each model is easily seen to be equal to the entire cone of positive definite 2×2 matrices.

From an applied perspective, two different graphs G and G' encode different scientific/causal hypotheses. If $\mathcal{M}_G = \mathcal{M}_{G'}$, then the two hypotheses cannot be distinguished based on data from a linear and Gaussian structural equation model. It is thus useful to be able to decide whether two graphs G and G' are covariance equivalent, that is, we would like to be able to answer:

Question 3.7. *When do two maps ϕ_G and $\phi_{G'}$ have the same image?*

Existing results addressing this questions make comparisons between certain types of relations among the entries of the covariance matrices in each model. This ties into the basic problem of implicitization:

Question 3.8. *What are the algebraic relations among the coordinates of ϕ_G?*

Such relations are also of interest for statistical tests that assess whether the model given by G is compatible with available data [6, 11, 17]. We review results on relations among the covariances in Sections 9-12. An important role is played by determinants that represent probabilistic conditional independence in a Gaussian random vector. We note that models can, in principle, also be distinguished using inequality constraints. However, as less is know about inequalities, we do not treat them here. Examples of models with latent variables for which a full semi-algebraic description is available can be found in [25, 53].

Remark 3.9. As defined above, covariance equivalence is based on data that is observational, i.e., it is collected by merely observing the considered physical system. The situation is different when experimental data is available, i.e., (some) data is collected in settings in which the system is subject to various experimental interventions. We will not treat such interventional data in this paper. Interested readers may find discussions of the problem in [43, 49, 57]. Similarly, even for observational data, questions of equivalence differ from Question 3.7 in non-linear models or linear models with non-Gaussian errors [26, 54].

3.4. Maximum likelihood

The parameters of linear structural equation models are most commonly estimated using the technique of maximum likelihood. Suppose we observe a sample $X^{(1)}, \ldots, X^{(n)}$ drawn independently from the multivariate normal distribution $\mathcal{N}(\mu, \Sigma)$, where μ is the mean vector and Σ is the covariance matrix. The joint distribution of the random vectors $X^{(1)}, \ldots, X^{(n)}$ is the n-fold product of $\mathcal{N}(\mu, \Sigma)$. The likelihood of the sample is the value of the joint density of the product distribution at $(X^{(1)}, \ldots, X^{(n)})$. The *likelihood function* maps the pair (μ, Σ) to the likelihood of the sample. The *maximum likelihood estimator* (MLE) of (μ, Σ) under the model given by a mixed graph G is the maximizer of the likelihood function when restricting Σ to be in $\mathcal{M}_G$.

Define the sample mean vector and the sample covariance matrix as

$$(3.1)\qquad \bar{X}_n = \frac{1}{n}\sum_{i=1}^{n} X^{(i)} \quad \text{and} \quad S_n = \frac{1}{n}(X^{(i)} - \bar{X}_n)(X^{(i)} - \bar{X}_n)^T,$$

respectively. It is convenient to treat the likelihood function on the log-scale. With an additive constant omitted and $n/2$ divided out, the *log-likelihood function* is

$$(\mu, \Sigma) \mapsto -\log\det(\Sigma) - \operatorname{trace}\left(\Sigma^{-1}S_n\right) - (\bar{X}_n - \mu)^T\Sigma^{-1}(\bar{X}_n - \mu).$$

Because the considered models place no constraint on the mean vector, its MLE is always $\bar{X}_n$. The MLE of Σ maximizes the function

$$(3.2)\qquad \ell(\Sigma) = -\log\det(\Sigma) - \operatorname{trace}\left(\Sigma^{-1}S_n\right)$$

subject to $\Sigma \in \mathcal{M}_G$. Using the covariance parametrization, the MLE is found by maximizing $\ell \circ \phi_G$. A key problem is then understanding the existence and uniqueness of the MLE. We record:

Question 3.10. *For which sample covariance matrices S_n does the likelihood function $\ell \circ \phi_G$ achieve its maximum?*

Graphical models theory solves Question 3.10 when $G = (V, D, B)$ is an acyclic digraph, i.e., has $B = \emptyset$ and D without directed cycles [46]. More generally, it is well known that $\ell \circ \phi_G$ is bounded when S_n is positive definite but this is not necessary [32]. An issue that is not well explored is the fact that even if $\ell \circ \phi_G$ is bounded it may fail to achieve its maximum as the model $\mathcal{M}_G$ need not be closed. For instance, the model in Example 3.14 is not closed. We remark that Question 3.10 is closely related to a positive definite matrix completion problem that arises in ML estimation for other types of graphical models [9, 63, 67].

Fig. 3.1. Mixed graph for a bivariate seemingly unrelated regressions model.

In some models, the MLE is known to admit a closed-form expression as a rational function of the data. Such models have *maximum likelihood (ML) degree* equal to one, in the sense of the following:

Question 3.11. *The MLE of Σ in model $\mathcal{M}_G$ is an algebraic function of the data. What is the degree of this function?*

An introduction to the notion of ML degree is given in [21, Chapter 2]. Here, we merely not that the ML degree is one when G is an acyclic digraph. More general models may have higher ML degree and a log-likelihood function with more than one local maximum. We exemplify this for a model discussed in detail in [19].

Example 3.12. Suppose the sample covariance matrix is

$$S_n = \begin{array}{c} \\ X_1 \\ X_2 \\ X_3 \\ X_4 \end{array} \begin{array}{c} \begin{array}{cccc} X_1 & X_2 & X_3 & X_4 \end{array} \\ \begin{bmatrix} 8 & -5 & 10 & 3 \\ -5 & 27 & 4 & 49 \\ 10 & 4 & 21 & 24 \\ 3 & 49 & 24 & 114 \end{bmatrix} \end{array}.$$

The matrix is positive definite such that the log-likelihood function ℓ from (3.2) is bounded above on the entire cone of positive definite matrices. Moreover, ℓ has compact level sets, i.e., for any constant $c \in \mathbb{R}$ the set of positive definite matrices Σ with $\ell(\Sigma) \geq c$ is compact [2].

Let G be the graph from Figure 3.1. It is not difficult to show that the parametrization ϕ_G admits a rational inverse. Let $\Sigma = (\sigma_{ij})$ satisfy $\Sigma = \phi_G(\Lambda, \Omega)$ with $\Lambda = (\lambda_{ij}) \in \mathbb{R}^D$ and $\Omega = (\omega_{ij}) \in PD(B)$. Then

$$\lambda_{12} = \frac{\sigma_{12}}{\sigma_{11}}, \qquad \lambda_{43} = \frac{\sigma_{34}}{\sigma_{44}}, \tag{3.3}$$

and the entries of $\Omega = (I - \Lambda)^T \Sigma (I - \Lambda)$ are rational functions of Σ as well. All the rational functions are defined on the entire cone of positive definite matrices because $\sigma_{11}, \sigma_{44} > 0$. It is also clear that the considered map ϕ_G is proper, that is, compact subsets of the positive definite cone have compact preimages under ϕ_G. It follows that $\ell \circ \phi_G$ has compact level sets and, thus, achieves its maximum on the open set $\mathbb{R}^D \times PD(B)$.

The critical points of $\ell \circ \phi_G$ satisfy a rational equation system, in which the determinant of $\phi_G(\Lambda, \Omega)$ appears in the denominator. Since G is acyclic the determinant is equal to the determinant of Ω. Clearing the denominator yields a polynomial equation system. Saturating the system with respect to the determinant removes infeasible solutions with $\det(\Omega) = 0$. Computing a lexicographic Gröbner basis after the saturation shows that the critical points (Λ, Ω) solve the equation

$$10583160\,\lambda_{12}^5 + 43115307\,\lambda_{12}^4 + 72738452\,\lambda_{12}^3 \\ + 55482894\,\lambda_{12}^2 + 8437660\,\lambda_{12} - 4703765 \;=\; 0.$$

All other entries of Λ and also Ω solve linear equations whose coefficients depend on λ_{12} and the data. We conclude that the MLE of Σ is an algebraic function of degree 5. The displayed equation for λ_{12} has three real roots and, thus, is not solvable by radicals.

3.5. Model singularities

As noted in Section 3.2, the distributions of test statistics are frequently approximated using asymptotic theory. For so-called likelihood ratio tests, this asymptotic theory can be thought of as assessing infinitesimal distances between a positive semidefinite data matrix and the given model $\mathcal{M}_G$. The data matrix is generated from a distribution that corresponds to a particular point in $\mathcal{M}_G$. At a smooth point of $\mathcal{M}_G$, the squared infinitesimal distance follows a chi-square distribution, which is the distribution of the squared Euclidean distance between a Gaussian random vector and a linear space. At singular points, the distribution is determined by the tangent cone at the considered point [15]. Singularities also impact other approaches such as Wald tests [24], and it is important to clarify:

Question 3.13. *Is the image of ϕ_G a smooth manifold? If not, what are the tangent cones of the image?*

We do not address the question explicitly in this paper. However, whenever ϕ_G is injective (see Section 7) its image is smooth. Indeed, when ϕ_G is injective it has a rational inverse whose domain of definition includes the cone of all positive definite matrices [16]. The next example illustrates that not all models are smooth.

Example 3.14. Consider the mixed graph $G = (V, D, B)$ from Figure 3.2. Let $\Sigma \in \mathbb{R}^{4\times 4}$ be a positive definite matrix. Define the matrix

$$\Sigma_{\{3,4\}.\{1,2\}} = \Sigma_{\{3,4\},\{1,2\}} \left(\Sigma_{\{1,2\},\{1,2\}}\right)^{-1},$$

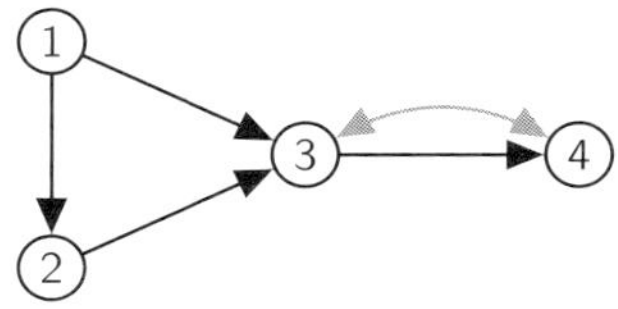

Fig. 3.2. Mixed graph of a model with two instruments.

and the Schur complement

$$\Sigma_{\{3,4\},\{3,4\}.\{1,2\}} = \Sigma_{\{3,4\},\{3,4\}} - \Sigma_{\{3,4\},\{1,2\}} \left(\Sigma_{\{1,2\},\{1,2\}}\right)^{-1} \Sigma_{\{1,2\},\{3,4\}}.$$

Change coordinates to the triple of 2×2 matrices

$$\left(\Sigma_{\{1,2\},\{1,2\}},\ \Sigma_{\{3,4\}.\{1,2\}},\ \Sigma_{\{3,4\},\{3,4\}.\{1,2\}}\right).$$

If $\Sigma = \phi_G(\Lambda, \Omega)$ for $\Lambda = (\lambda_{ij}) \in \mathbb{R}^D$ and $\Omega = (\omega_{ij}) \in PD(B)$, then

$$\Sigma_{\{1,2\},\{1,2\}} = \begin{pmatrix} \omega_{11} & \lambda_{12}\omega_{11} \\ \lambda_{12}\omega_{11} & \omega_{22} \end{pmatrix},$$
$$\Sigma_{\{3,4\}.\{1,2\}} = \begin{pmatrix} \lambda_{13} & \lambda_{23} \\ \lambda_{13}\lambda_{34} & \lambda_{23}\lambda_{34} \end{pmatrix},$$
$$\Sigma_{\{3,4\},\{3,4\}.\{1,2\}} = \begin{pmatrix} \omega_{33} & \omega_{34} + \lambda_{34}\omega_{33} \\ \omega_{34} + \lambda_{34}\omega_{33} & \omega_{44} + 2\lambda_{34}\omega_{34} + \lambda_{34}^4\omega_{33} \end{pmatrix}.$$

We observe that Σ is in the (topological) closure of $\mathcal{M}_G$ if and only if $\Sigma_{\{3,4\}.\{1,2\}}$ is a matrix of rank at most one. For Σ to be in $\mathcal{M}_G$, the second row of $\Sigma_{\{3,4\}.\{1,2\}}$ may be zero only if the first row is zero.

Geometrically, the closure of $\mathcal{M}_G$ is equivalent to the product of two cones of positive definite 2×2 matrices and the set of 2×2 matrices of rank at most one. The latter set is singular at the zero matrix. For more details see the related example in [21, Exercise 6.4].

3.6. Singularities of fibers

Finally, without going into any detail, we note that it is also of statistical interest to study the geometry of the fibers $\mathcal{F}_G(\Lambda, \Omega)$. In particular, the resolution of singularities of $\mathcal{F}_G(\Lambda, \Omega)$ is connected to asymptotic approximations in Bayesian approaches to model selection.

Bayesian methods assess the goodness-of-fit of a model by integrating the likelihood function with respect to a prior distribution. In models with many parameters, the integration is over a domain of larger dimension and may constitute a difficult numerical problem. While carefully tuned Monte Carlo methods can be effective, it can also be useful to invoke asymptotics. For large sample size n, the integrated likelihood function behaves like a Laplace integral. Under regularity conditions, a Laplace approximation can yield accurate approximations that have been used in many applications [42]. However, the models considered here may also lead to singular Laplace integrals for which asymptotic approximations are more involved.

Asymptotic expansions for singular Laplace integrals are well-studied [3]. The work of Sumio Watanabe brings the ideas to bear in the statistical context [70]. For several practically relevant settings, it has become tractable to determine or bound the *real log-canonical threshold* and its multiplicity, which determine how the integrated likelihood scales with the sample size n. This information can be used in model selection [18]. Computing real log-canonical thresholds for data generated under the distribution with covariance matrix $\phi_G(\Lambda, \Omega)$ requires careful study of the singularities of the fiber $\mathcal{F}_G(\Lambda, \Omega)$. Bounds on the thresholds can be obtained from cruder information such as the dimension of the fiber.

Part II. Treks, Subgraphs and Decomposition

§4. Trek Rule

In solving the problems from Section 3, it is desirable to exploit the connection between the covariance parametrization ϕ_G of a structural equation model and the underlying mixed graph $G = (V, D, B)$. The trek rule that we present in this section makes the connection precise and is behind results that allow one to answer some of the questions we posed with efficient algorithms.

It is natural to expect the covariance between random variables X_i and X_j to be determined by the semi-walks between the nodes i and j in the graph G. A semi-walk is an alternating sequence of nodes from V and edges from either D or B such that the endpoints of each edge are the nodes immediately preceding and succeeding the edge in the sequence. In other words, a semi-walk is a walk that uses bidirected or directed edges but is allowed to traverse directed edges in the 'wrong direction'. As we will see, only special semi-walks contribute to the covariance.

Definition 4.1. *A* trek *τ from initial node i to target node j is a semi-walk from i to j whose consecutive edges do not have any colliding arrowheads. In other words, τ is a sequence of the form*

(a) $i \leftarrow i_l \leftarrow \cdots \leftarrow i_1 \leftarrow i_0 \longleftrightarrow j_0 \rightarrow j_1 \rightarrow \cdots \rightarrow j_r \rightarrow j$, *or*

(b) $i \leftarrow i_l \leftarrow \cdots \leftarrow i_1 \longleftarrow i_0 \longrightarrow j_1 \rightarrow \cdots \rightarrow j_r \rightarrow j$.

A trek has a left- *and a* right-hand side, *denoted* left (τ) *and* right (τ), *respectively. We have* left $(\tau) = \{i_0, \ldots, i_l, i\}$ *and* right $(\tau) = \{j_0, \ldots, j_r, j\}$ *in case* (a), *and* left $(\tau) = \{i_0, \ldots, i_l, i\}$ *and* right $(\tau) = \{i_0, j_1, \ldots, j_r, j\}$ *in case* (b). *In case* (b) *the* top *node* i_0 *belongs to both sides.*

In an acyclic graph, if we think of directed edges pointing 'downward', then a trek takes us up and/or down a 'mountain'. A trek τ from i to i may have no edges, in which case i is the top node, left (τ) = right (τ) = $\{i\}$. We call such a trek trivial. Any directed path is a trek, in which case $|\text{left}(\tau)| = 1$ or $|\text{right}(\tau)| = 1$ depending on the direction in which the path is traversed. A trek may contain the same node on both its left- and right-hand sides. If the graph contains a cycle, then the left- or right-hand side of τ may contain this cycle, possibly repeated.

For a trek τ that contains no bidirected edge and has top node i_0, define a trek monomial as

$$\tau(\Lambda, \Omega) = \omega_{i_0 i_0} \prod_{k \to l \in \tau} \lambda_{kl}.$$

For a trek τ that contains a bidirected edge $i_0 \leftrightarrow j_0$, define the trek monomial as

$$\tau(\Lambda, \Omega) = \omega_{i_0 j_0} \prod_{k \to l \in \tau} \lambda_{kl}.$$

The following rule expresses the covariance matrix as a summation over treks [57, 73, 74]. We write $\mathcal{T}(i, j)$ for the set of all treks from i to j.

Theorem 4.2 (Trek rule). *Let $G = (V, D, B)$ be any mixed graph, and let $\Lambda \in \mathbb{R}^D$ and $\Omega \in PD(B)$. Then the covariances are*

$$\phi_G(\Lambda, \Omega)_{ij} = \sum_{\tau \in \mathcal{T}(i,j)} \tau(\Lambda, \Omega), \qquad i, j \in V. \tag{4.1}$$

Some clarification is in order. If G is acyclic, then the summation in (4.1) is finite and yields a polynomial. If G contains a directed cycle, then the right-hand side of (4.1) may yield a power series as shown in Example 4.4 below. Under suitable assumptions on the spectrum of Λ, the power series converges and yields the value of $\phi_G(\Lambda, \Omega)_{ij}$. Such spectral conditions are also needed to give cyclic models an interpretation of representing observation of an equilibrium [31, 45]. This said, it is also useful to treat the right-hand side of (4.1) as a formal power series. If so desired, a combinatorial description can also be given for a rational expression for $\phi_G(\Lambda, \Omega)_{ij}$; see [14].

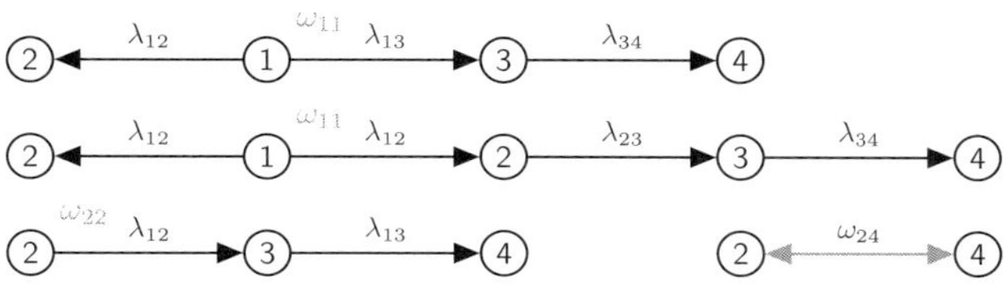

Fig. 4.1. Four treks in the graph from Figure 2.1.

Proof of the trek rule. Writing $(I-\Lambda)^{-1} = I + \Lambda + \Lambda^2 + \dots$, we observe that

$$((I-\Lambda)^{-1})_{ij} = \sum_{\tau \in \mathcal{P}(i,j)} \prod_{k \to l \in \tau} \lambda_{kl}, \tag{4.2}$$

where $\mathcal{P}(i,j)$ is the set of directed paths from i to j in G. If G is acyclic, then $\Lambda^m = 0$ for all $m \geq |V|$, and the geometric series of matrices has only finitely many nonzero terms. If G is cyclic the geometric series is infinite and converges if and only if all eigenvalues of Λ have magnitude less than 1. Now, observe that a product of three entries of $(I-\Lambda)^{-T}$, Ω, and $(I-\Lambda)^{-1}$, respectively, corresponds to the concatenation of two directed paths at a common top node or by joining them with a bidirected edge. A top node represents a diagonal entry of Ω, and a bidirected edge an off-diagonal entry of Ω. Q.E.D.

Example 4.3. In Example 2.4, the coordinate function $(\phi_G)_{24}$ is a polynomial with four terms; see (2.8). The terms correspond to the four treks shown in Figure 4.1.

Example 4.4. Let G be the graph from Figure 4.2, which contains the directed cycle $2 \to 3 \to 4 \to 2$. Due to this cycle, $\det(I-\Lambda) = 1 - \lambda_{23}\lambda_{34}\lambda_{42}$. As an example of a coordinate of ϕ_G we select

$$\phi_G(\Lambda,\Omega)_{24} = \\ \frac{1}{(1-\lambda_{23}\lambda_{34}\lambda_{42})^2}\left[\lambda_{12}^2\lambda_{23}\lambda_{34}\omega_{11} + \lambda_{12}\lambda_{13}\lambda_{34}\omega_{11}(\lambda_{23}\lambda_{34}\lambda_{42}+1)\right. \\ \left. +\lambda_{13}^2\lambda_{34}^2\lambda_{42}\omega_{11} + \lambda_{23}\lambda_{34}\omega_{22} + \lambda_{34}^2\lambda_{42}\omega_{33} + 2\lambda_{34}\lambda_{42}\omega_{34} + \lambda_{42}\omega_{44}\right].$$

To understand how this rational formula relates to the trek rule, let us focus on the treks from 2 to 4 that use the bidirected edge $2 \leftrightarrow 4$. There are then two treks for which both left and right side are self-avoiding

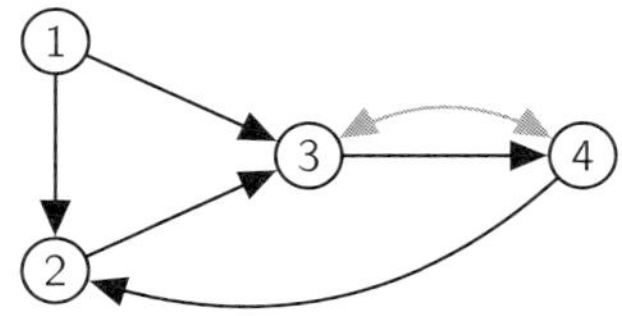

Fig. 4.2. A cyclic mixed graph.

paths, namely,

$$\tau_1 : 2 \leftarrow 4 \leftrightarrow 3 \rightarrow 4, \qquad \tau_2 : 2 \leftarrow 4 \leftarrow 3 \leftrightarrow 4.$$

Both of these treks yield the same monomial and together contribute the term $2\lambda_{34}\lambda_{42}\omega_{34}$ to $(\phi_G)_{24}$. All other treks from 2 to 4 that use edge $2 \leftrightarrow 4$ are obtained by inserting directed cycles into τ_1 or τ_2. For instance, inserting one cycle on the left- and one on the right-hand side of τ_1 gives

$$2 \leftarrow 4 \leftarrow 3 \leftarrow 2 \leftarrow 4 \leftrightarrow 3 \rightarrow 4 \rightarrow 2 \rightarrow 3 \rightarrow 4.$$

The monomials associated with these treks are

$$\lambda_{34}\lambda_{42}\omega_{34}\left(\lambda_{23}\lambda_{34}\lambda_{42}\right)^k, \qquad k = 1, 2, \dots.$$

The monomial for exponent k arises from $k+1$ different treks; $l = 0, 1, \dots, k$ cycles are inserted on the left, the remaining $k - l$ cycles are inserted on the right. Hence, the contribution to $(\phi_G)_{24}$ made by all treks from 2 to 4 that use edge $2 \leftrightarrow 4$ is

$$2\sum_{k=0}^{\infty}(k+1)\lambda_{34}\lambda_{42}\omega_{34}\left(\lambda_{23}\lambda_{34}\lambda_{42}\right)^k = \frac{2\lambda_{34}\lambda_{42}\omega_{34}}{(1-\lambda_{23}\lambda_{34}\lambda_{42})^2},$$

assuming that $|\lambda_{23}\lambda_{34}\lambda_{42}| < 1$. This explains one of the terms in the rational expression for $(\phi_G)_{24}$. The reasoning for the other terms is analogous.

§5. Induced Subgraphs and Principal Submatrices

Suppose $X = (X_i : i \in V)$ follows the linear structural equation model given by mixed graph $G = (V, D, B)$, so $\mathrm{Var}[X] = \phi_G(\Lambda, \Omega)$ for some $\Lambda \in \mathbb{R}^D_{\mathrm{reg}}$ and $\Omega \in PD(B)$. Let $A \subseteq V$ be a subset of nodes. Then the covariance matrix of the subvector $X_A = (X_i : i \in A)$ is obtained by projecting to the relevant principal submatrix, that is,

$$\text{Var}[X_A] \;=\; \phi_G(\Lambda, \Omega)_{A,A}. \tag{5.1}$$

The resulting map $(\Lambda, \Omega) \mapsto \phi_G(\Lambda, \Omega)_{A,A}$ may be complicated, even when ϕ_G is not.

Example 5.1. Suppose $G = (V, D, \emptyset)$ is a directed graph with $i \to j \in D$ if and only if $i \notin A$ and $j \in A$; the graph is thus bipartite. Then the image of $\phi_G(\Lambda, \Omega)_{A,A}$ is the set of covariance matrices of a factor analysis model with $|V \setminus A|$ factors. Factor analysis models have complicated geometric structure, particularly when $V \setminus A$ has more than two elements [8, 20, 23, 62]. Open problems remain even for $|V \setminus A| = 1$ when allowing additional directed edges among the nodes in A [47].

For a general mixed graph $G = (V, D, B)$, let $D_A = D \cap (A \times A)$ be the set of directed edges with both endpoints in A. Similarly, let $B_A \subset B$ be the set of bidirected edges that have both endpoints in A. The subgraph induced by A is the mixed graph $G_A = (A, D_A, B_A)$. Example 5.1 and also already Example 1.1 show that the covariance matrices obtained from ϕ_{G_A} generally differ from those obtained by projecting ϕ_G onto the $A \times A$ submatrix. However, as we now emphasize, induced subgraphs are relevant in a special case.

Define the set of *parents* of a node $i \in V$ as

$$\text{pa}(i) \;=\; \{j \in V \;:\; j \to i\}.$$

A set $A \subseteq V$ is *ancestral* if $i \in A$ implies $\text{pa}(i) \subseteq A$. The terminology indicates that such a set contains all its ancestors, where an ancestor is a node from which there is a directed path to some node in A. Ancestral sets are obtained by recursively removing sink nodes from V. A *sink* is a node that is a head on all edges it is incident to. By our convention, both endpoints of bidirected edges are heads. Hence, node i is a sink of $G = (V, D, B)$ if and only if i is a sink of the directed part (V, D).

Theorem 5.2. *Let $G = (V, D, B)$ be a mixed graph, and let G_A be the subgraph induced by an ancestral set $A \subset V$. Then for all $\Lambda \in \mathbb{R}^D_{\text{reg}}$ and $\Omega \in PD(B)$, we have*

$$\phi_{G_A}(\Lambda_{A,A}, \Omega_{A,A}) \;=\; [\phi_G(\Lambda, \Omega)]_{A,A}\,.$$

Proof. Let $i, j \in A$. By the trek-rule, the (i, j) entry of $\phi_G(\Lambda, \Omega)$ is given by summing the monomials associated to treks from i to j in G. Because A is ancestral, a trek from i to j in G cannot leave A. Hence, G

and G_A have the same sets of treks from i to j. Applying the trek-rule to G_A yields the claim. Q.E.D.

Example 5.3. Take up Example 2.1. Node 3 is a sink in the graph from Figure 1.2 and the set $\{1, 2\}$ is ancestral. Inspecting the matrix displayed in (2.5), we see that removing the third row and column yields the matrix for the induced subgraph $1 \to 2$.

§6. Graph Decomposition

We now present an important decomposition, which allows one to address several questions of interest by treating smaller subgraphs. The decomposition for acyclic mixed graphs was introduced by Jin Tian [65, 66]. Here we give the natural extension to graphs with directed cycles.

We consider two partitions of the vertex set of a mixed graph $G = (V, D, B)$. The first is given by the connected components of the bidirected part (V, B). The second is given by the strongly connected components of the digraph (V, D). Two distinct nodes belong to the same strongly connected component if there are directed paths in either direction between them. Let $\mathcal{C}(G)$ be the finest common coarsening of the two partitions. Two nodes are in the same block of $\mathcal{C}(G)$ if and only if they are connected by a path that uses only edges that are bidirected or part of some directed cycle. Note that G is acyclic if and only if all strongly connected components are singleton sets. The blocks of $\mathcal{C}(G)$ are then simply the connected components of the bidirected part (V, B).

For a block $C \in \mathcal{C}(G)$, define

$$V[C] := C \cup \bigcup_{i \in C} \mathrm{pa}(i)$$

to be the union of the block and all parents of nodes in the block. Let $D[C] = D \cap (V[C] \times C)$ be the set of directed edges with head in C. Let $B[C]$ be the set of bidirected with both endpoints in C.

Definition 6.1. *The graphs $G[C] = (V[C], D[C], B[C])$, $C \in \mathcal{C}(G)$, form a decomposition of G, and we refer to them as the* mixed components *of G.*

Graph decompositions partition edge sets. As we are working with mixed graphs both edge sets are partitioned in the decomposition. We note that the set $V[C] \setminus C$ contains the sources of $G[C]$. A source node is a tail on all edges it is incident to, with the convention that both endpoints of a bidirected edge are heads.

Example 6.2. The graph in Figure 6.1 has bidirected components $\{1,4\}$, $\{3\}$, and $\{2,5\}$. The strongly connected directed components are $\{1\}$, $\{2,3\}$, $\{4\}$, and $\{5\}$. The finest common coarsening of the two partitions is $\{\{1,4\},\{2,3,5\}\}$. The graph thus has two mixed components with vertex sets $V[\{2,3,5\}] = \{1,2,3,4,5\}$ and $V[\{1,4\}] = \{1,2,3,4\}$. The mixed component with vertex set $V[\{2,3,5\}]$ contains all edges with a head in $\{2,3,5\}$, and the second mixed component contains all edges with a head in $\{1,4\}$. The components are depicted in Figure 6.1.

For $C \in \mathcal{C}(G)$, define the projection $\pi_C^{\rightarrow} : \mathbb{R}^{V\times V} \to \mathbb{R}^{V[C]\times V[C]}$ by

$$\pi_C^{\rightarrow}(\Lambda)_{ij} = \begin{cases} \lambda_{ij} & \text{if } j \in C, \\ 0 & \text{if } j \in V[C] \setminus C. \end{cases}$$

Define a second map $\pi_C^{\leftrightarrow} : \mathbb{R}^{V\times V} \to \mathbb{R}^{V[C]\times V[C]}$ by

$$\pi_C^{\leftrightarrow}(\Omega)_{ij} = \begin{cases} \omega_{ij} & \text{if } i,j \in C, \\ 1 & \text{if } i = j \in V[C] \setminus C, \\ 0 & \text{otherwise.} \end{cases}$$

So $\pi_C^{\leftrightarrow}$ projects to the $C \times C$ submatrix and then adds the identity as diagonal block over $V[C] \setminus C$. Let $PD(B[C]) \subset \mathbb{R}^{V[C]\times V[C]}$ be the set of positive definite matrices supported over $B[C]$ and define the subset

$$PD_I(B[C]) \;=\; \{\Omega = (\omega_{ij}) \in PD(B[C]) \,:\, \omega_{ii} = 1 \;\text{ if }\; i \in V[C] \setminus C\}.$$

Then we have

$$\pi_C^{\rightarrow} : \mathbb{R}^D \to \mathbb{R}^{D[C]} \quad \text{and} \quad \pi_C^{\leftrightarrow} : PD(B) \to PD_I(B[C])$$

because $G[C]$ is a subgraph of G. With $\pi_C = (\pi_C^{\rightarrow}, \pi_C^{\leftrightarrow})$, we obtain the isomorphism

$$\pi = (\pi_C)_{C\in\mathcal{C}(G)} \;:\; \mathbb{R}^D \times PD(B) \;\to\; \prod_{C\in\mathcal{C}(G)} \mathbb{R}^{D[C]} \times PD_I(B[C]).$$

Example 6.3. Let G be the graph from Figure 6.1, which has $\mathcal{C}(G) = \{\{1,4\},\{2,3,5\}\}$. A matrix in $\mathbb{R}^D$ is of the form

$$\Lambda = \begin{bmatrix} 0 & \lambda_{12} & \lambda_{13} & 0 & 0 \\ 0 & 0 & \lambda_{23} & \lambda_{24} & \lambda_{25} \\ 0 & \lambda_{32} & 0 & \lambda_{34} & 0 \\ 0 & 0 & 0 & 0 & \lambda_{45} \\ 0 & 0 & 0 & 0 & 0 \end{bmatrix}.$$

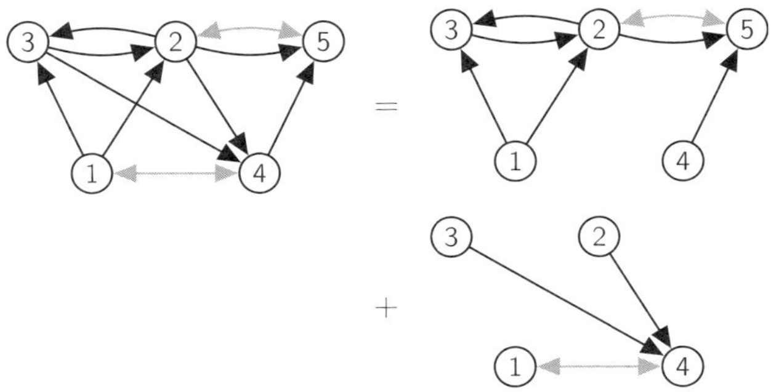

Fig. 6.1. A mixed graph is decomposed into its two mixed components.

Its projections are

$$\pi^{\rightarrow}_{\{2,3,5\}}(\Lambda) = \begin{bmatrix} 0 & \lambda_{12} & \lambda_{13} & 0 & 0 \\ 0 & 0 & \lambda_{23} & 0 & \lambda_{25} \\ 0 & \lambda_{32} & 0 & 0 & 0 \\ 0 & 0 & 0 & 0 & \lambda_{45} \\ 0 & 0 & 0 & 0 & 0 \end{bmatrix}, \quad \pi^{\rightarrow}_{\{1,4\}}(\Lambda) = \begin{bmatrix} 0 & 0 & 0 & 0 \\ 0 & 0 & 0 & \lambda_{24} \\ 0 & 0 & 0 & \lambda_{34} \\ 0 & 0 & 0 & 0 \end{bmatrix}.$$

A (symmetric) error covariance matrix in $PD(B)$ has the form

$$\Omega = \begin{bmatrix} \omega_{11} & 0 & 0 & \omega_{14} & 0 \\ 0 & \omega_{22} & 0 & 0 & \omega_{25} \\ 0 & 0 & \omega_{33} & 0 & 0 \\ \omega_{14} & 0 & 0 & \omega_{44} & 0 \\ 0 & \omega_{25} & 0 & 0 & \omega_{55} \end{bmatrix},$$

and we have

$$\pi^{\leftrightarrow}_{\{2,3,5\}}(\Omega) = \begin{bmatrix} 1 & 0 & 0 & 0 & 0 \\ 0 & \omega_{22} & 0 & 0 & \omega_{25} \\ 0 & 0 & \omega_{33} & 0 & 0 \\ 0 & 0 & 0 & 1 & 0 \\ 0 & \omega_{25} & 0 & 0 & \omega_{55} \end{bmatrix},$$

$$\pi^{\leftrightarrow}_{\{1,4\}}(\Omega) = \begin{bmatrix} \omega_{11} & 0 & 0 & \omega_{14} \\ 0 & 1 & 0 & 0 \\ 0 & 0 & 1 & 0 \\ \omega_{14} & 0 & 0 & \omega_{44} \end{bmatrix}.$$

With this preparation, Tian's theorem may be stated as follows. Recall that PD_V denotes the cone of positive definite $V \times V$ matrices.

Theorem 6.4. *Let $G = (V, D, B)$ be a mixed graph with mixed components $G[C] = (V[C], D[C], B[C])$ for $C \in \mathcal{C}(G)$. Then there is an invertible map τ such that the following diagram commutes:*

$$\begin{array}{ccc} \mathbb{R}^{D}_{\mathrm{reg}} \times PD(B) & \xrightarrow{\quad \phi_G \quad} & PD_V \\ \uparrow\downarrow \pi & & \uparrow\downarrow \tau \\ \prod_{C\in\mathcal{C}(G)} \mathbb{R}^{D[C]}_{\mathrm{reg}} \times PD_I(B[C]) & \xrightarrow{(\phi_{G[C]})_{C\in\mathcal{C}(G)}} & \prod_{C\in\mathcal{C}(G)} PD_{V[C]} \end{array}$$

In other words, $\tau \circ \phi_G = (\phi_{G[C]} \circ \pi_C)_{C\in\mathcal{C}(G)}$. Both τ and its inverse are rational maps, defined on all of PD_V and all of $\prod_{k=1}^m PD_{V_k}$, respectively.

Below we give a linear algebraic proof that makes τ and its rational nature explicit. Alternatively, a proof in probabilistic notation could be given by generalizing the proof of [66, Lemma 1]. The strongly connected components of (V, D) would play the role of nodes in the setup of [66].

Theorem 6.4 is a very useful result as questions about ϕ_G can be answered by studying, one by one, the maps $\phi_{G[C]}$ for the mixed components. The fact that τ and τ^{-1} are rational is important. For instance, it allows one to obtain precise algebraic information about parameter identifiability in the sense of Definition 3.5.

Corollary 6.5. *The degree of identifiability of a mixed graph G is the product of the degrees of identifiability of its mixed components $G[C]$, $C \in \mathcal{C}(G)$. In particular, ϕ_G is (generically) injective if and only if each $\phi_{G[C]}$ is so, for $C \in \mathcal{C}(G)$.*

Our proof of Theorem 6.4 is presented in terms of Cholesky decompositions. When applied to a Gaussian covariance matrix, the Cholesky decomposition corresponds to factoring the multivariate normal density into a product of conditional densities, which is the connection to the probabilistic setting of [66]. We begin by stating a lemma on uniqueness and sparsity in block-Cholesky decomposition.

If $\mathcal{C}$ is a partition of a finite set V, then we write $\mathrm{Diag}(\mathcal{C})$ for the space of matrices that are block-diagonal with respect to $\mathcal{C}$. So, $A = (a_{ij}) \in \mathbb{R}^{V\times V}$ is in $\mathrm{Diag}(\mathcal{C})$ if and only if $a_{ij} = 0$ whenever i and j are in distinct blocks of $\mathcal{C}$. If we order the blocks of the partition as $\mathcal{C} = \{C_1, \ldots, C_k\}$, then we may define a space of strictly block upper-triangular matrices $\mathrm{Upper}(C)$, which contains $A = (a_{ij}) \in \mathbb{R}^{V\times V}$ if and only if $a_{ij} = 0$ whenever $i \in C_u$ and $j \in C_v$ with $u \geq v$.

Lemma 6.6. *Let $\Sigma \in PD_V$, and let $\mathcal{C}$ be a partition of V, with ordered blocks.*

(i) *There exist unique matrices $A \in \mathrm{Upper}(\mathcal{C})$ and $\Delta \in \mathrm{Diag}(\mathcal{C})$ such that*

$$\Sigma = (I - A)^{-T} \Delta (I - A)^{-1}.$$

The matrix Δ has positive definite diagonal blocks.

(ii) *Let $\mathcal{C}'$ be a second partition of V that is coarser than $\mathcal{C}$. If $\Sigma \in \mathrm{Diag}(\mathcal{C}')$, then the matrix A from* (i) *satisfies $A \in \mathrm{Diag}(\mathcal{C}')$.*

Proof. (i) A block-LDL decomposition yields $\Sigma = (I + L)\Delta(I + L)^T$ for unique $L^T \in \mathrm{Upper}(\mathcal{C})$ and $\Delta \in \mathrm{Diag}(\mathcal{C})$. Unit block upper-triangular matrices form a group and, thus, $(I + L)^{-T} = I - A$ for $A \in \mathrm{Upper}(\mathcal{C})$. (ii) The claim is a consequence of the way fill-in occurs in the Cholesky decomposition of a sparse matrix [68, Section 4.1]. Q.E.D.

Proof of Theorem 6.4. We first show that, for every block $C \in \mathcal{C}(G)$, there exists a map τ_C such that $\tau_C \circ \phi_G = \phi_{G[C]} \circ \pi_C$. Let $\Lambda \in \mathbb{R}^D_{\mathrm{reg}}$, $\Omega \in PD(B)$, and $\Sigma = \phi_G(\Lambda, \Omega)$. Then the claim is that

(6.1)

$$\tau_C(\Sigma) = \left[I - \begin{pmatrix} 0 & \Lambda_{\overline{C},C} \\ 0 & \Lambda_{C,C} \end{pmatrix}\right]^{-T} \begin{pmatrix} I & 0 \\ 0 & \Omega_{C,C} \end{pmatrix} \left[I - \begin{pmatrix} 0 & \Lambda_{\overline{C},C} \\ 0 & \Lambda_{C,C} \end{pmatrix}\right]^{-1}.$$

Let $\mathcal{C}(D)$ be the partition of V given by the strongly connected components of (V, D). Order the blocks of $\mathcal{C}(D)$ topologically as $W_1, \ldots, W_k$ such that the existence of a directed path from a node in W_u to a node in W_v implies that $v \geq u$. By Lemma 6.6(i), there are $A \in \mathrm{Upper}(\mathcal{C}(D))$ and $\Delta \in \mathrm{Diag}(\mathcal{C}(D))$ such that

$$\Sigma = (I - A)^{-T} \Delta (I - A)^{-1}. \tag{6.2}$$

Letting $\overline{C} = V[C] \setminus C$, define

(6.3)

$$\tau_C(\Sigma) = \left[I - \begin{pmatrix} 0 & A_{\overline{C},C} \\ 0 & A_{C,C} \end{pmatrix}\right]^{-T} \begin{pmatrix} I & 0 \\ 0 & \Delta_{C,C} \end{pmatrix} \left[I - \begin{pmatrix} 0 & A_{\overline{C},C} \\ 0 & A_{C,C} \end{pmatrix}\right]^{-1}.$$

If G is acyclic then Λ is strictly upper-triangular under a topological ordering and, thus, $\Lambda \in \mathrm{Upper}(\mathcal{C}(D))$. When G has directed cycles, then Λ is block upper-triangular but not strictly so. Hence, we consider the block-diagonal matrix

$$\Delta_\Lambda = \mathrm{diag}(I - \Lambda_{W,W} : W \in \mathcal{C}(D)),$$

which is invertible because $\det(I-\Lambda)=\det(\Delta_\Lambda)$ and $\Lambda\in\mathbb{R}^D_{\text{reg}}$. Hence,

$$\Sigma=\left[(I-\Lambda)\Delta_\Lambda^{-1}\right]^{-T}\left[\Delta_\Lambda^{-1}\Omega\Delta_\Lambda^{-1}\right]\left[(I-\Lambda)\Delta_\Lambda^{-1}\right]^{-1}. \tag{6.4}$$

Because $\Delta_\Lambda,\Omega\in\text{Diag}(\mathcal{C}(G))$, we have

$$\tilde\Omega \;=\; \Delta_\Lambda^{-T}\Omega\Delta_\Lambda^{-1} \;\in\; \text{Diag}(\mathcal{C}(G)).$$

Moreover, due to the block upper-triangular shape of Λ,

$$\tilde\Lambda \;=\; I-(I-\Lambda)\Delta_\Lambda^{-1} \;\in\; \text{Upper}(\mathcal{C}(D)).$$

By Lemma 6.6(i) and (ii), there are matrices $\Delta_\Omega\in\text{Diag}(\mathcal{C}(D))$ and $U\in\text{Upper}(\mathcal{C}(D))\cap\text{Diag}(\mathcal{C}(G))$ such that

$$\tilde\Omega=(I-U)^{-T}\Delta_\Omega(I-U)^{-1}. \tag{6.5}$$

Combining (6.4) and (6.5) gives

$$\Sigma=\left[(I-\tilde\Lambda)(I-U)\right]^{-T}\Delta_\Omega\left[(I-\tilde\Lambda)(I-U)\right]^{-1}, \tag{6.6}$$

where $(I-\tilde\Lambda)(I-U)=I-(\tilde\Lambda+U-\tilde\Lambda U)$ with $\tilde\Lambda+U-\tilde\Lambda U\in\text{Upper}(\mathcal{C}(D))$.

By the uniqueness in Lemma 6.6(i), equations (6.2) and (6.6) imply

$$\Delta=\Delta_\Omega, \tag{6.7}$$

$$A=\tilde\Lambda+U-\tilde\Lambda U. \tag{6.8}$$

Since $U\in\text{Diag}(\mathcal{C}(G))$, we have that

$$U_{V\times C}=\begin{pmatrix}U_{(V\setminus C)\times C}\\ U_{C\times C}\end{pmatrix}=\begin{pmatrix}0\\ U_{C\times C}\end{pmatrix}. \tag{6.9}$$

Therefore, by (6.8), $A_{C,C}=\tilde\Lambda_{C,C}+U_{C,C}-\tilde\Lambda_{C,C}U_{C,C}$. We deduce that

$$I-A_{C,C}=(I-\tilde\Lambda_{C,C})(I-U_{C,C}) \quad\text{and}\quad A_{\overline{C},C}=\tilde\Lambda_{\overline{C},C}(I-U_{C,C}). \tag{6.10}$$

Moreover,

$$\tilde\Omega_{C,C}=(I-U_{C,C})^{-T}\Delta_{C,C}(I-U_{C,C})^{-1}, \tag{6.11}$$

which follows from (6.7) and the fact that

$$\left[(I-U)^{-1}\right]_{V,C}=\begin{pmatrix}0\\ \left[(I-U)^{-1}\right]_{C,C}\end{pmatrix}=\begin{pmatrix}0\\ (I-U_{C,C})^{-1}\end{pmatrix},$$

which in turn follows from U being in $\text{Diag}(\mathcal{C}(G))$.

Substituting the formulas from (6.10) into (6.3) yields that

$$\tau_C(\Sigma) = \left[\begin{pmatrix} I & -\tilde{\Lambda}_{\overline{C},C} \\ 0 & I - \tilde{\Lambda}_{C,C} \end{pmatrix} \begin{pmatrix} I & 0 \\ 0 & I - U_{C,C} \end{pmatrix}\right]^{-T} \times \begin{pmatrix} I & 0 \\ 0 & \Delta_{C,C} \end{pmatrix} \left[\begin{pmatrix} I & -\tilde{\Lambda}_{\overline{C},C} \\ 0 & I - \tilde{\Lambda}_{C,C} \end{pmatrix} \begin{pmatrix} I & 0 \\ 0 & I - U_{C,C} \end{pmatrix}\right]^{-1}. \tag{6.12}$$

Using (6.11), we get that

$$\begin{pmatrix} I & 0 \\ 0 & I - U_{C,C} \end{pmatrix}^{-T} \begin{pmatrix} I & 0 \\ 0 & \Delta_{C,C} \end{pmatrix} \begin{pmatrix} I & 0 \\ 0 & I - U_{C,C} \end{pmatrix}^{-1} = \begin{pmatrix} I & 0 \\ 0 & \tilde{\Omega}_{C,C} \end{pmatrix}. \tag{6.13}$$

Recalling the definition of Δ_Λ and $\tilde{\Lambda}$, we have

$$\begin{pmatrix} I & -\tilde{\Lambda}_{\overline{C},C} \\ 0 & I - \tilde{\Lambda}_{C,C} \end{pmatrix} = \begin{pmatrix} I & -\Lambda_{\overline{C},C} \\ 0 & I - \Lambda_{C,C} \end{pmatrix} \begin{pmatrix} I & 0 \\ 0 & (\Delta_\Lambda^{-1})_{C,C} \end{pmatrix} \tag{6.14}$$

Plugging (6.13) and (6.14) into (6.12), we obtain the claim from (6.1).

The entries of the matrices A and Δ in (6.2) are rational functions of Σ that are defined on all of PD_V. Hence, the same is true for the map τ_C defined in (6.3).

The value of $\tau_C(\Sigma)$ uniquely determines the matrices $A_{C,C}$, $A_{\overline{C},C}$, and $\Delta_{C,C}$ in (6.3). They are determined through a block LDL decomposition and, thus, rational functions of $\tau_C(\Sigma)$. Knowing the three matrices for all $C \in \mathcal{C}(G)$, we can form A and Δ and recover Σ using (6.2). We conclude that τ is invertible and the inverse is rational. Q.E.D.

Part III. Parameter Identification

§7. Global Identifiability

This section treats Question 3.1, which asks for a characterization of the mixed graphs $G = (V, D, B)$ for which the map ϕ_G is injective. In the statistical literature a model with injective parametrization is also called *globally identifiable*.

Example 7.1. If G is the graph from Figure 3.1, then ϕ_G is injective. Indeed, the coefficients for the two directed edges of G satisfy $\sigma_{11}\lambda_{12} = \sigma_{12}$ and $\sigma_{44}\lambda_{43} = \sigma_{34}$; recall (3.3). Since every positive definite matrix $\Sigma = (\sigma_{ij}) \in \mathbb{R}^{4\times 4}$ has $\sigma_{11}, \sigma_{44} > 0$, these two equations always have a unique solution. Hence, all fibers $\mathcal{F}_G(\Lambda, \Omega)$ are singleton sets. In contrast, if G is the graph from Figure 1.2, then only generic fibers are singleton sets and ϕ_G is not injective; recall Example 3.2.

Our first observation ties in with classical linear algebra.

Theorem 7.2. *If $G = (V, D, \emptyset)$ is an acyclic digraph, then ϕ_G is injective and has a rational inverse.*

We give two proofs. The first one emphasizes the connection to Cholesky decomposition.

Proof A. Suppose $V = \{1, \dots, m\}$ is enumerated in reversed topological order such that $i \to j \in D$ implies that $j < i$. Then Λ is a strictly lower-triangular matrix, and $\phi_G(\Lambda, \Omega)$ has matrix inverse $(I - \Lambda)\Omega^{-1}(I - \Lambda)^T$. This is the product of a unit lower-triangular matrix, a positive diagonal matrix and a unit upper-triangular matrix. We may compute $I - \Lambda$ and Ω^{-1} by an LDL decomposition. Q.E.D.

The second proof emphasizes the graphical nature of the problem and possible sparsity of Λ. It shows more explicitly that the inverse of ϕ_G is rational.

Proof B. Letting $\Sigma = (\sigma_{ij}) = \phi_G(\Lambda, \Omega)$, we have that

$$\Sigma_{\mathrm{pa}(i),i} = \Sigma_{\mathrm{pa}(i),\mathrm{pa}(i)} \Lambda_{\mathrm{pa}(i),i} \tag{7.1}$$

because if $j \in \mathrm{pa}(i)$, then every trek from j to i ends with an edge $k \to i$ for $k \in \mathrm{pa}(i)$. Indeed, a trek from j to i for which this fails has to be a directed path from i to j. Adding the edge $j \to i$ to this path would yield a directed cycle. Similarly, every nontrivial trek from i to i begins and ends with a directed edge whose tail is a parent of i. Hence,

$$\sigma_{ii} = \omega_{ii} + \Lambda_{\mathrm{pa}(i),i}^T \Sigma_{\mathrm{pa}(i),\mathrm{pa}(i)} \Lambda_{\mathrm{pa}(i),i}. \tag{7.2}$$

The matrix $\Sigma_{\mathrm{pa}(i),\mathrm{pa}(i)}$ is a principal submatrix of the positive definite matrix Σ and, thus, invertible. Therefore,

$$\Lambda_{\mathrm{pa}(i),i} = \left(\Sigma_{\mathrm{pa}(i),\mathrm{pa}(i)}\right)^{-1} \Sigma_{\mathrm{pa}(i),i}, \tag{7.3}$$

$$\omega_{ii} = \sigma_{ii} - \Sigma_{i,\mathrm{pa}(i)} \left(\Sigma_{\mathrm{pa}(i),\mathrm{pa}(i)}\right)^{-1} \Sigma_{\mathrm{pa}(i),i}. \tag{7.4}$$

Q.E.D.

Proof B shows that the formula from (7.3) holds more generally. It merely needs to hold that every trek from a node $j \in \mathrm{pa}(i)$ to i ends with a directed edge that has i as its head, so an edge of the form $k \to i$. This holds for every node in the graph if and only if the graph is *ancestral* [50]. A mixed graph is ancestral if the presence of a directed path from node i to node j implies that $i \neq j$ and $i \leftrightarrow j \notin B$. An ancestral graph is in particular acyclic.

The next easy lemma is crucial for the understanding of injectivity of ϕ_G.

Lemma 7.3. *If ϕ_G is injective and $H \subseteq G$ is a* subgraph, *then ϕ_H is injective.*

Proof. Any subgraph can be obtained by removing edges one at a time, and then removing isolated nodes. If H is obtained from $G = (V, D, B)$ by removing the edge $i \to j$, then ϕ_H is the restriction of ϕ_G to the subset of matrices $\Lambda \in \mathbb{R}^D_{\text{reg}}$ that have $\lambda_{ij} = 0$. If we instead remove the edge $i \leftrightarrow j$, then the restriction is to matrices $\Omega \in PD(B)$ with $\omega_{ij} = 0$. If H is obtained by removing the isolated node i, then

$$\phi_G(\Lambda, \Omega) = \begin{pmatrix} \phi_H(\Lambda, \Omega) & 0 \\ 0 & \omega_{ii} \end{pmatrix}.$$

In either case non-injectivity of ϕ_H implies non-injectivity of ϕ_G. Q.E.D.

Corollary 7.4. *If ϕ_G is injective, then G is simple, that is, for any two distinct vertices i and j at most one of the three edges $i \leftrightarrow j$, $i \to j$ and $i \leftarrow j$ may appear in G.*

Proof. If G is not simple then it contains a subgraph H with two nodes and two edges. The map ϕ_H is infinite-to-one as it maps a four-dimensional domain into the three-dimensional set of symmetric 2×2 matrices. Now apply Lemma 7.3. Q.E.D.

Theorem 7.5. *If ϕ_G is injective, then G is acyclic.*

Proof sketch. By Lemma 7.3, we may restrict to studying directed cycles $1 \to 2 \to \ldots \to m \to 1$. The case of $m = 2$ is covered by Corollary 7.4. If $m \geq 3$, then it is possible to show that ϕ_G is generically 2-to-1, that is, the fiber $\mathcal{F}_G(\Lambda, \Omega)$ is generically of size two [16]. Q.E.D.

It remains to characterize injectivity for acyclic graphs $G = (V, D, B)$. The next theorem shows that injectivity can be decided in polynomial time by alternatingly decomposing the bidirected part (V, B) into connected components and removing sink nodes of the directed part (V, D).

Theorem 7.6. *Suppose G is an acyclic mixed graph. Then:*

(a) *ϕ_G is injective if and only if G does not contain a subgraph whose bidirected part is connected and whose directed part has a unique sink.*

(b) *If ϕ_G is injective, then its inverse is rational and $\mathcal{M}_G$ smooth.*

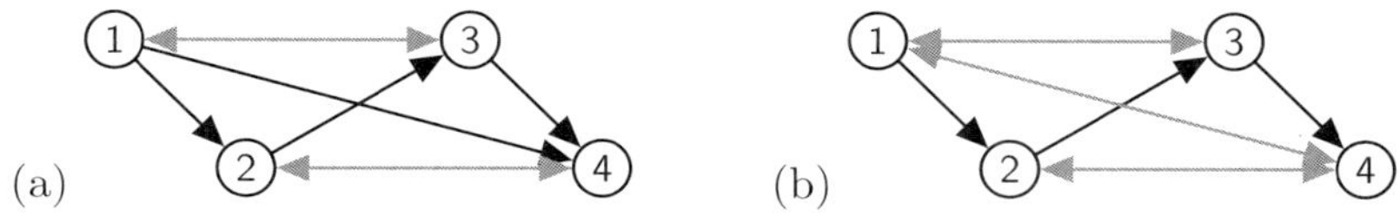

Fig. 7.1. (a) A mixed graph for which the parametrization ϕ_G is injective. (b) A graph for which ϕ_G is not injective.

Figure 7.1 illustrates the characterization in part (a) of the theorem. A full proof of the theorem can be found in [16]. The fact that ϕ_G is not injective if the combinatorial condition in (a) fails can be shown by a counterexample for the particular subgraph and then invoking Lemma 7.3. The sufficiency of the condition for injectivity can be proven by repeatedly applying the graph decomposition result in Theorem 6.4 and the result on ancestral subgraphs from Theorem 5.2. These results as well as Theorem 7.6 have generalizations to nonlinear structural equation models [56, 66].

§8. Generic Identifiability

The difference between injectivity and generic injectivity of ϕ_G may appear minute. However, the two properties are quite different, and failure of generic injectivity cannot be argued by studying subgraphs (as in Lemma 7.3). According to Corollary 7.4, a mixed graph G can have the map ϕ_G injective only if it is acyclic and simple. The deeper issue is then to find out which simple acyclic mixed graphs have ϕ_G injective. In contrast, the next result shows that all simple acyclic mixed graphs are generically injective. The deeper issue for generic injectivity is thus the treatment of graphs that contain directed cycles or are not simple.

Theorem 8.1. *If $G = (V, D, B)$ is acyclic and simple, then ϕ_G is generically injective and algebraically one-to-one.*

The theorem is due to [7]. It shows that the graph from Figure 7.1(b) has ϕ_G generically injective, but not injective. A short proof of Theorem 8.1 is obtained from the following observation.

Lemma 8.2. *Let $G = (V, D, B)$ be a mixed graph, and let $\Sigma = \phi_G(\Lambda_0, \Omega_0)$ for $\Lambda_0 \in \mathbb{R}^D_{\mathrm{reg}}$ and $\Omega_0 \in PD(B)$. The fiber $\mathcal{F}_G(\Lambda_0, \Omega_0)$ is isomorphic to the set of matrices $\Lambda \in \mathbb{R}^D_{\mathrm{reg}}$ that solve the equation system*

$$\left[(I-\Lambda)^T\Sigma(I-\Lambda)\right]_{ij} = 0, \qquad i \neq j,\ i \leftrightarrow j \notin B. \tag{8.1}$$

Proof. The projection $(\Lambda, \Omega) \mapsto \Lambda$ maps $\mathcal{F}_G(\Lambda_0, \Omega_0)$ to the set of matrices $\Lambda \in \mathbb{R}^D_{\text{reg}}$ that solve the equations in (8.1). Indeed, as $I - \Lambda$ is invertible for $\Lambda \in \mathbb{R}^D_{\text{reg}}$,

$$\Sigma = \phi_G(\Lambda, \Omega) = (I - \Lambda)^{-T}\Omega(I - \Lambda) \implies \Omega = (I - \Lambda)^T\Sigma(I - \Lambda).$$

If $i \neq j$ and $i \leftrightarrow j \notin B$, then the (i, j) entry of Ω is zero. Conversely, if Λ solves (8.1), then $(\Lambda, (I - \Lambda)^T\Sigma(I - \Lambda)) \in \mathcal{F}_G(\Lambda, \Omega)$. Q.E.D.

We emphasize that the equations in (8.1) are bilinear as

$$\begin{aligned}&\left[(I - \Lambda)^T\Sigma(I - \Lambda)\right]_{ij}\\ &\quad = \sigma_{ij} - \sum_{k \in \text{pa}(i)} \lambda_{ki}\sigma_{ki} - \sum_{l \in \text{pa}(j)} \sigma_{il}\lambda_{lj} + \sum_{k \in \text{pa}(i)} \sum_{l \in \text{pa}(j)} \lambda_{ki}\sigma_{kl}\lambda_{lj}.\end{aligned}$$

Proof of Theorem 8.1. Because G is acyclic, we may enumerate the vertex set in a topological order as $V = \{1, \dots, m\}$. Then $\text{pa}(i) \subseteq \{1, \dots, i-1\}$ for $i = 1, \dots, m$. Moreover, because G is simple, $j \in \text{pa}(i)$ implies that $j \leftrightarrow i \notin B$. By Lemma 8.2,

$$\left[(I - \Lambda)^T\Sigma(I - \Lambda)\right]_{\text{pa}(i),i} = 0, \qquad i = 1, \dots, m.$$

These equations can be rewritten as

$$\left[(I - \Lambda)^T\Sigma\right]_{\text{pa}(i),\text{pa}(i)} \Lambda_{\text{pa}(i),i} = \left[(I - \Lambda)^T\Sigma\right]_{\text{pa}(i),i}, \quad i = 1, \dots, m. \tag{8.2}$$

By the topological order, if $j \in \text{pa}(i)$, then the j-th row of $(I - \Lambda)^T\Sigma$ depends only on the first $i-1$ columns of Λ. The system in (8.2) can thus be solved recursively, each step requiring solution of a linear system.

To show that ϕ_G is generically injective, it remains to argue that the equations in (8.2) generically have a unique solution. It suffices to exhibit a single pair (Λ, Ω) for which this is true. We may choose $\Lambda = 0$ and $\Omega = I$, so $\Sigma = I$. Then the matrix for the i-th group of equations in (8.2) is $\Sigma_{\text{pa}(i),\text{pa}(i)}$, which is invertible. Q.E.D.

Although a combinatorial characterization of the graphs with generically injective parametrization ϕ_G is not known, Gröbner basis techniques can be used to determine the degree of identifiability from Definition 3.5 and, thus, decide whether ϕ_G is algebraically one-to-one. Gröbner bases are computationally tractable for non-trivial examples and have been used for a classification of all graphs with up to 5 nodes [34]. For larger graphs, algebraic methods can be applied after decomposition according to Theorem 6.4.

We describe two options for the computation. In either case, we advocate working with the equation system from (8.1) as opposed to the fiber equation $\Sigma = \phi_G(\Lambda, \Omega)$. System (8.1) has Ω eliminated and may be far more compact as it avoids inversion of $I - \Lambda$. This said, although system (8.1) is polynomial also for graphs that contain directed cycles, care must be taken to avoid spurious solutions with $I - \Lambda$ non-invertible.

The first possibility is to perform a parametric Gröbner basis computation. We introduce a matrix Λ whose nonzero entries λ_{ij}, $i \to j \in D$, are indeterminates and a pair of matrices (Λ_0, Ω_0) that are parameters. We form the matrix $(I - \Lambda)^T \phi_G(\Lambda_0, \Omega_0)(I - \Lambda)$ and set to zero the off-diagonal entries indexed by non-edges of the bidirected part (V, B). We then compute a Gröbner basis for the resulting system in the polynomial ring with coefficients in the field of rational fractions $\mathbb{R}(\Lambda_0, \Omega_0)$. The Gröbner basis readily yields the dimension of the generic fibers. If the dimension is finite we may also find the algebraic degree of the generic fibers, which is what we referred to as degree of identifiability. When the graph G contains directed cycles, we first saturate our equation system with respect to $\det(I - \Lambda)$ in order to remove solutions that have $I - \Lambda$ non-invertible.

Example 8.3. The following code for the system SINGULAR [13] implements the approach just described for a directed 3-cycle:

```
LIB "linalg.lib"; option(redSB);
ring R = (0,l012,l023,l031,w011,w022,w033),(l12,l23,l31),dp;
matrix L[3][3] = 1,-l12,0,
                 0,1,-l23,
                 -l31,0,1;
matrix L0[3][3] = 1,-l012,0,
                  0,1,-l023,
                  -l031,0,1;
matrix W0[3][3] = w011,0,0,
                  0,w022,0,
                  0,0,w033;
matrix W[3][3] =
  transpose(L)*inverse(transpose(L0))*W0*inverse(L0)*L;
ideal GB = sat(ideal(W[1,2],W[1,3],W[2,3]), det(L))[1];
dim(GB); mult(GB);
```

The output first certifies that the fibers are generically zero-dimensional, that is, contain finitely many points. The multiplicity computed with the last command shows the degree of identifiability to be two.

The second possibility is to consider only polynomials with real-valued coefficients but to introduce as polynomial variables the nonzero entries of Λ as well as a symmetric matrix Σ. These variables are ordered with respect to a block monomial order in which the variables in Λ are larger than the variables in Σ. Let $\mathcal{I}$ be the ideal generated by the off-diagonal entries of $(I-\Lambda)^T\Sigma(I-\Lambda)$ that are indexed by the non-edges of (V,B). Saturate $\mathcal{I}$ with respect to $\det(I-\Lambda)$. Let $\mathcal{J}$ be the reduced Gröbner basis of the resulting ideal. Elimination theory yields the following fact [34, Section 8 of the supplemental material].

Proposition 8.4. *A mixed graph $G=(V,D,B)$ has ϕ_G algebraically one-to-one if and only if for each $i\to j\in D$, the reduced Gröbner basis $\mathcal{J}$ contains an element with leading monomial $a(\Sigma)\lambda_{ij}$.*

In comparison to the first approach, the second method yields relations that show how to identify coefficients λ_{ij} from Σ. By analyzing the monomials under the staircase of the initial ideal of $\mathcal{J}$ [12, Chapter 9], we may also determine the generic dimension and degree of the fiber $\mathcal{F}_G(\Lambda,\Omega)$. This way we may find the degree of identifiability of G.

Example 8.5. Treating again a directed 3-cycle, we give an example of the second type of computation in Singular:

```
LIB "linalg.lib"; option(redSB);
ring R = 0,(l12,l23,l31,s11,s12,s13,s22,s23,s33),(dp(3));
matrix L[3][3] = 1,-l12,0,
                 0,1,-l23,
                 -l31,0,1;
matrix S[3][3] = s11,s12,s13,
                 s12,s22,s23,
                 s13,s23,s33;
matrix W[3][3] = transpose(L)*S*L;
ideal GB = sat(ideal(W[1,2],W[1,3],W[2,3]), det(L))[1]; GB;
```

The output is a list of 9 polynomials whose leading terms are, in our usual notation,

$$\begin{array}{ccc}\lambda_{23}\lambda_{31}\sigma_{23}, & \lambda_{12}\lambda_{31}\sigma_{13}, & \lambda_{12}\lambda_{23}\sigma_{12},\\ \lambda_{12}\lambda_{31}\sigma_{12}\sigma_{33}, & \lambda_{12}\sigma_{11}\sigma_{13}\sigma_{23}, & \lambda_{12}\lambda_{23}\sigma_{11}\sigma_{23},\\ \lambda_{23}\sigma_{12}\sigma_{13}\sigma_{22}, & \lambda_{23}\lambda_{31}\sigma_{13}\sigma_{22}, & \lambda_{31}^2\sigma_{13}\sigma_{22}\sigma_{33}.\end{array}$$

By Proposition 8.4, ϕ_G is not algebraically one-to-one because there is no leading term of the form $\lambda_{31}a(\Sigma)$. The last leading term belongs to a polynomial that shows that λ_{31} is algebraic function of degree 2 of the covariance matrix Σ because it solves the equation

$$\lambda_{31}^2\sigma_{33}(\sigma_{13}\sigma_{22}-\sigma_{12}\sigma_{23})-\lambda_{31}(\sigma_{13}^2\sigma_{22}-\sigma_{11}\sigma_{23}^2-\sigma_{12}^2\sigma_{33}+\sigma_{11}\sigma_{22}\sigma_{33})\\ +\sigma_{11}(\sigma_{13}\sigma_{22}-\sigma_{12}\sigma_{23}) = 0.$$

The equations with leading terms $\lambda_{23}\sigma_{12}\sigma_{13}\sigma_{22}$ and $\lambda_{12}\sigma_{11}\sigma_{13}\sigma_{23}$ show that λ_{23} and λ_{12} are rational functions of Σ and λ_{31}. Altogether, we have verified that ϕ_G is algebraically 2-to-one. Checking this by counting monomials under the staircase means considering the leading monomials while focusing only the variables $\lambda_{12}, \lambda_{23}, \lambda_{31}$ we seek to solve for. The monomials are

$$(8.3) \qquad \lambda_{23}\lambda_{31}, \qquad \lambda_{12}\lambda_{31}, \qquad \lambda_{12}\lambda_{23}, \qquad \lambda_{12}, \qquad \lambda_{23}, \qquad \lambda_{31}^2.$$

They generate the ideal $\mathcal{I} = \langle \lambda_{12}, \lambda_{23}, \lambda_{31}^2 \rangle$. The monomials under the staircase are the monomials in $\mathbb{R}[\lambda_{12}, \lambda_{23}, \lambda_{31}] \setminus \mathcal{I}$. The fact that there are two, namely, 1 and λ_{31}, implies that ϕ_G is algebraically 2-to-one.

Although Gröbner basis methods can be effective, it is desirable to obtain combinatorial methods that are efficient also for large-scale problems. The half-trek criteria of [34] are state-of-the-art methods whose conditions can be checked in time that is polynomial in the size of the vertex set of the considered graph. They provide a sufficient as well as a necessary condition for generic injectivity of ϕ_G. More precisely, there is a condition that is sufficient for ϕ_G to be algebraically one-to-one and a related condition that is necessary for ϕ_G to be generically finite-to-one. The conditions are implemented in a package for the R project for statistical computing [4]. We begin our discussion of the half-trek criteria by introducing some needed terminology.

A *half-trek* from initial node i to target node j is a trek τ from i to j whose left-hand side is a singleton set, so $\text{left}(\tau) = \{i\}$. In other words, a half-trek is of the form

$$i \to j_1 \to \ldots \to j_r \to j \quad \text{or} \quad i \leftrightarrow j_1 \to \ldots \to j_r \to j.$$

Let $X, Y \subseteq V$ be two sets of nodes of equal cardinality $|X| = |Y| = k$. Let Π be a set of k treks. Then Π is a system of treks from X to Y, denoted $\Pi : X \rightrightarrows Y$, if X is the set of initial nodes of the treks in Π and Y is the set of target nodes. Note that we allow $X \cap Y \neq \emptyset$. The system Π is a system of half-treks if every trek π_i is a half-trek. Finally, the system Π has no sided intersection if

$$\text{left}(\pi) \cap \text{left}(\pi') = \emptyset = \text{right}(\pi) \cap \text{right}(\pi')$$

for all pairs of treks $\pi, \pi' \in \Pi$.

Definition 8.6. *A set $Y \subseteq V$ satisfies the* half-trek criterion *with respect to node i if* (i) $|Y| = |\text{pa}(i)|$, (ii) *$j = i$ or $j \leftrightarrow i$ implies that $j \notin Y$, and* (iii) *there exists a system of half-treks $\Pi : Y \rightrightarrows \text{pa}(i)$ that has no sided intersection.*

Theorem 8.7. *Let $G = (V, D, B)$ be a mixed graph.*

(i) *Suppose for every $i \in V$ there exists a set $Y_i \subseteq V$ that satisfies the half-trek criterion with respect to i. If there exists a total ordering $\prec$ such that $j \prec i$ whenever $j \in Y_i$ and there is a half-trek from i to j, then ϕ_G is generically injective and algebraically one-to-one.*

(ii) *For ϕ_G to be generically finite-to-one it is necessary that there exists a family of sets $Y_i \subseteq V$, $i \in V$, such that Y_i satisfies the half-trek criterion with respect to i and $j \in Y_i$ implies $i \notin Y_j$.*

We merely outline the proof of the theorem; for details see [34]. Some of the arguments are further illustrated in Example 8.8. Note also that Theorem 8.1 is obtained from Theorem 8.7(i) by taking $Y_i = \mathrm{pa}(i)$ and $\prec$ as a topological order.

Outline of proof of Theorem 8.7. (i) Let $\Sigma = \phi_G(\Lambda_0, \Omega_0)$ for $\Lambda_0 \in \mathbb{R}^D_{\mathrm{reg}}$ and $\Omega_0 \in PD(B)$. Suppose $(\Lambda, \Omega) \in \mathcal{F}_G(\Lambda_0, \Omega_0)$. To show that $(\Lambda, \Omega) = (\Lambda_0, \Omega_0)$, we visit the nodes $i \in V$ from smallest to largest in the order $\prec$ and iteratively find a linear equation system that is uniquely solved by the i-th column of Λ. The starting point is Lemma 8.2, by which we have

$$\big[(I - \Lambda)^T \Sigma (I - \Lambda)\big]_{Y_i, i} = 0, \quad i \in V. \tag{8.4}$$

This is true because Definition 8.6 yields that $j \neq i$ and $j \leftrightarrow i \notin B$ when $j \in Y_i$. Similar to the proof of Theorem 8.1, we may rearrange (8.4) to

$$A_i(\Lambda, \Sigma)\Lambda_{\mathrm{pa}(i),i} \;=\; b_i(\Lambda, \Sigma),$$

with $A_i(\Lambda, \Sigma) = \big[(I - \Lambda)^T \Sigma\big]_{Y_i, \mathrm{pa}(i)}$ and $b_i(\Lambda, \Sigma) = \big[(I - \Lambda)^T \Sigma\big]_{Y_i, i}$. Both $A_i(\Lambda, \Sigma)$ and $b_i(\Lambda, \Sigma)$ can be shown to depend only on those columns of Λ that are indexed by nodes j with a half-trek from i to j. Thus, the proof is complete if we can show that $A_i(\Lambda_0, \Sigma)$ is invertible for generic choices of Λ_0 and Ω_0. To verify this, we may use the existence of a half-trek system without sided intersection from Y_i to $\mathrm{pa}(i)$ to argue that the determinant of $A_i(\Lambda_0, \Sigma)$ is not the zero polynomial. This last step is in the spirit of the Lindström-Gessel-Viennot lemma.

(ii) The Jacobian of the equations from Lemma 8.2 has its rows indexed by the non-edges of the bidirected part (V, B) and its columns indexed by the directed edges in D. For ϕ_G to be generically finite-to-one, it is necessary that the Jacobian has full column rank D. It can be shown that the Jacobian contains an invertible $|D| \times |D|$ submatrix only if the given condition holds. Let J_i be the submatrix of the Jacobian

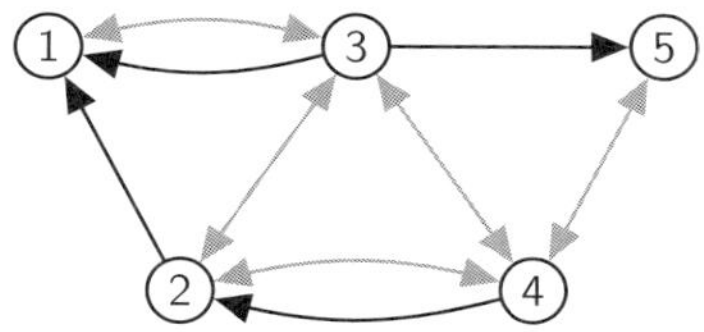

Fig. 8.1. Illustration of Theorem 8.7.

obtained by selecting the columns corresponding to directed edges with head at i. Then, more specifically, J_i has full column rank only if there exists a subset $Y_i \subseteq V$ that satisfies the half-trek criterion with respect to i. Moreover, if $j \in Y_i$ and $i \in Y_j$, then the same row, namely, that corresponding to $i \leftrightarrow j \notin B$, would be used to get an invertible square submatrix of J_i and J_j. Q.E.D.

The conditions from Theorem 8.7 can be checked in polynomial time. For condition (i), we may use recursively that the existence of a suitable set satisfying the half-trek criterion is equivalent to a condition on the maximum flow in a certain network-flow problem. Condition (ii) can be checked via a single larger network-flow problem [34, Section 6].

Example 8.8. Let G be the graph in Figure 8.1. The sets

$$Y_1 = \{2,5\}, \qquad Y_2 = \{5\}, \qquad Y_3 = \emptyset, \qquad Y_4 = \emptyset, \qquad Y_5 = \{3\}$$

each satisfy the half-trek criterion with respect to the node they are indexed by. This is least evident for Y_1, and we highlight the half-treks $2 \leftrightarrow 3$ and $5 \leftrightarrow 4 \to 2$, which have no sided intersection. Choosing the ordering as $3 \prec 4 \prec 5 \prec 1 \prec 2$, Theorem 8.7(i) shows that ϕ_G is algebraically one-to-one. Other possible orderings are obtained by permuting $\{3,4,5\}$ or $\{1,2\}$.

To illustrate ideas from the proof of Theorem 8.7(i), we focus on node 1, with $\text{pa}(1) = \{2,3\}$. Since $Y_1 = \{2,5\}$, we work with the equations

$$\left[(I-\Lambda)^T\Sigma(I-\Lambda)\right]_{51} = 0, \qquad \left[(I-\Lambda)^T\Sigma(I-\Lambda)\right]_{21} = 0.$$

Expanding out the matrix product, the equations become

$$\sigma_{51} - \sigma_{52}\lambda_{21} - \sigma_{53}\lambda_{31} - \lambda_{35}\sigma_{31} + \lambda_{35}\sigma_{32}\lambda_{21} + \lambda_{35}\sigma_{33}\lambda_{31} = 0, \tag{8.5}$$

$$\sigma_{21} - \sigma_{22}\lambda_{21} - \sigma_{23}\lambda_{31} - \lambda_{42}\sigma_{41} + \lambda_{42}\sigma_{42}\lambda_{21} + \lambda_{42}\sigma_{43}\lambda_{31} = 0, \tag{8.6}$$

and we wish to solve for λ_{21} and λ_{31}. With $5 \prec 1$, we have already solved for λ_{35}; since $Y_5 = \mathrm{pa}(5)$ we have $\lambda_{35} = \sigma_{35}/\sigma_{55}$ as is also clear from the discussion after Proof B for Theorem 7.2. Substituting the ratio for λ_{35} turns (8.5) into a linear equation in λ_{21} and λ_{31}. The equation in (8.6) could be linearized similarly, except that now the relevant coefficient λ_{42} has not yet been determined in an ordering with $1 \prec 2$. However, if Λ is part of a pair (Λ, Ω) in the fiber given by Σ, then

$$-\lambda_{42}\sigma_{41} + \lambda_{42}\sigma_{42}\lambda_{21} + \lambda_{42}\sigma_{43}\lambda_{31} \;=\; 0$$

because there is no half-trek from 1 to 2. To see this note that the term $\lambda_{42}\sigma_{41}$ corresponds to treks from 2 to 1 that start with the edge $2 \leftarrow 4$, whereas the sum $\lambda_{42}\sigma_{42}\lambda_{21} + \lambda_{42}\sigma_{43}\lambda_{31}$ corresponds to treks from 2 to 1 that start with the edge $2 \leftarrow 4$ and end in either $2 \rightarrow 1$ or $3 \rightarrow 1$. These two sets of treks coincide when there is no half-trek from 1 to 2.

Condition (i) in Theorem 8.7 can be improved by applying the graph decomposition from Section 6 and checking the condition in each subgraph. No such strengthening is possible for the necessary condition from part (ii) of the theorem [34]. Further strengthening of the sufficient condition is possible by first removing sink nodes from the graph using the observation from Theorem 5.2. When a sink node is removed a more refined graph decomposition may become possible; we refer the reader to [10, 22]. While a specific polynomial-time algorithm using this idea is given in [22], it is still unclear how to best design algorithms based on recursive graph decomposition and removal of sink nodes.

We conclude our discussion of parameter identification with two examples from the exhaustive computational study of graphs with up to 5 nodes in [34]. Both graphs in Figure 8.1 satisfy the necessary condition in Theorem 8.7(ii) and, thus, have ϕ_G generically finite-to-one. Neither graph satisfies the sufficient condition from Theorem 8.7(i). The graph in panel (a) indeed does not have ϕ_G generically injective. Instead, ϕ_G is algebraically 3-to-one. The graph in panel (b), however, is algebraically one-to-one but Theorem 8.7(i) fails to recognize it. Decomposition and removal of sink nodes do not help and other ideas are needed; see [71].

Part IV. Relations Among Covariances

§9. Implicitization

Let $\mathcal{M}_G$ be the set of covariance matrices of the structural equation model given by a mixed graph $G = (V, D, B)$. Motivated, in particular, by the covariance equivalence problem from Question 3.7, we now discuss polynomial relations among the entries of the matrices in $\mathcal{M}_G$. Let

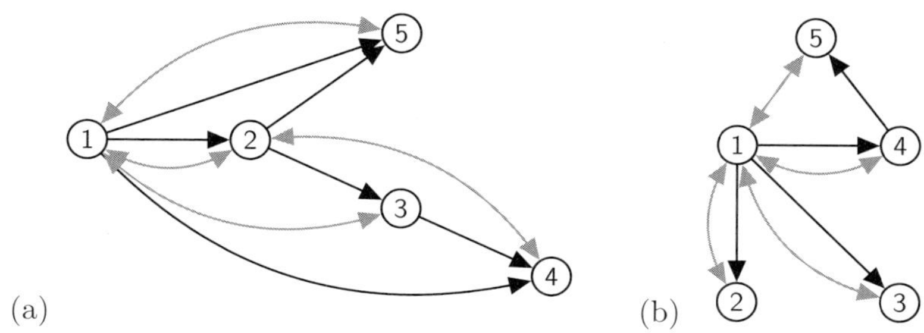

Fig. 8.2. Graphs that satisfy the necessary but not the sufficient condition from Theorem 8.7: (a) ϕ_G is algebraically 3-to-one, (b) ϕ_G is algebraically one-to-one and, thus, generically injective.

$\Sigma = (\sigma_{ij})$ be a symmetric $V \times V$ matrix of variables, and let $\mathbb{R}[\Sigma]$ be the ring of polynomials in the indeterminates σ_{ij} with real coefficients. The relations we seek to understand make up the *vanishing ideal*

$$\mathcal{I}(G) \;=\; \{f \in \mathbb{R}[\Sigma] \;:\; f(\Sigma) = 0 \text{ for all } \Sigma \in \mathcal{M}_G\}\,.$$

Since $\mathcal{M}_G$ is the image of the rational map ϕ_G, we may compute a generating set for $\mathcal{I}(G)$ by implicitization. Assume for simplicity that G is acyclic and, thus, ϕ_G polynomial. Define $\mathbb{R}[\Sigma, \Lambda, \Omega]$ to be the polynomial ring with the additional indeterminates from a $V \times V$ matrix $\Lambda = (\lambda_{ij})$ and a symmetric $V \times V$ matrix $\Omega = (\omega_{ij})$ whose supports are determined by D and B, respectively. Then $\langle \sigma_{ij} - \phi_G(\Lambda, \Omega)_{ij} : i \le j \rangle \subset \mathbb{R}[\Sigma, \Lambda, \Omega]$ is the ideal for the graph of ϕ_G, and

$$\mathcal{I}(G) \;=\; \langle\, \sigma_{ij} - \phi_G(\Lambda, \Omega)_{ij} : i \le j \,\rangle \;\cap\; \mathbb{R}[\Sigma].$$

A better approach, however, is to start with the equations from Lemma 8.2, which have Ω already eliminated. We compute

$$\mathcal{I}(G) \;=\; \Big\langle \big[(I - \Lambda)^T \Sigma (I - \Lambda)\big]_{ij} \;:\; i \ne j,\; i \leftrightarrow j \notin B \Big\rangle \;\cap\; \mathbb{R}[\Sigma].$$

If G is not acyclic, we saturate with respect to $\det(I - \Lambda)$ before intersecting with $\mathbb{R}[\Sigma]$; compare Examples 8.3 and 8.5.

Example 9.1. To illustrate the use of a different software, we change to Macaulay2 [40]. The following code computes a generating set for the vanishing ideal $\mathcal{I}(G)$ when G is the graph from Figure 9.1:

```
R = QQ[l12,l13,l24,l34, s11,s12,s13,s14,s22,s23,s24,s33,s34,s44,
     MonomialOrder => Eliminate 4];
```

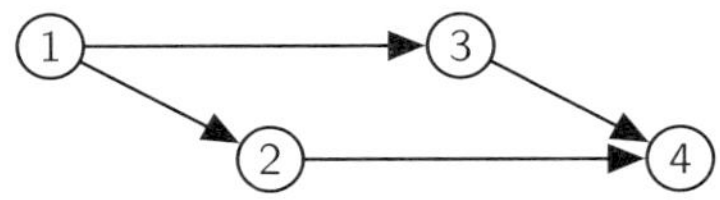

Fig. 9.1. An acyclic digraph.

```
Lambda = matrix{{1, -l12, -l13, 0},
                {0, 1, 0, -l24},
                {0, 0, 1, -l34},
                {0, 0, 0, 1}};
S = matrix{{s11, s12, s13, s14},
           {s12, s22, s23, s24},
           {s13, s23, s33, s34},
           {s14, s24, s34, s44}};
W = transpose(Lambda)*S*Lambda;
I = ideal{W_(0,1),W_(0,2),W_(0,3),W_(1,2),W_(1,3),W_(2,3)};
eliminate({l12,l13,l24,l34},I)
```

The 'GraphicalModels' package [37] automates the computation:

```
needsPackage "GraphicalModels";
G = digraph {{1,{2,3}},{2,{4}},{3,{4}}};
R = gaussianRing G;
gaussianVanishingIdeal R
```

Reproduced in our notation, the output shows that the ideal $\mathcal{I}(G)$ is generated by the two polynomials

$$f_1 = \sigma_{12}\sigma_{13} - \sigma_{11}\sigma_{23}, \tag{9.1}$$

$$\begin{aligned} f_2 = \sigma_{14}\sigma_{23}^2 - \sigma_{13}\sigma_{23}\sigma_{24} - \sigma_{14}\sigma_{22}\sigma_{33} \\ + \sigma_{12}\sigma_{24}\sigma_{33} + \sigma_{13}\sigma_{22}\sigma_{34} - \sigma_{12}\sigma_{23}\sigma_{34}. \end{aligned} \tag{9.2}$$

Computing $\mathcal{I}(G)$ using Gröbner bases is a method that applies to any mixed graph but can be computationally prohibitive for graphs with more than 5 or 6 nodes. In order to solve model equivalence problems combinatorial insight on particular types of relations is needed.

Example 9.2. Continuing with Example 9.1, observe that f_1 and f_2 from (9.1) and (9.2) are determinants of submatrices of the covariance matrix Σ. The following two displays locate the submatrices:

$$\begin{bmatrix} \boxed{\sigma_{11}} & \sigma_{12} & \sigma_{13} & \sigma_{14} \\ \sigma_{12} & \sigma_{22} & \sigma_{23} & \sigma_{24} \\ \sigma_{13} & \sigma_{23} & \sigma_{33} & \sigma_{34} \\ \sigma_{14} & \sigma_{24} & \sigma_{34} & \sigma_{44} \end{bmatrix} \qquad \begin{bmatrix} \sigma_{11} & \sigma_{12} & \sigma_{13} & \sigma_{14} \\ \sigma_{12} & \boxed{\sigma_{22}} & \boxed{\sigma_{23}} & \sigma_{24} \\ \sigma_{13} & \boxed{\sigma_{23}} & \boxed{\sigma_{33}} & \sigma_{34} \\ \sigma_{14} & \sigma_{24} & \sigma_{34} & \sigma_{44} \end{bmatrix}$$

The boxes emphasize that each submatrix contains a principal submatrix along with one additional row and column. The determinants f_1 and f_2 are thus almost principal minors of Σ. As we discuss in Section 10, the vanishing of almost principal minors of a Gaussian covariance matrix has the probabilistic meaning of conditional independence [60].

§10. Conditional Independence

Let $X = (X_i : i \in V)$ be a Gaussian random vector with covariance matrix $\Sigma \in PD_V$. Let $i, j \in V$ be two distinct indices, and let $S \subseteq V \setminus \{i, j\}$. The random variables X_i and X_j are conditionally independent given X_S if and only if $\det(\Sigma_{iS\times jS}) = 0$; see [21, Chapter 3] and [48].

It is fully understood which conditional independence relations hold in the covariance matrices of a linear structural equation model. The following concepts are needed to state the result. Let π be a semi-walk from node i to node j in the considered mixed graph $G = (V, D, B)$, and let node k be a non-endpoint of π. A *collider* on π is a node k that is an internal node on π, and a head on the two edges that precede and succeed k on π. We recall our convention that the two nodes incident to a bidirected edge are both heads. Pictorially, π includes the segment $\to k \leftarrow$, $\to k \leftrightarrow$, $\leftrightarrow k \leftarrow$ or $\leftrightarrow k \leftrightarrow$. A *non-collider* on π is an internal node of π that is not a collider on π.

Definition 10.1. *Fix a set $S \subseteq V$. Two nodes $i, j \in V$ are* d-connected *by S if G contains a semi-walk from i to j that has all colliders in S and all non-colliders outside S. If no such semi-walk exists, then i and j are* d-separated *by S.*

The following theorem was first proven for acyclic digraphs [39] and later extended to cover arbitrary mixed graphs [58].

Theorem 10.2. *Let i and j be two distinct nodes of a mixed graph $G = (V, D, B)$, and let $S \subseteq V \setminus \{i, j\}$. Then $\det(\phi_G(\Lambda, \Omega)_{iS\times jS}) = 0$ for all $\Lambda \in \mathbb{R}^D_{\mathrm{reg}}$, $\Omega \in PD(B)$ if and only if i and j are d-separated by S.*

For acyclic digraphs, the theorem can be derived in a probabilistic setup that extends to non-Gaussian conditional independence [46]. Examples with three binary variables, such as the one in Figure 10.1, serve well to illustrate the intuition behind d-separation. In our linear Gaussian setting, Theorem 10.2 is a special case of Theorem 11.3 that we treat in more detail.

Define the *conditional independence ideal*

$$\mathcal{I}_{\mathrm{CI}}(G) \;=\; \langle\, \det(\Sigma_{iS\times jS}) \;:\; i, j \text{ d-separated by } S \,\rangle.$$

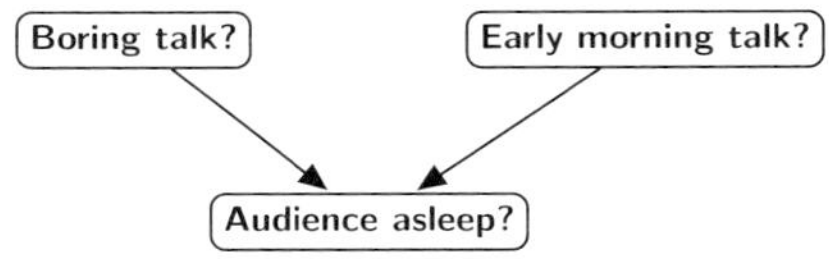

Fig. 10.1. A graph for three binary variables.

By Theorem 10.2, $\mathcal{I}_{\mathrm{CI}}(G) \subseteq \mathcal{I}(G)$ for any mixed graph G. In Example 9.1, $\mathcal{I}_{\mathrm{CI}}(G) = \mathcal{I}(G)$ but this may be false even for acyclic digraphs [61]. Nevertheless, if G is an acyclic digraph, then the set of covariance matrices $\mathcal{M}_G$ is cut out by conditional independence relations, that is,

$$\mathcal{M}_G \;=\; V\left(\mathcal{I}_{\mathrm{CI}}(G)\right) \cap PD_V. \tag{10.1}$$

Here, $V(\mathcal{I}_{\mathrm{CI}}(G))$ is the variety of the conditional independence ideal, i.e., the set of symmetric matrices that have all the determinants generating $\mathcal{I}_{\mathrm{CI}}(G)$ zero. The equality in (10.1) also holds when G is an ancestral graph, as defined in Section 7. However, it is false in general as can be seen in Example 3.14. We remark that for acyclic digraphs it has been proven that saturating the conditional independence with respect to all principal minors yields the vanishing ideal [52]:

$$\mathcal{I}(G) \;=\; \mathcal{I}^{\mathrm{CI}}(G) \;:\; \Big(\prod_{A \subset V} \det(\Sigma_{A\times A})\Big)^{\infty}.$$

For acyclic digraphs and more generally ancestral graphs, the equality from (10.1) allows us to answer Question 3.7 on covariance equivalence by checking whether two graphs have the same d-separation relations. The latter comparison can be accomplished in polynomial time. We state the result for acyclic digraphs [35, 69]. A generalization for ancestral graphs was given more recently [1]; see also [75].

Theorem 10.3. *Let G and G' be two acyclic digraphs. Then $\mathcal{M}_G = \mathcal{M}_{G'}$ if and only if G and G' have the same adjacencies and the same unshielded colliders. An unshielded collider is an induced subgraph of the form $i \to j \leftarrow k$.*

We remark that it can also be decided in polynomial time whether two digraphs that are not necessarily acyclic have the same d-separation relations [51]. When the graphs have directed cycles then d-separation equivalence is necessary but not sufficient for covariance equivalence; see, e.g., the example in [15].

§11. Trek-Separation

We now turn to the characterization of the vanishing of general minors of the covariance matrices in a linear structural equation model. Our first example clarifies the importance of minors that are not almost principal.

Example 11.1. If G is the graph from Figure 3.2 and Example 3.14, then $\mathcal{I}(G)$ is generated by $\det(\Sigma_{12,34})$. Such off-diagonal 2×2 minors are known as tetrads in the statistical literature; see e.g. [20].

The tetrad representation theorem gives a combinatorial characterization of the vanishing of any 2×2 determinant [57]. The theorem has been greatly generalized, and we now have a full combinatorial understanding of when a minor of the covariance matrix vanishes based on the notion of trek-separation [64]. Moreover, the non-vanishing determinants can be described precisely [14].

Definition 11.2. *Let $A, C, S_A, S_C \subseteq V$ be subsets of the vertex set of the mixed graph $G = (V, D, B)$. The pair (S_A, S_C)* trek-separates *A and C if every trek from $i \in A$ to $j \in C$ intersects S_A on its left side or S_C on its right side.*

Theorem 11.3. *Let G be a mixed graph. Then the $A \times C$ submatrix of $\phi_G(\Lambda, \Omega)$ has generic rank $\leq r$ if and only if there exist sets S_A and S_C with $|S_A| + |S_C| \leq r$ such that (S_A, S_C) trek-separates A and C.*

The theorem for acyclic mixed graphs is proven in [64]. The cases with directed cycles are covered by the results in [14]. We describe the ideas behind the proof.

Proof outline. The problem can be reduced to the case where G is a digraph by subdividing bidirected edges. For each edge $i \leftrightarrow j \in B$ we introduce a new node $\{i, j\}$ and two edges $\{i, j\} \to i$ and $\{i, j\} \to j$. The new digraph G' thus has $|V| + |B|$ nodes and $|D| + 2|B|$ edges. For instance, if G is the graph from Figure 1.2, then G' is the digraph in Figure 1.1. Every trek in G corresponds to a trek in G' where an edge $i \leftrightarrow j$ in G corresponds to $i \leftarrow \{i, j\} \to j$ in G'. The entries of $(\phi_G)_{A,C}$ and those of $(\phi_{G'})_{A,C}$ can be shown to admit the same set of relations.

In the sequel, assume that G itself is a digraph. Then $PD(B)$ contains diagonal matrices with positive entries. The rank of a submatrix of $\phi_G(\Lambda, \Omega)$ for $\Lambda \in \mathbb{R}^D_{\text{reg}}$ and $\Omega \in PD(B)$ is then the same as the rank of the corresponding submatrix of $\Sigma = \phi_G(\Lambda, I)$.

To establish the claim, we may study the vanishing of the determinants of square submatrices. So assume that $|A| = |C| = r + 1$. The

Cauchy-Binet formula gives that

(11.1)
$$\det(\Sigma_{A\times C}) = \sum_E \det\left(\left[(I-\Lambda)^{-1}\right]_{E\times A}\right)\det\left(\left[(I-\Lambda)^{-1}\right]_{E\times C}\right),$$

where the sum is over subsets $E \subseteq V$ of cardinality $r+1$. By the Lindström-Gessel-Viennot lemma for general digraphs, it holds that

$$\det\left(\left[(I-\Lambda)^{-1}\right]_{E\times A}\right) = 0 \quad \text{for all } \Lambda \in \mathbb{R}^D_{\text{reg}}, \tag{11.2}$$

if and only if no system of $r+1$ directed paths from E to A is vertex-disjoint. Applying this to all terms in (11.1) shows that $\det(\Sigma_{A\times C})$ vanishes if and only if every system of treks from A to C has a sided intersection. Here, an intersection on the left side of a trek corresponds to the vanishing of determinants of the matrices $\left[(I-\Lambda)^{-1}\right]_{E\times A}$ and, similarly, intersections on the right side are related to the determinants of the matrices $\left[(I-\Lambda)^{-1}\right]_{E\times C}$. The characterization by sided intersections in trek systems can be turned into the claimed statement about trek-separation via Menger's theorem. To account for the distinct role played by the left and the right sides of the treks, Menger's theorem is applied in a digraph $\tilde{G}$ that results from duplicating the nodes and edges of G. Each node and each edge of G has a left and a right side version in $\tilde{G}$. Menger's theorem yields a cut set S of cardinality $|S| \le r$ in $\tilde{G}$. Partitioning S according to the left and right side yields the pair (S_A, S_C) for the claimed trek-separation. Q.E.D.

Example 11.4. When G is the graph from Figure 3.2, then the submatrix $[\phi_G(\Lambda,\Omega)]_{12,34}$ has generic rank 1; recall Example 11.1. Theorem 11.3 shows that the rank is at least 1 because $(\emptyset,\emptyset)$ does not trek-separate $\{1,2\}$ and $\{3,4\}$. For instance, there is the trek $1\to 3$. That the rank is no larger than 1 follows from $(\emptyset,\{3\})$ trek-separating $\{1,2\}$ and $\{3,4\}$. For instance, the node 3 is on the right side of the two treks $1\to 3$ and $2\to 3\to 4$.

Example 11.5. What is the generic rank of the $A\times C$ submatrix of $\phi_G(\Lambda,\Omega)$ when G is the 'spider graph' from Figure 11.1, $A=\{1,2,3,4\}$ and $C=\{5,6,7\}$? Clearly, node c is on every trek between A and C. Hence, $(\{c\},\{c\})$ trek-separates A and C and the rank is at most two. In fact, the rank is two. Consider the two treks

$$\pi_1 : 1 \leftrightarrow \circ \to c \to 5,$$
$$\pi_2 : 3 \leftarrow c \leftarrow \circ \leftrightarrow 6.$$

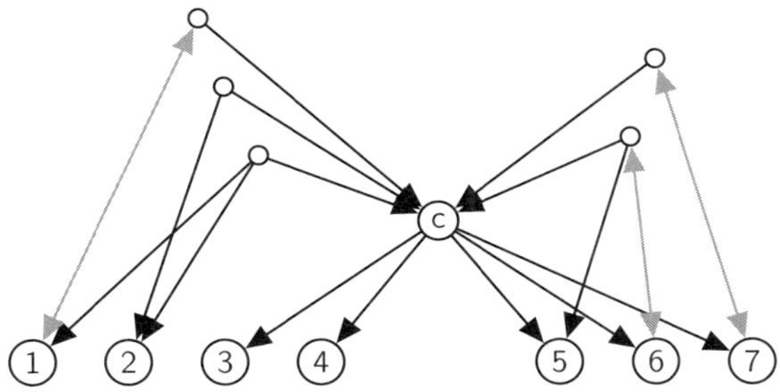

Fig. 11.1. A 'spider graph' with $\{1,2,3,4\} \times \{5,6,7\}$ submatrix of rank two.

They have only node c in common but $c \in \text{right}\,(\pi_1)$ and $c \in \text{left}\,(\pi_2)$. Hence, a pair of trek-separating sets must use at least two nodes.

Trek-separation solves the problem of characterizing the vanishing of determinants. However, there is currently no efficient method for deciding when two mixed graphs are trek-separation equivalent.

§12. Verma Constraints

Much interesting ground lies beyond determinants of the covariance matrix. We content ourselves with two examples concerning a relation first presented in [69].

Example 12.1. Let G be the graph from Figure 2.1. The graph has no trek-separation relations as can be checked with the package 'GraphicalModels' for Macaulay2 [37]. The commands

```
needsPackage "GraphicalModels";
G = mixedGraph(digraph {{1,{2,3}},{2,{3}},{3,{4}}}, bigraph {{2,4}});
R = gaussianRing G;
trekIdeal(R,G)
```

return the zero ideal. Issuing the command

```
gaussianVanishingIdeal R
```

shows that $\mathcal{I}(G)$ is generated by

(12.1)
$$f_{\text{Verma}} = \sigma_{11}\sigma_{13}\sigma_{22}\sigma_{34} - \sigma_{11}\sigma_{13}\sigma_{23}\sigma_{24} - \sigma_{11}\sigma_{14}\sigma_{22}\sigma_{33} + \sigma_{11}\sigma_{14}\sigma_{23}^2 - \sigma_{12}^2\sigma_{13}\sigma_{34} + \sigma_{12}^2\sigma_{14}\sigma_{33} + \sigma_{12}\sigma_{13}^2\sigma_{24} - \sigma_{12}\sigma_{13}\sigma_{14}\sigma_{23}.$$

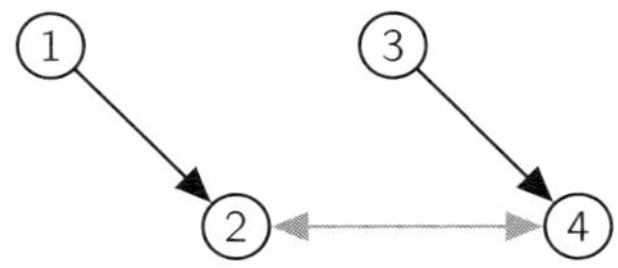

Fig. 12.1. The largest mixed component of the Verma graph.

Clearly, f_{Verma} is not a determinant. Its vanishing can be explained as follows. Decompose G into its mixed components. The largest component $G[\{2,4\}]$ is depicted in Figure 12.1. By Theorem 6.4, there is a rational function $\tau_{\{2,4\}} : \mathcal{M}_G \to \mathcal{M}_{G'}$, so $\tau_{\{2,4\}}$ is the covariance matrix for $G[\{2,4\}]$. In $G[\{2,4\}]$, there is no trek from node 1 to node 4. Hence, the $(1,4)$ entry of $\tau_{\{2,4\}}(\Sigma)$ is zero. Clearing the denominator yields $f_{\text{Verma}}(\Sigma) = 0$ for $\Sigma \in \mathcal{M}_G$.

The example suggests that new relations can be discovered by decomposing the graph and studying d-separation or trek-separation relations in the mixed components. However, the matter is not as simple as applying a single decomposition.

Example 12.2. In order to prevent decomposition of the Verma graph, add a fifth node and bidirected edges such that (V, B) becomes connected. Specifically, consider the graph G from Figure 12.2. The new graph G is simple and acyclic and, thus, $\mathcal{M}_G$ has expected dimension 13. Because the nodes 2 and 5 are d-separated by node 1, we have $\sigma_{12}\sigma_{15} - \sigma_{11}\sigma_{25} \in \mathcal{I}(G)$. Other relations must exist, and indeed a Gröbner basis computation shows that

$$\mathcal{I}(G) \;=\; \langle \sigma_{12}\sigma_{15} - \sigma_{11}\sigma_{25},\; f_{\text{Verma}} \rangle : \sigma_{11}^{\infty}$$

with f_{Verma} being the polynomial from (12.1). The fact that $f_{\text{Verma}} \in \mathcal{I}(G)$ is explained by Theorem 5.2. Since no directed edge of G has a tail at node 5, the theorem allows us to consider the subgraph induced by the set of nodes $\{1,2,3,4\}$. We are back in Example 12.1 and decomposition yields f_{Verma} as a relation.

It is clear that more contrived examples can be constructed in which d-/trek-separation applies only after several rounds of alternating between graph decomposition and forming a subgraph induced by an ancestral set. An overview of what is known about such a recursive approach can be found in [55], where the focus is on non-linear models and manipulation of probability density functions.

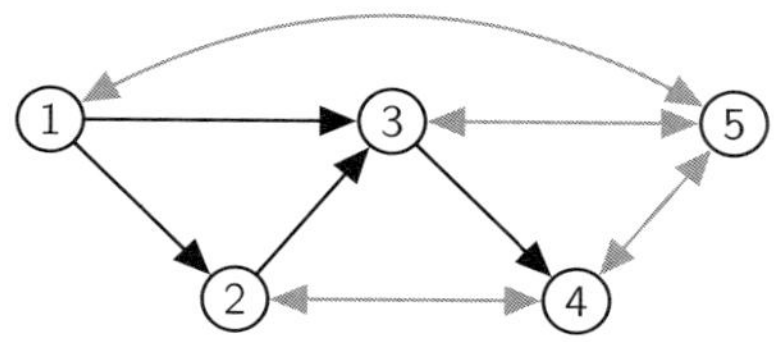

Fig. 12.2. A graph that can be decomposed only after removing the sink node 5.

§13. Conclusion

Linear structural equation models have covariance matrices with rich algebraic structure. As we showed in Section 3, statistical considerations motivate a host of different algebraic problems. In this review, we focused on methods for parameter identification as well as relations among covariances. While much progress has been made, and continues to be made [30], we have encountered a plethora of open problems of algebraic and combinatorial nature.

In our review, we focused exclusively on linear and Gaussian models. As noted repeatedly, many of the questions have interesting generalizations to non-linear or non-Gaussian models. In particular, in settings with discrete random variables, as considered in [27, 28], algebra and Gröbner basis techniques continue to be useful [21]. Similarly, many additional challenges arise in models with explicit latent variables, which motivate studying maps that result from projecting onto a principal submatrix of ϕ_G (recall Section 5).

References

[1] R. Ayesha Ali, Thomas S. Richardson, and Peter Spirtes, *Markov equivalence for ancestral graphs*, Ann. Statist. **37** (2009), no. 5B, 2808–2837.

[2] Theodore W. Anderson, *An introduction to multivariate statistical analysis*, third ed., Wiley Series in Probability and Statistics, Wiley-Interscience [John Wiley & Sons], Hoboken, NJ, 2003. MR 1990662

[3] Vladimir I. Arnol′d, Sabir M. Guseĭn-Zade, and Aleksandr N. Varchenko, *Singularities of differentiable maps. Vol. II*, Monographs in Mathematics, vol. 83, Birkhäuser Boston, Inc., Boston, MA, 1988, Monodromy and asymptotics of integrals, Translated from the Russian by Hugh Porteous, Translation revised by the authors and James Montaldi. MR 966191

[4] Rina Foygel Barber, Mathias Drton, and Luca Weihs, *Semid: Identifiability of linear structural equation models*, 2015, R package version 0.2.

[5] Kenneth A. Bollen, *Structural equations with latent variables*, Wiley Series in Probability and Mathematical Statistics: Applied Probability and Statistics, John Wiley & Sons Inc., New York, 1989, A Wiley-Interscience Publication. MR 996025 (90k:62001)

[6] Kenneth A. Bollen and Kwok-Fai Ting, *A tetrad test for causal indicators*, Psychological methods **5** (2000), no. 1, 3–22.

[7] Carlos Brito and Judea Pearl, *A new identification condition for recursive models with correlated errors*, Struct. Equ. Model. **9** (2002), no. 4, 459–474. MR 1930449

[8] Andries E. Brouwer and Jan Draisma, *Equivariant Gröbner bases and the Gaussian two-factor model*, Math. Comp. **80** (2011), no. 274, 1123–1133. MR 2772115

[9] Søren L. Buhl, *On the existence of maximum likelihood estimators for graphical Gaussian models*, Scand. J. Statist. **20** (1993), no. 3, 263–270.

[10] Bryant Chen, *Decomposition and identification of linear structural equation models*, ArXiv e-prints (2015), 1508.01834.

[11] Bryant Chen, Jin Tian, and Judea Pearl, *Testable implications of linear structural equations models*, Proceedings of the Twenty-Eighth AAAI Conference on Artificial Intelligence (Palo Alto, CA) (Carla E. Brodley and Peter Stone, eds.), AAAI Press, 2014, pp. 2424–2430.

[12] David Cox, John Little, and Donal O'Shea, *Ideals, varieties, and algorithms*, third ed., Undergraduate Texts in Mathematics, Springer, New York, 2007, An introduction to computational algebraic geometry and commutative algebra. MR 2290010 (2007h:13036)

[13] Wolfram Decker, Gert-Martin Greuel, Gerhard Pfister, and Hans Schönemann, SINGULAR *4-0-2 — A computer algebra system for polynomial computations*, `http://www.singular.uni-kl.de`, 2015.

[14] Jan Draisma, Seth Sullivant, and Kelli Talaska, *Positivity for Gaussian graphical models*, Adv. in Appl. Math. **50** (2013), no. 5, 661–674. MR 3044565

[15] Mathias Drton, *Likelihood ratio tests and singularities*, Ann. Statist. **37** (2009), no. 2, 979–1012. MR 2502658

[16] Mathias Drton, Rina Foygel, and Seth Sullivant, *Global identifiability of linear structural equation models*, Ann. Statist. **39** (2011), no. 2, 865–886.

[17] Mathias Drton, Hélène Massam, and Ingram Olkin, *Moments of minors of Wishart matrices*, Ann. Statist. **36** (2008), no. 5, 2261–2283. MR 2458187

[18] Mathias Drton and Martyn Plummer, *A Bayesian information criterion for singular models*, J. R. Stat. Soc. Ser. B. Stat. Methodol. **79** (2017), 323–380.

[19] Mathias Drton and Thomas S Richardson, *Multimodality of the likelihood in the bivariate seemingly unrelated regressions model*, Biometrika **91** (2004), no. 2, 383–392.

[20] Mathias Drton, Bernd Sturmfels, and Seth Sullivant, *Algebraic factor analysis: tetrads, pentads and beyond*, Probab. Theory Related Fields **138** (2007), no. 3-4, 463–493. MR 2299716

[21] ______, *Lectures on algebraic statistics*, Oberwolfach Seminars, vol. 39, Birkhäuser Verlag, Basel, 2009. MR 2723140 (2012d:62004)

[22] Mathias Drton and Luca Weihs, *Generic identifiability of linear structural equation models by ancestor decomposition*, Scand. J. Stat. **43** (2016), no. 4, 1035–1045. MR 3573674

[23] Mathias Drton and Han Xiao, *Finiteness of small factor analysis models*, Ann. Inst. Statist. Math. **62** (2010), no. 4, 775–783. MR 2652316

[24] ______, *Wald tests of singular hypotheses*, Bernoulli **22** (2016), no. 1, 38–59. MR 3449776

[25] Mathias Drton and Josephine Yu, *On a parametrization of positive semidefinite matrices with zeros*, SIAM J. Matrix Anal. Appl. **31** (2010), no. 5, 2665–2680. MR 2740626 (2011j:15057)

[26] Jan Ernest, Dominik Rothenhäusler, and Peter Bühlmann, *Causal inference in partially linear structural equation models: identifiability and estimation*, ArXiv e-prints (2016), 1607.05980.

[27] Robin J. Evans, *Graphs for margins of Bayesian networks*, Scand. J. Stat. **43** (2016), no. 3, 625–648. MR 3543314

[28] Robin J. Evans and Thomas S. Richardson, *Smooth, identifiable supermodels of discrete DAG models with latent variables*, ArXiv e-prints (2015), 1511.06813.

[29] William N. Evans and Jeanne S. Ringel, *Can higher cigarette taxes improve birth outcomes?*, Journal of Public Economics **72** (1999), no. 1, 135–154.

[30] Alex Fink, Jenna Rajchgot, and Seth Sullivant, *Matrix Schubert varieties and Gaussian conditional independence models*, J. Algebraic Combin. **44** (2016), no. 4, 1009–1046. MR 3566228

[31] Franklin M. Fisher, *A correspondence principle for simultaneous equation models*, Econometrica **38** (1970), 73–92.

[32] Christopher Fox, *Interpretation and inference of linear structural equation models*, Ph.D. thesis, University of Chicago, 2014.

[33] Christopher J. Fox, Andreas Käufl, and Mathias Drton, *On the causal interpretation of acyclic mixed graphs under multivariate normality*, Linear Algebra Appl. **473** (2015), 93–113. MR 3338327

[34] Rina Foygel, Jan Draisma, and Mathias Drton, *Half-trek criterion for generic identifiability of linear structural equation models*, Ann. Statist. **40** (2012), no. 3, 1682–1713.

[35] Morten Frydenberg, *The chain graph Markov property*, Scand. J. Statist. **17** (1990), no. 4, 333–353. MR 1096723

[36] Luis D. Garcia-Puente, Sarah Spielvogel, and Seth Sullivant, *Identifying causal effects with computer algebra*, Proceedings of the 26th Conference on Uncertainty in Artificial Intelligence (UAI) (Peter Grünwald and Peter Spirtes, eds.), AUAI Press, 2010.

[37] Luis David García-Puente, Sonja Petrović, and Seth Sullivant, *Graphical models*, J. Softw. Algebra Geom. **5** (2013), 1–7. MR 3073716

[38] Dan Geiger, David Heckerman, Henry King, and Christopher Meek, *Stratified exponential families: graphical models and model selection*, Ann. Statist. **29** (2001), no. 2, 505–529. MR 1863967

[39] Dan Geiger, Thomas Verma, and Judea Pearl, *Identifying independence in Bayesian networks*, Networks **20** (1990), no. 5, 507–534, Special issue on influence diagrams. MR 1064736

[40] Daniel R. Grayson and Michael E. Stillman, *Macaulay2, a software system for research in algebraic geometry*, Available at http://www.math.uiuc.edu/Macaulay2/.

[41] Frank Harary, *The determinant of the adjacency matrix of a graph*, SIAM Rev. **4** (1962), 202–210. MR 0144330

[42] Dominique M. A. Haughton, *On the choice of a model to fit data from an exponential family*, Ann. Statist. **16** (1988), no. 1, 342–355. MR 924875

[43] Alain Hauser and Peter Bühlmann, *Jointly interventional and observational data: estimation of interventional Markov equivalence classes of directed acyclic graphs*, J. R. Stat. Soc. Ser. B. Stat. Methodol. **77** (2015), no. 1, 291–318. MR 3299409

[44] Jan T. A. Koster, *Marginalizing and conditioning in graphical models*, Bernoulli **8** (2002), no. 6, 817–840. MR 1963663

[45] Gustavo Lacerda, Peter Spirtes, Joseph Ramsey, and Patrik Hoyer, *Discovering cyclic causal models by independent components analysis*, Proceedings of the Twenty-Fourth Conference Annual Conference on Uncertainty in Artificial Intelligence (UAI-08) (Corvallis, Oregon), AUAI Press, 2008, pp. 366–374.

[46] Steffen L. Lauritzen, *Graphical models*, Oxford Statistical Science Series, vol. 17, The Clarendon Press, Oxford University Press, New York, 1996, Oxford Science Publications. MR 1419991

[47] Dennis Leung, Mathias Drton, and Hisayuki Hara, *Identifiability of directed Gaussian graphical models with one latent source*, Electron. J. Stat. **10** (2016), no. 1, 394–422. MR 3466188

[48] Radim Lněnička and František Matúš, *On Gaussian conditional independent structures*, Kybernetika (Prague) **43** (2007), no. 3, 327–342. MR 2362722

[49] Judea Pearl, *Causality*, second ed., Cambridge University Press, Cambridge, 2009, Models, reasoning, and inference. MR 2548166 (2010i:68148)

[50] Thomas Richardson and Peter Spirtes, *Ancestral graph Markov models*, Ann. Statist. **30** (2002), no. 4, 962–1030. MR 1926166 (2003h:60017)

[51] Thomas S. Richardson, *A characterization of Markov equivalence for directed cyclic graphs*, International Journal of Approximate Reasoning **17** (1997), no. 2/3, 107–162.

[52] Hajir Roozbehani and Yury Polyanskiy, *Algebraic methods of classifying directed graphical models*, 2014 IEEE International Symposium on Information Theory, June 2014, pp. 2027–2031.

[53] N. Shiers, P. Zwiernik, J. A. D. Aston, and J. Q. Smith, *The correlation space of Gaussian latent tree models and model selection without fitting*, Biometrika **103** (2016), no. 3, 531–545. MR 3551782

[54] Shohei Shimizu, *LiNGAM: Non-Gaussian methods for estimating causal structures*, Behaviormetrika **41** (2014), no. 1, 65–98.

[55] Ilya Shpitser, Robin Evans, Thomas Richardson, and James Robins, *Introduction to nested Markov models*, Behaviormetrika **41** (2014), no. 1, 3–39.

[56] Ilya Shpitser and Judea Pearl, *Identification of joint interventional distributions in recursive semi-Markovian causal models*, Proceedings of the Twenty-First National Conference on Artificial Intelligence (Menlo Park, CA), AAAI Press, 2006, pp. 1219–1226.

[57] Peter Spirtes, Clark Glymour, and Richard Scheines, *Causation, prediction, and search*, second ed., Adaptive Computation and Machine Learning, MIT Press, Cambridge, MA, 2000, With additional material by David Heckerman, Christopher Meek, Gregory F. Cooper and Thomas Richardson, A Bradford Book. MR 1815675 (2001j:62009)

[58] Peter Spirtes, Thomas Richardson, Chris Meek, Richard Scheines, and Clark Glymour, *Using path diagrams as a structural equation modelling tool*, Sociological Methods and Research **27** (1998), 182–225.

[59] Paul D. Stolley, *When genius errs: R. A. Fisher and the lung cancer controversy*, Am. J. Epidemiol. **133** (1991), no. 5, 416–425.

[60] Milan Studený, *Probabilistic conditional independence structures*, Information Science and Statistics, Springer, London, 2005. MR 3183760

[61] Seth Sullivant, *Algebraic geometry of Gaussian Bayesian networks*, Adv. in Appl. Math. **40** (2008), no. 4, 482–513. MR 2412156

[62] ———, *A Gröbner basis for the secant ideal of the second hypersimplex*, J. Commut. Algebra **1** (2009), no. 2, 327–338. MR 2504939

[63] Seth Sullivant and Elizabeth Gross, *The maximum likelihood threshold of a graph*, Bernoulli **24** (2018), no. 1, 386–407.

[64] Seth Sullivant, Kelli Talaska, and Jan Draisma, *Trek separation for Gaussian graphical models*, Ann. Statist. **38** (2010), no. 3, 1665–1685. MR 2662356 (2011f:62076)

[65] Jin Tian, *Identifying direct causal effects in linear models*, Proceedings of the Twentieth National Conference on Artificial Intelligence (AAAI), 2005, pp. 346–353.

[66] Jin Tian and Judea Pearl, *A general identification condition for causal effects*, AAAI/IAAI, 2002, pp. 567–573.

[67] Caroline Uhler, *Geometry of maximum likelihood estimation in Gaussian graphical models*, Ann. Statist. **40** (2012), no. 1, 238–261. MR 3014306

[68] Lieven Vandenberghe and Martin S. Andersen, *Chordal graphs and semidefinite optimization*, Foundations and Trends in Optimization **1** (2014), no. 4, 241–433.

[69] Thomas S. Verma and Judea Pearl, *Equivalence and synthesis of causal models*, Uncertainty in Artificial Intelligence 6, Elsevier, 1991, UCLA Cognitive Systems Laboratory, Technical Report (R-150), pp. 255–268.
[70] Sumio Watanabe, *Algebraic geometry and statistical learning theory*, Cambridge Monographs on Applied and Computational Mathematics, vol. 25, Cambridge University Press, Cambridge, 2009. MR 2554932
[71] Luca Weihs, Bill Robinson, Emilie Dufresne, Jennifer Kenkel, Kaie Kubjas, Reginald McGee, II, Nhan Nguyen, Elina Robeva, and Mathias Drton, *Determinantal generalizations of instrumental variables*, Journal of Causal Inference **6** (2018), no 1.
[72] Nanny Wermuth, *Probability distributions with summary graph structure*, Bernoulli **17** (2011), no. 3, 845–879. MR 2817608
[73] Sewall Wright, *Correlation and causation*, J. Agricultural Research **20** (1921), 557–585.
[74] ______, *The method of path coefficients*, Ann. Math. Statist. **5** (1934), no. 3, 161–215.
[75] Hui Zhao, Zhongguo Zheng, and Baijun Liu, *On the Markov equivalence of maximal ancestral graphs*, Sci. China Ser. A **48** (2005), no. 4, 548–562. MR 2157690 (2006e:62069)

Department of Statistics, University of Washington, Seattle, WA 98195-4322, U.S.A.
E-mail address: md5@uw.edu

Advanced Studies in Pure Mathematics 77, 2018
The 50th Anniversary of Gröbner Bases
pp. 87–107

BM algorithms for noisy data and implicit regression modelling

Claudia Fassino, Hans Michael Möller and Eva Riccomagno

Abstract.

In this paper we consider the problem of finding a set of monomials $\mathcal{O}$ and a polynomial f whose support is contained in $\mathcal{O}$, such that (1) f is almost vanishing at a set of points $\mathbb{X}$ whose coordinates are not known exactly and (2) $\mathcal{O}$ exhibits structural stability, that is the model/design matrix associated to $\mathcal{O}$ is full rank for each set of points differing only slightly from $\mathbb{X}$. We review some numerical versions of the Buchberger-Möller (BM) algorithm for computing the set $\mathcal{O}$ and the polynomial f and we present a variant, called LDP-LP, which integrates one of these methods with a classical statistical least squares algorithm for implicit regression from [1]. To illustrate the usefulness of these numerical BM algorithms, we review some of their application in the analyses of data sets for which standard techniques did not yield satisfactory results.

§1. Introduction

For n-distinct points $\mathbb{X} \subset \mathbb{R}^k$ we consider the generic problem of finding a polynomial function f which almost vanishes at $\mathbb{X}$. The Buchberger-Möller (BM) algorithm would return polynomials whose zero sets include the zero-dimensional variety $\mathbb{X}$. Nevertheless such polynomials might be too complex, e.g. of too high degree, to be useful for some practical purposes in some areas such as statistical modelling [10, 11]. For example from the four points $\mathbb{X} = (\pm 1, \pm 1)$, the two polynomials $x_1^2 - 1, x_2^2 - 1$ are the generators of the polynomial ideal of $\mathbb{X}$ and the vector space basis of the quotient space $\mathbb{R}[x_1, x_2]/\langle x_1^2 - 1, x_2^2 - 1\rangle$ returned by the algorithm is $\mathcal{O} = \{1, x_2, x_1, x_1x_2\}$ for any term ordering. Both generators are simple and the model/design matrix (also called evaluation

---

Received August 9, 2016.
2010 *Mathematics Subject Classification.* Primary: 13P, 62J, 65F. Secondary: 13P10, 65F20.
Key words and phrases. least squares estimates, vanishing ideals, Gröbner bases.

matrix) $M = [s(d)]_{d\in\mathbb{X},s\in\mathcal{O}}$ is not only full rank, but also well conditioned. Nevertheless slightly perturbing the point $(1,1)$ into $(0.9,1.1)$ leads to a non unique nor nice choice of generators, for example a generator set with respect to the DegRevLex term ordering is given by the three polynomials

$$x_1x_2 - 19x_2^2 + x_1 + x_2 + 20,\ x_1^2 + \frac{19}{21}x_2^2 - \frac{40}{21},\ x_2^3 - \frac{11}{10}x_2^2 - x_2 + \frac{11}{10}$$

with corresponding $\mathcal{O} = \{1, x_1, x_2, x_2^2\}$. The model/design matrix is ill-conditioned. In statistics this is known as collinearity of the explanatory functions labelling the columns of M. These functions depend on the variables representing the characteristics of the population under study and of which $\mathbb{X}$ can be considered a collection of instances. Regression models based on $\mathcal{O}$ would be unstable, specifically stability of algorithms commonly used in statistics cannot be guaranteed.

To overcome this, numerical versions of the BM algorithm are available in the literature which take into account the fact that the coordinates of the points in $\mathbb{X}$ might be effected by random error. We discuss some of them in this paper. The general strategy is to consider a sequence of embedded $\mathbb{R}$-vector spaces of increasing dimension $(V_m)_{m=1,2,\ldots}$ such that V_m is a subset of polynomials. At step m the objective is to determine a polynomial $p \in V_m$ such that $\sum_{u\in\mathbb{X}} p^2(u)$ is minimal. A variant requires to find p such that not only this sum is minimised, but also the variety $\{y : p(y) = 0\}$ admits a n-point subset, each point of which can be paired to a point in $\mathbb{X}$ so that the distances between paired points is smaller than a given threshold ε, namely for each $u \in \mathbb{X}$ find a y in $\{y : p(y) = 0\}$ such that $\|u - y\|$ is smaller than ε. In Section 6 p and the y's are chosen so that $\min\{\|u - y\| \mid p(y) = 0\}$ is reached by following a classical algorithm for implicit regression analysis.

The paper is organised as follows. In Section 2 the classical BM algorithm is recalled together with an example illustrating the issues dealt with in the paper. In Section 3 two classical notions of numerical independence of vectors used to avoid collinearity are outlined. The first one is based on singular values and the second one on least squares. The latter leads to the NBM algorithm in Section 4 and tightened into the LDP algorithm (Section 5) which in Section 6 is integrated with the classical Britt-Luecke method for implicit regression from Statistics.

§2. BM algorithm for zero dimensional varieties and noisy data

The BM algorithm [2] is a classical method for computing the Gröbner basis of the ideal $\mathcal{I}(\mathbb{X})$ of a set of points $\mathbb{X} \subset \mathbb{R}^k$ with respect to a term ordering τ. It can be shortly described as follows.

The BM algorithm (BM)

- **Input:** a set of points $\mathbb{X} \subset \mathbb{R}^k$ and a term ordering τ.
- **Output:** the τ-Gröbner basis of $\mathcal{I}(\mathbb{X})$.
- **Core:** stepwise construction of a monomial basis $\mathcal{O}$ of the quotient space $\mathbb{R}[x_1, \ldots, x_k]/\mathcal{I}(\mathbb{X})$
- **Step Zero:** set $\mathcal{O} = \{1\}$.
- **Generic Step:** given $\mathcal{O}$ and $t >_\tau s$ for $s \in \mathcal{O}$, check if the vector $t(\mathbb{X}) = [t(d)]_{d\in\mathbb{X}}$ and the columns of $M(\mathbb{X}) = [s(u)]_{u\in\mathbb{X},s\in\mathcal{O}}$ are linearly dependent.
  If the answer is **Yes**, the polynomial $g = t - \sum_{s\in\mathcal{O}} \alpha_s s$, where α is s.t. $M(\mathbb{X})\alpha = t(\mathbb{X})$, is added to the Gröbner basis.
  If the answer is **No**, the monomial t is added to $\mathcal{O}$.

The main check of the BM algorithm is strongly effected by data perturbations, since small variations of the coordinates of linear dependent vectors can turn them into linear independent vectors. For this reason small perturbations of the points in $\mathbb{X}$ can correspond to a very different Gröbner bases, as illustrated in the following example.

Example 2.1. The set $\widetilde{\mathbb{X}} = \{(1,1),(2,3),(3,5.1)\}$ is obtained by slightly changing a coordinate of a point in $\mathbb{X} = \{(1,1),(2,3),(3,5)\}$. The DegLex-Gröbner bases $\widetilde{\mathcal{G}}$ and $\mathcal{G}$ of their vanishing ideals are very different, since $\mathcal{G} = \{x - 0.5y - 0.5,\ y^3 - 9y^2 + 23y - 15\}$ and $\widetilde{\mathcal{G}} = \{y^2 - 80x + 36y + 43,\ xy - 45x + 20y + 24,\ x^2 - 26.1x + 11.55y + 13.55\}$ and the corresponding algebraic varieties are very different as well.

In order to avoid this drawback and to obtain an $\mathcal{O}$ set robust with respect to small perturbations, the check of exact linear dependence of vectors can be relaxed and substituted with that of numerical linear dependence. In literature [4, 7, 9] there are several ways for testing the dependence of a set of vectors from the numerical point of view. Exploiting these ideas of numerical dependence of a set of vectors, it is possible to design different numerical versions of the BM algorithm, for which at each step the main check consists in testing whether $t(\mathbb{X})$ is numerically, and not exactly, dependent on the columns of $M(\mathbb{X}) = [s(\mathbb{X})]_{s\in\mathcal{O}}$.

§3. Numerical dependence of vectors

3.1. An approach based on singular value decomposition

A classical strategy for testing the numerical dependence of a set of vectors is based on the Singular Values Decomposition (SVD) [7]. Given a set W of vectors, let $\sigma_1 \geq \ldots \sigma_p > 0$ be the singular values of the matrix $A = [v]_{v \in W}$ whose columns are the elements of W and p is the rank of A. Given a threshold ε, if $\sigma_1 \geq \cdots \geq \sigma_r \geq \varepsilon > \sigma_{r+1} \geq \cdots \geq \sigma_p$, then r is called the numerical rank of A w.r.t. ε. The numerical rank gives information about the numerical dependence of the vectors of W. Indeed, it is well known that there exists a rank deficient matrix B such that the rank of B is r and $\|A - B\|_2 \leq \sigma_{r+1} < \varepsilon$, where, for a matrix M, $\|M\|_2$ is its largest singular value. In this case the columns of A can be considered numerically dependent. Some numerical versions of the BM algorithm exploit this notion of numerical dependence, for example the BM Approximation algorithm in [9].

Nevertheless, the main check of the BM algorithm, which involves the monomial t, the set of monomials $\mathcal{O}$ and the matrix $M(\mathbb{X})$, tests if the vector $t(\mathbb{X})$ is a linear combination of the columns of $M(\mathbb{X})$ in order to compute a polynomial vanishing at $\mathbb{X}$. If the points in $\mathbb{X}$ are effected by data errors, then it is of interest to check if there exists a small perturbation $\widetilde{\mathbb{X}}$ of $\mathbb{X}$ such that $g(\widetilde{\mathbb{X}}) = 0$, for some polynomial g with support contained in $\mathcal{O} \cup \{t\}$. For this reason the generic step of the algorithm should test if there exists a set $\widetilde{\mathbb{X}}$ close to $\mathbb{X}$ such that the vectors $\{s(\widetilde{\mathbb{X}})\}_{s \in \mathcal{O} \cup \{t\}}$ are linearly dependent. The computation of the singular values of the matrix $[M(\mathbb{X}), t(\mathbb{X})]$ in the general case does not give such information usually, as illustrated in the following example.

Example 3.1. For the three points set $\mathbb{X} = \{1, 3, 3.1\}$ in $\mathbb{R}$, the set of monomials $\mathcal{O} = \{1, x\}$ and the monomial $t = x^2$, the evaluation matrix is

$$\widehat{M}(\mathbb{X}) = [M(\mathbb{X}), t(\mathbb{X})] = \begin{bmatrix} 1 & 1 & 1 \\ 1 & 3 & 9 \\ 1 & 3.1 & 3.1^2 \end{bmatrix}.$$

For $\varepsilon = 0.03$, since $\{13.9931, 1.0744, 0.0279\}$ are the singular values of $\widehat{M}(\mathbb{X})$ and since $\sigma_2 > \varepsilon > \sigma_3$, the numerical rank of $\widehat{M}(\mathbb{X})$ w.r.t. ε is equal to 2 and thus there exists a rank deficient matrix B such that $\|\widehat{M}(\mathbb{X}) - B\|_2 < \varepsilon$. A possible polynomial almost vanishing at $\mathbb{X}$ is $f(x) = x^2 - 4.0549x + 3.0599$.

Nevertheless, there does not exist any small perturbation $\widetilde{\mathbb{X}}$ of $\mathbb{X}$ such that $t(\widetilde{\mathbb{X}})$ is a linear combination of the columns of $M(\widetilde{\mathbb{X}})$. In fact, for a generic perturbation $\widetilde{\mathbb{X}} = \{1+\delta_1, 3+\delta_2, 3.1+\delta_3\}$, the evaluation matrix is

$$\widehat{M}(\widetilde{\mathbb{X}}) = \begin{bmatrix} 1 & 1+\delta_1 & (1+\delta_1)^2 \\ 1 & 3+\delta_2 & (3+\delta_2)^2 \\ 1 & 3.1+\delta_3 & (3.1+\delta_3)^2 \end{bmatrix}.$$

Since $\det(\widehat{M}(\widetilde{\mathbb{X}})) = (2+\delta_2-\delta_1)(2.1+\delta_3-\delta_1)(0.1+\delta_3-\delta_2)$, does not vanish for any $\delta_i < \varepsilon$, $i = 1, 2, 3$, it follows that there does not exist any set close to $\mathbb{X}$ by less than ε such that the columns of $\widehat{M}(\widetilde{\mathbb{X}})$ are linearly dependent. In conclusion there do not exist an $\widetilde{\mathbb{X}}$ small perturbation of $\mathbb{X}$ and a polynomial f with support $\{1, x, x^2\}$ such that f vanishes at $\widetilde{\mathbb{X}}$.

3.2. An approach based on least squares method

A different definition of numerical linear dependence is presented in [4] and [5], which formalises and exploits the concept of *admissible perturbation* of $\mathbb{X}$. Given a set of points $\mathbb{X}$ known to be effected by experimental error and an estimation ε of the maximum componentwise error, a set $\widetilde{\mathbb{X}}$ is an admissible perturbation of $\mathbb{X}$ if for each $u \in \mathbb{X}$ there exists $\delta_u \in \mathbb{R}^k$ such that $\|\delta_u\|_\infty = \max_{i=1,\dots,k} |\delta_i| < \varepsilon$ and $\widetilde{\mathbb{X}} = \{u + \delta_u \mid u \in \mathbb{X}\}$.

The sets $\mathbb{X}$ and $\widetilde{\mathbb{X}}$ are indistinguishable from the numerical points of view and so the vector $t(\mathbb{X})$ can be considered linearly dependent on the columns of the matrix $M(\mathbb{X})$ with respect to ε if there exists an admissible perturbation $\widetilde{\mathbb{X}}$ of $\mathbb{X}$ such that $t(\widetilde{\mathbb{X}})$ is exactly linearly dependent on the columns of $M(\widetilde{\mathbb{X}})$. According to this definition, the columns of the matrix $\widehat{M}(\mathbb{X})$ in Example 3.1 are linearly independent w.r.t. $\varepsilon = 0.03$, even if $\widehat{M}(\mathbb{X})$ is numerically rank deficient w.r.t. ε.

Let $\rho(\mathbb{X}) = t(\mathbb{X}) - M(\mathbb{X})\alpha$ be the residual of the least squares problem $M(\mathbb{X})\alpha = t(\mathbb{X})$. Two necessary conditions for $t(\mathbb{X})$ and the columns of $M(\mathbb{X})$ to be numerically dependent are

$$(1) \qquad |\rho(\mathbb{X})| \le \varepsilon |I - M(\mathbb{X})M^+(\mathbb{X})| \sum_{i=1}^{k} \left| \frac{\partial g}{\partial x_i}(\mathbb{X}) \right| + O(\varepsilon^2) \quad \text{and}$$

$$(2) \qquad \|\rho(\mathbb{X})\|_2^2 \le \varepsilon^2 \sum_{u \in \mathbb{X}} \|\nabla g(u)\|_2^2 + O(\varepsilon^3),$$

where the absolute value is intended componentwise and $M^+(\mathbb{X}) = (M^t(\mathbb{X})M(\mathbb{X}))^{-1}M^t(\mathbb{X})$ is the pseudo inverse of $M(\mathbb{X})$. The first upper bound is presented, in a slightly different formulation, in [4, Th. 3.5] while the second one is in [5, Prop. 4]. Thus if upper bound (1) or upper bound (2) are not satisfied, then there does not exist any polynomial with support contained in $\mathcal{O} \cup \{t\}$ which vanishes at any admissible perturbation of $\mathbb{X}$. If both upper bounds (1) and (2) are satisfied, we cannot conclude that there is a polynomial with support in $\mathcal{O} \cup \{t\}$ vanishing at some $\widetilde{\mathbb{X}}$. Nevertheless the polynomial $g = t - \sum_{s \in \mathcal{O}} \alpha_s s$ assumes small values at $\mathbb{X}$, since $g(\mathbb{X}) = \rho(\mathbb{X})$, with $\|\rho(\mathbb{X})\|_2 \sim O(\varepsilon)$.

§4. The NBM algorithm

A numerical version of the BM algorithm should check, in its generic step, whether the vector $t(\mathbb{X})$ and the columns of $M(\mathbb{X})$ are linearly dependent from a numerical point of view. Obviously, in the lucky case when a polynomial vanishing on $\mathbb{X}$ is available, no check is required, both in the cases where $\mathbb{X}$ is exactly known or includes noisy data.

For not noisy data, when the residual $\rho(\mathbb{X}) = t(\mathbb{X}) - M(\mathbb{X})\alpha$ of the least squares problem $M(\mathbb{X})\alpha = t(\mathbb{X})$ is different from zero, then the polynomial $g = t - \sum_{s \in \mathcal{O}} \alpha_s s$ does not vanish at $\mathbb{X}$ and thus the BM algorithm inserts t into $\mathcal{O}$.

For noisy data, when neither the upper bound (1) or (2) are satisfied, then the monomial t can be inserted in $\mathcal{O}$ to give a normal set for each admissible perturbation of $\mathbb{X}$. If instead (1) and (2) are both satisfied, then, as already mentioned, a polynomial g almost vanishing at $\mathbb{X}$ can be computed.

Exploiting the above, we present below a slight modification of the Numerical BM algorithm (NBM) in [4], which is a numerical version of the BM algorithm.

The Numerical BM algorithm

- **Input:** A set of points $\mathbb{X}$, an error estimation ε and a term ordering τ.
- **Output:** A set $\mathcal{O}$ of monomials and a set $\mathcal{G}$ of polynomials.
- **Core:** stepwise construction of a set $\mathcal{O}$ of monomials.
- **Step Zero:** Set $\mathcal{O} = \{1\}$.
- **Generic Step:** Given $\mathcal{O}$ and $t >_\tau s$ for $s \in \mathcal{O}$, solve the least squares problem $M(\mathbb{X})\alpha = t(\mathbb{X})$, with $M(\mathbb{X}) = [s(\mathbb{X})]_{s \in \mathcal{O}}$. Check if $\rho(\mathbb{X})$ satisfies upper bound (2).
  If the answer is **No**, the monomial t is added to $\mathcal{O}$.
  If the answer is **Yes**, then check upper bound (1) and
  - if the answer is **No**, the monomial t is added to $\mathcal{O}$;
  - if the answer is **Yes**, $g = t - \sum_{s \in \mathcal{O}} \alpha_s s$ is added to $\mathcal{G}$.

Importantly, the set $\mathcal{O}$ computed by the NBM algorithm is a basis of the quotient space $\mathbb{R}[x_1, \ldots, x_n]/\mathcal{I}(\widetilde{\mathbb{X}})$ for *each* admissible perturbation $\widetilde{\mathbb{X}}$. In fact, if at least one coordinate of $\rho(\mathbb{X})$ does not satisfy the upper bound (1) or if the 2-norm of $\rho(\mathbb{X})$ does not satisfy the upper bound (2), then there are no admissible perturbations $\widetilde{\mathbb{X}}$ of $\mathbb{X}$ such that $\rho(\widetilde{\mathbb{X}}) = 0$, that is such that the vector $t(\widetilde{\mathbb{X}})$ and the columns of $M(\widetilde{\mathbb{X}})$ are dependent. This implies that if $M(\mathbb{X})$ is a full rank matrix then the matrix $[M(\widetilde{\mathbb{X}}), t(\widetilde{\mathbb{X}})]$ is a full rank matrix too, for each admissible perturbation $\widetilde{\mathbb{X}}$ and thus $\mathcal{O}$ is a basis of the quotient space. Furthermore, since by construction, $\mathcal{O}$ is closed under taking divisors, we conclude that $\mathcal{O}$ is a normal set.

Furthermore, if each point $u \in \mathbb{X}$ belongs to $[-1, 1]^k$, each polynomial $g \in \mathcal{G}$ is almost vanishing, that is $\|g(\mathbb{X})\|_2/\|c\|_2 = O(\varepsilon)$, where c is the coefficient vector of g. Such result is shown in [4, Th. 5.1] if the upper bound (1) is used, while it follows from the fact that $\|\nabla g(u)\|_2 \leq \sqrt{v}\sqrt{\sum_{j=1}^{k} D_j^2}\|c\|_2$, where v is the cardinality of the support of g and D_j is the maximum degree of each x_j in g (see Lemma 7.1), monomials of g, if the upper bound (2) is implemented.

In the algorithm, we choose to check first upper bound (2) since it is easier to compute than (1), because it does not require the computation of a pseudo inverse matrix.

§5. The LDP algorithm

Example 5.1. The polynomial $f = y^2 + 0.01x^2 - 1$ has coefficient vector $c = [1, 0.01, 1]$ and it is almost vanishing at the point $u = (12, 0)$ for all $\varepsilon > 0.312$ since $|f(u)|/\|c\|_2 = 0.312$. Nevertheless, the minimum euclidean distance between u and the zero set of f is equal to 2, as shown in Figure 1. Thus even if f is almost vanishing at u, its affine variety does not pass close to u.

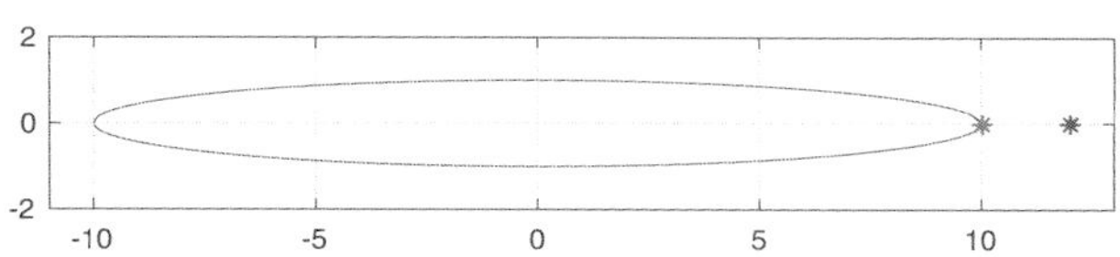

Fig. 1. The variety $Z(f) = \{(x, y) \mid y^2 + 0.01x^2 - 1 = 0\}$ does not pass close to $(12, 0)$, w.r.t. $\varepsilon = 0.4$, even if $|f(u)|/\|c\|_2 = 0.312 < \varepsilon$.

Given $\mathbb{X}$ and ε, NBM returns a normal set $\mathcal{O}$ such that $M(\mathbb{X})$ and $M(\widetilde{\mathbb{X}})$ are full rank matrices for all $\widetilde{\mathbb{X}}$ admissible perturbation of $\mathbb{X}$. Moreover, it returns polynomials g such that $g(\mathbb{X})$ is almost zero. Now to workaround the issue in Example 1, the LDP algorithm below seeks an admissible perturbation $\widetilde{\mathbb{X}}$ and a polynomial $\widetilde{g}$ whose coefficients are a perturbation of those of g such that the zero set of $\widetilde{g}$ includes $\widetilde{\mathbb{X}}$. LDP or LPA stands for Low Degree Polynomial algorithm, first presented in [6].

As in NBM algorithm, the main check of LDP analyses the numerical dependence of a set of vectors using upper bounds (1) and (2). But LDP and NBM differ in the last step, where the g polynomials are constructed. In theory, with LDP we want to find a set $\widetilde{\mathbb{X}}$ such that $\rho(\widetilde{\mathbb{X}}) = 0$, i.e. $t(\widetilde{\mathbb{X}}) - M(\widetilde{\mathbb{X}})\alpha(\widetilde{\mathbb{X}}) = 0$. If such $\widetilde{\mathbb{X}}$ is an admissible perturbation of $\mathbb{X}$ then the polynomial $\widetilde{g} = t - \sum_{s\in\mathcal{O}} \alpha_s(\widetilde{\mathbb{X}})s$ is returned and LDP stops. Otherwise, t is added to $\mathcal{O}$. The algorithm ends because in the worst case it computes an element of the exact Gröwner basis of $\mathcal{I}(\mathbb{X})$. The LDP algorithm can be summarised as follows.

The Low Degree Polynomial Algorithm (LDP)

- **Input:** As NBM.
- **Output:** A set $\mathcal{O}$ of monomials, an admissible perturbation $\widetilde{\mathbb{X}}$ and *a* low degree polynomial $\widetilde{g}$ such that $\widetilde{g}(\widetilde{\mathbb{X}}) = 0$.

All equal to NBM with the exception of the last row which changes to

- **Key step:** if the answer is **Yes**, compute a set $\widetilde{\mathbb{X}}$ such that $\rho(\widetilde{\mathbb{X}}) = 0$.
  * If $\widetilde{\mathbb{X}}$ is an admissible perturbation of $\mathbb{X}$, the polynomial $\widetilde{g} = t - \sum_{s\in\mathcal{O}} \alpha(\widetilde{\mathbb{X}})_s s$ is formed.
  * Otherwise, the monomial t is added to $\mathcal{O}$.

In practice, solving $\rho(\widetilde{\mathbb{X}}) = 0$ is a difficult task. It can be reformulated introducing a vector of additive errors $e = (e_u)_{u\in\mathbb{X}}$, where each e_u is a vector whose coordinates represent the perturbation of the point u in $\mathbb{X}$. Since each point $\widetilde{u}$ belonging to a generic admissible perturbation $\widetilde{\mathbb{X}}$ of $\mathbb{X}$ can be expressed as $\widetilde{u} = u + e_u$, then $\widetilde{\mathbb{X}}$ and also $\rho(\widetilde{\mathbb{X}})$ can be expressed in function of e. To keep track of this, we denote a generic perturbation of $\mathbb{X}$ by $\mathbb{X}(e)$ and the residual of the least squares problem $M(\mathbb{X}(e))\alpha(\mathbb{X}(e)) = t(\mathbb{X}(e))$ by $\rho(e) = t(e) - M(e)\alpha(e)$.

In this formulation, to solve $\rho(\widetilde{\mathbb{X}}) = 0$ is equivalent to find a vector $\widetilde{e}$ such that $\rho(\widetilde{e}) = 0$ and this gives $\widetilde{\mathbb{X}} = \mathbb{X}(\widetilde{e})$. Since the exact solution of a nonlinear system of equations can be found in very few cases, LDP does not compute $\widetilde{e}$, but an approximation of $\widetilde{e}$. Obviously, different techniques can be adopted to approximate $\widetilde{e}$. The original version of the LDP, presented in [6], finds an approximation $\widehat{e}$ of $\widetilde{e}$ by means of a root finding method based on the Normal Flow Algorithm [14], a classical iterative method for approximating a solution of a non linear system. Section 6 presents an alternative method. Theorem 4.2 in [6] shows that the polynomial $\widehat{g}$ computed by the LDP algorithm is almost vanishing at $\mathbb{X}(\widehat{e})$ even if it does vanish at it. Furthermore, if $\widehat{g}$ satisfies some simple conditions similar to the hypotheses of the Kantorovich theorem on the convergence of Newton method, then Theorem 4.3 in [6] shows that the zero set of $\widehat{g}$ lies close to $\mathbb{X}$ by less than ε, that is $\widehat{g}$ vanishes at an admissible perturbation of $\mathbb{X}$.

Example 5.2. In the left hand side plot of Figure 2, for the set of points

$$\mathbb{X} = \{(0.95, 1), (5.05, 2.95), (5.05, -0.95), (9.98, 4), (10.05, -2), (17.01, 5)\},$$

obtained by perturbing less than 0.1 the coordinates of six points on $f = y^2 - x - 2y + 2 = 0$, the dot-dashed curve is the zero set of the minimal degree polynomial in the Gröbner basis returned by the BM algorithm with DegRevLex ordering; the dashed curve is the zero set of the minimal degree polynomial returned by the NBM algorithm with $\varepsilon = 0.1$ and the black curve is the zero set of the polynomial returned by the LDP algorithm with the same ε. The three polynomials with the coefficients rounded to the first 4 decimal digits are

$$\begin{aligned} f_{BM} &= y^3 + 0.012541x^2 - 1.0439xy - 4.1372y^2 + 2.0233x \\ &\quad + 6.7437y - 4.5483 \\ f_{NBM} &= y^2 - 1.004x - 2.009y + 2.128 \\ f_{LDP} &= y^2 - 0.975x - 2.005y + 1.977 \end{aligned}$$

The right hand side plot of Figure 2 zooms around the vertex of the parabola. The polynomials f, f_{NBM} and f_{LDP} have the same support and, in this case, denoting with α, α_{NBM} and α_{LDP} the coefficients of f, f_{NBM} and f_{LDP}, respectively, we have that

$$\frac{\|\alpha - \alpha_{NBM}\|_2}{\|\alpha\|_2} = \frac{\|[0, 0.004, 0.009, -0.128]\|_2}{\|[1, -1, -2, 2]\|_2} = 0.0406 < \varepsilon$$

$$\frac{\|\alpha - \alpha_{LDP}\|_2}{\|\alpha\|_2} = \frac{\|[0, -0.025, -0.005, 0.023]\|_2}{\|[1, -1, -2, 2]\|_2} = 0.0109 < \varepsilon$$

By construction, there are six points in the zero set of f_{LDP} which are an admissible perturbation of $\mathbb{X}$, specifically they are

$$\begin{aligned} \widetilde{\mathbb{X}} = \{&(0.9969, 1), (5.0892, 3), (5.0688, -0.99), (9.9680, 3.96), \\ &(10.0592, -1.97), (17.0603, 4.96)\}. \end{aligned}$$

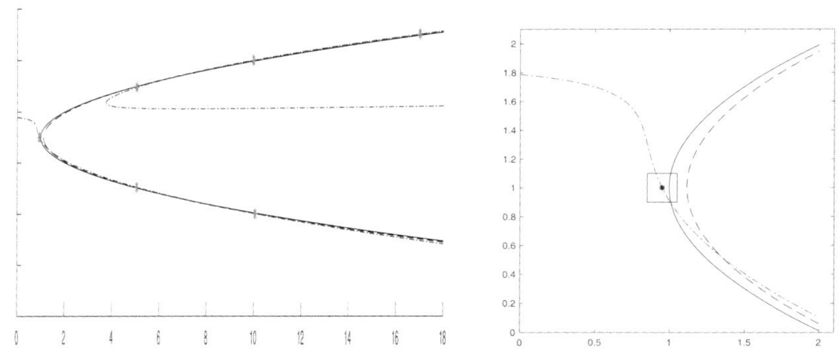

Fig. 2. The left plot shows the polynomials computed by BM (-·-), NBM (- - -) and LDP (black). The right plot shows around the point closest to the origin.

In summary, the vector $\widehat{e}$ found above is such that $\mathbb{X}(\widehat{e})$ is an admissible perturbation of $\mathbb{X}$ and it is an approximation of the zero set of ρ. By further assuming normally distributed random error with zero mean and known covariance matrix, the algorithm could be adapted so that $\widehat{e}$ is the solution a least squares problem as shown in Section 6 and a maximum likelihood problem.

§6. The LDP-BL algorithm

For the long vectors $z = (u+e_u)_{u\in\mathbb{X}}$, $e = (e_u)_{u\in\mathbb{X}}$ and a vector function $f(z,\theta)$ on the noisy data z and depending on a parameter vector θ, the problem of finding a solution to the following constrained minimisation problem: $\min_e\{e^t R^{-1} e\}$ subject to $f(z,\theta) = 0$, is the subject of [1]. The authors assume that the measurement error e is normally distributed with zero mean and known covariance matrix R. The solution to the constrained minimisation problem is thus the least squares estimate of θ which can be shown to be the maximum likelihood estimate when the model is given in implicit form ($f = 0$).

Under some regularity assumptions on f the authors of [1] propose an iterative method based on Lagrange multipliers. Let $z_0 = (u)_{u\in\mathbb{X}}$, z_c and θ_c be the current estimates of z and θ, F_θ and F_z be the Jacobian matrix of $f(z,\theta)$ with respect to θ and to z evaluated at z_c and θ_c, respectively. The vectors θ and z are updated as $\theta+\delta\theta$ and $z+\delta z$ where

$$\delta\theta = -\left(F_\theta^t(F_zRF_z^t)^{-1}F_\theta\right)^{-1} F_\theta^t(F_zRF_z^t)^{-1}\left(f(z_c,\theta_c) + F_z\cdot(z_0 - z_c)\right)$$
$$\delta z = -RF_z^t(F_zRF_z^t)^{-1}\left(f(z_c,\theta_c) + F_\theta\cdot\delta\theta + F_z\cdot(z_0 - z_c)\right)$$

A new step is performed until $\|\delta\theta\|_2$ and $\|\delta z\|_2$ satisfy a smallness criterion. Let us call this the Britt-Luecke algorithm.

It can be implemented inside the **key step** of the LDP algorithm, where the implicit function $f(z,\theta)$ is the evaluation at an admissible perturbation $\mathbb{X}(e)$ of the polynomial $g = t - \sum_{s\in\mathcal{O}} \alpha_s s$, namely $\rho(\mathbb{X}(e))$ and the parameter vector θ is the coefficient vector of g. At each step the Jacobian matrix F_θ coincides with the evaluation matrix of the monomials in $\mathcal{O}$ at the current set of points derived from z and the updating steps are rather simple as the function is a polynomial. We call the LDP algorithm with the strategy due to Britt and Luecke for approximating the zeros of ρ the LDP-BL algorithm.

In general, the admissible perturbations of $\mathbb{X}$ computed by the LDP and the LDP-BL algorithms are different and the set of points computed by the LDP-BL algorithm is closer to $\mathbb{X}$ than the set computed by the LDP algorithm, since the Britt-Luecke procedure minimises the 2-norm of the vector e. Example 6.1 illustrates the different behaviour of the LDP and of the LDP-BL algorithm.

Example 6.1. For $\varepsilon = 0.55$, R the identity matrix and the ten point set

$$\begin{aligned}\mathbb{X} = \{&(1,2),\ (0.5,1.3),\ (1.98,2.05),\ (0,2.08),\ (-0.48,3.18),\\ &(2.95,5.05),\ (-0.95,5.05),\ (-1.45,7.2),\ (4,9.98),\ (-2,10.05)\},\end{aligned}$$

the NBM algorithm computes $g = y - 0.9668x^2 + 1.9308x - 2.2780$, which is almost vanishing at $\mathbb{X}$, since $\|g(\mathbb{X})\|_2/\|[1,-0.9668,1.9308,-2.2780]\|_2 = 0.2527 < \varepsilon$. But g does not vanish at any admissible perturbation of the point $(1,2)$, since the zero set of g does not intersect the ball with centre in $(1,2)$ and radius ε (see Figure 3).

The original LDP algorithm computes $g_1 = y - 0.9657x^2 + 1.8292x - 2.3703$ and the admissible perturbation

$$\begin{aligned}\mathbb{X}_1 = \{&(1.007,1.508),\ (0.719,1.554),\ (1.772,2.162),\ (0.129,2.150),\\ &(-0.385,3.218),\ (2.869,5.070),\ (-0.967,5.044),\ (-1.480,7.192),\\ &(3.913,9.997),\ (-2.027,10.045)\}\end{aligned}$$

while the LDP-BL algorithm computes the polynomial $g_2 = y - 0.9467x^2 + 1.7542x - 2.2715$ and the admissible perturbation

$$\begin{aligned}\mathbb{X}_2 = \{&(1.301,1.592),\ (0.630,1.542),\ (1.789,2.163),\ (0.083,2.132),\\ &(-0.429,3.200),\ (2.879,5.069),\ (-1.016,5.032),\ (-1.532,7.182),\\ &(3.929,9.992),\ (-2.083,10.03)\}.\end{aligned}$$

The polynomials g_1 and g_2 have similar behaviour: they are almost vanishing at $\mathbb{X}$ since

$$\frac{\|g_1(\mathbb{X})\|_2}{\|[1, -0.9657, 1.8292, -2.3703]\|_2} = 0.3372 < \varepsilon$$
$$\frac{\|g_2(\mathbb{X})\|_2}{\|[1, -0.9467, 1.7542, -2.2715]\|_2} = 0.3694 < \varepsilon,$$

they do not vanish at $\mathbb{X}_1$ and $\mathbb{X}_2$ exactly, nevertheless they take very small values at $\mathbb{X}_1$ and $\mathbb{X}_2$ as $\|g_1(\mathbb{X}_1)\|_2 = 0.00024$ and $\|g_2(\mathbb{X}_2)\|_2 = 0.00039$ and their zero sets contain admissible perturbations of $\mathbb{X}$ (see Figure 3). However note that the sum of squared euclidean distances between associated pairs of points in $\mathbb{X}$ and $\mathbb{X}_2$ is 0.6512 which is smaller than 0.6782, the analogue quantity for $\mathbb{X}$ and $\mathbb{X}_1$, because the LDP-BL algorithm aims at minimising this quantity.

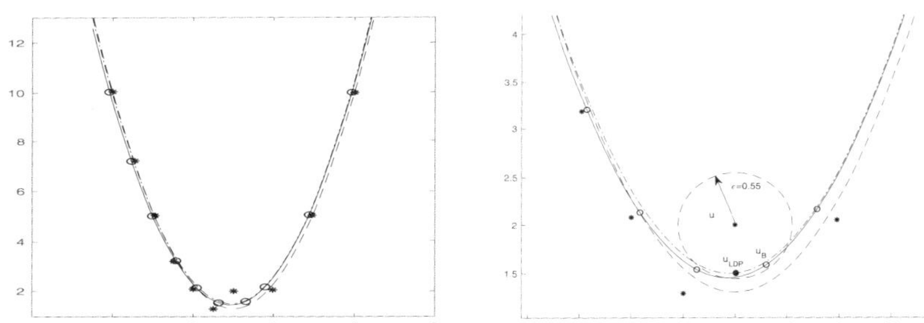

Fig. 3. The left plot shows the polynomials computed by NBM (dashed curve), LDP (dot-dashed curve) and LDP-BL (black curve). The right plot zooms the polynomials around the vertex. The LDP algorithm moves from the original point u to u_{LDP}, while the LDP-BL algorithm moves from u to u_B.

§7. Properties of almost vanishing polynomials

Let f be a monic polynomial almost vanishing at $\mathbb{X}$. If there exists a monic polynomial $\widetilde{f}$ with the same support as f and vanishing at an admissible perturbation $\widetilde{\mathbb{X}}$ of $\mathbb{X}$, then Theorem 7.1 shows that f and $\widetilde{f}$ have similar coefficients.

Theorem 7.1. *Let $\mathbb{X}$ be a set of points, ε a data error estimation, $\mathcal{O}$ a finite set of monomials, t a monomial not in $\mathcal{O}$,* $M(\mathbb{X}) = [s(u)]_{u \in \mathbb{X}, s \in \mathcal{O}}$

be full rank and $t(\mathbb{X}) = [t(u)]_{u\in\mathbb{X}}$. *Let* $f = t - \sum_{s\in\mathcal{O}} \alpha_s s$ *be a polynomial whose coefficient vector solves the least squares problem* $M(\mathbb{X})\alpha = t(\mathbb{X})$, *and let* $\widetilde{f} = t - \sum_{s\in\mathcal{O}} \widetilde{\alpha}_s s$ *be a polynomial vanishing at an admissible perturbation* $\widetilde{\mathbb{X}}$ *of* $\mathbb{X}$. *Then*

$$\|\alpha - \widetilde{\alpha}\|_2 \leq \frac{1}{\sigma}\left(\|f(\mathbb{X})\|_2 + \varepsilon\sqrt{\sum_{u\in\mathbb{X}} \|\nabla\widetilde{f}(u)\|_2^2 + O(\varepsilon^2)}\right) \tag{3}$$

where σ *is the smallest singular value of* $M(\mathbb{X})$.

Proof. Let x be a non zero vector in $\mathbb{R}^{\#\mathcal{O}}$. By the min-max theorem, the smallest eigenvalue λ of $M(\mathbb{X})^t M(\mathbb{X})$ is such that $\lambda = \min_{x\neq 0} \frac{\|M(\mathbb{X})x\|_2^2}{\|x\|_2^2}$. Since the square roots of the eigenvalues of $M(\mathbb{X})^t M(\mathbb{X})$ are the singular values of $M(\mathbb{X})$, the smallest singular value σ of $M(\mathbb{X})$ is equal to $\min_{x\neq 0} \frac{\|M(\mathbb{X})x\|_2}{\|x\|_2}$.

For each $x \neq 0$, we have $\|x\|_2 \leq \frac{\|M(\mathbb{X})x\|_2}{\sigma}$ and in particular

$$\begin{aligned}
\|\alpha - \widetilde{\alpha}\|_2 &\leq \frac{1}{\sigma}\|M(\mathbb{X})(\alpha - \widetilde{\alpha})\|_2 = \frac{1}{\sigma}\|M(\mathbb{X})\alpha - M(\mathbb{X})\widetilde{\alpha}\|_2 \\
&= \frac{1}{\sigma}\|M(\mathbb{X})\alpha - t(\mathbb{X}) + t(\mathbb{X}) - M(\mathbb{X})\widetilde{\alpha}\|_2 \\
&= \frac{1}{\sigma}\|f(\mathbb{X}) - \widetilde{f}(\mathbb{X})\|_2 \leq \frac{1}{\sigma}(\|f(\mathbb{X})\|_2 + \|\widetilde{f}(\mathbb{X})\|_2)\,.
\end{aligned}$$

The thesis follows since $\widetilde{f}$ vanishes at $\widetilde{\mathbb{X}}$ and so, from Proposition 4 in [5], $\|\widetilde{f}(\mathbb{X})\|_2^2 \leq \varepsilon^2 \sum_{u\in\mathbb{X}} \widetilde{M}_u^2$, where $\widetilde{M}_u^2 = \|\nabla\widetilde{f}(u)\|_2^2 + O(\varepsilon^2)$. Q.E.D.

Sufficient conditions for the existence of $\widetilde{f}$ vanishing at $\widetilde{\mathbb{X}}$ are shown in Theorem 4.3 in [6].

Lemma 7.1 below gives a rough estimate of the sum in the square root of Equation (3).

Lemma 7.1. *Let* $p = \sum_{s\in\mathcal{T}} \beta_s s \in \mathbb{R}[x_1, \ldots, x_k]$ *be a polynomial with coefficients vector* β, v *the cardinality of the support of* p, D_j *the maximum degree of* x_j *in* p *(for* $j = 1, \ldots, k$*) and* $u = (u_1, \ldots, u_k) \in [-1, 1]^k$, *then*

$$\left(\frac{\|\nabla p(u)\|_2}{\|\beta\|_2}\right)^2 \leq v\sum_{j=1}^{k} D_j^2.$$

Proof. For a monomial $s = x_1^{n_1} \cdots x_k^{n_k}$, it holds $\nabla s(u) = [n_j u_1^{n_1} \cdots u_j^{n_j-1} \cdots u_k^{n_k}]_{j=1,\ldots,k}$ and for $u \in [-1,1]^k$ it holds

$$\|\nabla s(u)\|_2^2 = \sum_{j=1}^{k} n_j^2 u_1^{2n_1} \cdots u_j^{2(n_j-1)} \cdots u_k^{2n_k} \leq \sum_{j=1}^{k} n_j^2 \leq \sum_{j=1}^{k} D_j^2.$$

Since $\nabla p(u) = \sum_{s\in\mathcal{T}} \beta_s \nabla s(u)$, it holds

$$\|\nabla p(u)\|_2 \leq \sum_{s\in\mathcal{T}} |\beta_s| \, \|\nabla s(u)\|_2 \leq \sum_{s\in\mathcal{T}} |\beta_s| \sqrt{\sum_{j=1}^{k} D_j^2}$$

and the thesis follows since $\sum_{s\in\mathcal{T}} |\beta_s| = \|\beta\|_1 \leq \sqrt{v}\|\beta\|_2$. Q.E.D.

Now, substituting in Equation (3) and dividing by $\|\widetilde{\alpha}\|_2$ yields

$$\frac{\|\widetilde{\alpha}-\alpha\|_2}{\|\widetilde{\alpha}\|_2} \leq \frac{1}{\sigma}\left(\frac{\|f(\mathbb{X})\|_2}{\|\alpha\|_2}\frac{\|\alpha\|_2}{\|\widetilde{\alpha}\|_2} + \varepsilon\sqrt{n(m+1)\sum_{j=1}^{k} D_j^2 + \frac{O(\varepsilon^2)}{\|\widetilde{\alpha}\|_2^2}}\right)$$

indeed $v = \#\mathcal{O} + 1$ is the cardinality of $\mathcal{O}$ plus one. If f is almost vanishing at $\mathbb{X}$ w.r.t. ε, namely $\frac{\|f(\mathbb{X})\|_2}{\|\alpha\|_2} = O(\varepsilon)$, for example if it has been computed with NBM or LDP algorithms, and since $\|\widetilde{\alpha}\|_2 \geq 1$ because the leading coefficient of $\widetilde{f}$ is one, then

$$\frac{\|\widetilde{\alpha}-\alpha\|_2}{\|\widetilde{\alpha}\|_2} \leq \frac{1}{\sigma}\left(O(\varepsilon)\left(\frac{\|\widetilde{\alpha}-\alpha\|_2}{\|\widetilde{\alpha}\|_2}+1\right) + \varepsilon\sqrt{n(m+1)\sum_{j=1}^{k} D_j^2 + O(\varepsilon^2)}\right)$$

equivalently

$$\frac{\|\widetilde{\alpha}-\alpha\|_2}{\|\widetilde{\alpha}\|_2}\left(1-\frac{O(\varepsilon)}{\sigma}\right) \leq \frac{O(\varepsilon)}{\sigma} + \frac{\varepsilon}{\sigma}\sqrt{n(m+1)\sum_{j=1}^{k} D_j^2 + O(\varepsilon^2)}$$

In our typical applications σ is order of magnitude larger than ε, $\mathcal{O}$ does not involve many terms and the D_j's are small integers, e.g. $0, 1, 2$. Thus the right hand side is small showing that the coefficients of the polynomials obtained with NBM, LDP, LDP-Britt are similar. Examples 5.2 and 6.1 show this in details.

§8. Numerical aspects

The test for numerical independence of the vector $t(\mathbb{X})$ and the columns of the matrix $M(\mathbb{X})$ in the generic step of the NBM and LDP algorithms is performed by computing the residual $\rho(\mathbb{X})$ of the least squares problem $M(\mathbb{X})\alpha = t(\mathbb{X})$. If $t(\mathbb{X})$ is numerically independent of the columns of $M(\mathbb{X})$, then t is included into $\mathcal{O}$ and, at the next step, a new monomial $\widehat{t}$ and a new least squares problem $\widehat{M}(\mathbb{X})\alpha = \widehat{t}(\mathbb{X})$ are considered, where the coefficient matrix $\widehat{M}(\mathbb{X})$ is obtained augmenting $M(\mathbb{X})$ with the column vector $t(\mathbb{X})$ as the last column. In essence, the NBM and LDP algorithms are based on a sequence of least squares problems where, at each step, the coefficient matrix is an update of the coefficient matrix at the previous step.

Using the QR decomposition of the coefficient matrices, the residual at the current step can be obtained exploiting the computation performed at the previous step, without solving explicitly the least squares problem. As shown in [5, Th. 3], if at the previous step $M(\mathbb{X})$ is an $n \times m$ matrix, $n \geq m$, and if $M(\mathbb{X}) = QR$ is its QR decomposition, where Q is an orthogonal $n \times n$ matrix and R is an upper triangular $n \times m$ matrix with positive diagonal entries, then the QR decomposition of $\widehat{M}(\mathbb{X}) = [M(\mathbb{X}), t(\mathbb{X})]$ at the current step is such that the first m columns of $\widehat{Q}$ coincide with the first m columns of Q, the $(m+1)$-th column of $\widehat{Q}$ is $\frac{\rho(\mathbb{X})}{\|\rho(\mathbb{X})\|_2}$, where $\rho(\mathbb{X})$ is the residual vector at the current step, and nothing can be said on the remaining columns of $\widehat{Q}$. The first m columns of $\widehat{R}$ are given by R and the first m elements of the last column of $\widehat{R}$ are given by the first m elements of $R\alpha$, the $(m+1)$-th element is given by $\|\rho(\mathbb{X})\|_2$ and the remaining positions of the last column are filled with zeros.

Thus for evaluating the residual $\rho(\mathbb{X})$ at the current step, it is sufficient to update the QR decomposition of $M(\mathbb{X}) = QR$ for obtaining the QR decomposition of $[M(\mathbb{X}), t(\mathbb{X})] = \widehat{Q}\widehat{R}$. If the last diagonal element of $\widehat{R}$, which is equal to $\|\rho(\mathbb{X})\|_2$, does not satisfy the upper bound (1) or the upper bound (2), then t is added to $\mathcal{O}$. Otherwise, both NBM and LDP algorithms compute a polynomial f whose coefficient vector α is the solution of the linear system whose coefficient matrix consists of the first m rows of R and whose right hand side is given by the first m elements of the last column of $\widehat{R}$.

A stable algorithm with low computational cost for updating the QR decomposition of $M(\mathbb{X})$ consists in the computation, at each step, of a Householder matrix H such that the last $(n - m - 1)$ elements of

$HQ^t t(\mathbb{X})$ are equal to zero. It follows that $HQ^t t(\mathbb{X})$ is the last column of $\widehat{R}$ and that HQ is the matrix $\widehat{Q}$ [7].

§9. Applications

The above algorithms have been employed for a number of analyses of data sets of various complexity for which standard techniques did not yield satisfactory results. For example they have been employed as part of a real time statistical classifier used in colour recognition [3]; as a tool to determine identifiable polynomial regression models from non-standard, noisy experiments used in a thermal spraying process producing particle coatings of surfaces [11]; in marine robotics for the approximation, through a polynomial curve, of the target path an unmanned marine vehicle should follow [12, 13]; for the modelling the trajectories of near-Earth asteroids [8]. Some of these applications have peculiarities which make a numerical version of the BM algorithm more suitable than another one.

9.1. Colour recognition

The first application we consider concerns the problem of colour recognition. Training surgeons can practice by collecting little scattered objects, i.e. coloured chickpeas, inside a laparoscopic training box, and group them according to their colour by looking at their movements in a monitor screen. The performance of the student is assessed in realtime through a computer system. This requires accurate and instantaneous colour identification of images such as the one shown in the left plot of Figure 4. Various sources of uncertainty and subjectivity effect the identification. For example colour is a personal interpretation of the reflected light and are influenced by random factors (light, texture, position, temperature, humidity, etc.), also objects reflect different colours when exposed to different sources of light.

In [3], the problem of colour recognition is addressed by applying a semi-parametric algebraic approach to statistical classification, based on a polynomial approximation to probability density functions. The left plot in Figure 4 is mapped into the right one using the CIELAB model [15] and for each colour 500 points are sampled. Out of these 500 points a subset representing the boundary is extracted, e.g. by using a method based on the Mahalanobis distance. This subset is the $\mathbb{X}$ for a colour given in input to the NBM algorithm. The output of NBM is a polynomial p used to construct a probability density function supported on the interior of the "chickpea".

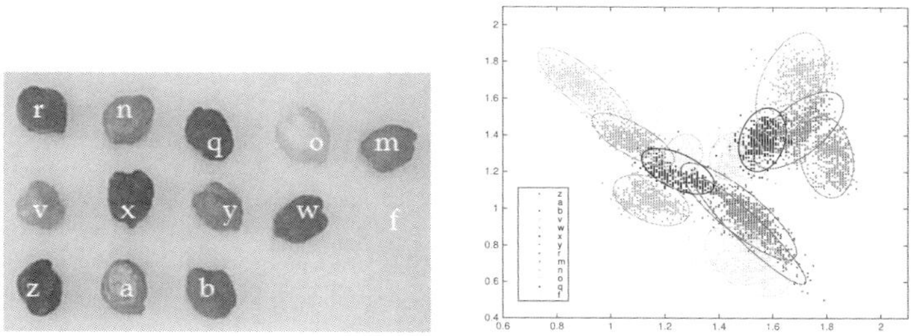

Fig. 4. Image of and representation on the colour system CIELAB.

The objective is thus to determine a smooth approximation p to the boundary of a closed bounded two-dimensional set. To ensure that p is a polynomial whose zero set is indeed a closed curve, only degree two polynomials are used in NBM. As the estimation of the probability densities supported on the clusters were robust to slight misclassification of the cluster boundaries, the precision given by LDP was not paying off for extra computational cost even if the algorithm was run only once for each colour. Anyway we tried and verified that the polynomials returned by the LDP and NBM algorithm were leading to 'ellipses' whose differences were not effecting the classification.

9.2. Industrial coating of surfaces

This project was carried out under the umbrella of the collaborative research center SFB823 of Dortmund University and published in [11]. Its objective was on determining a large set of $\mathcal{O}$ sets, called numerical fan, all of which were identifiable with observations or predictions from an experimental design. Thus, contrary to the standard set up in linear regression where outputs Y are observed at inputs X and a linear model f is sought in the form $Y = f(X) +$ random error, here the data generating process is from X to Z via a third set Y of observables which are easier to measure than Z. The interest is in determining a linear model g either from Y to Z or from X to Z via Y.

A regular design is often chosen on the controlled variables X which induces some canonical regression model f, but the regularity of the design on X is not transferred into Y because either the observed Y or the predicted $f(X)$ are affected by random error. The instability in the observed or predicted designs Y can naturally be dealt with some numerical version of the BM algorithms. Varying the term-ordering in the NBM algorithms yields a finite (possibly large depending on the coordinates of the Y) collection of $\mathcal{O}$ sets. As the issue is in avoiding collinearity any of NBM, LDP, LDP-BL could be safely employed.

In the specific application described in [11], related to a thermal spraying application, the design on the control variables X was a full factorial with central point on four variables in two levels, for a total of 17 points. An identifiable, robust to misspecification, linear regression model f from X to Y is supported on the set of square free monomials in four variables of degree not larger than four plus a simple quadratic term. Either the 17 values $\hat{\mathbb{Y}} = f(\mathbb{X})$ or the observed values at $\mathbb{Y}$ of the in-flight particles properties, are input to the NBM algorithm. In either cases the $\mathbb{Y}$ sets is a seemingly random design. By varying the term-ordering a set of 72 possible identifiable models is obtained. Each of the 72 possible identifiable models is used as maximal model in a forward backward selection based on the AIC criterion to identify a robust model for the coating properties Z in function of Y.

A leave-one-out cross validation analysis was performed based on the PRESS statistics in order to compare with standard methods. It showed the worthwhileness of the computational effort required to compute (a part of) the numerical fan of a design for model search.

9.3. Performance assessment in marine robotics

A topic of interest in marine robotics is the evaluation of the performance on Unmanned Marine Vehicles (UMVs) during a path-following or a path-tracking exercise based solely on the knowledge of the positions of the vehicle in time and of the reference target. Both reference and actual positions are give as sets of points $\mathcal{R}$ and $\mathcal{V}$ respectively, in a two or three dimensional space according to whether the vehicle is an Unmanned Surface Vehicle (USV) or an Unmanned Underwater Vehicle.

In [12, 13] a new criterion for evaluating the capability of an USV to follow a desired generic curvilinear path is presented and tested based on the LDP/LPA algorithm. By using the LDP algorithm a polynomial approximation f to $\mathcal{R}$ is computed with respect to a certain tolerance ε_1 and the value $f(p)$ is used to test whether the point $p \in \mathcal{V}$ is closed to $\mathcal{R}$. This test requires that f is near to $\mathcal{R}$ in the L_2-norm. This requirement is the reason why the LDP algorithm (and ideally its LPD-Britt version) is used and why the NBM algorithm, which only guarantees values of f small in a region around $\mathcal{R}$, might give misleading results. Other norms can be used and are discussed in [13], although the Euclidean norm seems a natural choice for the context.

The measure of closeness is quantified in two bounds called B_1 and B_2 whose formulæ are provided in [13] and which depend on the Jacobian vector and Hessian matrix of f. If $|f(p)| > B_1$ then the curve $f = 0$

does not cross an L_2-ball centred in $p \in \mathcal{V}$ and of radius ε_2 chosen by the experimenter; if $|f(p)| < B_2$ then it crosses it; otherwise it cannot be decided. The associated performance indices are the percentage of points in $\mathcal{V}$ crossing the ball and the percentage of points far from the path. Both indices can be computed efficiently online.

In summary, the requirements of a good estimation of the error, of high precision and of a speedy test based on close-by-ness, make the applicative scenarios of this section a good test bed for the LPD algorithm when used real time and online. The method proposed in [13] to evaluate when an UMV is close to the reference path is thus

(1) computation of an algebraic curve $f = 0$ that approximates the points in $\mathcal{R}$ within a tolerance ε_1;
(2) identification of the points in $\mathcal{V}$ far from the reference path $f = 0$ for more than a tolerance ε_2.

9.4. Near-Earth asteroid modelling

In [8] the trajectories of near-Earth asteroids (NEA), that is the whole set of heliocentric orbital elements with perihelion distance $q \leq 1.3\,au$ and eccentricity $e \leq 1$, are considered with the aim of studying their distances from the trajectory of the Earth. The authors analyse the distribution of the NEA asteroids with respect to some of their orbital elements. Their reconstruction is based on analytical knowledge of the phenomenon and consists of the computation of the minimum of specific analytic functions which at times can be hard to compute analytically.

We take a different perspective, not dissimilar to the one in Section 9.3, and starting from the available data we construct an implicit regression model for the distribution of the NEA asteroids. We have 300 noisy points in the two variables (q, ω) which we write as $\{(q, y) \mid y = \cos\omega,\ \omega \in [0, 2\pi]\} \subset \mathbb{R}^2$. The variable q is the perihelion distance and ω is the perihelion argument of a set of NEAs with absolute magnitude $H \geq 26$, the measurement error is estimated to be $\varepsilon = 0.002$.

The NBM algorithm can be applied for obtaining information about the distribution of NEAs directly from some measured parameters, without requiring, for each selected orbital parameters, the solution of a specific different analytical problem as it is done in [8]. The 300 input points have to be preprocessed in order to obtain a much smaller set of points excluding those in low density areas of the (q, ω) plane and taking representatives for high density areas (see Figure 5). This gives a set of 21 points along a curve which exhibits symmetries and singularities. Only the seven points for which $\omega \in [0, \pi/2]$ are used as input in the NBM algorithm which returns the polynomial $g(q, y) = q - 0.306687y - 0.708222$, and thus $g(q, \omega) = q - 0.306687\cos(\omega) - 0.708222$. By reflection and translation of g, one obtains the red (non affine) curve in Figure 5 which almost vanishes at the selected 21 points and is very close to the curve $\gamma(q,\omega) = 0$ presented in [8].

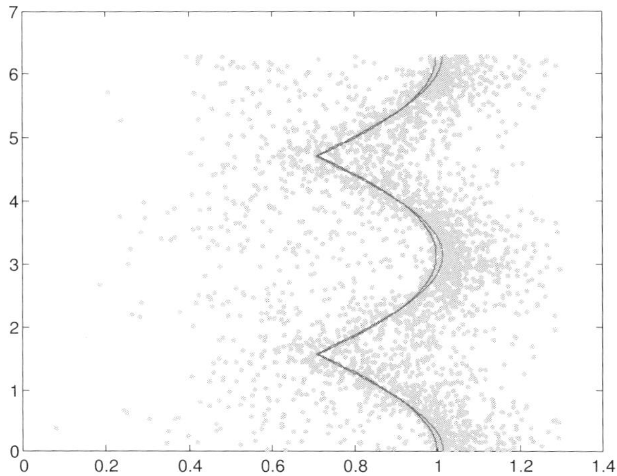

Fig. 5. The set of points $\mathbb{X}$, the (red) curve $g(q,\omega) = 0$ and the (blue) curve $\gamma(q,\omega) = 0$

References

[1] H. I. Britt and R. H. Luecke (1973). *The Estimation of Parameters in Nonlinear, Implicit Models.* Technometrics, 15 (2), 233–247.

[2] B. Buchberger and H. M. Möller (1982). *The construction of multivariate polynomials with preassigned zeros.* Proc. EUROCAM '82, LNCS, 144, 24–31.

[3] C. Cuevas Covarrubias, C. Fassino, E. Riccomagno and C. Villar-Patiño (2016). *Probability density functions on star domains with an application to classification.* (Submitted)

[4] C. Fassino (2010). *Almost vanishing polynomials for sets of limited precision points.* J. Symbolic Comput., 45 (1), 19–37.

[5] C. Fassino and H. M. Möller(2016). *Multivariate polynomial interpolation with disturbed data.* Numerical Algorithms, 71 (2), 273–292.

[6] C. Fassino and M.-L. Torrente (2013). *Simple varieties for limited precision points.* Theoretical Computer Science, 479, 174–186.

[7] G. H. Golub and C. F. Van Loan (1996). *Matrix Computations. 3rd Ed.*, John Hopkins University Press, Baltimore.

[8] G. F. Gronchi and G. B. Valsecchi (2013). *On the possible values of the orbit distance between a near-Earth asteroid and the Earth.* Monthly Notices of the Royal Astronomical Society, 429, 2687–2699.

[9] D. Heldt, M. Kreuzer, S. Pokutta, H. Poulisse (2009). *Approximate computation of zero-dimensional polynomial ideals.* J. Symbolic Comput., 44, 156–1591.

[10] G. Pistone, E. Riccomagno and H. P. Wynn (2001). Algebraic statistics: computational commutative algebra in statistics, Volume 89 of *Monographs on Statistics and Applied Probability.* Chapman & Hall/CRC, Boca Raton.

[11] N. Rudak, S. Kuhnt and E. Riccomagno (2016). *Numerical algebraic fan of a design for statistical model building.* Statistica Sinica, 26(3), 1021–1035.

[12] E. Saggini, M.-L. Torrente, E. Riccomagno, M. Bibuli, G. Bruzzone, M. Caccia and E. Zereik (2014). *Assessing path-following performance for Unmanned Marine Vehicles with algorithms from Numerical Commutative Algebra.* MED'14, 22nd Mediterranean Conference on Control and Automation, 16–19 June, Palermo, Italy.

[13] E. Saggini and M.-L. Torrente (2016). *An Euclidean norm based criterion to assess robots' 2D path-following performance.* Journal of Algebraic Statistics, 7, 45–71.

[14] H. F. Walker and L. T. Watson (1990). *Least-change secant update methods for underdetermined systems.* SIAM J. Numer. Anal., 27 (5), 1227–1262.

[15] Carmen Villar Patiño, Carlos Cuevas Covarrubias (2016). *Controlled condensation in k-NN and its application for real time color identification.* Revista de Matemática: teoriá y aplicationes, 23(1) : 143–154 (in Spanish).

(Claudia Fassino) *Department of Mathematics, University of Genova, via Dodecaneso 35, 16146 Genova, Italy*
E-mail address: `fassino@dima.unige.it`

(Hans Michael Möller) *Universität Dortmund, Institut für Angewandte Mathematik, Vogelpothsweg 87, D-44221 Dortmund, Germany*
E-mail address: `moeller@mathematik.tu-dortmund.de`

(Eva Riccomagno) *Department of Mathematics, University of Genova, via Dodecaneso 35, 16146 Genova, Italy*
E-mail address: `riccomagno@dima.unige.it`

Advanced Studies in Pure Mathematics 77, 2018
The 50th Anniversary of Gröbner Bases
pp. 109–119

The multiple roots phenomenon in maximum likelihood estimation for factor analysis

Elizabeth Gross, Sonja Petrović, Donald Richards and Despina Stasi

Abstract.

Multiple root estimation problems in statistical inference arise in many contexts in the literature. In maximum likelihood estimation, the existence of multiple roots causes uncertainty in the computation of maximum likelihood estimators using hill-climbing algorithms, and consequent difficulties in the resulting statistical inference.

In this paper, we study the multiple roots phenomenon in maximum likelihood estimation for factor analysis. We prove that the corresponding likelihood equations have uncountably many feasible solutions even in the simplest cases. For the case in which the observed data are two-dimensional and the unobserved factor scores are one-dimensional, we prove that the solutions to the likelihood equations form a one-dimensional real curve.

§1. Introduction

Systems of equations which have multiple roots, such as critical equations in estimation problems, arise in many contexts in statistical inference and there is now a large literature on these problems. We refer to Small, et al. [15] for an extensive account of the history and methods for investigating numerous multiple roots problems; Buot, et al. [5, 6] for enumeration of the roots of likelihood equations arising from mixture models or from the Behrens-Fisher problem; Catanese, et al. [7] for algebraic aspects of some of these problems and Gross, et al. [10] for applications to variance component models; Hoferkamp and Peddada

---

Received September 1, 2016.
2010 *Mathematics Subject Classification.* 62F10, 62N02, 62H25, 62H12, 62H20, 62F30.
Key words and phrases. maximum likelihood estimation, factor analysis, multiple roots, inference, small sample.

[12] for order-restricted inference; and Zwiernik, et al. [17] for aspects of linear Gaussian covariance models.

In the context of estimation by the method of maximum likelihood, the existence of multiple roots for the likelihood equations causes uncertainty in the computation of maximum likelihood estimators using hill-climbing algorithms. Consequently, the resulting statistical inference faces the difficulty that it is necessary to determine which of the multiple roots produces the maximum value of the likelihood function.

It is well-known that maximum likelihood estimators (MLEs) enjoy many appealing asymptotic properties. Subject to various regularity conditions and as the sample size tends to infinity, the MLE of a parameter converges in probability to the true value of the parameter, a property known as *asymptotic consistency*. MLEs also are *asymptotically efficient*: The variance of the MLE converges to the minimal value among all consistent estimators. Moreover, the asymptotic distribution of MLEs are known, and this enables standard statistical procedures, such as the construction of confidence regions and tests of hypotheses.

Despite the large-sample advantages of MLEs, there are instances in which large samples are unavailable, e.g., in the statistical analyses of clinical trials data for rare diseases. On the other hand, from a purely inferential perspective, one must bear in mind that, as the sample size increases, the likelihood of rejecting the null hypothesis increases simply as an artifact of said increasing sample size. Consequently, it is important to investigate maximum likelihood estimation problems in which sample sizes are small.

In this paper, we study the multiple roots phenomenon in maximum likelihood estimation for the problem of factor analysis; cf., Anderson [3], Harman [11]. The factor analysis problem is old and venerable, with many applications to the social sciences, and its difficult mathematical nature is well-known; cf., Adachi [1], Anderson and Rubin [2], and Rubin and Thayer [14]. Starting with the formulation of the factor analysis model in [14], we prove that the corresponding likelihood equations have uncountably many feasible solutions even in the simplest cases. For the case in which the observed data are two-dimensional and the unobserved factor scores are one-dimensional, we prove that the solutions to the likelihood equations form a one-dimensional *real* curve.

§2. A model for factor analysis

Rubin and Thayer [14] considered a problem in factor analysis, as follows. Let Y denote a $n \times p$ matrix of observed data. Let Z be a $n \times q$ matrix of unobserved factor scores, where $q < p$, and each row of

Z is assumed to be marginally distributed as $N_q(0, R)$, a q-dimensional multivariate normal distribution with mean vector 0 and $q \times q$ *correlation* matrix $R = (r_{ij})$, i.e., $r_{jj} = 1$ for all $j = 1, \ldots, q$. It also is assumed that the rows of the matrix (Y, Z) are mutually independent and identically distributed.

For $i = 1, \ldots, n$, denote by Y_i and Z_i the ith rows of Y and Z, respectively. We suppose that the conditional distribution of Y_i, given Z is given by $Y_i|Z \sim N_p(\alpha + Z_i\beta, \tau^2)$, where $\tau^2 = \operatorname{diag}(\tau_1^2, \ldots, \tau_p^2)$ is a diagonal matrix and $\beta = (\beta_{ij})$ is a $q \times p$ matrix called the *factor-loading matrix*. By centering the matrix Y it can be assumed, with no loss of generality that $\alpha = 0$. Then the problem of factor analysis is to calculate the maximum likelihood estimators of the parameters β, τ^2, and R.

It was noted in [14, p. 75] that the likelihood function for this factor analysis model generally has multiple local maxima. The issue of the number of critical points, i.e., the number of roots of the likelihood equations, was left open, however, and this is the focus of the present paper.

§3. The derivatives of the likelihood function

Following [14], we write the log-likelihood function in the form

$$\ell(\tau^2, \beta, R) = -\tfrac{1}{2} n\, f(\tau^2, \beta, R)$$

with $C_{yy} = (c_{ij})$ denoting the sample covariance matrix constructed from Y, and

$$f(\tau^2, \beta, R) = \log\det(\tau^2 + \beta' R\beta) + \operatorname{tr} C_{yy}(\tau^2 + \beta' R\beta)^{-1}. \tag{1}$$

Thus, we wish to *minimize* the function f with respect to the parameters (β, τ^2, R).

To calculate the derivatives of f with respect to τ and β, we will apply repeatedly the following result provided by Magnus and Neudecker [13, p. 169 (Theorem 1) and p. 171 (Theorem 3)].

Lemma 3.1. *Suppose that the square matrix $A(t)$ is a differentiable function of a real parameter t. Then*

$$\frac{\partial}{\partial t} \log\det(A(t)) = \operatorname{tr}\left[A(t)^{-1} \frac{\partial}{\partial t} A(t)\right] \tag{2}$$

and

$$\frac{\partial}{\partial t} A(t)^{-1} = -A(t)^{-1} \left[\frac{\partial}{\partial t} A(t)\right] A(t)^{-1}. \tag{3}$$

Denote by E_{kk} the $p \times p$ matrix with entry 1 in the (k,k)th position and zeros elsewhere. It is straightforward to verify that

$$\frac{\partial}{\partial \tau_k}(\tau^2 + \beta' R \beta) = 2\tau_k E_{kk},$$

and then it follows from formula (2) that

$$\begin{aligned}\frac{\partial}{\partial \tau_k} \log\det(\tau^2 + \beta' R \beta) &= 2\tau_k \operatorname{tr}[(\tau^2 + \beta' R \beta)^{-1} E_{kk}] \\ &= 2\tau_k \det(\tau^2 + \beta' R \beta)^{-1} \operatorname{Cof}_{kk}(\tau^2 + \beta' R \beta),\end{aligned}$$

where $\operatorname{Cof}_{ij}(A)$ is the (i,j)th cofactor of the matrix A.

Next, we apply the formula (3) to obtain

$$\begin{aligned}&\frac{\partial}{\partial \tau_k} \operatorname{tr} C_{yy}(\tau^2 + \beta' R \beta)^{-1} \\ &\qquad = \operatorname{tr}[C_{yy} \frac{\partial}{\partial \tau_k}(\tau^2 + \beta' R \beta)^{-1} \\ &\qquad = -2\tau_k \operatorname{tr} C_{yy}(\tau^2 + \beta' R \beta)^{-1} E_{kk}(\tau^2 + \beta' R \beta)^{-1}.\end{aligned}$$

This latter expression seems formidable initially, but it can be rewritten in terms of the entries of C_{yy} and the cofactors of $\tau^2 + \beta' R \beta$, as follows: Since

$$(\tau^2 + \beta' R \beta)^{-1} = \det(\tau^2 + \beta' R \beta)^{-1} \Big(\operatorname{Cof}_{ij}(\tau^2 + \beta' R \beta) \Big)$$

then

$$\begin{aligned}&(\tau^2 + \beta' R \beta)^{-1} E_{kk} (\tau^2 + \beta' R \beta)^{-1} \\ &\quad = \det(\tau^2 + \beta' R \beta)^{-2} \Big(\operatorname{Cof}_{ij}(\tau^2 + \beta' R \beta) \Big) E_{kk} \Big(\operatorname{Cof}_{ij}(\tau^2 + \beta' R \beta) \Big) \\ &\quad = \det(\tau^2 + \beta' R \beta)^{-2} \Big(\operatorname{Cof}_{ik}(\tau^2 + \beta' R \beta) \cdot \operatorname{Cof}_{kj}(\tau^2 + \beta' R \beta) \Big).\end{aligned}$$

Therefore,

$$\begin{aligned}&\operatorname{tr} C_{yy}(\tau^2 + \beta' R \beta)^{-1} E_{kk} (\tau^2 + \beta' R \beta)^{-1} \\ &\quad = \det(\tau^2 + \beta' R \beta)^{-2} \sum_{i,j=1}^{p} c_{ij} \operatorname{Cof}_{jk}(\tau^2 + \beta' R \beta) \cdot \operatorname{Cof}_{ki}(\tau^2 + \beta' R \beta).\end{aligned}$$

Collecting together the formulas above, we obtain

$$\begin{aligned}
&\frac{\partial}{\partial \tau_k} f(\tau^2, \beta, R) \\
&= 2\tau_k \det(\tau^2 + \beta' R\beta)^{-1} \ \mathrm{Cof}_{kk}(\tau^2 + \beta' R\beta) \\
&\quad - 2\tau_k \det(\tau^2 + \beta' R\beta)^{-2} \sum_{i,j=1}^{p} c_{ij} \,\mathrm{Cof}_{jk}(\tau^2 + \beta' R\beta) \cdot \mathrm{Cof}_{ki}(\tau^2 + \beta' R\beta).
\end{aligned}$$

Next, we calculate the derivative of $f(\tau^2, \beta, R)$ with respect to β. Let E_{kl} be a $q \times p$ matrix with entry 1 in the (k, l)th position and zeros elsewhere; that is, the (i, j)th entry of E_{kl} is $\delta_{ik}\delta_{jl}$, where δ_{ij} denotes Kronecker's delta. Then

$$\begin{aligned}
\frac{\partial}{\partial \beta_{kl}}(\beta' R\beta) &= (\frac{\partial}{\partial \beta_{kl}}\beta)' R\beta + \beta' R \frac{\partial}{\partial \beta_{kl}}\beta \\
&= E'_{kl} R\beta + \beta' R E_{kl}.
\end{aligned}$$

Since

$$\begin{aligned}
E'_{kl} R\beta = (\delta_{ik}\delta_{jl})' R\beta &= (\delta_{il}\delta_{jk}) R\beta \\
&= \Big(\sum_{m=1}^{q} \delta_{il}\delta_{mk}(R\beta)_{mj} \Big) = \big(\delta_{il}(R\beta)_{kj}\big),
\end{aligned}$$

then

$$\beta' R E_{kl} = (E'_{kl} R\beta)' = \big(\delta_{jl}(R\beta)_{ki}\big),$$

and hence

$$\frac{\partial}{\partial \beta_{kl}}(\beta' R\beta) = \big(\delta_{il}(R\beta)_{kj} + \delta_{jl}(R\beta)_{ki}\big).$$

Therefore, by formula (2),

$$\begin{aligned}
&\frac{\partial}{\partial \beta_{kl}} \log\det(\tau^2 + \beta' R\beta) \\
&= \mathrm{tr}\Big[(\tau^2 + \beta' R\beta)^{-1} \frac{\partial}{\partial \beta_{kl}}(\tau^2 + \beta' R\beta)\Big] \\
&= \mathrm{tr}\Big[(\tau^2 + \beta' R\beta)^{-1} \big(\delta_{il}(R\beta)_{kj} + \delta_{jl}(R\beta)_{ki}\big)\Big] \\
&= \det(\tau^2 + \beta' R\beta)^{-1} \cdot \sum_{i,j=1}^{p} \mathrm{Cof}_{ij}(\tau^2 + \beta' R\beta) \cdot (\delta_{jl}(R\beta)_{ki} + \delta_{il}(R\beta)_{kj}).
\end{aligned}$$

Next, by formula (3),

$$\begin{aligned}
&\frac{\partial}{\partial\beta_{kl}}(\tau^2+\beta'R\beta)^{-1}\\
&\qquad=-(\tau^2+\beta'R\beta)^{-1}\Big[\frac{\partial}{\partial\beta_{kl}}(\tau^2+\beta'R\beta)\Big](\tau^2+\beta'R\beta)^{-1}\\
&\qquad=-(\tau^2+\beta'R\beta)^{-1}\big(\delta_{il}(R\beta)_{kj}+\delta_{jl}(R\beta)_{ki}\big)(\tau^2+\beta'R\beta)^{-1},
\end{aligned}$$

Therefore,

$$\begin{aligned}
&\frac{\partial}{\partial\beta_{kl}}\operatorname{tr}C_{yy}(\tau^2+\beta'R\beta)^{-1}\\
&\qquad=-\operatorname{tr}C_{yy}(\tau^2+\beta'R\beta)^{-1}\big(\delta_{il}(R\beta)_{kj}+\delta_{jl}(R\beta)_{ki}\big)(\tau^2+\beta'R\beta)^{-1}.
\end{aligned}$$

Consequently,

$$\begin{aligned}
&\frac{\partial}{\partial\beta_{kl}}f(\tau^2,\beta,R)\\
&=\det(\tau^2+\beta'R\beta)^{-1}\cdot\sum_{i,j=1}^{p}\operatorname{Cof}_{ij}(\tau^2+\beta'R\beta)\cdot(\delta_{jl}(R\beta)_{ki}+\delta_{il}(R\beta)_{kj})\\
&\quad-\operatorname{tr}C_{yy}(\tau^2+\beta'R\beta)^{-1}\big(\delta_{il}(R\beta)_{kj}+\delta_{jl}(R\beta)_{ki}\big)(\tau^2+\beta'R\beta)^{-1}.
\end{aligned}$$

Finally, we calculate the derivatives with respect to the parameters r_{kl}, $k<l$. On writing

$$\beta'R\beta=\Big(\sum_{m,n=1}^{q}\beta_{mi}r_{mn}\beta_{nj}\Big)=\Big(\sum_{m=1}^{q}\beta_{mi}\beta_{mj}+2\sum_{1\le m<n\le q}r_{mn}\beta_{mi}\beta_{nj}\Big),$$

we obtain

$$\frac{\partial}{\partial r_{kl}}\beta'R\beta=2r_{kl}\big(\beta_{ki}\beta_{lj}\big)$$

where, for fixed (k,l), $(\beta_{ki}\beta_{lj})$ denotes the matrix with generic (i,j)th entry $\beta_{ki}\beta_{lj}$. Consequently,

$$\begin{aligned}
\frac{\partial}{\partial r_{kl}}\log\det(\tau^2+\beta'R\beta)&=\operatorname{tr}\Big[(\tau^2+\beta'R\beta)^{-1}\frac{\partial}{\partial r_{kl}}(\tau^2+\beta'R\beta)\Big]\\
&=2r_{kl}\operatorname{tr}\big[(\tau^2+\beta'R\beta)^{-1}\big(\beta_{ki}\beta_{lj}\big)\big].
\end{aligned}$$

Next,

$$\begin{aligned}
\frac{\partial}{\partial r_{kl}}(\tau^2+\beta'R\beta)^{-1}&=-(\tau^2+\beta'R\beta)^{-1}\Big[\frac{\partial}{\partial r_{kl}}(\tau^2+\beta'R\beta)\Big](\tau^2+\beta'R\beta)^{-1}\\
&=-2r_{kl}(\tau^2+\beta'R\beta)^{-1}\big(\beta_{ki}\beta_{lj}\big)(\tau^2+\beta'R\beta)^{-1},
\end{aligned}$$

and therefore

$$
\begin{aligned}
&\frac{\partial}{\partial r_{kl}} f(\tau^2, \beta, R) \\
&\quad = 2r_{kl}\, \mathrm{tr}\left[(\tau^2 + \beta' R\beta)^{-1}(\beta_{ki}\beta_{lj})\right] \\
&\quad\quad - 2r_{kl}\, \mathrm{tr}\, C_{yy}(\tau^2 + \beta' R\beta)^{-1}(\beta_{ki}\beta_{lj})(\tau^2 + \beta' R\beta)^{-1} \\
&\quad = 2r_{kl}\, \mathrm{tr}\Big[(\tau^2 + \beta' R\beta)^{-1}(\beta_{ki}\beta_{lj})\big(I_p - (\tau^2 + \beta' R\beta)^{-1} C_{yy}\big)\Big].
\end{aligned}
$$

§4. The solutions of the likelihood equations

The likelihood equations derived in the previous section are rational equations whose denominators are $(\det(\tau^2 + \beta' R\beta))^2$. In order to work with polynomial equations, we will substitute γ for $(\det(\tau^2 + \beta' R\beta))^{-1}$ in the likelihood equations, add to the system of equations the new equation

$$\gamma \det(\tau^2 + \beta' R\beta) - 1 = 0,$$

and denote by $\mathbf{f}_{p,q,C_{yy}} = 0$ the resulting expanded system of equations.

The set $V_{p,q,C_{yy}} \subset \mathbb{R}^{p+pq+q(q-1)/2+1}$ of real solutions to the system $\mathbf{f}_{p,q,C_{yy}} = 0$ is equal to the set

(4)

$$
\begin{aligned}
V_{p,q,C_{yy}} = \Big\{ &(\tau, \beta, R, \gamma) \ : \ (\tau, \beta, R) \text{ is a critical point of } f(\tau^2, \beta, R), \\
&\det(\tau^2 + \beta' R\beta) \neq 0, \text{ and } \gamma = \frac{1}{\det(\tau^2 + \beta' R\beta)} \Big\}.
\end{aligned}
$$

We now set $p = 2$ and $q = 1$. Let us consider the polynomials in $\mathbf{f}_{2,1,C_{yy}}$ and treat the entries of $C_{yy} = (c_{ij})$ as unknown variables. Furthermore, for simplification, we make the substitution,

$$\begin{pmatrix} x_{11} & x_{12} \\ x_{12} & x_{22} \end{pmatrix} = \tau^2 + \beta' R\beta, \tag{5}$$

and add to $\mathbf{f}_{2,1,C_{yy}}$ the equations from (5). This new set of polynomials defines an ideal $I_{2,1} \subseteq \mathbb{Q}[c_{11}, c_{12}, c_{22}, \tau_1, \tau_2, \beta_{11}, \beta_{12}, \gamma, x_{11}, x_{12}, x_{22}]$. As we are interested in the solution set associated with these polynomials, we study the variety $V(I_{2,1}) \subseteq \mathbb{C}^{11}$ defined to the set of all points $p = (c_{11}, c_{12}, c_{22}, \tau_1, \tau_2, \beta_{11}, \beta_{12}, \gamma, x_{11}, x_{12}, x_{22}) \in \mathbb{C}^{11}$ such that $g(p) = 0$ for all polynomials $g \in I_{2,1}$. To compute the dimension of the set $V(I_{2,1})$, we first decompose it into a union of irreducible varieties. The corresponding algebraic object is the intersection of the primary ideals. For an introduction to ideals, varieties, and primary decompositions see [8].

Using `Macaulay2` (Grayson and Stillman [9]), we obtain the decomposition $I_{2,1} = J_1 \cap J_2$ of $I_{2,1}$ into an intersection of primary ideals J_1 and J_2, where

$$J_1 = \langle x_{22} - c_{22}, x_{12} - c_{12}, x_{11} - c_{11}, \beta_{11}\beta_{12} - c_{12}, \tau_2^2 + \beta_{12}^2 - c_{22}, \\ \tau_1^2 + \beta_{11}^2 - c_{11}, \gamma c_{12}^2 - \gamma c_{11}c_{22} + 1\rangle$$

and

$$J_2 = \langle x_{22} - c_{22}, x_{12}, x_{11} - c_{11}, \beta_{12}, \beta_{11}, \tau_2^2 - c_{22}, \tau_1^2 - c_{11}, \gamma c_{11}c_{22} - 1\rangle.$$

Moreover, we obtain from `Macaulay2` that the dimensions of the components $V(J_1)$ and $V(J_2)$ corresponding to the ideals J_1 and J_2 are 4 and 3, respectively. These computations can be confirmed using numerical algebraic geometry software such as `PHCpack` (Verschelde [16]). Since there are no relations between the variables c_{11}, c_{12}, and c_{22}, in other words, they are free variables, it follows that if we choose C_{yy} generically from the set of $p \times p$ positive definite matrices then the solution set $V_{2,1,C_{yy}}$ contains a curve (i.e., a one-dimensional component) and a set of points (i.e., a zero-dimensional component).

Moreover, the curve in $V_{2,1,C_{yy}}$ can be described parametrically in terms of t for all values of t such that $c_{12}^2/c_{22} \leq t^2 \leq c_{11}$:

$$\beta_{11} = t \qquad \beta_{12} = \frac{c_{12}}{t} \tag{6}$$

$$\tau_1^2 = c_{11} - t^2 \qquad \tau_2^2 = c_{22} - \frac{c_{12}^2}{t^2} \tag{7}$$

Since the matrix C_{yy} is positive definite then the interval $\{t \in \mathbb{R} : c_{12}^2/c_{22} \leq t^2 \leq c_{11}\}$ has positive measure.

The isolated points in $V_{2,1,C_{yy}}$ also can be described analytically:

$$\beta_{11} = 0 \qquad \beta_{12} = 0$$
$$\tau_1 = \pm\sqrt{c_{11}} \qquad \tau_2 = \pm\sqrt{c_{22}}$$

In summary, we have proved the following result.

Theorem 4.1. *Suppose that $p = 2$ and $q = 1$. For a generic sample covariance matrix C_{yy}, the likelihood equations for the factor analysis model in Section 2 have an infinite number of real solutions.*

For $p \geq 2$ and $q \geq 1$, we can apply the same approach as in the case $(p, q) = (2, 1)$ to construct a sample covariance matrix C_{yy} such that $V_{p,q,C_{yy}}$ is positive dimensional.

Theorem 4.2. *For all $p \geq 2$ and $q \geq 1$, there exists a covariance matrix C_{yy} such that $V_{p,q,C_{yy}}$ is positive dimensional.*

Proof. Let C_{yy} be a positive definite matrix of the following form:

$$C_{yy} = \begin{pmatrix} c_{11} & c_{12} & 0 & \cdots & 0 \\ c_{12} & c_{22} & 0 & \cdots & 0 \\ 0 & 0 & c_{33} & \cdots & 0 \\ \vdots & \vdots & \vdots & \ddots & \vdots \\ 0 & 0 & 0 & \cdots & c_{pp} \end{pmatrix}.$$

The set of all (τ, β, R) such that $r_{kl} = 0$ for $1 \leq k < l \leq q$ and

$$C_{yy} = \tau^2 + \beta'\beta$$

is a one-dimensional subset of $V_{p,q,C_{yy}}$. Indeed, this set is obtained by solving for $\beta_{11}, \beta_{12}, \tau_1, \tau_2$ in equations (6)-(7) and then setting $\beta_{kl} = 0$ for $(k,l) \neq (1,1), (1,2)$ and $\tau_k = \pm\sqrt{c_{kk}}$. Q.E.D.

Remark 4.3. We note that the set of solutions in Theorem 4.1 is the set of solutions to the equation

$$C_{yy} = \tau^2 + \beta'\beta. \tag{8}$$

Since $f(\tau^2, \beta, R)$ is a function of $\tau^2 + \beta'\beta$ for the case in which $(p,q) = (2,1)$ then the likelihood function is constant on $V_{2,1,C_{yy}}$. Thus, this set of solutions forms a ridge on the likelihood hypersurface. Although it follows immediately from (1) that the log-likelihood function attains the same value at every solution to (8), one cannot similarly conclude immediately that these solutions are critical points, the reason being that the first partial derivatives of $f(\tau^2, \beta, R)$ are not functions of $\tau^2 + \beta'\beta$.

While Theorem 4.2 describes some sample covariance matrices which result in infinite number of solutions to the likelihood equations, we believe that there are more. However, once we increase p and q, even modestly, the computations become infeasible with current methodology. One approach would be to use numerical algebraic geometry, but such approaches would require creative manipulation of the system.

§5. Acknowledgments

The research of Gross was supported by the National Science Foundation award DMS-1304167 and DMS-1620109. The research of Richards was partially supported by the National Science Foundation under grant

DMS-1309808. The research of Stasi and Petrović was partially supported by Air Force Office for Scientific Research grant FA9550-14-1-0141. We thank Daniel Brake for helping us to visualize the curve in Theorem 4.1 using `bertini_real` [4].

References

[1] Adachi, K. (2013). Factor analysis with EM algorithm never gives improper solutions when sample covariance and initial parameter matrices are proper. *Psychometrika*, **78**, 380–394.

[2] Anderson, T. W., and Rubin, H. (1956). Statistical inference in factor analysis. In: *Proc. Third Berkeley Symposium on Mathematical Statistics and Probability*, **5**, 11–150.

[3] Anderson, T. W. (2002). *An Introduction to Multivariate Statistical Analysis*, third edition. Wiley, New York.

[4] Brake, D. A., Bates, D. J., Hao, W., Hauenstein, J. D., Sommese, A. J., and Wampler, C. W. `Bertini_real`: Software for One- and Two-Dimensional Real Algebraic Sets. *Proceedings of ICMS 2014*, extended abstract.

[5] Buot, M.-L. G., and Richards, D. St. P. (2006). Counting and locating the solutions of polynomial systems of maximum likelihood equations, I. *J. Symbolic Computation*, **41**, 234–244.

[6] Buot, M.-L. G., Hoşten, S., and Richards, D. St. P. (2007). Counting and locating the solutions of polynomial systems of maximum likelihood equations, II: The Behrens-Fisher problem. *Statist. Sinica*, **17**, 1343–1354.

[7] Catanese, F., Hoşten, S., Khetan, A. and Sturmfels, B. (2006). The maximum likelihood degree. *Amer. J. Math.*, **128** no. 3, 671–697.

[8] Cox, D., Little, J. and O'Shea, D. (2007). *Ideals, Varieties, and Algorithms: An Introduction to Computational Algebraic Geometry and Commutative Algebra*, 3/E (Undergraduate Texts in Mathematics), Springer-Verlag New York Secaucus, NJ.

[9] Grayson, D. and Stillman, M. *Macaulay2*, a software system for research in algebraic geometry. Available at `http://www.math.uiuc.edu/Macaulay2/`.

[10] Gross, E., Drton, M., and Petrović, S. (2012). ML degree of variance component models. *Electron. J. Statist.*, **6**, 993–1016.

[11] Harman, H. H. (1976). *Modern Factor Analysis*, third edition revised. University of Chicago Press, Chicago, IL.

[12] Hoferkamp, C., and Peddada, S. D. (2002). Parameter estimation in linear models with heteroscedastic variances subject to order restrictions. *J. Multivariate Anal.*, **82**, 65–87.

[13] Magnus, J. R., and Neudecker, H. (2007). *Matrix Differential Calculus with Applications in Statistics and Econometrics*, third edition. Wiley, New York.

[14] Rubin, D. B., and Thayer, D. T. (1982). EM algorithms for ML factor analysis. *Psychometrika*, **47**, 69–76.

[15] Small, C. G., Wang, J., and Yang, Z. (2000). Eliminating multiple root problems in estimation, *Statist. Sci.*, **15**, 313–343.

[16] Verschelde, J. (1999). Algorithm 795: PHCpack: A general-purpose solver for polynomial systems by homotopy continuation. *ACM Trans. Math. Softw.*, **25** no. 2, 251–276. Software available at `http://www.math.uic.edu/~jan/download.html`.

[17] Zwiernik, P., Uhler, C., and Richards, D. (2017). Maximum likelihood estimation for linear Gaussian covariance models, *Journal of the Royal Statistical Society, Series B*, **79**,1269–1292.

(Elizabeth Gross) *Department of Mathematics, San José State University, San José, CA 95192, USA*
E-mail address: `elizabeth.gross@sjsu.edu`

(Sonja Petrović) *Department of Applied Mathematics, Illinois Institute of Technology, Chicago, IL 60616, USA*
E-mail address: `sonja.petrovic@iit.edu`

(Donald Richards) *Department of Statistics, Pennsylvania State University, University Park, PA 16802, USA*
E-mail address: `richards@stat.psu.edu`

(Despina Stasi) *Department of Applied Mathematics, Illinois Institute of Technology, Chicago, IL 60616, USA*
E-mail address: `stasdes@iit.edu`

Advanced Studies in Pure Mathematics 77, 2018
The 50th Anniversary of Gröbner Bases
pp. 121–128

A necessary condition for an edge ring to satisfy Serre's condition (S_2)

Akihiro Higashitani and Kyouko Kimura

Abstract.

Let G be a finite simple connected graph. We give a necessary condition for the edge ring of G to satisfy Serre's condition (S_2) in terms of a graph.

§1. Introduction

Let G be a finite simple connected graph. We denote by $V = V(G)$ the vertex set of G and $E(G)$ the edge set of G. We set $V = [d] := \{1, 2, \ldots, d\}$ and $E(G) = \{e_1, \ldots, e_r\}$, where $e_\ell = \{j_{\ell 1}, j_{\ell 2}\}$. Let K be a field. Consider the polynomial ring $K[\mathbf{t}]$ with d variables $t_1, t_2, \ldots, t_d$. The edge ring of G, denoted by $K[G]$ is the subring of $K[\mathbf{t}]$ generated by $\mathbf{t}^{e_1}, \ldots, \mathbf{t}^{e_r}$, where $\mathbf{t}^{e_\ell} := t_{j_{\ell 1}} t_{j_{\ell 2}}$. Let $\rho(e_\ell)$ be the $(0,1)$-vector of $\mathbb{R}^d$ which has 1 only $j_{\ell 1}$-, $j_{\ell 2}$-entries, and let S_G be the affine semigroup generated by $\rho(e_1), \ldots, \rho(e_r)$. Then $K[G]$ is the affine semigroup ring of S_G.

In [6], Ohsugi and Hibi characterized when $K[G]$ is normal in terms of the graph; see Section 3 for detail. Since $K[G]$ is an affine semigroup ring, by Hochster [5, Theorem 1], it is known that if $K[G]$ is normal then $K[G]$ is Cohen–Macaulay. Hence it is natural to ask when $K[G]$ is Cohen–Macaulay. For a general affine semigroup ring, the characterization of its Cohen–Macaulayness has been investigated by Goto and Watanabe [3] and Trung and Hoa [8, Theorem 4.1]. The purpose of our study is to illustrate the characterization in terms of the graph. Note that $K[G]$ is normal if and only if $K[G]$ satisfies Serre's conditions (R_1) and (S_2); see [2, Theorem 2.2.22]. Hibi and Katthän [4] give the characterization for $K[G]$ satisfying (R_1). On the other hand we try

Received August 30, 2016.
2010 *Mathematics Subject Classification.* 13H10.
Key words and phrases. Serre's condition (S_2), edge ring, toric ideal, finite graph.

to characterize $K[G]$ which satisfies (S_2). Note that Serre's condition (S_2) is a necessary condition for $K[G]$ to be Cohen–Macaulay. The main result of the present report is a necessary condition for $K[G]$ to satisfy (S_2) in terms of a graph; see Theorem 4.1.

The organization of the present report is as follow. In Section 2, we recall the characterization of a Cohen–Macaulay affine semigroup ring. In Section 3, we recall the properties of an edge ring. Finally, in Section 4, we give a necessary condition for an edge ring to satisfy Serre's condition (S_2) in terms of a graph.

§2. Cohen–Macaulayness of affine semigroup rings

Recall that $K[G]$ is an affine semigroup ring, as stating at Introduction. In this section, we recall the characterization of a Cohen–Macaulay affine semigroup ring investigated by Goto and Watanabe [3] and Trung and Hoa [8].

Let $S \subset \mathbb{N}^d$ be an affine semigroup. Then the affine semigroup ring of S, denoted by $K[S]$, is the subring of the polynomial ring $K[\mathbf{t}]$ with d variables $t_1, t_2, \dots, t_d$ generated by $\{\mathbf{t}^x : x \in S\}$, where $\mathbf{t}^x = t_1^{x_1} t_2^{x_2} \cdots t_d^{x_d}$ for $x = (x_1, x_2, \dots, x_d)$. Let C_S be the convex rational polyhedral cone spanned by S in $\mathbb{Q}^d$. We may assume that C_S is d-dimensional. Let $F_1, \dots, F_m$ be all the facets of C_S. Also let $\mathcal{G}$ be the group generated by S. We set

$$\begin{aligned} S_i :&= S - S \cap F_i \\ &= \{x \in \mathcal{G} \ : \ \text{there exists } y \in S \cap F_i \text{ such that } x + y \in S\} \end{aligned}$$

and $S' = \bigcap_{i=1}^m S_i$. Let J be a subset of $[m] := \{1, 2, \dots, m\}$. Set

$$G_J := \bigcap_{i \notin J} S_i \setminus (\bigcup_{j \in J} S_j).$$

Also set

$$\pi_J := \left\{ I \subset J \ : \ I \neq \emptyset, \bigcap_{i \in I} S \cap F_i \neq (0) \right\} \cup \{\emptyset\}.$$

Then π_J is a simplicial complex.

Theorem 2.1 (Trung and Hoa [8, Theorem 4.1])**.** *Let S be an affine semigroup ring. Assume that C_S has m facets. Then $K[S]$ is Cohen–Macaulay if and only if the following two conditions are satisfied:*

(i) $S' = S$;

(ii) $G_J = \emptyset$ *or* π_J *is acyclic for all proper subsets* J *of* $[m]$.

Remark 2.2. *The first condition of Theorem 2.1 corresponds to Serre's condition* (S_2); *see* [1, Exercise 4.17].

§3. Edge rings

As noted in Remark 2.2, the condition (i) of Theorem 2.1 corresponds to Serre's condition (S_2). Thus we investigate which graph G satisfies the condition. In order to do this, we need the facets of C_{S_G}. Fortunately, these have been studied by Ohsugi and Hibi [6]. In this section, we recall their result.

Let G be a finite simple connected graph. We denote $V(G)$ its vertex set and $E(G)$ its edge set. Let $W \subset V(G)$. The induced subgraph of G on W, denoted by G_W, is the graph whose vertex set is W and whose edge set is given by $\{e \in E(G) : e \subset W\}$. Also we set

$$N(G; W) := \{v \in V(G) : \{v, w\} \in E(G) \text{ for some } w \in W\}.$$

A subset T of $V(G)$ is said to be independent if $\{v_1, v_2\}$ is not an edge of G for any distinct vertices $v_1, v_2 \in T$. Let T be an independent set of G. Then we can consider the bipartite graph on $T \cup N(G; T)$ whose edge set is given by

$$\{\{v, w\} \in E(G) : v \in T,\ w \in N(T; G)\}.$$

We call such a graph the *bipartite graph induced by* T.

A pair (C_1, C_2) of odd cycles of G with no common vertex is said to be *exceptional* if it has no bridge, that is, for any vertices $v_1 \in V(C_1)$ and $v_2 \in V(C_2)$, $\{v_1, v_2\}$ is not an edge of G. The graph G is said to satisfy the *odd cycle condition* if it has no exceptional pair. We have already known a graph G whose edge ring is normal:

Theorem 3.1 (Ohsugi and Hibi [6]). *Let* G *be a finite simple connected graph. Then* $K[G]$ *is normal if and only if* G *satisfies the odd cycle condition.*

Since we have already known a normal graph, we focus on a non-normal graph, namely, the graph whose edge ring is not normal, and in what follows, we always assume that the graph G is non-normal. Note that G is not bipartite.

In order to illustrate the facets of C_{S_G}, we need some notion.

Definition 3.2. *The vertex* $i \in V$ *is said to be regular in* G *if each connected component of* $G_{V \setminus \{i\}}$ *has at least one odd cycle.*

Definition 3.3. *A non-empty subset T of V is said to be fundamental in G if the following three conditions are satisfied:*

(i) *T is an independent set;*
(ii) *the bipartite graph induced by T is connected;*
(iii) *if $T \cup N(G;T) \neq V(G)$, then each connected component of $G_{V(G)\setminus(T\cup N(G;T))}$ has at least one odd cycle.*

The facets of C_{S_G} are given by the intersection of supporting hyperplanes of C_{S_G}, which are defined by the regular vertices in G and fundamental sets in G.

Theorem 3.4 (Ohsugi and Hibi [6, Theorem 1.7]). *Let G be a non-normal finite connected simple graph. Then all the supporting hyperplanes of C_{S_G} are the following ones:*

(i) $\mathcal{H}_i = \{(x_1, x_2, \dots, x_d) \in \mathbb{R}^d : x_i = 0\}$, *where i is a regular vertex in G.*
(ii) $\mathcal{H}_T = \{(x_1, x_2, \dots, x_d) \in \mathbb{R}^d : \sum_{i\in T} x_i = \sum_{j\in N(G;T)} x_j\}$, *where T is a fundamental set in G.*

We close the present section by recalling the characterization of the graph G whose edge ring satisfies Serre's condition (R_1) given by Hibi and Katthän [4].

Theorem 3.5 (Hibi and Katthän [4, Theorem 2.1]). *Let G be a finite simple connected graph. Assume that G is not bipartite. Then $K[G]$ satisfies Serre's condition (R_1) if and only if G satisfies the following two conditions:*

(i) *$G \setminus \{i\}$ is connected for all regular vertex i in G;*
(ii) *for each fundamental set T in G with $T \cup N(G;T) \neq V(G)$, the induced subgraph $G_{V(G)\setminus(T\cup N(G;T))}$ is connected.*

We note that if a non-normal graph G satisfies Serre's condition (R_1), then G does not satisfy Serre's condition (S_2).

§4. Main result

Our purpose of the study is to characterize a Cohen–Macaulay edge ring $K[G]$ in terms of the graph G. As the first step of our study, we try to characterize the Serre's condition (S_2); see Theorem 2.1 and Remark 2.2. In this section, we give a necessary condition for G satisfying the Serre's condition (S_2), namely, satisfying $S_G = S'_G$.

Before stating our result, we define one notation. Let C be an odd cycle. Set

$$\mathbf{e}_C := \sum_{i\in V(C)} \mathbf{e}_i \in \mathbb{N}^d,$$

where $\mathbf{e}_i$ is the i-th elementary vector of $\mathbb{R}^d$.

The following theorem is the main result of the present report.

Theorem 4.1. *Let G be a finite simple connected graph. Suppose that there exists an exceptional pair (C_1, C_2) satisfying following two conditions*:

(i) *for each regular vertex $i \in V(G) \setminus (V(C_1) \cup V(C_2))$ in G, both C_1 and C_2 belong to the same connected components of $G \setminus \{i\}$*;

(ii) *for each fundamental set T in G with $(V(C_1) \cup V(C_2)) \cap (T \cup N(G;T)) = \emptyset$, both C_1 and C_2 belong to the same connected components of $G_{V(G)\setminus(T\cup N(G;T))}$.*

Then $\mathbf{e}_{C_1} + \mathbf{e}_{C_2} \in S'_G$. In particular, $S_G \neq S'_G$.

Theorem 4.1 is restated as follows:

Theorem 4.2. *Let G be a finite simple connected graph. If $S_G = S'_G$, then for each exceptional pair (C_1, C_2), one of the following is satisfied*:

(i) *there exists a regular vertex i in G such that both C_1 and C_2 belong to different connected components of $G \setminus \{i\}$.*

(ii) *there exists a fundamental set T in G such that both C_1 and C_2 belong to different connected components of $G_{V(G)\setminus(T\cup N(G;T))}$.*

Before proving Theorem 4.1, we consider which elements can be the elements of S_G or S'_G. Note that $S_G \subset S'_G$ holds in general. Hence we would like to know which element belongs to $S'_G \setminus S_G$.

We need some more notation. Let $\Gamma = (e_{j_1}, \dots, e_{j_q})$ be a walk. We set

$$\mathbf{e}_{\Gamma,o} = \sum_{\ell} \mathbf{e}_{j_{2\ell+1}}, \qquad \mathbf{e}_{\Gamma,e} = \sum_{\ell} \mathbf{e}_{j_{2\ell}}.$$

Note that $\mathbf{e}_\ell$ is the ℓ-th elementary vector of $\mathbb{R}^r$ and $\mathbf{e}_{\Gamma,o}, \mathbf{e}_{\Gamma,e} \in \mathbb{N}^r$. Also for $f = (f_1, \dots, f_r) \in \mathbb{N}^r$, we set

$$\rho(f) := \sum_{j=1}^{r} f_j \rho(\mathbf{e}_j).$$

Let i be a regular vertex in G. Then we denote by Γ_i, the subsequence of Γ whose edges contain i. Let T be a fundamental set in G. Then we denote by Γ_T, the subsequence of Γ whose edges contain an element of $N(G;T)$ and do not contain any element of T.

For $f = (f_1, \dots, f_r), g = (g_1, \dots, g_r) \in \mathbb{N}^r$, we denote $g \leq f$ if $g_i \leq f_i$ for all i. Katthän proved the following proposition.

Proposition 4.3 (Katthän). *Let G be a finite simple connected non-normal graph. Let (C_1, C_2) be an exceptional pair of odd cycles of G and $f \in \mathbb{N}^r$. Then $\mathbf{e}_{C_1} + \mathbf{e}_{C_2} + \rho(f) \in S_G$ if and only if there exists a walk Γ which combine C_1 and C_2 with either $\mathbf{e}_{\Gamma,o} \leq f$ or $\mathbf{e}_{\Gamma,e} \leq f$.*

We give a proof of this proposition for the convenience of the reader.

Proof. "If" part is obvious. We consider "Only if" part.

Take $v_1 \in V(C_1)$ and $v_2 \in V(C_2)$. Set $e' := e_{r+1} := \{v_1, v_2\}$. (Recall that $E(G) = \{e_1, \ldots, e_r\}$.) Then $e' \notin E(G)$ since (C_1, C_2) is exceptional. Let G' be the graph obtained by adding an edge e' to the graph G. Then there exists a $(0,1)$-vector $g = (g_1, \ldots, g_{r+1}) \in \mathbb{N}^{r+1}$ with $g_{r+1} = 1$ such that $\mathbf{e}_{C_1} + \mathbf{e}_{C_2} = \rho(g)$ for the graph G'.

On the other hand, since $\mathbf{e}_{C_1} + \mathbf{e}_{C_2} + \rho(f) \in S_G$, there exists $h \in \mathbb{N}^r$ with $\mathbf{e}_{C_1} + \mathbf{e}_{C_2} + \rho(f) = \rho(h)$ for the graph G.

Identifying an element $\mathbb{N}^r$ with the element $\mathbb{N}^{r+1}$ whose $(r+1)$-th entry is 0, we have $\rho(g + f) = \rho(h)$ for the graph G'. Then the binomial $\mathbf{t}^{\rho(g+f)} - \mathbf{t}^{\rho(h)}$ belongs to the toric ideal of G'. By [7, Lemma 1.1], there exists an even closed walk $\Gamma' = (e', e_{j_2}, \ldots, e_{j_q})$ with $\mathbf{e}_{\Gamma',o} \leq g + f$. Let e_{j_α} be the last edge of Γ' which contains the vertex of C_2 and let e_{j_β} be the first edge of Γ' which contains the vertex of C_1 after e_{j_α}. Then the subwalk Γ of Γ' starting at e_{j_α} and ending at e_{j_β} has the desired property. Q.E.D.

On the other hand, we have the following proposition.

Proposition 4.4. *Let G be a graph with the same conditions as in Proposition 4.3. Let (C_1, C_2) be an exceptional pair and $f \in \mathbb{N}^r$. Then $\mathbf{e}_{C_1} + \mathbf{e}_{C_2} + \rho(f) \in S'_G$ if and only if the following two conditions are satisfied:*

(i) *for any regular vertex i in G such that C_1 and C_2 belong to different connected components of $G \setminus \{i\}$, there exists a walk $\Gamma^{(i)}$ which combine C_1 and C_2 with either $\mathbf{e}_{\Gamma_i^{(i)},o} \leq f$ or $\mathbf{e}_{\Gamma_i^{(i)},e} \leq f$;*

(ii) *for any fundamental set T in G such that C_1 and C_2 belong to different connected components of $G_{V(G)\setminus(T \cup N(G;T))}$, there exists a walk $\Gamma^{(T)}$ which combine C_1 and C_2 with either $\mathbf{e}_{\Gamma_T^{(T)},o} \leq f$ or $\mathbf{e}_{\Gamma_T^{(T)},e} \leq f$.*

Proof. "If" part is obvious. We consider "Only if" part.

Let i be a regular vertex such that C_1 and C_2 belong to different connected components of $G \setminus \{i\}$. Since $\mathbf{e}_{C_1} + \mathbf{e}_{C_2} + \rho(f) \in S'_G \subset$

$S_G - S_G \cap \mathcal{H}_i$, we can write

$$\mathbf{e}_{C_1} + \mathbf{e}_{C_2} + \rho(f) = \sum_{e \in E(G)} a_e \rho(e) - \sum_{e' \in E(G),\, i \notin e'} b_{e'} \rho(e').$$

Then $\mathbf{e}_{C_1} + \mathbf{e}_{C_2} + \rho(f) + \sum_{e'} b_{e'}\rho(e') = \sum_{e \in E(G)} a_e \rho(e) \in S_G$. Let $f' = (f'_1, \ldots, f'_r)$ be the element of $\mathbb{N}^r$ such that $\rho(f') = \sum_{e'} b_{e'}\rho(e')$. By Proposition 4.3, there exists a walk $\Gamma^{(i)}$ combining C_1 and C_2 with either $\mathbf{e}_{\Gamma^{(i)},o} \leq f + f'$ or $\mathbf{e}_{\Gamma^{(i)},e} \leq f + f'$. Note that $f'_\ell = 0$ if $i \in e_\ell$. Then focusing on the entry ℓ with $i \in e_\ell$, we see that $\mathbf{e}_{\Gamma_i^{(i)},o} \leq f$ or $\mathbf{e}_{\Gamma_i^{(i)},e} \leq f$.

Let T be a fundamental set such that C_1 and C_2 belong to different connected components of $G_{V(G)\setminus(T \cup N(G;T))}$. Since $\mathbf{e}_{C_1} + \mathbf{e}_{C_2} + \rho(f) \in S'_G \subset S_G - S_G \cap \mathcal{H}_T$, we can write

$$\begin{aligned}&\mathbf{e}_{C_1} + \mathbf{e}_{C_2} + \rho(f)\\ &= \sum_{e \in E(G)} a_e \rho(e) - \sum_{e' \in E(G),\, e' \cap (T \cup N(G;T)) = \emptyset \text{ or } e' \cap T \neq \emptyset} b_{e'} \rho(e').\end{aligned}$$

Then $\mathbf{e}_{C_1} + \mathbf{e}_{C_2} + \rho(f) + \sum_{e'} b_{e'}\rho(e') = \sum_{e \in E(G)} a_e \rho(e) \in S_G$. Let $f' = (f'_1, \ldots, f'_r)$ be the element of $\mathbb{N}^r$ such that $\rho(f') = \sum_{e'} b_{e'}\rho(e')$. By Proposition 4.3, there exists a walk $\Gamma^{(T)}$ combining C_1 and C_2 with either $\mathbf{e}_{\Gamma^{(T)},o} \leq f + f'$ or $\mathbf{e}_{\Gamma^{(T)},e} \leq f + f'$. Focus on entries ℓ with $e_\ell \cap T = \emptyset$ and $e' \cap N(G;T) \neq \emptyset$. Since $f'_\ell = 0$, we have $\mathbf{e}_{\Gamma_T^{(T)},o} \leq f$ or $\mathbf{e}_{\Gamma_T^{(T)},e} \leq f$. Q.E.D.

Propositions 4.3 and 4.4 show the gap between the sets S_G and S'_G. In these propositions, we consider the elements which are of the form $e_{C_1} + e_{C_2} + \rho(f)$. Although these look special, in the proof of the characterization of a normal edge ring given by Ohsugi and Hibi [6] (Theorem 3.1), they reduced the discussion to the case of only one exceptional pair. Hence we believe that we can also reduce the discussion to such a case and the above proposition would be the first step for the characterization of a non-normal edge ring which satisfies Serre's condition (S_2).

Theorem 4.1 is an easy consequence of these propositions.

Proof of Theorem 4.1. By Proposition 4.3, we have $\mathbf{e}_{C_1} + \mathbf{e}_{C_2} = \mathbf{e}_{C_1} + \mathbf{e}_{C_2} + \rho(\mathbf{0}) \notin S_G$. On the other hand, we have $\mathbf{e}_{C_1} + \mathbf{e}_{C_2} \in S'_G$ by Proposition 4.4. Q.E.D.

Acknowledgment. The first author is partially supported by JSPS Grant-in-Aid for Young Scientists (B) 26800015. The Second author is partially supported by JSPS Grant-in-Aid for Young Scientists (B) 15K17507.

References

[1] W. Bruns and J. Gubeladze, *Polytopes, rings, and K-theory*, Monographs in Mathematics, Springer, 2009.

[2] W. Bruns and J. Herzog, *Cohen–Macaulay rings*, revised ed., Cambridge University press, 1998.

[3] S. Goto and K. Watanabe, On graded rings II ($\mathbf{Z}^n$-graded rings), *Tokyo J. Math.* **1** (1978), 237–261.

[4] T. Hibi and L. Katthän, Edge rings satisfying Serre's condition (R_1), *Proc. Amer. Math. Soc.* **142** (2014), 2537–2541.

[5] M. Hochster, Rings of invariants of tori, Cohen-Macaulay rings generated by monomials, and polytopes, *Ann. of Math.* (2) **96** (1972), 318–337.

[6] H. Ohsugi and T. Hibi, Normal polytopes arising from finite graphs, *J. Algebra* **207** (1998), 409–426.

[7] H. Ohsugi and T. Hibi, Toric ideals generated by quadratic binomials, *J. Algebra* **218** (1999), 509–527.

[8] N. V. Trung and L. T. Hoa, Affine semigroups and Cohen–Macaulay rings generated by monomials, *Trans. Amer. Math. Soc.* **298** (1986), 145–167.

(Akihiro Higashitani) *Department of Mathematics, Kyoto Sangyo University, Motoyama, Kamigamo, Kita-ku, Kyoto 603-8555, Japan*
E-mail address: `ahigashi@cc.kyoto-su.ac.jp`

(Kyouko Kimura) *Department of Mathematics, Faculty of Science, Shizuoka University, 836 Ohya, Suruga-ku, Shizuoka 422-8529, Japan*
E-mail address: `kimura.kyoko.a@shizuoka.ac.jp`

Advanced Studies in Pure Mathematics 77, 2018
The 50th Anniversary of Gröbner Bases
pp. 129–154

Equivariant Gröbner bases

Christopher J. Hillar, Robert Krone and Anton Leykin

Abstract.

Algorithmic computation in polynomial rings is a classical topic in mathematics. However, little attention has been given to the case of rings with an infinite number of variables until recently when theoretical efforts have made possible the development of effective routines. Ability to compute relies on finite generation up to symmetry for ideals invariant under a large group or monoid action, such as the permutations of the natural numbers. We summarize the current state of theory and applications for equivariant Gröbner bases, develop several algorithms to compute them, showcase our software implementation, and close with several open problems and computational challenges.

§1. Introduction

1.1. History

The theory of polynomial rings is an old and well-studied subject. However, as far as we can tell, a rigorous set of tools for algorithmic computation in such rings was only first developed starting in 1913 [27] by the Russian/Soviet mathematician N. Gjunter. This project culminated with Gjunter's review of the theory in 1941 [28] but went unnoticed until recently [44]. Outside of this rather newly discovered reference, general algorithmic theory in (possibly non-commutative) rings has a long history of independent thinkers. For instance, Bergman's work [5] (see also [6] as influenced by [48]) was inspired by an algorithmic proof of the Poincaré-Birkhoff-Witt theorem.

Received October 5, 2016.
Revised March 31, 2017.
2010 *Mathematics Subject Classification.* 13E05, 13E15, 20B30, 06A07.
Key words and phrases. Invariant ideal, well-quasi-ordering, symmetric group, Gröbner basis, generating sets, infinite dimensional algebra.
Research of AL is supported in part by NSF grant DMS-1151297.
Part of this work took place during the "Free Resolutions, Representations, and Asymptotic Algebra" workshop at the BANFF International Research Station, April, 2016.

Attribution of an algorithmic theory of polynomial rings and ideals is usually given to Buchberger [8], who named the main tools "Gröbner bases" after his Ph.D. advisor. Hironaka [35] used a similar concept called "standard bases" in power series rings to prove his theorem on resolutions of singularities.

The main consequences of these projects are effective procedures for polynomial equation solving over fields such as the complex numbers $\mathbb{C}$. Practical questions of ideal membership or equation feasibility were now answerable (provably) using a finite programmable set of computations.

Since these early efforts, much progress has been made on the mathematical and computational theory of polynomial algebra involving a finite number of indeterminates. In this article, we consider computation in rings with infinite numbers of indeterminates, a topic that is part of a burgeoning new field called "asymptotic algebra". At first, such a concept seems at odds with the non-Noetherianity of even simple ideals such as the maximal ideal:

$$I = \langle x_0, x_1, \ldots \rangle \subset \mathbb{C}[x_0, x_1, \ldots].$$

However, if extra structure is imposed on the class of ideals under consideration, such as a large group action, then it is possible to develop a theory of algorithmic computation. For instance, the ideal I above has a single generator up to the action of permuting indices on polynomials.

The concept of *equivariant Gröbner bases* (EGB) was first used in an application to meta-abelian group theory [10] and later developed into an algorithmic theory [25, 11]. Similar to the story of Gröbner bases (in finitely many variables), the concept was rediscovered several decades later in [3, 4] and applied to solve in a unified manner several problems in algebraic statistics [34]. The theory was also useful in other applications such as those to algebraic chemistry [15] and asymptotic tensor geometry [21] (see [16] for an elegant survey of these techniques).

In the meantime, several works have started to make practical use of this effective computational machinery. As a simple example, consider the following classical theorem in toric algebra that has been a starting point for several investigations into finiteness in asymptotic algebra.

Theorem 1.1. *Let $i > j$ run over natural numbers. The kernel of $\mathbb{C}[y_{ij}] \to \mathbb{C}[x_i]$, $y_{ij} \mapsto x_i x_j$, is generated by the 2×2 minors (not containing diagonal entries) of the symmetric matrix y.*

This result can be proved using equivariant Gröbner bases, as first demonstrated by J. Draisma with the following rather innocuous-looking Input/Output pair on a computer:

```
Input:  { y_{10} - x_1 x_0 }.
Output:  { x_0 x_1 - y_{10}, x_2 y_{10} - x_1 y_{20},
  x_2 y_{10} - x_0 y_{21}, x_1 y_{20} - x_0 y_{21},
  x_0^2 y_{21} - y_{20} y_{10},
  y_{32} y_{10} - y_{30} y_{21},
  y_{31} y_{20} - y_{30} y_{21} }.
```

Specifically, the ideal $\ker(y_{ij} \mapsto x_i x_j)$ for $i > j$ is generated by a finite set of 2×2 minors up to symmetry, which is witnessed by the last two polynomials of the EGB output above.

The first equivariant Gröbner basis computation to prove a new theorem that we are aware of occurs in [7]. This leads us to the next subsection.

1.2. Applications

As a prelude, we start with an application of classical Gröbner bases that deals with a seemingly infinite problem; here, of course, a recurrence helps to "control infinity".

The Fibonacci sequence $F_n = 1, 1, 2, 3, 5, \ldots$ $(n = 1, 2, \ldots)$ is a *strong divisibility sequence*, in that we have $\gcd(F_n, F_m) = F_{\gcd(n,m)}$; in particular, F_m divides F_n if $m \mid n$. This surprising fact was used by Édouard Lucas for Mersenne prime testing. Is there a direct explanation for the integrality of $F_{3n}/F_n \in \mathbb{Z}$? It turns out that there is an identity:

$$(F_{3n} - 5F_n^3 - 3F_n)(F_{3n} - 5F_n^3 + 3F_n) = 0, \tag{1}$$

which explains in an explicit manner strong divisibility for this case. Is it possible to use Gröbner bases to derive this relation? The following Macaulay 2 code does exactly that:

```
i1: R = QQ[z, x, y, t, MonomialOrder => Eliminate 2]
i2: I = ideal(x + y - z, (x*z - y^2)^2 - 1, t - z^3 - y^3 + x^3)
i3: toString groebnerBasis I
o3 = matrix {{25*y^6-10*y^3*t-9*y^2+t^2, z-x-y, ...
```

In this computation, the variables z, x, y, t represent the recurrence values $F_{n+1}, F_{n-1}, F_n, F_{3n}$, respectively. The first generator of I defines the recurrence, the second is Cassini's identity, and the third is Lucas'.

One can check that factoring the first polynomial in the list above gives (1). Bootstrapping with extra equations, we can also discover that:

$$(F_{5n} - 25F_n^5 - 25F_n^3 - 5F_n)(F_{5n} - 25F_n^5 + 25F_n^3 - 5F_n) = 0. \tag{2}$$

In turn, these findings incite conjectures and proofs. For instance, we leave it to the reader to use modular arithmetic to verify from (2) that the integer $\frac{F_{5n}}{5F_n}$ always has unit digit 1 base ten. More generally, the following natural problem arises from this investigation: Given ℓ, find a nonzero polynomial $P(y,t) \in \mathbb{Z}[y,t]$ satisfying an identity of the form $P(F_n, F_{\ell n}) = 0$ (see [32] for more on "polynomial recurrences").

The above is classical. Here, we are interested in problems with not four or even twenty-four indeterminates, but rather an infinite number of them. Take, for instance, the following basic ideal membership question. Let $I \subset \mathbb{C}[x_0, x_1, \ldots]$ be the ideal generated by all permutations acting on the polynomial $f = x_0x_1 - x_1x_2^2 + x_1^2$. Is the following in I?

$$h = x_0x_4^2 + x_0x_1^2 + x_1x_0^2 - 2x_1x_0 + x_0x_3x_4 - x_0x_5^2 - x_0x_3x_5 - 2x_1^2.$$

The difference in this question from classical problems of polynomial algebra is that *a priori* there is no guarantee a particular computation, say, with a truncated polynomial ring $\mathbb{C}[x_0, x_1, \ldots, x_N]$ will do the job. Nonetheless, the following code gives us an answer to our question [39].

```
i1: needsPackage "EquivariantGB"
i2: R = buildERing({symbol x}, {1}, QQ, 6);
i3: h = x_0*x_4^2+x_0*x_1^2+x_1*x_0^2-2*x_1*x_0+x_0*x_3*x_4- ...
i4: G = egb({x_0*x_1 - x_1*x_2^2 + x_1^2}, Algorithm=>Incremental)
          2         2         2              3        2        2      2
o4 = {x x   - 2x   + x x   - 2x x , x   - x x , x x   - x   - x x ,
       1 0       1    1 0      1 0   1     1 0   2 0     1     1 0
                    2            2
       x x  - x x , x   + x x  - x   - x x }
        2 1    2 0   2     2 0    1     1 0
i5: reduce(h, G)
o5 = 0
```

With the equivariant Gröbner basis produced above, we can solve ideal membership problems and much more, just as we can use classical Gröbner bases in numerous applications.

Developing the machinery to solve such questions is more than an intellectual curiosity. Not only can basic facts now be proved by computer such as Theorem 1.1 but also cutting edge conjectures. For example, using [39], it is possible to verify [19, 40] the first nontrivial case of a basic finiteness conjecture for toric ideals [3].

Theorem 1.2 (Proved by computer). *For $n > 1$, let $I_n = \ker(y_{ij} \mapsto x_i^2x_j)$, $1 \le i \neq j \le n$. The invariant chain of toric ideals $I_2 \subset I_3 \subset \cdots$ stabilizes up to the symmetric group. That is, there is some N such that all elements of I_m, $m > N$, are polynomial consequences of relabellings of a finite set of generators of I_N.*

We next provide a summary of applications of equivariant Gröbner bases in rings with infinite numbers of indeterminates.

1.2.1. *Group theory and Chemistry.* The first use of the concept "finite up to symmetry" for polynomial rings that we are aware of is in the group theory work of Cohen in [10]. Independently, it was problems in algebraic chemistry [45], brought to the attention of the authors of [3] by Andreas Dress, that motivated further applications of asymptotic polynomial algebra in chemistry [15].

1.2.2. *Toric Algebra and Algebraic Statistics.* A major inspiration for asymptotic algebra arises from studying chains of toric ideals, many of which arise naturally in algebraic statistics. The series of works [31, 33, 19, 38, 40] have developed fundamental finiteness properties of these structures, but many questions remain open, as we outline in Section 5.

In this regard, one of the major motivations for equivariant Gröbner bases and infinite symbolic algebra are their application to the problem of sampling from conditional distributions by algebraic methods [14]. At its essence, the strategy is to find a collection of elementary moves through model space that preserves the sufficient statistics of the data. The idea then is to consider growing families of model classes and show that, up to obvious symmetries, only a finite set of moves suffices for all infinite numbers of models (e.g., [2, 47, 36, 22, 15, 7, 20, 34, 17]). Typically, these moves correspond to elements of a Gröbner basis or at least a generating set for some ideal.

1.2.3. *Invariants.* Recently, Nagel and Römer [42] have introduced Hilbert series for Noetherian infinite-dimensional rings. Their original theoretical treatment that leads to a proof of rationality of the Hilbert series, in principle, also leads to an effective procedure to compute the series. For an ideal generated up to symmetry by one monomial, such computation was carried out in [30]. An alternative approach of [41] computes the Hilbert series given an equivariant Gröbner basis as the generating function counting words in a regular language.

1.3. Finiteness up to symmetry in general.

Although the equivariant Gröbner bases described in this article may not directly apply, finiteness up to symmetry plays the central role in the following results (this list is by no means exhaustive).

It appears in homological stability [43, 9], the moduli space of n points in a line [37], geometry as the positivity of the embedding line bundle grows [24], syzygies of Segre embeddings [49], Betti tables as their length goes to infinity [23], tensor geometry [21, 17], and limiting Grassmannians [18]. Gröbner methods have also been used to understand representations of combinatorial categories [46].

1.4. Goals and structure

Since their introduction, Gröbner bases techniques have improved immensely. We believe that in this new setting of infinite-dimensional polynomial algebras, which demands far more computational power, algorithmic development is at the beginning of a similar road, with similar advances ahead. Our aims here are to outline the current state of effective computation in this setting and to provide a background for researchers to start tackling problems in this exciting domain.

After some preliminaries in Section 2, we quickly move on to describing equivariant Gröbner bases algorithms in Section 3. Section 4 goes on to explain a modern signature-based approach and a strategy inspired by it for an equivariant Buchberger's algorithm. The final Section 5 outlines computational and theoretical challenges for future exploration.

§2. Preliminaries

Let R be a commutative K-algebra equipped with a left action of monoid Π (a Π-algebra structure). We mainly consider the case where R has the structure of a monoid algebra; that is, for some abelian monoid M, the elements of R consist of formal sums of elements of M with coefficients in the field K. An example of such a monoid algebra is the polynomial ring $R = K[X]$ with variables from the set X. In this case, M is the free abelian monoid generated by X, which we will denote by $[X]$. To make our notation consistent with the polynomial case, we will denote the monoid algebra of M over K by KM even though this is not standard (often it is written as $K[M]$, but this creates ambiguity with polynomial rings). We also generally refer to elements of M as "monomials" in analogy to the polynomial case. Additionally, we will assume that Π acts on monoid algebra R through a Π-action on M by monoid homomorphisms.

Our particular focus in this paper is when Π is an infinite symmetric group $\mathfrak{S}_\infty$ or certain related monoids. For our purposes, we take $\mathfrak{S}_\infty$ to be the group of all finite permutations of $\mathbb{N}$ (i.e., permutations that fix all but a finite number of elements).

Example 2.1. *Let*[1] $R = K[x_1, x_2, x_3, \ldots]$ *with* $\mathfrak{S}_\infty$ *acting on* R *by permuting the variables, so that* $\sigma x_i = x_{\sigma(i)}$.

Definition 2.2. *An ideal* $I \subseteq R$ *is a* Π-invariant ideal *if* $\sigma I \subseteq I$ *for all* $\sigma \in \Pi$.

[1] Here the subscripts of the generators start with 1. Note that we start with subscript 0 in the introduction, software examples, and several other places in this article.

The ring R is both an R-algebra and a Π-algebra, and there is a ring $R * \Pi$ which captures both of these actions, and which will be referred to as the *twisted monoid ring* of Π with coefficients in R. The elements of $R * \Pi$ are of the form $\sum_{\sigma \in \Pi} f_\sigma \cdot \sigma$ with each $f_\sigma \in R$ and only a finite number nonzero. The additive structure is the same as the usual monoid ring, but multiplication is "twisted":

$$(f \cdot \sigma)(g \cdot \tau) = f\sigma(g) \cdot \sigma\tau,$$

where $\sigma(g)$ denotes the element of R obtained by acting on g by σ.

The ring R is a $R*\Pi$-module, and the definition of Π-invariant ideals can be restated as the collection of $R * \Pi$-submodules of R.

When $R = KM$ with Π acting on M, we can define a monoid $M * \Pi$ whose elements are pairs in $M \times \Pi$ with monoid operation:

$$(m, \sigma)(n, \tau) = (m\sigma(n), \sigma\tau).$$

There is a left action of $M * \Pi$ on M, and the elements of $M * \Pi$ are the "monomials" of $R * \Pi$.

Definition 2.3. *A Π-invariant ideal $I \subseteq R$ is* Π-finitely generated *if there is a finite set $F \subseteq I$ such that the Π-orbits of the elements of F generate I. The ring R is called* Π-Noetherian *if every Π-invariant ideal in R is Π-finitely generated.*

If a Π-invariant ideal I is generated by the Π-orbits of a set F, we shall write:

$$I = \langle F \rangle_\Pi.$$

Such a set F generates I as an $R * \Pi$-module.

We can also say that monoid M with Π-action is Π-finitely generated if it is generated by the Π-orbits of a finite number of elements. Then, $R = KM$ is Π-finitely generated as a K-algebra.

Example 2.4. *Continuing the example of $R = K[x_1, x_2, x_3, \ldots]$ with $\mathfrak{S}_\infty$ action, the ideal $\mathfrak{m} = \langle x_1, x_2, x_3, \ldots \rangle$ is a $\mathfrak{S}_\infty$-invariant ideal. Moreover, it is $\mathfrak{S}_\infty$-finitely generated because $\mathfrak{m} = \langle x_1 \rangle_{\mathfrak{S}_\infty}$. Also, R is a $\mathfrak{S}_\infty$-finitely generated K-algebra with generator x_1.*

Definition 2.5. *Let R be a $\mathfrak{S}_\infty$-algebra. For $f \in R$, the* width *of f is the smallest integer n such that for every $\sigma \in \mathfrak{S}_\infty$ that fixes $\{1, \ldots, n\}$, σ also fixes f. The width of f is denoted $w(f)$. If no such integer n exists, then $w(f) := \infty$. For a set $F \subseteq R$, its width is $w(F) := \max_{f \in F}\{w(f)\}$.*

If every element of R has finite width, we say that R satisfies the *finite width condition.* This is primarily the situation we want to address in this paper, and so we shall assume from here forward that all rings with $\mathfrak{S}_\infty$-action satisfy the finite width condition unless stated otherwise. For a $\mathfrak{S}_\infty$-invariant ideal $I \subseteq R$ and an integer n, we can define the nth truncation of I as:

$$I_n := \{f \in I \mid w(f) \leq n\}.$$

The set I_n is naturally a $\mathfrak{S}_n$-invariant ideal of R_n. If R satisfies the finite width condition, then I is the union of all its truncations. Moreover, if I is $\mathfrak{S}_\infty$-finitely generated, there is sufficiently large $n \in \mathbb{N}$ such that $I = \langle I_n \rangle_{\mathfrak{S}_\infty}$.

The definition of width also applies to $\Pi = \mathrm{Inc}(\mathbb{N})$, the monoid of strictly increasing functions, which is introduced below.

Definition 2.6. *Given $R = KM$ with Π acting on M, there is a natural partial order $|_\Pi$ on M called the* Π-divisibility partial order *defined by $a|_\Pi b$ if there exists $\sigma \in \Pi$ such that σa divides b. Equivalently, $a|_\Pi b$ iff $b \in \langle a \rangle_\Pi$.*

Recall that a monomial order on $R = KM$ is a total order $\leq$ on M that is a well-order and that respects multiplication (i.e., if $a \leq b$ then $ac \leq bc$ for all $c \in M$).

Definition 2.7. *A monomial order $\leq$ on $R = KM$ is said to* respect *Π if whenever $a \leq b$, then $\sigma a \leq \sigma b$ for all $\sigma \in \Pi$.*

Therefore, order $\leq$ is a Π-respecting monomial order on R if $\leq$ is a total well-order on M that respects the action of $M * \Pi$. We now have all the tools to describe the Π-equivariant version of Gröbner bases.

Definition 2.8. *Let $R = KM$ be a monoid ring with Π action on M, and let $\leq$ be a Π-respecting monomial order. Given a Π-invariant ideal $I \subseteq R$, a* Π-equivariant Gröbner basis *of I is a set $G \subseteq I$ such that the Π orbits of G form a Gröbner basis of I:*

$$\langle \mathrm{in}_\leq \Pi G \rangle = \mathrm{in}_\leq I.$$

We require $\leq$ to be a Π-respecting order because it is equivalent to the condition that:

$$\mathrm{in}_\leq \sigma f = \sigma \, \mathrm{in}_\leq f,$$

for all $f \in R$ and $\sigma \in \Pi$. Therefore, with such an order, we have:

$$\langle \mathrm{in}_\leq G \rangle_\Pi = \langle \mathrm{in}_\leq \Pi G \rangle = \mathrm{in}_\leq I.$$

This also implies that $\text{in}_{\leq} I$ is a Π-invariant ideal. Note that since Π orbits of G are a Gröbner basis of I, we naturally have $\langle G \rangle_\Pi = I$.

Proposition 2.9 (Remark 2.1 of [7]). *Let Π be a group which acts nontrivially on M. Then KM has no Π-respecting monomial orders.*

Proof. Suppose that $\leq$ is a Π-respecting order and choose $\sigma \in \Pi$ and $m \in M$ such that $m \neq \sigma m$. If $m > \sigma m$, then $\sigma^n m > \sigma^{n+1} m$ for all n, and thus it follows that:

$$m > \sigma m > \sigma^2 m > \cdots$$

is an infinite descending chain of monomials, contradicting the fact that $\leq$ is a well-order. If $m < \sigma m$, then $m > \sigma^{-1} m > \sigma^{-2} m > \cdots$ is an infinite descending chain. Q.E.D.

In particular, this means that R with nontrivial $\mathfrak{S}_\infty$ action has no $\mathfrak{S}_\infty$-respecting monomial orders. To deal with this problem, a related monoid is introduced to replace $\mathfrak{S}_\infty$ that allows for monomial orders but is somehow large enough compared to $\mathfrak{S}_\infty$ not to break properties like finite generation.

Define the *monoid of strictly increasing functions* as:

$$\text{Inc}(\mathbb{N}) := \{\rho : \mathbb{N} \to \mathbb{N} \mid \text{ for all } a < b, \rho(a) < \rho(b)\}.$$

For any $\mathfrak{S}_\infty$-algebra R with the finite width property, there is a natural action of $\text{Inc}(\mathbb{N})$ on R as follows. Fixing $f \in R$, for any $\sigma \in \mathfrak{S}_\infty$ the value of σf depends only on the restriction $\sigma|_{[w(f)]}$ considering σ as a function $\mathbb{N} \to \mathbb{N}$. For any $\rho \in \text{Inc}(\mathbb{N})$, there exists $\sigma \in \mathfrak{S}_\infty$ such that $\sigma|_{[w(f)]} = \rho|_{[w(f)]}$ and defines $\rho f = \sigma f$. It can be checked that this gives a well-defined action of $\text{Inc}(\mathbb{N})$ on R.

It immediately follows from this definition that $\text{Inc}(\mathbb{N}) f \subseteq \mathfrak{S}_\infty f$. Despite the fact that $\text{Inc}(\mathbb{N})$ is not a submonoid of $\mathfrak{S}_\infty$, it behaves like one in terms of its action on R. An injective map $\sigma|_{[w(f)]} : [w(f)] \to \mathbb{N}$ can always be factored into $\rho' \circ \tau$ with $\tau \in \mathfrak{S}_{w(f)}$ and $\rho' : [w(f)] \to \mathbb{N}$ a strictly increasing function. The map ρ' can be extended to some $\rho \in \text{Inc}(\mathbb{N})$, and then $\sigma f = \rho(\tau f)$. Thus, we have:

$$\mathfrak{S}_\infty f = \bigcup_{\tau \in \mathfrak{S}_{w(f)}} \text{Inc}(\mathbb{N})(\tau f).$$

The fact that the $\mathfrak{S}_\infty$-orbit of any f is a finite union of $\text{Inc}(\mathbb{N})$-orbits implies the following statements.

Proposition 2.10. *Let R be a $\mathfrak{S}_\infty$-algebra satisfying the finite width condition, and let $I \subseteq R$ be a $\mathfrak{S}_\infty$-invariant ideal.*

- *I is* Inc($\mathbb{N}$)*-invariant.*
- *I is $\mathfrak{S}_\infty$-finitely generated if and only if I is* Inc($\mathbb{N}$)*-finitely generated.*
- *If R is* Inc($\mathbb{N}$)*-Noetherian then R is $\mathfrak{S}_\infty$-Noetherian.*

Remark 2.11. *For practical purposes, we may replace* Inc($\mathbb{N}$) *with* $\mathrm{Inc}^\tau(\mathbb{N})$*, the monoid of all increasing maps $\pi : \mathbb{N} \to \mathbb{N}$ such that* $\mathrm{im}(\pi)$ *has a finite complement. For $i \in \mathbb{N}$, let τ_i denote the element of* Π *defined by*

$$\tau_i(j) = \begin{cases} j & \text{if } j < i, \text{ and} \\ j+1 & \text{if } j \geq i. \end{cases}$$

The maps τ_i generate Π*, and they satisfy relations*

$$\tau_{j+1}\tau_i = \tau_i\tau_j \text{ if } j \geq i.$$

This gives a presentation of Π*, and any element of* Π *has a unique expression of the form $\tau_{i_1} \cdots \tau_{i_d}$ with $i_1 \leq \cdots \leq i_d$.*

When computing Gröbner bases of $\mathfrak{S}_\infty$-invariant ideals, we will work with the Inc($\mathbb{N}$) action instead. If G is an Inc($\mathbb{N}$)-equivariant Gröbner basis for $\mathfrak{S}_\infty$-invariant ideal I, then the $\mathfrak{S}_\infty$-orbits of G also form a Gröbner basis of I. Generally, the rings we are interested in will have Inc($\mathbb{N}$)-respecting monomial orders.

Example 2.12. *Let $R = K[x_1, x_2, \ldots]$ with* Inc($\mathbb{N}$)*-action defined by $\rho \cdot x_i = x_{\rho(i)}$. The lexicographic order $\leq$ on the monomials of R with $x_1 < x_2 < x_3 < \cdots$ is a* Inc($\mathbb{N}$)*-respecting monomial order. This is the only possible lexicographic order on R that respects* Inc($\mathbb{N}$)*. There are also a graded lexicographic and a graded reverse lexicographic order on R that respect* Inc($\mathbb{N}$)*. There is no* Inc($\mathbb{N}$)*-respecting monomial order on R that is defined by a single weight vector in $\mathbb{R}^{\mathbb{N}}$.*

It is an open question to characterize all possible Inc($\mathbb{N}$)-respecting monomial orders on a given ring KM with Inc($\mathbb{N}$) action. We can make the following statement about such orders.

Proposition 2.13. *If $\leq$ is a* Π*-respecting monomial order on KM, then $\leq$ refines the* Π*-divisibility quasi-order $|_\Pi$.*

Proof. Suppose a and b are monomials with $a|_\Pi b$, so there is some pair $\sigma \in \Pi$, $c \in M$ such that $c\sigma a = b$. From the proof of Proposition 2.9, we see that $a \leq \sigma a$. Since $1 \leq c$ and $\leq$ respects multiplication, it follows that $\sigma a \leq c\sigma a = b$. Q.E.D.

One implication of this proposition is that if KM has a Π-respecting monomial order, then the Π-divisibility quasi-order must be a partial order (i.e., it has the *anti-symmetry* property: if $a \geq b$ and $a \leq b$ then $a = b$). If anti-symmetry fails for $|_\Pi$, it will also fail for any refinement.

If R is Π-Noetherian with a Π-respecting monomial order, then any Π-invariant ideal $I \subseteq R$ will have a finite Π-equivariant Gröbner basis. This follows from the fact that $\text{in}_\leq I$ is Π-finitely generated. We recount two previous results that give examples of $\text{Inc}(\mathbb{N})$-Noetherian rings, and they will be directly relevant to the results of this paper.

Theorem 2.14 (Theorem 1.1 of [34]). *Let $X = \{x_{ij} \mid i \in [k], j \in \mathbb{N}\}$, and let $\mathfrak{S}_\infty$ act on $[X]$ by permuting the second index: $\sigma x_{ij} = x_{i\sigma(j)}$ for $\sigma \in \mathfrak{S}_\infty$. Then, $K[X]$ is* $\text{Inc}(\mathbb{N})$*-Noetherian.*

Theorem 2.15 (Theorem 1.1 of [19]). *Let $K[Y]$ be a $\mathfrak{S}_\infty$-algebra with $\mathfrak{S}_\infty$ action on variable set Y. Suppose Y has a finite number of $\mathfrak{S}_\infty$-orbits, and $K[Y]$ satisfies the finite width condition. For $K[X]$ defined as in Theorem 2.14, let ϕ be a monomial map:*

$$\phi : K[Y] \to K[X].$$

Then, the following hold:

- $\ker \phi$ *is* $\text{Inc}(\mathbb{N})$*-finitely generated,*
- $\text{im}\, \phi$ *is* $\text{Inc}(\mathbb{N})$*-Noetherian.*

The conditions on the ring $K[Y]$ in Theorem 2.15 are quite general although [34] proves that such rings are generally not $\mathfrak{S}_\infty$-Noetherian. They give the example of $K[Y]$ where $Y = \{y_{ij} \mid i, j \in \mathbb{N}\}$ with $\sigma y_{ij} = y_{\sigma(i)\sigma(j)}$ for $\sigma \in \mathfrak{S}_\infty$ and prove that Noetherianity fails.

When R is not Π-Noetherian, we do not know in general if a Π-finitely generated ideal $I \subseteq R$ has a finite Π-equivariant Gröbner basis, or if so, for which monomial orders. However, [40] shows that the $\mathfrak{S}_\infty$-invariant toric ideal $\ker \phi$ as in Theorem 2.15 does have finite $\text{Inc}(\mathbb{N})$-equivariant Gröbner bases for specifically chosen monomial orders. This allows for an algorithm to compute a Gröbner basis of $\ker \phi$ given ϕ.

§3. Equivariant Buchberger algorithm

3.1. Description of the algorithm

First proposed in [3] and formalized in [7], the classical Buchberger's algorithm [8] may be adapted to the equivariant setting in a straightforward way.

Let $R = KM$ with Π acting on M, and let $\leq$ be a Π-respecting monomial order. For $f, g \in R$, we say that g Π-*reduces* f if $\text{in}_\leq g|_\Pi \,\text{in}_\leq f$

and the reduction is $f - \frac{\mathrm{LC}(f)}{\mathrm{LC}(g)} mg$ where $m \in M * \Pi$ is such that $\mathrm{in}_\leq f = m\,\mathrm{in}_\leq g$ (and $\mathrm{LC}(f)$ denotes the lead coefficient of f). For $G \subseteq R$, a Π-*normal form* of f with respect to G, denoted $\mathrm{NF}_{\Pi G}(f)$, is the result of repeated Π-reductions of f by elements of G until no more reductions are possible. Equivalently, $\mathrm{NF}_{\Pi G}(f)$ is a normal form of f with respect to ΠG.

The equivariant Buchberger's algorithm is described below, which departs from the conventional Buchberger's algorithm only at the step of adding new S-pairs to the list S. The necessity and extent of this departure becomes clear with the definition of $O_{f,g}$ and the *finite S-pair condition* (Definition 3.3) that are given after the description of the algorithm.

Algorithm 3.1 (Brouwer–Draisma [7]). $G = \mathrm{Buchberger}(F)$

Require: *F is a finite set of elements in $R = KM$ with Π acting on M and satisfying the finite S-pair condition.*

Ensure: *G is Π-equivariant Gröbner basis of $\langle F \rangle_\Pi$.*

$G \leftarrow F$
$S \leftarrow \bigcup_{f,g\in G} O_{f,g}$
while $S \neq \emptyset$ ***do***
  pick $(h_1, h_2) \in S$
  $S \leftarrow S \setminus \{(h_1, h_2)\}$
  $h \leftarrow \mathrm{NF}_{\Pi G}(h_1 - \frac{\mathrm{LC}(h_1)}{\mathrm{LC}(h_2)} h_2)$
  if $h \neq 0$ ***then***
    $G \leftarrow G \cup \{h\}$
    $S \leftarrow S \cup \left(\bigcup_{g\in G} O_{g,h}\right)$
  end if
end while

Given $f, g \in R$, define:

$$\mathcal{S}_{f,g} := \{(m_1 f, m_2 g) \mid m_1, m_2 \in M * \Pi \text{ such that } \mathrm{in}_\leq m_1 f = \mathrm{in}_\leq m_2 g\}.$$

This collection is closed under the diagonal action of $M * \Pi$, making $\mathcal{S}_{f,g}$ a $M * \Pi$-module.

Definition 3.2. *A set $G \subseteq R$ satisfies the* equivariant Buchberger criterion *if for all $(h_1, h_2) \in \bigcup_{f,g\in G} \mathcal{S}_{f,g}$:*

$$\mathrm{NF}_{\Pi G}(h_1 - \frac{\mathrm{LC}(h_1)}{\mathrm{LC}(h_2)} h_2) = 0.$$

The set G is a Π-equivariant Gröbner basis of $\langle G \rangle_\Pi$ if and only if it satisfies the equivariant Buchberger criterion. The proof of this fact follows by applying the usual Buchberger criterion to the set ΠG (see Theorem 2.5 of [7]).

For each pair $f, g \in G$, we need not check the criterion on every pair in the infinite set $\mathcal{S}_{f,g}$. It is instead sufficient to check on a $M * \Pi$ generating set of $\mathcal{S}_{f,g}$, which we denote $O_{f,g}$. Still, in general, it may be that no finite generating set of $\mathcal{S}_{f,g}$ exists, in which case we cannot apply the algorithm in finite time.

Definition 3.3. *A* Π*-algebra* $R = KM$ *has the* finite S-pair condition *if for any* $f, g \in R$*, the set* $\mathcal{S}_{f,g}$ *is finitely generated as a* $M * \Pi$*-module. In* [7]*, this condition is referred to as "EGB4."*

When Π is trivial and R is a polynomial ring (the setting of the conventional Buchberger's algorithm), $\mathcal{S}_{f,g}$ is generated by a single pair $(m_1 f,\ m_2 g)$ where:

$$m_1 = \operatorname{lcm}(\mathrm{in}_\leq f, \mathrm{in}_\leq g)/\,\mathrm{in}_\leq(f), \quad m_2 = \operatorname{lcm}(\mathrm{in}_\leq f, \mathrm{in}_\leq g)/\,\mathrm{in}_\leq(g)\,.$$

This generator is typically referred to as the *S-pair* of f, g. Therefore, R in this case satisfies the finite S-pair condition, and the equivariant Buchberger's algorithm specializes to the conventional Buchberger's algorithm.

Proposition 3.4. *If* R *is a polynomial ring* $R = K[Y]$ *with* $\mathrm{Inc}(\mathbb{N})$*-action on* $[Y]$ *satisfying the finite width condition, then* R *has the finite S-pair condition.*

Proof. Fix $f, g \in R$. Since R is a polynomial ring, for fixed $\sigma_1, \sigma_2 \in \mathrm{Inc}(\mathbb{N})$, all elements of $\mathcal{S}_{f,g}$ of the form $(m_1\sigma_1 f, m_2\sigma_2 g)$ with $m_1, m_2 \in M$ are monomial multiplies of the usual S-pair of $\sigma_1 f, \sigma_2 g$:

$$\left(\frac{m}{\mathrm{in}_\leq \sigma_1 f}\sigma_1 f,\ \frac{m}{\mathrm{in}_\leq \sigma_2 g}\sigma_2 g\right),$$

where $m = \operatorname{lcm}(\mathrm{in}_\leq \sigma_1 f, \mathrm{in}_\leq \sigma_2 g)$.

Any $f, g \in R$ have finite width so that $\sigma_1 f$ depends only on $\sigma_1|_{[w(f)]}$, and similarly for $\sigma_2 g$. In fact, we can always factor the pair as:

$$(\sigma_1 f, \sigma_2 g) = \rho(\sigma_1' f, \sigma_2' g),$$

for some $\rho \in \mathrm{Inc}(\mathbb{N})$, while $\sigma_1' : [w(f)] \to [w(f)+w(g)]$ and $\sigma_2' : [w(g)] \to [w(f)+w(g)]$ are strictly increasing functions. Here σ_1' and σ_2' are chosen to "interlace" the variables of f and g in the same way as σ_1, σ_2. (To consider σ_1', σ_2' as elements of $\mathrm{Inc}(\mathbb{N})$, take any choice of extensions to maps on $\mathbb{N}$.)

Then, $\mathcal{S}_{f,g}$ is generated by the finite set of pairs of the form:

$$\left(\frac{m}{\mathrm{in}_\leq \sigma_1' f} \sigma_1' f,\ \frac{m}{\mathrm{in}_\leq \sigma_2' g} \sigma_2' g \right),$$

with $\sigma_1' : [w(f)] \to [w(f) + w(g)]$ and $\sigma_2' : [w(g)] \to [w(f) + w(g)]$ where $m = \mathrm{lcm}(\mathrm{in}_\leq \sigma_1' f, \mathrm{in}_\leq \sigma_2' g)$. Q.E.D.

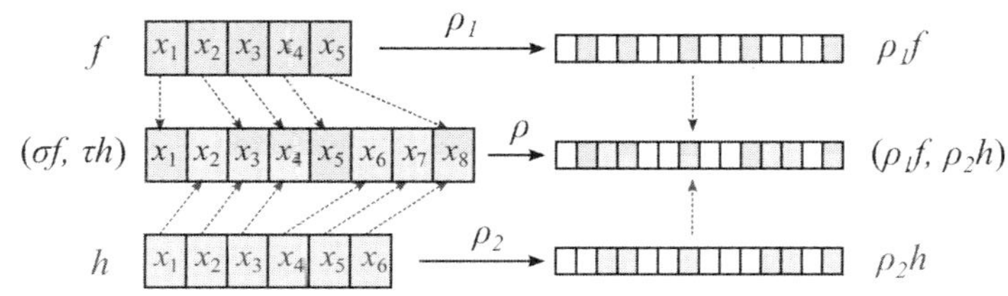

Fig. 1. For f and g of width 5 and 6 respectively, any S-pair, $(\rho_1 f, \rho_2 g)$, is in the orbit of some S-pair obtained from an "interlacing" of [5] and [6], $(\sigma f, \tau g)$.

We note that Algorithm 3.1 is guaranteed to terminate when R is Π-Noetherian. Let $G_0, G_1, \ldots$ be the value of G at each step. The initial ideals of these sets form a strictly increasing chain of Π-invariant monomial ideals:

$$\langle \mathrm{in}_\leq G_0 \rangle_\Pi \subsetneq \langle \mathrm{in}_\leq G_1 \rangle_\Pi \subsetneq \cdots,$$

which must terminate. However, without Noetherianity, we offer no termination guarantee of the algorithm as stated above, even when a finite equivariant Gröbner basis for the ideal exists. Algorithm 3.5 is a modification of the algorithm which repairs this when $\Pi = \mathrm{Inc}(\mathbb{N})$, a finite equivariant Gröbner basis exists, and the truncated rings R_n are Noetherian.

3.2. Termination of Inc($\mathbb{N}$)-equivariant Buchberger

Let $R = KM$ with $\mathrm{Inc}(\mathbb{N})$ action on M, with R satisfying the finite width and finite S-pair conditions, and with each truncation R_n a Noetherian ring. Let $I \subseteq R$ be a $\mathrm{Inc}(\mathbb{N})$-invariant ideal which is $\mathrm{Inc}(\mathbb{N})$-generated by finite set F, and, moreover, has finite $\mathrm{Inc}(\mathbb{N})$-equivariant Gröbner basis G. Define the *generator truncation* of I to be $\tilde{I}_{F,n} := \langle \mathrm{Inc}(\mathbb{N})F \cap R_n \rangle \cap R_n$. Note that $\tilde{I}_{F,n} \subseteq I_n$, but, in general, equality does not hold. For $f \in I$, define $w_F(f)$ to be the minimum value of n for which $f \in \tilde{I}_{F,n}$.

The truncated EGB algorithm takes a finite generating set F as its input. For each successive $n \geq w(F)$, it computes a set G_n such that $\mathrm{Inc}(\mathbb{N})G_n \cap R_n$ is a Gröbner basis for $\tilde{I}_{F,n}$. Then it checks if G_n is a $\mathrm{Inc}(\mathbb{N})$-equivariant Gröbner basis of I using the equivariant Buchberger criterion (Definition 3.2), and if so returns G_n.

Algorithm 3.5. $G = \mathrm{TruncatedEGB}(F)$

Require: *F is a finite set of elements in $R = KM$ with $\mathrm{Inc}(\mathbb{N})$ acting on M, R satisfies the finite width and finite S-pair conditions, and each R_n is Noetherian.*

Ensure: *G is a $\mathrm{Inc}(\mathbb{N})$-equivariant Gröbner basis of $I := \langle F \rangle_{\mathrm{Inc}(\mathbb{N})}$.*

```
G ← F
n ← w(F)
while G not a Inc(ℕ)-equivariant Gröbner basis of I do
    G ← Gröbner basis of Ĩ_{F,n}
    n ← n + 1
end while
```

Proof of termination (supposing a finite EGB for I). For each n, let G_n denote the value of G after that step. Computing G_n is a finite process since it takes place in R_n, which is Noetherian. G_n is a finite set, and so it has a finite number of S-pairs to be checked. Therefore, testing whether G_n is a $\mathrm{Inc}(\mathbb{N})$-equivariant Gröbner basis is finite.

It remains to be proved that G_n is a $\mathrm{Inc}(\mathbb{N})$-equivariant Gröbner basis for some value of n. If H is a $\mathrm{Inc}(\mathbb{N})$-equivariant Gröbner basis of I, for any $h \in H$ we have $h \in \tilde{I}_{F,n}$ for all $n \geq w_F(h)$, so $\mathrm{in}_{\leq}(h) \in \mathrm{in}_{\leq}(\tilde{I}_{F,n})$. Thus, there is some $g \in G_n$ with $\mathrm{in}_{\leq}(g)|_{\mathrm{Inc}(\mathbb{N})}\, \mathrm{in}_{\leq}(h)$. For $n = \max_{h\in H} w_F(h)$, the initial ideal $\langle \mathrm{in}_{\leq}(G_n)\rangle_{\mathrm{Inc}(\mathbb{N})}$ then necessarily contains $\langle \mathrm{in}_{\leq}(H)\rangle_{\mathrm{Inc}(\mathbb{N})}$, and so G_n is a $\mathrm{Inc}(\mathbb{N})$-equivariant Gröbner basis of I. Q.E.D.

In practice, G_n can be computed either using a traditional Gröbner basis algorithm on input $\mathrm{Inc}(\mathbb{N})F \cap R_n$ or using an equivariant Buchberger's algorithm on input F with the following two caveats:

- Consider only S-pairs $(m_1 f, m_2 g)$ with $m_1 f$ and $m_2 g$ both having width $\leq n$,
- Perform only reductions such that the outcome has width $\leq n$.

Moreover, we do not need to restart the algorithm from scratch for each n: $G_{n-1} \cup F$ can be used as the input for the nth step instead of F.

Suppose R has the form $K[Y]$ and each $R_n = K[Y_n]$ for some $Y_n \subseteq Y$. If $\leq$ is a width order (a monomial order such that $w(a) < w(b)$ implies $a < b$), the second condition is satisfied automatically since reductions cannot increase the width. Therefore, the normal form of a given S-pair does not depend on n and only needs to be computed once. As a result, we can use Algorithm 3.1, queuing S-pairs by width so that the smallest width S-pairs are considered first. The algorithm terminates once the queue is empty. A separate check for whether G_n is a Inc($\mathbb{N}$)-equivariant Gröbner basis for I is not needed since this is equivalent to reducing all S-pairs in the queue.

3.3. Macaulay2 package

We have implemented several strategies for computing equivariant Gröbner bases in a package:

`EquivariantGB` (see `http://rckr.one/EquivariantGB.html`)

for Macaulay2 [29], a software system for computational algebraic geometry and commutative algebra.

The main command of our Macaulay2 package, `egb`, has an optional argument that determines how the computation is done:

- `egb(...,Algorithm=>Buchberger)` uses Algorithm 3.1,
- `egb(...,Algorithm=>Incremental)` uses Algorithm 3.5,
- `egb(...,Algorithm=>Signature)` uses the approach in §4.

Remark 3.6. *With the assumptions of Theorem 2.15, one can operate with truncated toric ideals and use specialized lattice based Gröbner bases algorithms to improve performance. Our Macaulay2 command for that,* `egbToric`, *outsources heavy computation to* `4ti2` (*see* [1]), *a special software package for algebraic, geometric, and combinatorial problems on linear spaces.*

§4. A signature-based approach

In this section, we describe an approach to computing equivariant Gröbner bases that utilizes the information stored in *signatures*.

Signature-based algorithms for computing Gröbner bases in the most common (finite-dimensional, commutative) setting acquired popularity due to Faugere's F5 (see a short description in §4 of Chapter 10 of the new edition of Cox, Little, and O'Shea [12]). We give a description of one of the signature-based approaches due to Gao *et al.* in [26], followed by its modification needed to compute equivariant Gröbner bases.

4.1. Strong Gröbner basis

Let $I = \langle F \rangle \subset R = K[x_1, \ldots, x_n]$, where $|F| = r \in \mathbb{N}$.

A subset G of:

$$P = \left\{ (s, f) \in R^r \times R \mid f = s \cdot F = \sum_{i=1}^{r} s_i F_i \right\}$$

is called a *strong Gröbner basis* if every nonzero pair is *top-reducible* by some pair in G.

A pair (s_f, f) is *top-reducible* by (s_g, g) if $\operatorname{LM} g | \operatorname{LM} f$ and for some a with $\operatorname{LM} f = a \operatorname{LM} g$, we have $a \operatorname{LM} s_g \leq \operatorname{LM} s_f$. If the reduction:

$$(s_{f'}, f') := (s_f - a s_g, f - ag)$$

has $\operatorname{LM} s_{f'} = \operatorname{LM} s_f$, then it is *regular top-reducible.*

If G is a strong Gröbner basis, then by Proposition 2.2 of [26]:

(1) $\{f \mid (s, f) \in G\}$ is a Gröbner basis of I, and
(2) $\{s \mid (s, 0) \in G\}$ is a Gröbner basis of the module of syzygies $\operatorname{Syz}(F) \subset R^r$.

Take two pairs $p_f = (s_f, f)$ and $p_g = (s_g, g)$. For monomials a and b such that $a \operatorname{LM} f = b \operatorname{LM} g \in \operatorname{lcm}(\operatorname{LM} f, \operatorname{LM} g)$, form a *J-pair* by taking the "larger side" of the corresponding S-polynomial; e.g., if $a \operatorname{LM} s_f \geq b \operatorname{LM} s_g$, then the J-pair is $(a s_f, a f)$.

We denote the set of all J-pairs of p_f and p_g as J_{p_f, p_g}. Note that $\operatorname{lcm}(\operatorname{LM} f, \operatorname{LM} g)$, the *set* of lowest common multiples, has one element in our current setting as does J_{p_f, p_g}.

Example 4.1. *If we have:*

$$p_f = (e_1 + \cdots, x_1^2 x_2 + \cdots), \; p_g = (x_2 e_1 + \cdots, x_1 x_2^2 + \cdots),$$

then, since $x_2 \operatorname{LM} s_f < x_1 \operatorname{LM} s_g$, *we have:*

$$J_{p_g, p_f} = \{x_1 p_g\} = \{(x_1 x_2 e_1 + \cdots, x_1^2 x_2^2 + \cdots)\}.$$

A pair (s_f, f) is *covered* by (s_g, g) if $\operatorname{LM} g | \operatorname{LM} f$ and for some a such that $\operatorname{LM} f = a \operatorname{LM} g$, we have $a \operatorname{LM} s_g < \operatorname{LM} s_f$.

Algorithm 4.2. StrongBuchberger(F)

Require: $F \subset R$.

Ensure: $G \cup S$ *is a strong Gröbner basis for* F.

$G \leftarrow \emptyset$, $S \leftarrow \emptyset$
$J \leftarrow \{(e_i, F_i) : i \in r = |F|\} \subset R^r \times R$
while $J \neq \emptyset$ **do**
  pick $p_f = (s_f, f) \in J$; $J \leftarrow J \setminus \{p_f\}$
  $p_h = (s_h, h) \leftarrow$ regular top-reduction *of* (s_f, f) *with respect to* G

if $h \neq 0$ ***then***
  $G \leftarrow G \cup \{p_h\}$
  append to J *all J-pairs* $\bigcup_{(p_g)\in G} J_{p_g,p_h}$ *not* covered *by* $G \cup S$
else
  $S \leftarrow S \cup \{(s_h, 0)\}$
end if
end while

Proof of termination relies on Noetherianity of the free module R^r.

4.2. Translation to an equivariant setting

Let us return to an infinite-dimensional polynomial ring $R = K[X]$ with some Π-action. As a running example, take $R = K[x_i,\ i \in \mathbb{N}]$ with a Π-compatible order, $\Pi = \mathrm{Inc}(\mathbb{N})$.

To draw parallels with the approach of the previous section, we need to work with pairs:

$$P = \{(s, f) \in (R * \Pi)^r \times R \mid f = s \cdot F\}.$$

Recall that, for instance, in our running example, one can think of $\Pi = \mathrm{Inc}^{\tau}(\mathbb{N}) = [\tau_i | i \in \mathbb{N}] \subset \mathrm{Inc}(\mathbb{N})$ in Remark 2.11, so:

$$R * \Pi = K[X] * \Pi = K([X] * \Pi).$$

The semidirect product $[X] * \Pi$ is a *non-Noetherian noncommutative* monoid where every element can be written in a *left standard form*:

$$x_{i_1} \cdots x_{i_c} \cdot \tau_{j_1} \cdots \tau_{j_d},$$

with $i_1 \leq \cdots \leq i_c$ and $j_1 \leq \cdots \leq j_d$.

Since the Π-divisibility order on $[X] * \Pi$ is not a well-partial-order (indeed, τ_i are pairwise not comparable), the (left) free module $(R*\Pi)^r$, $r \in \mathbb{N}$, is not Noetherian. In the presence of F (the vector of generators of the given ideal) and with a fixed order on R, we define the *Schreyer order* on $(R*\Pi)^r$ as follows. For two terms me_i and $m'e_j$, with $m, m' \in [X]*\Pi$:

- compare $m \,\mathrm{LM}\, f_i$ and $m' \,\mathrm{LM}\, f_j$ using the order on R,
- then break the ties according to the position (i.e., compare i and j).

While we see the Schreyer order as natural in some sense, any term order compatible with the order on R may be used.

A strong equivariant Gröbner basis, which can be defined similarly to a strong Gröbner basis in the previous section, is infinite (for a nonzero Π-invariant ideal). For instance, $I = R = K[x_1, x_2, \ldots]$ has a Gröbner basis $\{1\}$. However, a strong Gröbner basis has to include the elements $\{(\tau_i - 1)e_1 \mid i \in \mathbb{N}\} \subset (R * \Pi)^1$.

We found a way to modify Algorithm 4.2 to compute an equivariant Gröbner basis. It, of course, falls short of computing a strong equivariant Gröbner basis, but the partial information computed about the syzygies and the mechanism of top-reduction of J-pairs eliminate a large number of unnecessary iterations in a naïve implementation of an equivariant Buchberger's algorithm (Algorithm 3.1).

Algorithm 4.3. EquivariantSignatureBuchberger(F)

Require: $F \subset R$.

Ensure: *G such that $\pi_2(G)$ is an equivariant Gröbner basis of $\langle F\rangle_\Pi$.*

$r \leftarrow |F|$.
$G \leftarrow \emptyset$, $S \leftarrow \emptyset$
$J \leftarrow s\{(e_i, F_i) : i \in r = |F|\} \subset R^r \times R$
while $J \neq \emptyset$ ***do***
    pick $p_f = (s_f, f) \in J$; $J \leftarrow J \setminus \{p_f\}$
    $p_h = (s_h, h) \leftarrow$ regular top-reduction *of* (s_f, f) *with respect to* G
    if $h \neq 0$ ***then***
        $h' \leftarrow \mathrm{NF}_{\Pi\pi_2(G)}\, h$
        if $h' \neq 0$ ***then***
            if $h' \neq h$ ***then***
                $r \leftarrow r + 1$
                $p_h \leftarrow (e_r, h')$
            end if
            $G \leftarrow G \cup \{p_h\}$
            append to J *all J-pairs* $\bigcup_{(p_g)\in G} J_{p_g,p_h}$ *not* covered *by* $G \cup S$
        end if
    else
        $S \leftarrow S \cup \{(s_h, 0)\}$
    end if
end while

The highlighted part of the algorithm ensures that it terminates for an input for which Algorithm 3.1 terminates. Note that the rank r (recall: G and S are contained in $(R * \Pi)^r \times R$) may grow as the algorithm progresses.

Example 4.4. *Consider the ideal $I = \langle F\rangle_{\mathrm{Inc}^\tau(\mathbb{N})}$ in the ring $R = K[x_i, y_{ij} \mid i, j \in \mathbb{N}, i > j]$ where $F = y_{21} - x_2x_1$.*

The implementation of Algorithm 4.3 produces the following output:

```
i1:  needsPackage "EquivariantGB";
i2 : -- QQ[x_0,x_1,...; y_(0,1),y(1,0),...]
```

```
     -- (NOTE: indices start with 0, not 1)
     R = buildERing({symbol x, symbol y}, {1,2}, QQ, 2,
                        MonomialOrder=>Lex, Degrees=>{1,2});
i3:  egbSignature(y_(1,0) - x_0*x_1)
...
...
-- 95th syzygy: (0, y_(6,0)*y_(4,3)*y_(2,1)*{2, 5, 6, 7, 8}*[0])
...
...
-- TOTAL covered pairs = 1528
o3 = {- x x  + y   , ...  ...  ...
         1 0    1,0
       - y   y    + y   y   , - y   y    + y   y   }
          3,2 1,0    3,1 2,0     3,1 2,0    3,0 2,1
```

In particular, this computation shows that the kernel of the monomial map induced by $y_{ij} \mapsto x_i x_j$ *is* $\langle y_{43}y_{21} - y_{42}y_{31}, y_{42}y_{31} - y_{41}y_{32}\rangle_{\mathrm{Inc}^{\tau}(\mathbb{N})}$.

The number of times a polynomial corresponding to a J-pair in the queue J was reduced to zero is **95**. *However, in this signature-based algorithm, the knowledge of 95 syzygies is still useful as their signatures are stored and may "cover" some J-pairs in the queue. The total number of covered J-pairs, 1528, could be taken as a measure of how many useless reductions are avoided.*

There is an optional parameter:

`egbSignature(...,PrincipalSyzygies=>true)`,

that instructs the algorithm to construct the so-called principal syzygies, *the syzygies that correspond to the trivial commutation relations on the generators:* $(\sigma F_i)(\sigma' F_j) - (\sigma' F_j)(\sigma F_i) = 0$, $i \neq j$, *where* $\sigma, \sigma' \in \Pi$ *are extensions of the maps* $[w(F_i)] \to [w(F_i) + w(F_j)]$ *and* $[w(F_j)] \to [w(F_i) + w(F_j)]$. *With this option, the previous computation produces a much larger number of syzygies,* **1114**; *however, there is no improvement obtained in terms of covered J-pairs, and the improvement in the number of J-pairs that need to be stored is insignificant.*

It is our understanding that in the usual setting (where the results of [26] *apply in their entirety), the introduction of principal syzygies leads to a significant speedup. While we can find examples where the effect of principal syzygies is nontrivial, it still seems to be negligible in the setting of this paper.*

Our general conclusion at the moment of writing is that signature-based approaches are applicable for computing EGBs, however, the savings produced by eliminating unnecessary reductions are largely offset by the amount of J-pairs needed to be stored. Perhaps with a more careful implementation of what we have proposed and some new ideas, one could overcome the bottlenecks of the required space complexity and the complexity of looking up J-pairs.

At the moment, implementations of algorithms that fall back onto highly optimized Gröbner bases routines in the finite-dimensional setting (such as Algorithm 3.5 and its variation in Remark 3.6) seem to be the best practical choice.

§5. Open questions and challenges

In this final section, we raise several computational challenges and theoretical problems arising from equivariant Gröbner bases and asymptotic symbolic algebra. Often these challenge problems can serve as benchmark tests for sharpening the methods of practitioners who are improving and implementing these new classes of algorithms.

Problem 5.1 (Chains induced by a monomial). *Compute symbolically an EGB for the chain of toric ideals* $I_n = \ker(y_{ij} \mapsto x_i^a x_j^b)$, $1 \leq i \neq j \leq n$, *for small* $a > b$ *with* $\gcd(a, b) = 1$. (*Compare to* [31, 33, 38, 19, 40]).

The case $a = 2, b = 1$ is the only one explicitly computed (Theorem 1.2). A variant of this problem has the same statement apart from considering a smaller subset of indices: $1 \leq j < i \leq n$ (see [19, Remark 6.3]) and also more indices:

Problem 5.2. *Develop combinatorial methodology to understand kernels with more than two indices such as* $\ker(y_{ijk} \mapsto x_i^3 x_j^2 x_k)$*?*

There are also some basic questions in the theory of EGBs that remain open. For instance, it is not so well understood exactly which classes of ideals have finite generation, much less an equivariant Gröbner basis. While [19] gives a definitive answer for a large class of invariant toric ideals (i.e., the kernels of equivariant monomial maps), the following question is open (but see [13] for recent work on the cubic case).

Question 5.3 (Kernel of a polynomial map). *Is there a finite set of generators (up to symmetry) for the chain* $I_n = \ker(y_{ij} \mapsto f(x_i, x_j))$, $1 \leq i \neq j \leq n$, *for a given polynomial* $f \in \mathbb{C}[s, t]$*?*

Even when finite generation is known, other problems still remain open. We know that the kernels of monomial maps stabilize [3, 38, 19] and have EGBs with respect to certain monomial orders [40]. What if the monomial order is not particularly nice? What if the map is a general equivariant polynomial map?

Question 5.4. *If the answer to the previous question is positive, is there a finite equivariant Gröbner basis with respect to an arbitrary order?*

One largely unexplored aspect of research efforts to date is the structure of term orders for equivariant Gröbner bases. In the classical application of Gröbner bases, term orders play a significant role and such concepts as the *Gröbner fan* and techniques such as the *Gröbner walk* arise. These seem not to have equivalents in the equivariant setting in view of the following question.

Question 5.5. *For* $R = \mathbb{C}[x_0, x_1, x_2, \ldots]$, *there are several natural monomial orders respecting* Inc($\mathbb{N}$)*-action and refining the* Inc($\mathbb{N}$)*-divisibility partial order: namely, lexicographic and graded lexicographic orders. Are there any others?*

In classical computational algebra, Gröbner bases do more than simply answer ideal membership questions. They also are used as input by other algorithms to find invariants describing the underlying geometry and algebra such as dimension, degree, Hilbert series, etc.

Question 5.6. *What is a good notion of the variety defined by an* $\mathfrak{S}_\infty$*-invariant ideal (of an infinite-dimensional ring)? How should one define its dimension?*

Question 5.7. *Is there a better (alternative) notion of Hilbert series, one that would be suitable for* Inc($\mathbb{N}$)*-invariant modules? (See issues discussed in the last section of* [41].)

While it is easily observed that, in practice, computations of EGBs tend to consume far more resources than in the classical case (per bit of input), there is no good understanding of theoretical complexity of an equivariant Buchberger's algorithm.

Question 5.8. *Given widths and degrees of a finite set of generators, is there an upper bound on widths and degrees of the elements of a reduced EGB?*

If so, then one could look for lower bounds (in the worst case).

One of the largest computations done so far is that of [7]; it is accomplished by a custom made program (not available publicly). The output gives a definitive algebraic-statistical description of the Gaussian two-factor model by means of EGBs. We propose the following difficult challenge.

Problem 5.9. *Use EGBs to study the Gaussian three-factor model; i.e., obtain the kernel of the map*:

$$\mathbb{C}[y_{ij} \mid i, j \in \mathbb{N}, i > j] \to \mathbb{C}[s_i, t_i, u_i \mid i \in \mathbb{N}],$$
$$y_{ij} \mapsto s_i s_j + t_i t_j + u_i u_j\,.$$

While this may be set up exactly with the same technique as in [7], the computation seems to present an insurmountable task for the current implementations of current algorithms executed on current computers.

References

[1] 4ti2 team. 4ti2—a software package for algebraic, geometric and combinatorial problems on linear spaces. Available at www.4ti2.de.

[2] S. Aoki and A. Takemura. Minimal basis for a connected markov chain over $3 \times 3 \times k$ contingency tables with fixed two-dimensional marginals. *Australian & New Zealand Journal of Statistics*, 45(2):229–249, 2003.

[3] M. Aschenbrenner and C. Hillar. Finite generation of symmetric ideals. *Trans. Amer. Math. Soc.*, 359(11):5171–5192, 2007.

[4] M. Aschenbrenner and C. Hillar. An algorithm for finding symmetric Gröbner bases in infinite dimensional rings. In *Proceedings of the international symposium on symbolic and algebraic computation* (*ISSAC*), pages 117–124. ACM, 2008.

[5] G. Bergman. The diamond lemma for ring theory. *Adv. Math.*, 29(2):178–218, 1978.

[6] L. Bokut'. Embeddings into simple associative algebras. *Algebra and Logic*, 15(2):73–90, 1976.

[7] A. Brouwer and J. Draisma. Equivariant Gröbner bases and the two-factor model. *Math. Comput.*, 80:1123–1133, 2011.

[8] B. Buchberger. Ein algorithmus zum auffinden der basiselemente des restklassenrings nach einem nulldimensionalen polynomideal. *Universitat Innsbruck, Austria, PhD Thesis*, 1965.

[9] T. Church. Homological stability for configuration spaces of manifolds. *Invent. Math.*, 188(2):465–504, 2012.

[10] D. Cohen. On the laws of a metabelian variety. *J. Algebra*, 5(3):267–273, 1967.

[11] D. Cohen. Closure relations, Buchberger's algorithm, and polynomials in infinitely many variables. In *Computation theory and logic*, volume 270 of *Lect. Notes Comput. Sci.*, pages 78–87, 1987.

[12] D. Cox, J. Little, and D. O'Shea. *Ideals, varieties, and algorithms: An introduction to computational algebraic geometry and commutative algebra.* Undergraduate Texts in Mathematics. Springer Publishing Company, Incorporated, 4th edition, 2015.

[13] H. Derksen, R. Eggermont, and A. Snowden. Topological noetherianity for cubic polynomials. *Algebra & Number Theory*, 11(9):2197–2212, 2017.

[14] P. Diaconis and B. Sturmfels. Algebraic algorithms for sampling from conditional distributions. *Ann. Stat.*, 26(1):363–397, 1998.

[15] J. Draisma. Finiteness for the k-factor model and chirality varieties. *Adv. Math.*, 223:243–256, 2010.

[16] J. Draisma. Noetherianity up to symmetry. In *Combinatorial algebraic geometry*, pages 33–61. Springer, 2014.

[17] J. Draisma and R. Eggermont. Finiteness results for abelian tree models. *J. Eur. Math. Soc.*, 17(4):711–738, 2015.

[18] J. Draisma and R. Eggermont. Plücker varieties and higher secants of Sato's Grassmannian. *Journal für die Reine und Angewandte Mathematik* (*Crelles Journal*), 2015.

[19] J. Draisma, R. Eggermont, R. Krone, and A. Leykin. Noetherianity for infinite-dimensional toric varieties. *Algebra & Number Theory*, 9(8):1857–1880, 2015.

[20] J. Draisma and J. Kuttler. On the ideals of equivariant tree models. *Math. Ann.*, 344(3):619–644, 2009.

[21] J. Draisma and J. Kuttler. Bounded-rank tensors are defined in bounded degree. *Duke Math. J.*, 163(1):35–63, 2014.

[22] M. Drton, B. Sturmfels, and S. Sullivant. Algebraic factor analysis: tetrads, pentads and beyond. *Probab. Theory Relat. Fields.*, 138(3-4):463–493, 2007.

[23] L. Ein, D. Erman, and R. Lazarsfeld. Asymptotics of random betti tables. *Journal für die Reine und Angewandte Mathematik* (*Crelles Journal*), 2015(702):55–75, 2015.

[24] L. Ein and R. Lazarsfeld. Asymptotic syzygies of algebraic varieties. *Invent. Math.*, 190(3):603–646, 2012.

[25] P. Emmott. *Some decision problems in group theory and ring theory*. PhD thesis, Queen Mary, University of London, 1987.

[26] S. Gao, F. Volny IV, and M. Wang. A new framework for computing Gröbner bases. *Math. Comp.*, 85(297):449–465, 2016.

[27] N. Gjunter. Ob odnom' neravenstja iz' teorij zjalych'racional'nych' funkcij [on an inequality from the theory of polynomials]. *Izdanie Inst. Inž. Putej Soobščenija Imp. Al. I.*, 84:1–18, 1913.

[28] N. Gjunter. Sur les modules des formes algébriques. *Trav. Inst. Math. Tbilissi* [*Trudy Tbiliss. Mat. Inst.*], 9:97–206, 1941.

[29] D. Grayson and M. Stillman. Macaulay 2: a software system for research in algebraic geometry. Available at `http://www.math.illinois.edu/Macaulay2/`.

[30] S. Gunturkun and U. Nagel. Equivariant Hilbert series of monomial orbits. *arXiv preprint arXiv:1608.06372*, 2016.

[31] C. Hillar and A. Martín del Campo. Finiteness theorems and algorithms for permutation invariant chains of Laurent lattice ideals. *J. Symb. Comput.*, 50:314–334, 2013.

[32] C. Hillar and L. Levine. Polynomial recurrences and cyclic resultants. *Proc. Amer. Math. Soc.*, 135(6):1607–1618, 2007.

[33] C. Hillar and A. Martin del Campo. Corrigendum to "Finiteness theorems and algorithms for permutation invariant chains of laurent lattice ideals", *J. Symb. Comput.*, 74(C):650–652, 2016.

[34] C. Hillar and S. Sullivant. Finite Gröbner bases in infinite dimensional polynomial rings and applications. *Adv. Math.*, 229(1):1–25, 2012.

[35] H. Hironaka. Resolution of singularities of an algebraic variety over a field of characteristic zero: II. *Ann. Math.*, pages 205–326, 1964.

[36] S. Hoşten and S. Sullivant. A finiteness theorem for markov bases of hierarchical models. *J. Comb. Theory A*, 114(2):311–321, 2007.

[37] B. Howard, J. Millson, A. Snowden, and R. Vakil. The equations for the moduli space of n points on the line. *Duke Math. J.*, 146(2):175–226, 2009.

[38] T. Kahle, R. Krone, and A. Leykin. Equivariant lattice generators and markov bases. In *Proceedings of the international symposium on symbolic and algebraic computation (ISSAC)*, pages 264–271, 2014.

[39] R. Krone. EquivariantGB: equivariant Gröbner bases in Macaulay2. `http://www.rckr.one/EquivariantGB.html`.

[40] R. Krone. Equivariant Gröbner bases of symmetric toric ideals. In *Proceedings of the international symposium on symbolic and algebraic computation (ISSAC)*, pages 311–318. ACM, 2016.

[41] R. Krone, A. Leykin, and A. Snowden. Hilbert series of symmetric ideals in infinite polynomial rings via formal languages. *J. Algebra*, to appear.

[42] U. Nagel and T. Römer. Equivariant Hilbert Series in non-Noetherian Polynomial Rings. *arXiv preprint arXiv:1510.02757*, 2015.

[43] O. Randal-Williams. Homological stability for unordered configuration spaces. *Q. J. Math.*, 64(1):303–326, 2013.

[44] B. Renschuch, H. Roloff, G. Rasputin, and M. Abramson. Contributions to constructive polynomial ideal theory XXIII: forgotten works of Leningrad mathematician N.M. Gjunter on polynomial ideal theory. *ACM SIGSAM Bulletin*, 37(2):35–48, 2003.

[45] E. Ruch, A. Schönhofer, and I. Ugi. Die Vandermondesche determinante als näherungsansatz für eine chiralitätsbeobachtung, ihre verwendung in der stereochemie und zur berechnung der optischen aktivität. *Theoretica Chimica Acta.*, 7(5):420–432, 1967.

[46] S. Sam and A. Snowden. Gröbner methods for representations of combinatorial categories. *J. Amer. Math. Soc.*, 30(1):159–203, 2017.

[47] F. Santos and B. Sturmfels. Higher Lawrence configurations. *J. Comb. Theory A*, 103(1):151–164, 2003.

[48] A. Shirshov. Some algorithmic problems for Lie algebras. *Sibirsk Mat. Zh*, 3(2):291–296, 1962.

[49] A. Snowden. Syzygies of Segre embeddings and Δ-modules. *Duke Math. J.*, 162(2):225–277, 2013.

University of California, Berkeley
E-mail address: chillar@msri.org

Queens University, Canada
E-mail address: rk71@queensu.ca

Georgia Institute of Technology
E-mail address: leykin@math.gatech.edu

Advanced Studies in Pure Mathematics 77, 2018
The 50th Anniversary of Gröbner Bases
pp. 155–170

Semigroups – A computational approach

Florian Kohl, Yanxi Li, Johannes Rauh and Ruriko Yoshida

Abstract.

The question whether there exists an integral solution to the system of linear equations with non-negativity constraints, $A\mathbf{x} = \mathbf{b}$, $\mathbf{x} \geq 0$, where $A \in \mathbb{Z}^{m \times n}$ and $\mathbf{b} \in \mathbb{Z}^m$, finds its applications in many areas such as operations research, number theory, combinatorics, and statistics. In order to solve this problem, we have to understand the semigroup generated by the columns of the matrix A and the structure of the "holes" which are the difference between the semigroup and its saturation. In this paper, we discuss the implementation of an algorithm by Hemmecke, Takemura, and Yoshida that computes the set of holes of a semigroup and we discuss applications to problems in combinatorics. Moreover, we compute the set of holes for the common diagonal effect model and we show that the nth linear ordering polytope has the integer-decomposition property for $n \leq 7$. The software is available at `http://ehrhart.math.fu-berlin.de/People/fkohl/HASE/`.

§1. Introduction

Consider the system of linear equations and inequalities

$$A\mathbf{x} = \mathbf{b}, \ \ \mathbf{x} \geq 0, \tag{1}$$

where $A \in \mathbb{Z}^{m \times n}$ and $\mathbf{b} \in \mathbb{Z}^m$. Suppose that the solution set over the real numbers $\{\mathbf{x} \in \mathbb{R}^n : A\mathbf{x} = \mathbf{b}, \mathbf{x} \geq 0\}$ is not empty.

Problem 1.1. *Decide whether there exists an integral solution to the system* (1) *or not.*

Problem 1.1 is called the *integer feasibility problem.* To decide whether a system of equations is feasible is the first step in integer programming, where the goal is to find an "optimal" solution. Therefore,

---

Received August 11, 2016.
2010 *Mathematics Subject Classification.* 11P21, 52B11, 90C08, 97K70.
Key words and phrases. feasibility problem, lattice points, polytopes, software.

problem 1.1 can be solved computationally using a linear programming system that can handle integer constraints, such as **lp_solve** [6]. However, this computational approach does not work if one wants to study a family of integer feasibility problems at the same time. Here, we consider the following problem:

Problem 1.2. *For fixed* $A \in \mathbb{Z}^{m\times n}$, *decide for which* $\mathbf{b} \in \mathbb{Z}^m$ *there exists an integral solution to the system* (1).

Problems 1.1 and 1.2 are of fundamental importance in many areas such as operations research, number theory, combinatorics, and statistics (see [18] and references within). For instance, the *Frobenius problem* is a simple-sounding yet wide-open integer feasibility problem, see e.g. [13] for an overview. For coprime integers $a_1, a_2, \ldots, a_n > 0$, the Frobenius problem asks to find the biggest positive integer that cannot be expressed as a non-negative linear combination of the a_i's with integral coefficients. Even for $n = 4$, this is an active area of research.

Feasibility can be described in terms of the *semigroup*

$$Q = Q(A) = \{\mathbf{a}_1 x_1 + \cdots + \mathbf{a}_n x_n \mid x_1, \ldots, x_n \in \mathbb{Z}_{\geq 0}\} \tag{2}$$

generated by the column vectors $\mathbf{a}_1, \ldots, \mathbf{a}_n$ of A. Here, $\mathbb{Z}_{\geq 0}$ denotes the set of non-negative integers, i.e., $\mathbb{Z}_{\geq 0} := \{0, 1, 2, \ldots\}$. Moreover, we need the *cone*

$$K = K(A) = \{\mathbf{a}_1 x_1 + \cdots + \mathbf{a}_n x_n \mid x_1, \ldots, x_n \in \mathbb{R}_{\geq 0}\}$$

generated by the columns of A, where $\mathbb{R}_{\geq 0} := [0, \infty)$. Throughout this paper, we assume that all cones are *pointed*, i.e., that they do not contain lines: if $\mathbf{v} \in K \setminus \{0\}$, then $-\mathbf{v} \notin K$. Finally, we need the *lattice*

$$L = L(A) = \{\mathbf{a}_1 x_1 + \cdots + \mathbf{a}_n x_n \mid x_1, \ldots, x_n \in \mathbb{Z}\}$$

generated by the columns of A. In this paper, we assume $L(A) = \mathbb{Z}^n$.

By definition, an integral solution to the system (1) exists if and only if $\mathbf{b} \in Q$. In general, it is difficult to check whether a given vector belongs to Q. However, it is much easier to check whether a given vector belongs to L or to K: To check whether $\mathbf{b} \in L$ is a problem of linear algebra (over the integers), and to check whether $\mathbf{b} \in K$ one can compute the inequality description of K and check whether $\mathbf{b}$ satisfies all linear inequalities. Admittedly, computing the inequalities can be a difficult problem in itself, but usually it is still easier than the integer feasibility problem. Therefore, it makes sense to compare Q to the larger semigroup $Q_{\text{sat}} = K \cap L$, which is called the *saturation* of Q. Clearly,

$Q \subset Q_{\text{sat}}$, and we call Q *saturated* (or *normal*) if $Q = Q_{\text{sat}}$. We define the set of *holes* H of the semigroup Q to be $H := Q_{\text{sat}} \setminus Q$.

If $\mathbf{b} \in H \subset Q_{\text{sat}}$, then the system

$$A\mathbf{x} = \mathbf{b}, \mathbf{x} \geq 0,$$

has a solution $\mathbf{x} \in \mathbb{R}^n$ over the *reals*, but no solution $\mathbf{x} \in \mathbb{Z}^n_{\geq 0}$ over the integers.

In general, the set H may be infinite, but it is possible to write H as a finite union of finitely generated (affine) monoids. The first step is to compute the fundamental holes, where we say that a hole $\mathbf{h} \in H$ is *fundamental* if there is no other hole $\mathbf{h}' \in H$ such that $\mathbf{h} - \mathbf{h}' \in Q$. In contrast to H, the set $F \subseteq H$ of fundamental holes is always finite, as it is contained in the bounded set

$$P := \left\{ \sum_{i=1}^{n} \lambda_i \mathbf{a}_i \mid 0 \leq \lambda_1, \ldots, \lambda_n < 1 \right\},$$

as shown in [18]. A finite algorithm to compute F is due to [10].

Once F is known, it is necessary to compute an explicit representation of the holes in $\mathbf{f} + Q$. Hemmecke et. al. [10] showed how the set of holes in $\mathbf{f} + Q$ can be expressed as a finite union of finitely generated monoids using ideas from commutative algebra. Together with an algorithm to compute the fundamental holes, this gives a *finite* algorithm to compute an *explicit* representation of H, even for an infinite set H.

As shown by [2, 18], computing the set of holes is polynomial in time in the input size of A if we fix the number of variables m and n (see the definition of input size in [3]). Once we compute Q for a particular matrix A, we do not have to compute it again as it does not depend on $\mathbf{b}$.

In this paper, we have implemented the algorithm introduced in [10] and we have applied our software to problems in combinatorics and statistics. We named the software HASE (**H**oles in **A**ffine **SE**migroups). It is available at `http://ehrhart.math.fu-berlin.de/People/fkohl/HASE/`. The homepage also contains the input files that are needed to reproduce the examples that are discussed in this paper.

This paper is organized as follows: In Section 2, we outline the algorithm. The performance of the algorithm and possible ways to speed up the process are described in Section 3. In Section 4, we compute the set of holes for the *common diagonal effect model* [17]. In Section 5, we show some computational experiments concerning the integer-decomposition property of polytopes and concerning a lifting algorithm for Markov bases, see [15]. We end with a discussion and open problems.

§2. Computing holes

In this section, we briefly describe our software and the implementation. The two main steps of the algorithm of [10] are:

(1) Compute the set F of fundamental holes.
(2) For each of the finitely many $\mathbf{f} \in F$, compute an explicit representation of the holes in $\mathbf{f} + Q$.

Our software outsources step 1 to **Normaliz**, see [7]. **Normaliz** is a computer program that computes the saturation (or **Normaliz**ation) of an affine semigroup. Usually, the saturation is output in the form of a matrix A' such that the saturation $Q_{\mathrm{sat}}(A) = Q(A')$ equals the semigroup generated by the columns of A'. Starting with version 3.0, **Normaliz** can also compute a second representation of $Q_{\mathrm{sat}}(A)$ by giving a minimal set F' of "*generators of $Q_{\mathrm{sat}}(A)$ as a $Q(A)$-module.*" Formally, this says that

$$Q_{\mathrm{sat}}(A) = \bigcup_{\mathbf{f} \in F'} (\mathbf{f} + Q(A)).$$

It is not difficult to see that $\mathbf{0} \in F'$ (since $Q \subseteq Q_{\mathrm{sat}}$) and that $F := F' \setminus \{\mathbf{0}\}$ is the set of fundamental holes of $Q(A)$. For details how **Normaliz** computes the set F', we refer to the documentation of **Normaliz**. As an illustration, Section 4 contains a description of the (fundamental and non-fundamental) holes of the common diagonal effect models.

It remains to determine the holes in $\mathbf{f} + Q$ for every fundamental hole $\mathbf{f} \in F$. Every **non**-hole belongs to $(\mathbf{f}+Q)\cap Q$ and if $\mathbf{z} \in (\mathbf{f}+Q)\cap Q$, then also $\mathbf{z} + A\boldsymbol{\lambda} \in (\mathbf{f} + Q) \cap Q$ for any $\boldsymbol{\lambda} \in \mathbb{Z}^n_{\geq 0}$. Consider the ideal

$$I_{A,\mathbf{f}} := \left\langle \mathbf{x}^{\boldsymbol{\lambda}} \mid \boldsymbol{\lambda} \in \mathbb{Z}^n_{\geq 0},\ \mathbf{f} + A\boldsymbol{\lambda} \in (\mathbf{f} + Q) \cap Q \right\rangle, \tag{3}$$

where $\mathbf{x}^{\boldsymbol{\lambda}} := \prod_{i=1}^n x_i^{\boldsymbol{\lambda}_i}$ is the monomial with exponent vector $\boldsymbol{\lambda}$. Then, $\mathbf{f} + A\boldsymbol{\lambda}$ is not a hole if and only if $\mathbf{x}^{\boldsymbol{\lambda}} \in I_{A,\mathbf{f}}$. So we need to find a description of the monomials not in $I_{A,\mathbf{f}}$. These monomials are called the *standard monomials*. There are algorithms for finding the standard monomials, once a generating set for the ideal $I_{A,\mathbf{f}}$ is known. A generating set of the ideal $I_{A,\mathbf{f}}$ is described by the following lemma:

Lemma 2.1 ([10], Lemma 4.1). *Let M be the set of $\leq$-minimal solutions $(\boldsymbol{\lambda}, \boldsymbol{\mu}) \in \mathbb{Z}^{2n}_{\geq 0}$ to $\mathbf{f} + A\boldsymbol{\lambda} = A\boldsymbol{\mu}$, where the partial order $\leq$ is given by coordinatewise comparison. Then*

$$I_{A,\mathbf{f}} = \left\langle \mathbf{x}^{\boldsymbol{\lambda}} \mid \exists \boldsymbol{\mu} \in \mathbb{Z}^n_{\geq 0} \text{ such that } (\boldsymbol{\lambda}, \boldsymbol{\mu}) \in M \right\rangle.$$

Therefore, we have to find minimal integral solutions to the above system of linear equations for every fundamental hole $\mathbf{f}$. For this task, we can again use **Normaliz**, or we can use the **zsolve** command of **4ti2**, see [1]. Usually, **zsolve** runs faster, so it is the default choice of HASE.

Once we have a generating set for $I_{A,\mathbf{f}}$, we can use a computer algebra software to find the standard monomials. In general, the set of standard monomials of a polynomial ideal can be infinite, but it has a finite representation in terms of *standard pairs*. HASE relies on **Macaulay2** [8], which has the command `standardPairs`. A standard pair is a pair that consists of a monomial $\mathbf{x}^\lambda$ and a set $\mathbf{x}^{\boldsymbol{\mu}_1}, \dots, \mathbf{x}^{\boldsymbol{\mu}_r}$ of monomials. Such a pair corresponds to the set of holes

$$\mathbf{f} + A(\boldsymbol{\lambda} + \textstyle\sum_{i=1}^r c_i \boldsymbol{\mu}_i), \quad c_i \in \mathbb{Z}_{\geq 0},$$

and the set of all such standard pairs gives all holes in $\mathbf{f} + Q$.

§3. Performance of the algorithm

As shown by [2, 18], computing the set of holes H for the semigroup Q is polynomial in time in the input size of A if we fix the number of variables m and n (see the definition of the input size in [3]). Still, computing H is a difficult problem, and our algorithm may fail to terminate due to limited memory or time even for reasonably-sized examples.

In the examples we computed, we experienced the following problems:

- **Normaliz** may fail to compute the set F of fundamental holes.
- For one of the fundamental holes $\mathbf{f} \in F$, **zsolve** or **Normaliz** may fail to find the $\leq$-minimal solutions to $\mathbf{f} + A\boldsymbol{\lambda} = A\boldsymbol{\mu}$.
- For one of the fundamental holes $\mathbf{f} \in F$, **Macaulay2** may fail to compute the standard pairs.

In this list, a failure means that either we ran out of memory or we ran out of time.

If **Normaliz** fails, there is not much we can do. We really need the fundamental holes, and if computing the fundamental holes overstrains our computational resources at hand, it is very probable that the problem is just too difficult. The only thing we could do is to ask the developpers of **Normaliz**, who are always up for a challenge, for advice.

If one of the later steps fails, there is much more that we could do. The translation of computing the holes of the form $\mathbf{f} + Q$ for a fundamental hole $\mathbf{f}$ into a problem of commutative algebra is not very direct, and there may be some room for improvements. We discuss one trick that we implemented in Section 3.1 below.

There may be fourth problem: Namely, the set F may be extremely large. Thus, even if **Normaliz** computes F within reasonable time and if **zsolve** and **Macaulay2** find the hole monoids reasonably fast for each single hole, the total running time may be unacceptable. However, at least in this case it is relatively easy to obtain a good estimate for the total running time that would be needed, since in this case the cardinality of F is known, and the running times of **zsolve** and **Macaulay2** per fundamental hole can be estimated by their performance on the first few holes.

If F is very large, a natural remedy is to look for symmetries of the problem. However, currently symmetries are not implemented in HASE.

3.1. Speeding up zsolve

Let $\mathbf{f} \in F$ be a fundamental hole. As explained in Section 2, we want to solve the linear system $\mathbf{f} + A\boldsymbol{\lambda} = A\boldsymbol{\mu}$. This system can be simplified considerably if certain non-holes are known in advance. The simplest case is to look at the vectors $\mathbf{f} + \mathbf{a}_i$, where $\mathbf{a}_i$ is a column of A.

Suppose that $\mathbf{f} + \mathbf{a}_i$ is not a hole. Then $\mathbf{f} + \mathbf{a}_i = A\boldsymbol{\mu}_0$ for some $\boldsymbol{\mu}_0$. Thus, $\mathbf{f} + \mathbf{a}_i + A\boldsymbol{\lambda} = A(\boldsymbol{\mu}_0 + \boldsymbol{\lambda})$. This shows that if $\mathbf{f} + \mathbf{a}_i$ is not a hole, then $\mathbf{f} + \mathbf{a}_i + Q$ contains no other holes. This implies that every minimal solution to $\mathbf{f} + A\boldsymbol{\lambda} = A\boldsymbol{\mu}$ has $\lambda_i = 0$. Let A' be the matrix A with the ith column $\mathbf{a}_i$ dropped. Then, instead of solving $\mathbf{f} + A\boldsymbol{\lambda} = A\boldsymbol{\mu}$, we may just as well solve $\mathbf{f} + A'\boldsymbol{\lambda}' = A\boldsymbol{\mu}$. Observe that this leads to a linear system with one variable fewer. If we can identify many columns $\mathbf{a}_i$ that we can drop, we can speed up the computation of the holes in $\mathbf{f} + Q$.

This idea is implemented in HASE and can be activated using the option `--trick`. With this option, HASE does the following instead of solving $\mathbf{f} + A\boldsymbol{\lambda} = A\boldsymbol{\mu}$:

(1) For each column $\mathbf{a}_i$ of A, check whether $\mathbf{f} + \mathbf{a}_i$ is a hole.
(2) Let A' be the matrix with columns those $\mathbf{a}_i$ for which $\mathbf{f} + \mathbf{a}_i$ is a hole.
(3) Compute the minimal solutions to $\mathbf{f} + A'\boldsymbol{\lambda}' = A\boldsymbol{\mu}$ (using either **zsolve** or **Normaliz**).
(4) Use **Macaulay2** to compute the standard pairs of the ideal $I_{A',\mathbf{f}}$.

Step 1 is an integer feasibility problem. HASE uses the open source (mixed-integer) linear programming system **lp_solve** [6] to solve this problem. Usually, this is a relatively quick step (and if it is not, it is again an indication that our original problem is too difficult).

In the last step, observe that the trick also leads to a smaller ideal $I_{A',\mathbf{f}}$ or, more precisely, an ideal in a smaller ambient ring ($I_{A,\mathbf{f}}$ and $I_{A',\mathbf{f}}$ will in fact have the same generators). This, however, should not lead to a big speed-up, since the command `standardPairs` in **Macaulay2** will usually realize when variables do not appear in the generating set of an ideal.

§4. Common diagonal effect models

In this section, we consider the common diagonal effect models (CDEM) introduced by [9]. The results were obtained by computing small examples using HASE to build a conjecture.

Let $A \in \mathbb{Z}^{(2d+1)\times d^2}$ be the matrix that computes the row sums, column sums, and also the diagonal sum of a $d \times d$ table. For instance, if $d = 3$, we have

$$A = \begin{pmatrix} 1 & 0 & 0 & 1 & 0 & 0 & 1 & 0 & 0 \\ 0 & 1 & 0 & 0 & 1 & 0 & 0 & 1 & 0 \\ 0 & 0 & 1 & 0 & 0 & 1 & 0 & 0 & 1 \\ 1 & 1 & 1 & 0 & 0 & 0 & 0 & 0 & 0 \\ 0 & 0 & 0 & 1 & 1 & 1 & 0 & 0 & 0 \\ 0 & 0 & 0 & 0 & 0 & 0 & 1 & 1 & 1 \\ 1 & 0 & 0 & 0 & 1 & 0 & 0 & 0 & 1 \end{pmatrix}.$$

The cone K generated by the columns of A lies in the hyperplane

$$\sum_{i=1}^{d} z_i = \sum_{i=d+1}^{2d} z_i, \tag{4}$$

since this linear equality is satisfied by all columns of A. Our goal is to describe the Hilbert basis of the saturation Q_{sat} of the semigroup Q generated by the columns of A. First, we define a set $\mathcal{F}$ that will later turn out to be the set of fundamental holes of Q.

Definition 4.1. *Let $\mathbf{a}_{ij}$ be the $((i-1)d+j)^{th}$ column of A, and let*

$$\mathbf{h}_{kl} := \frac{1}{2}\left(\mathbf{a}_{ll} + \mathbf{a}_{lk} + \mathbf{a}_{kl} + \mathbf{a}_{kk}\right).$$

Finally, let $\mathcal{F} := \{\mathbf{h}_{kl} : k, l \in [d], k < l\}$, where $[d] := \{1, 2, \ldots, d\}$.

There are $\binom{d}{2}$ choices for l and k. Since every choice yields a different vector, we get $|\mathcal{F}| = \binom{d}{2}$. The next lemma shows that $\mathcal{F}$ consists of holes.

Lemma 4.2. *$\mathcal{F} \subset K \cap \mathbb{Z}^{2d+1}$, and $\mathcal{F} \subseteq Q_{\text{sat}} \setminus Q$.*

Proof. $\mathbf{a}_{ij}$ has a 1 in the jth coordinate and in the $(d+i)^{\text{th}}$ coordinate. Moreover, if $i = j$, then there is a 1 in the $(2d+1)^{\text{th}}$ coordinate. So every vector of the form $2\mathbf{h}_{kl} = \mathbf{a}_{ll} + \mathbf{a}_{lk} + \mathbf{a}_{kl} + \mathbf{a}_{kk}$ has a 2 in the

l^{th}, k^{th}, $(d+l)^{\text{th}}$, $(d+k)^{\text{th}}$ and in the $(2d+1)^{\text{st}}$ coordinate, and all other coordinates are 0. Thus, $\mathcal{F} \subset K \cap \mathbb{Z}^{2d+1} = Q_{\text{sat}}$.

It remains to show that $\mathcal{F} \cap Q = \emptyset$. So suppose we have an non-negative integral linear combination of the $\mathbf{a}_{ij}$'s that lies in $\mathcal{F}$. To get a 1 in the lth coordinate, we need a generator $\mathbf{a}_{il}$ and to get a 1 in the kth coordinate we need a generator $\mathbf{a}_{i'k}$. Since there has to be a 1 in the $(d+l)^{\text{th}}$ and $(d+k)^{\text{th}}$ coordinate, we see that i, $i' \in \{l, k\}$. Note that we cannot use a different generator to obtain a 1 in these coordinates, since otherwise we would get another 1 in the first d coordinates. If there are more than 2 generators, then either there is an entry bigger than 1 or there are at least five $1'$s in the first $2d$ coordinates.

If $i = i'$, then without loss of generality our linear combination is $\mathbf{a}_{ll} + \mathbf{a}_{lk}$, which has a 2 in the $(d+l)^{\text{th}}$ coordinate. If $i \neq i'$, then either we have $\mathbf{a}_{kl} + \mathbf{a}_{lk}$, which has a 0 in the last entry, or we have $\mathbf{a}_{ll} + \mathbf{a}_{kk}$ which has a 2 in the last entry. Hence, we have $\mathcal{F} \cap Q = \emptyset$. To see that all elements in $\mathcal{F}$ are indeed fundamental holes, one checks that each vector of the form $\mathbf{h}_{kl} - \mathbf{a}_{ij}$ is not in Q_{sat}. Q.E.D.

We have now identified a set of fundamental holes. [18, proof of Proposition 3.1] have shown that the set of fundamental holes is contained in

$$P := \left\{ \sum_{i,j \in [d]} \lambda_{ij} \mathbf{a}_{ij} \mid 0 \leq \lambda_{ij} < 1 \text{ for } i, j \in [d] \right\}.$$

To identify the fundamental holes, we can focus on P. Moreover, this proposition also implies that the (minimal) Hilbert basis is contained in the closure of P. The next theorem describes the (minimal) Hilbert basis for Q_{sat}.

Theorem 4.3. *The minimal Hilbert basis for Q_{sat} is given by*

$$\mathcal{H} := \{\mathbf{a}_{ij}\}_{i,j \in [d]} \cup \mathcal{F}.$$

Proof. Let $\mathbf{z} = (z_1, z_2, \ldots, z_{2d+1}) \in P \cap \mathbb{Z}^{2d+1} \setminus \{0\}$. Let $S := z_1 + z_2 + \cdots + z_d = z_{d+1} + z_{d+2} + \cdots + z_{2d}$. For every $i, j \in [d]$, we have $z_i = \sum_{j'=1}^{d} \lambda_{ij'}$ and $z_{d+j} \geq \lambda_{ij}$. This implies

$$S - z_{d+i} = \sum_{j \in [d] \setminus \{i\}} z_{d+j} \geq \left\lceil \sum_{j \in [d] \setminus \{i\}} \lambda_{ij} \right\rceil \overset{z \text{ lattice point}}{=} z_i,$$

and hence

$$z_i + z_{d+i} \leq S. \tag{5}$$

We show that every non-negative integer vector $\mathbf{z}$ that satisfies (5) is a non-negative integer combination of $\mathcal{H}$. To do this, we show that if $\mathbf{z} \neq 0$, then there is an element $\mathbf{a} \in \mathcal{H}$ such that $\mathbf{z}-\mathbf{a}$ is non-negative and satisfies (5). Observe that subtracting $\mathbf{a}$ from $\mathbf{z}$ decreases the right-hand-side S, so we need to make sure that the left-hand-side also decreases for those $i \in [d]$ for which (5) holds with equality.

First, suppose that $z_{2d+1} > 0$.

If there are two indices l and k where equality holds in (5), then

$$z_k + z_{d+k} = S \geq z_k + z_l \quad \text{and} \quad z_l + z_{d+l} = S \geq z_{d+k} + z_{d+l}.$$

It follows that $z_{d+k} = z_l$ and $z_{d+l} = z_k$. Thus, z_l and z_k are the only nonzero entries among the first d coordinates. It is easy to check that in this case, $\mathbf{z} - \frac{1}{2}(\mathbf{a}_{ll} + \mathbf{a}_{lk} + \mathbf{a}_{kl} + \mathbf{a}_{kk})$ is non-negative and satisfies (5).

If there is only one index i for which equality holds in (5), we can check that $\mathbf{z} - \mathbf{a}_{ii}$ again is non-negative and satisfies (5). If there is nc i for which equality holds, we pick i such that the pair of indices $(i, d+i)$ with z_i, $z_{d+i} \neq 0$ contains the biggest entry and subtract $\mathbf{a}_{ii}$. Then $\mathbf{z} - \mathbf{a}_{ii}$ is non-negative and satisfies (5).

It remains to discuss the case $z_{2d+1} = 0$. To express $\mathbf{z}$ as a non-negative linear integral combination of $\mathbf{a}'_{ij}$s where $i \neq j$, we translate the problem to a matching problem. We have two labeled multi-sets $\bigcup_{i:z_i>0} \bigcup_{r=1}^{z_i} \{i\}$ and $\bigcup_{j:z_{d+j}>0} \bigcup_{s=1}^{z_{d+j}} \{j\}$, both of cardinality S. Writing $\mathbf{z}$ as a non-negative integer combination of the $\mathbf{a}_{ij}$ corresponds to a matching between the two sets. The matching has to be *proper* in the sense that we only match elements i, j with $i \neq j$. For example, if $\mathbf{z} = (2, 0, 2, 2, 0, 2, 0)$, the two multi-sets are both equal to $\{1, 1, 3, 3\}$. In this example, there is (up to symmetry) only one proper matching that matches 1 to 3 and 3 to 1, corresponding to the identity $\mathbf{z} = \mathbf{a}_{1,3} + \mathbf{a}_{3,1} + \mathbf{a}_{1,3} + \mathbf{a}_{3,1}$.

It remains to show that there always exist such a proper matching. This can either be seen directly by induction or by appealing to Hall's marriage theorem, noting that (5) always ensures that the marriage condition is satisfied.

This finishes the proof $\mathcal{H}$ is a Hilbert basis. It is straightforward to check that $\mathcal{H}$ is indeed minimal. Q.E.D.

Knowing the Hilbert basis for Q_{sat}, we can completely describe the set of fundamental holes.

Corollary 4.4. *$\mathcal{F}$ is the set of fundamental holes of Q.*

Proof. We have already seen that every element in $\mathcal{F}$ is a fundamental hole. It only remains to show that there are no other fundamental holes. Any fundamental hole is a non-negative integer combination of the Hilbert basis $\mathcal{H}$. Clearly, this combination cannot involve the columns $\mathbf{a}_{ij}$ (otherwise the combination would not be fundamental). Thus, it suffices to show that the sum of two holes in $\mathcal{H}$ is not a hole. This follows from the identity $\mathbf{h}_{ij} + \mathbf{h}_{kl} = \mathbf{a}_{kk} + \mathbf{a}_{ll} + \mathbf{a}_{ij} + \mathbf{a}_{ji}$. Q.E.D.

We have now seen that there are exactly $\binom{d}{2}$ *fundamental* holes. However, this semigroup has *infinitely* many holes:

Theorem 4.5. *The set of holes in the set* $\mathbf{h}_{kl} + Q$ *is the union of the two monoids*

$$\mathbf{h}_{kl} + \mathbb{Z}_{\geq 0}\mathbf{a}_{kk} + \mathbb{Z}_{\geq 0}\mathbf{a}_{kl} + \mathbb{Z}_{\geq 0}\mathbf{a}_{lk} + \mathbb{Z}_{\geq 0}\mathbf{a}_{ll} \tag{6a}$$

and

$$\mathbf{h}_{kl} + \sum_{i=1}^{d} \mathbb{Z}_{\geq 0}\mathbf{a}_{ii}. \tag{6b}$$

Proof. Fix $k, l \in [d]$, $k < l$. If $i \neq j$ and $i \notin \{k, l\}$, then

$$\mathbf{h}_{kl} + \mathbf{a}_{ij} = \mathbf{a}_{kk} + \mathbf{a}_{il} + \mathbf{a}_{lj}$$

assuming that $j \neq l$. If $j = l$, then we get

$$\mathbf{h}_{kl} + \mathbf{a}_{ij} = \mathbf{a}_{ll} + \mathbf{a}_{kl} + \mathbf{a}_{ik}.$$

Thus, if $i \neq j$ and $i \notin \{k, l\}$, then $\mathbf{h}_{kl} + \mathbf{a}_{ij}$ is not a hole. Similarly, if $j \notin \{k, l\}$, then $\mathbf{h}_{kl} + \mathbf{a}_{ij}$ is not a hole. Thus, if $\mathbf{h}_{kl} + \sum_{r=1}^{s} \mathbf{a}_{i_r j_r}$ is a hole, then either $i_r = j_r$ or $\{i_r, j_r\} = \{k, l\}$ for each r. We claim that either $i_r = j_r$ for all r, or $\{i_r, j_r\} = \{k, l\}$ for all r. This implies that each hole is as in the statement of the theorem. The claim follows from the computation

$$\mathbf{h}_{kl} + \mathbf{a}_{ii} + \mathbf{a}_{kl} = \mathbf{a}_{kk} + \mathbf{a}_{ll} + \mathbf{a}_{il} + \mathbf{a}_{ki},$$

which is valid whenever $i \notin \{k, l\}$ and $k \neq l$.

It remains to see that every integer vector in (6) is indeed a hole. Let $\mathbf{h}$ be of the form (6a), and suppose that $\mathbf{h} = \sum_{ij} \lambda_{ij}\mathbf{a}_{ij}$ with $\lambda_{ij} \geq 0$. Then $\lambda_{ij} \neq 0$ only for $\{i, j\} \subseteq \{k, l\}$, because $\mathrm{h}_i = 0$ or $\mathrm{h}_j = 0$ for $i, j \notin \{k, l\}$. The matrix A restricted to the columns $\mathbf{a}_{ij}$ with $\{i, j\} \subseteq \{k, l\}$

equals

$$\begin{pmatrix} 1 & 0 & 1 & 0 \\ 0 & 1 & 0 & 1 \\ 1 & 1 & 0 & 0 \\ 0 & 0 & 1 & 1 \\ 1 & 0 & 0 & 1 \end{pmatrix}$$

up to rows with only zeros. Since this matrix has rank four, the representation of $\mathbf{h}$ as a linear combination is unique. However, by assumption, $\mathbf{h}$ has a representation of the form $\mathbf{h} = \mathbf{h}_{kl} + \dots$ in which the coefficients are not integers (but half integers). Thus, $\mathbf{h}$ is a hole.

Finally, let $\mathbf{h}$ be of the form (6b). Suppose that $\mathbf{h} = \sum_{ij} \lambda_{ij} \mathbf{a}_{ij}$ with $\lambda_{ij} \in \mathbb{Z}_{\geq 0}$, and let $S = \sum_{i=1}^{d} \mathrm{h}_i = \sum_{ij} \lambda_{ij}$. Note that $\mathrm{h}_{2d+1} = S - 1$. Therefore, $\mathbf{h}$ is the sum of $S - 1$ "diagonal" columns $\mathbf{a}_{ii}$ and one "off-diagonal" column $\mathbf{a}_{ij}$. This is not possible, since, by assumption, the column sums and the row sums are the same. Q.E.D.

Remark 4.6. *Note the following two properties of the hole monoids:*

(1) *The two monoids corresponding to a single hole $\mathbf{h}_{kl}$ are not disjoint.*

(2) *The hole monoids corresponding to two different fundamental holes $\mathbf{h}_{kl}$, $\mathbf{h}_{k'l'}$ are not disjoint: For example,*

$$\mathbf{h}_{12} + \mathbf{a}_{33} = \mathbf{h}_{23} + \mathbf{a}_{11}.$$

§5. Computational experiments

Semigroups play an important role in combinatorics, in discrete geometry, and in combinatorial commutative algebra. The interplay between these areas is nicely exemplified by the theory of lattice polytopes. One can associate a semigroup to every lattice polytope. The Hilbert function of the corresponding graded semigroup ring turns out to be the Ehrhart function of the lattice polytope, see [12, Section 12.1] for more details about this connection and see [5] for a nice introduction to Ehrhart theory.

It is of particular interest to determine whether this semigroup has holes. For example, if the semigroup has holes, then there is no unimodular triangulation. Therefore, the algebraic structure is closely related to geometric properties. In Section 5.1, we briefly define what a (lattice) polytope is and how to construct the corresponding semigroup, and we present a computational result regarding the linear ordering polytope.

Holes of semigroups also play a role when computing Markov bases, as was recently shown by [15]. We give a brief example in Section 5.2.

5.1. The Integer-Decomposition Property and Linear Order Polytopes

A *polytope* $P \subset \mathbb{R}^d$ is the convex hull of finitely many vectors $\mathbf{v}_1$, $\mathbf{v}_2, \ldots, \mathbf{v}_n \in \mathbb{R}^d$, and we write

$$P = \operatorname{conv}(\mathbf{v}_1, \mathbf{v}_2, \ldots, \mathbf{v}_n),$$

where $\operatorname{conv}(\mathbf{v}_1, \mathbf{v}_2, \ldots, \mathbf{v}_n) = \left\{\sum_i \lambda_i \mathbf{v}_i \mid \lambda_i \geq 0, \sum_j \lambda_j = 1\right\}$.

The inclusion-minimal subset $V \subset \{\mathbf{v}_1, \mathbf{v}_2, \ldots, \mathbf{v}_n\}$ such that $P = \operatorname{conv}(V)$ is called the *vertex set* of P and an element of V is called a *vertex*. We say P is a *lattice polytope* if $V \subset \mathbb{Z}^d$, i.e., all coordinates of the vertices are integral. The *dimension* of a polytope P is the dimension of its affine span, i.e., the dimension of the smallest affine subspace containing P. A d-dimensional polytope is sometimes called a *d-polytope.*

A lattice d-polytope $P \in \mathbb{R}^d$ has the *integer-decomposition property* (IDP) if for every integer $k > 0$ and every integer point $\mathbf{z} \in kP \cap \mathbb{Z}^d$, there exist $\mathbf{x}_1, \mathbf{x}_2, \ldots, \mathbf{x}_k \in P \cap \mathbb{Z}^d$ with

$$\mathbf{z} = \mathbf{x}_1 + \mathbf{x}_2 + \cdots + \mathbf{x}_k. \tag{7}$$

Such a polytope is also called *integrally closed* by other authors.

We can also express the integer-decomposition property in the language of semigroups. Let $(P, 1) := \{(\mathbf{x}, 1) \mid \mathbf{x} \in P\} \subset \mathbb{R}^{d+1}$ be the polytope embedded in $\mathbb{R}^{d+1}$ at height 1. Moreover, let $K_P := \mathbb{R}_{\geq 0}(P, 1)$ denote the (pointed) cone generated by the points in $(P, 1)$ and let

$$Q_P := \left\{\mathbf{z} \mid \mathbf{z} = k_1\mathbf{x}_1 + \cdots + k_n\mathbf{x}_n, \text{ where } k_i \in \mathbb{Z}_{\geq 0},\ \mathbf{x}_i \in (P, 1) \cap \mathbb{Z}^{d+1}\right\}$$

be the semigroup generated by the integer points in $(P, 1)$. Then P has the IDP if and only if the semigroup Q_P is saturated. This means that we can check if a polytope has the integer-decomposition property by showing that the semigroup does not have any holes.

In a computational experiment, we examined whether the 7th linear ordering polytope has the integer decomposition property. Sturmfels and Welker already showed that for $n \leq 6$, the nth linear ordering polytope satisfies the IDP, see [16, Theorem 6.1].

For any permutation π of n elements, we define

$$v_{ij}(\pi) = \begin{cases} 1 & \text{if } \pi(i) > \pi(j) \\ 0 & \text{otherwise,} \end{cases}$$

where $1 \leq i < j \leq n$. We follow the definition of [11, Section 3.3] and define the nth linear ordering polytope P_n as the convex hull of the $n!$ vectors $\mathbf{v}(\pi) := (v_{ij}(\pi))_{1\le i<l\le n} \in \mathbb{R}^{\binom{n}{2}}$. Note that the vertices are the only integer points of P_n. A python program that generates the matrix can be downloaded from the HASE homepage. After a bit more than a month of computation time on a linux machine with 16 Intel processors (Intel(R) Xeon(R) CPU E5-2687W v2 at 3.40GHz), the program confirmed that P_7 also has the IDP.

Theorem 5.1. *The nth linear ordering polytope has the integer-decomposition property for $n \le 7$.*

The question whether or not P_n satisfies the IDP for all $n \in \mathbb{Z}_{\ge 1}$ is still open.

5.2. Lifting Markov bases and Gröbner bases

Recently, Rauh and Sullivant [15] have proposed a new iterative algorithm to compute Markov bases and Gröbner bases of toric ideals in which a key step is to understand the holes of an associated semigroup. We do not explain this theory here, but we summarize two examples that arose in this context and that can now be reproduced using HASE.

The first example is from the computation of the Markov basis of the binary complete bipartite graph $K_{3,N}$, as computed by [14]. The associated semigroup has two fundamental holes. Each fundamental hole has one associated monoid, generated by eight generators. The input file `K31codz.mat` for HASE can be downloaded from the HASE homepage. Within a few seconds, HASE produces the following output:

```
Normaliz found 2 fundamental holes.
Standard pairs of [1 1 1 1 1 1 1 1 1 1 1 1 0 1 1 0 1 0 0 1]:
   1: {x1, x2, x4, x7, x9, x10, x12, x15}
Standard pairs of [1 1 1 1 1 1 1 1 1 1 1 1 0 0 1 0 1 1 0]:
   1: {x0, x3, x5, x6, x8, x11, x13, x14}
```

The same method can be used to compute a Markov basis for the binary 3×3-grid. In this case, one needs to understand the holes of a larger semigroup. Again, the input file `3x3codz.mat` can be downloaded from the HASE homepage.

This problem turns out to be much more difficult for HASE, and in fact, after waiting 24 hours for HASE to finish we became impatient and aborted the program, even when the option `--trick` was activated.

Surprisingly, it turns out that the set of holes itself can be computed with some extra information: While the semigroup has 32 fundamental holes, there are only three symmetry classes. The time that HASE spends on a single hole varies greatly, even within a symmetry class. So all that is needed to finish the computation is to find representatives of the three symmetry classes such that the hole monoid computations run through relatively quickly. The current version of HASE cannot be used to run the algorithm on a subset of the fundamental holes (but it is not difficult to do this manually by looking at HASE's source code). This shows once again how important it is to take symmetry into account.

§6. Discussion and open problems

There are many open problems concerning semigroups and holes of semigroups. In this section, we just want to briefly mention a non-respresentative selection of open problems.

As mentioned in the beginning, the Frobenius problem is still open if there are more than two generators. There are several computational results, see e.g. [4]. It might be possible to use the structure of the holes, i.e. which hole is based on which fundamental hole, to say something about the Frobenius number. Alternatively, one could use a slightly modified version of HASE to compute the Frobenius number explicitly.

As briefly discussed in Section 5.1, holes in semigroups coming from lattice polytopes are of particular interest as they reflect geometric properties. Therefore, another application of HASE is to describe the semigroup coming from a user specified lattice polytope.

§7. Acknowledgements

R.Y. is supported by NSF Division of Mathematical Sciences: CDS&E-MSS program with the award number 1622369. The first author would like to thank Christian Haase for helpful discussions and for pointing his attention towards linear ordering polytopes. The first author was partially supported by a scholarship of the Berlin Mathematical School. We would also like to thank the **Normaliz** team for helpful discussions and suggestions.

References

[1] 4ti2 team. 4ti2—a software package for algebraic, geometric and combinatorial problems on linear spaces. Available at www.4ti2.de, 2008.

[2] K. Woods A. Barvinok. Short rational generating functions for lattice point problems. *J. Amer. Math. Soc.*, 16:957–979, 2003.

[3] A. Barvinok. Polynomial time algorithm for counting integral points in polyhedra when the dimension is fixed. *Math of Operations Research*, 19:769–779, 1994.

[4] Matthias Beck, David Einstein, and Shelemyahu Zacks. Some experimental results on the Frobenius problem. *Experiment. Math.*, 12(3):263–269, 2003.

[5] Matthias Beck and Sinai Robins. *Computing the Continuous Discretely*. Undergraduate Texts in Mathematics. Springer, New York, 2007. Integer-point enumeration in polyhedra.

[6] Michel Berkelaar, Kjell Eikland, and Peter Notebaert. *lp_solve : Open source (Mixed-Integer) Linear Programming system*, 2015.

[7] W. Bruns, B. Ichim, T. Roemer, R. Sieg, and C. Soeger. Normaliz. algorithms for rational cones and affine monoids. Available at `https://www.normaliz.uni-osnabrueck.de`.

[8] Daniel R. Grayson and Michael E. Stillman. Macaulay2, a software system for research in algebraic geometry. Available at `http://www.math.uiuc.edu/Macaulay2/`.

[9] H. Hara, A. Takemura, and R. Yoshida. A markov basis for conditional test of common diagonal effect in quasi-independence model for two-way contingency tables. *J of Computational Statistics and Data Analysis*, 53:1006–1014, 2009.

[10] R. Hemmecke, A. Takemura, and R. Yoshida. Computing holes in semigroups and its applications to transportation problems. *Contributions to Discrete Mathematics*, 4(1):81 – 91, 2009.

[11] Lukas Katthän. Decomposing sets of inversions. *Electron. J. Combin.*, 20:Paper 40, 16, 2013.

[12] E. Miller and B. Sturmfels. *Combinatorial Commutative Algebra*. Springer-Verlag, New York, 2005.

[13] J. L. Ramírez Alfonsín. *The Diophantine Frobenius problem*, volume 30 of *Oxford Lecture Series in Mathematics and its Applications*. Oxford University Press, Oxford, 2005.

[14] Johannes Rauh and Seth Sullivant. The Markov basis of $K_{3,N}$. *preprint*, 2014.

[15] Johannes Rauh and Seth Sullivant. Lifting Markov bases and higher codimension toric fiber products. *Journal of Symbolic Computation*, 74:276–307, 2015.

[16] Bernd Sturmfels and Volkmar Welker. Commutative algebra of statistical ranking. *J. Algebra*, 361:264–286, 2012.

[17] A. Takemura, P. Thomas, and R. Yoshida. Holes in semigroups and their applications to the two-way common diagonal effect model. *In: Proceedings of the 2008 International Conference on Information Theory and Statistical Learning, ITSL*, 2008.

[18] A. Takemura and R. Yoshida. A generalization of the integer linear infeasibility problem. *Discrete Optimization*, 5(1):36–52, 2008.

(Florian Kohl) *FB Mathematik and Informatik Institut fuer Mathematik, Arnimallee3, Room 033, Königin-Luise Str. 24-26, Berlin 14195, Germany*
E-mail address: fkohl@zedat.fu-berlin.de

(Yanxi Li) *Statistics Department, University of Kentucky, 725 Rose Street, Lexington KY 40536-0082, USA*
E-mail address: yli282@g.uky.edu

(Johannes Rauh) *Max Planck Institute for Mathematics in the Sciences, Inselstr. 21, 04103 Leipzig, Germany*
E-mail address: jrauh@mis.mpg.de

(Ruriko Yoshida) *Department of Operations Research, 1411 Cunningham Road, Naval Postgraduate School, Monterey, CA 93943-5219, USA*
E-mail address: ryoshida@np.sedu

Advanced Studies in Pure Mathematics 77, 2018
The 50th Anniversary of Gröbner Bases
pp. 171–205

Algebraic tools for the analysis of state space models

Nicolette Meshkat, Zvi Rosen and Seth Sullivant

Abstract.

We present algebraic techniques to analyze state space models in the areas of structural identifiability, observability, and indistinguishability. While the emphasis is on surveying existing algebraic tools for studying ODE systems, we also present a variety of new results. In particular: on structural identifiability, we present a method using linear algebra to find identifiable functions of the parameters of a model for unidentifiable models. On observability, we present techniques using Gröbner bases and algebraic matroids to test algebraic observability of state space models. On indistinguishability, we present a sufficient condition for distinguishability using computational algebra and demonstrate testing indistinguishability.

Contents

Received August 31, 2016.
2010 *Mathematics Subject Classification.* 13P10, 13P25, 05E40, 52B20.
Key words and phrases. identifiability, observability, indistinguishability, state space models.

§1. Introduction

Consider a dynamic systems model in the following state space form:

$$x'(t) = f(x(t), p, u(t)) \qquad y(t) = g(x(t), p) \tag{1}$$

Here $x(t)$ is the state variable vector, $u(t)$ is the input vector (or control vector), $y(t)$ is the output vector, and p is a parameter vector $(p_1, ..., p_n)$ composed of unknown real parameters $p_1, ..., p_n$. In this modeling framework the only observed quantities are the input and output trajectories, $u(t)$ and $y(t)$ (or more realistically, the trajectories observed at some finite number of time points $t_1, t_2, \ldots$), together with the underlying modeling structure (that is, the functions f and g). State space models are widely used throughout the applied sciences, including the areas of control [27, 53, 59, 68], systems biology [22], economics and finance [35, 77], and probability and statistics [11, 40].

A simple example of a state space model is a linear compartment model.

Example 1.1. Consider the following ODE:

$$\begin{pmatrix} x_1' \\ x_2' \end{pmatrix} = \begin{pmatrix} -(a_{01}+a_{21}) & a_{12} \\ a_{21} & -(a_{02}+a_{12}) \end{pmatrix} \begin{pmatrix} x_1 \\ x_2 \end{pmatrix} + \begin{pmatrix} u_1 \\ 0 \end{pmatrix} \qquad y_1 = x_1.$$

This model is called the linear 2-compartment model and will be referenced in later sections. Here $(x_1(t), x_2(t))$ is the state variable vector, $u_1(t)$ is the input (or control), $y_1(t)$ is the output, and $(a_{01}, a_{02}, a_{12}, a_{21})$ is the unknown parameter vector.

Although the analysis of the behavior and use of state space models falls under the dynamical systems research area umbrella, tools from algebra can be used to analyze these models when the functions f and g are rational functions. Algebraic methods typically focus on determining which key features the models satisfy *a priori* before the models are used to analyze data. The point of the present paper is to give an overview of these algebraic techniques to show how they can be applied to analyze state space models. We focus on three main problems where algebraic techniques can be helpful: determining structural identifiability, observability, and indistinguishability of the models. We provide an overview of techniques for these problems coming from computational algebra and we also introduce some new results coming from matroid theory.

§2. State Space Models

In this section, we provide a more detailed introduction to state space models, and the basic theoretical problems of identifiability, observability, and indistinguishability that we will address in this paper. We also provide a detailed introduction to the linear compartment models that will be an important set of examples that we use to illustrate the theory.

Consider a general state space model

$$x'(t) = f(x(t), p, u(t)) \qquad y(t) = g(x(t), p) \tag{2}$$

as in the introduction, with $x(t) \in \mathbb{R}^N$, $y(t) \in \mathbb{R}^M$, $u(t) \in \mathbb{R}^R$ and $p \in \mathbb{R}^n$.

The state space model (2) is called *identifiable* if the unknown parameter vector p can be recovered from observation of the input and output alone. The model is *observable* if the trajectories of the state space variables $x(t)$ can be recovered from observation of the input and output alone. Two state space models are *indistinguishable* if for any choice of parameters in the first model, there is a choice of parameters in the second model that will yield the same dynamics in both models, and vice versa. Before getting into the technical details of these definitions for state space models, we introduce some key examples of state space models that we will use to illustrate the main concepts throughout the paper.

Example 2.1 (SIR Model). A commonly used model in epidemiology is the Susceptible-Infected-Recovered model (SIR model) ([8], [9], [10], [47], [63]) which has the following form:

$$\begin{aligned} S' &= \mu N - \beta S \frac{I}{N} - \mu S \\ I' &= \beta S \frac{I}{N} - (\mu + \gamma) I \\ R' &= \gamma I - \mu R \\ y &= kI \end{aligned}$$

The interpretation of the state variables is that $S(t)$ is the number of susceptible individuals at time t, $I(t)$ is the number of infected individuals at time t, and $R(t)$ is the number of recovered individuals at time t. The unknown parameters are the birth/death rate μ, the transmission parameter β, the recovery rate γ, the total population N, and the proportion of the infected population measured k. In this model, we assume that we only observe the trajectory $y(t)$, an (unknown) proportion of the infected population. Note that this simple model has no input/control.

Identifiability and observability analysis in this model are concerned with determining which unmeasured quantities can be determined from only the observed output trajectory y. Identifiability specifically concerns the unobserved parameters μ, β, γ, N, and k, whereas observability specifically is concerned with the unobserved state variables S, I, and R.

A commonly used family of state space models are the linear compartment models. We outline these models here. Let $G = (V, E)$ be a directed graph with vertex set V and set of directed edges E. Each vertex $i \in V$ corresponds to a compartment in our model and each edge $j \to i$ corresponds to a direct flow of material from the jth compartment to the ith compartment. Let $In, Out, Leak \subseteq V$ be three sets of compartments: the set of input compartments, output compartments, and leak compartments respectively. To each edge $j \to i$ we associate an independent parameter a_{ij}, the rate of flow from compartment j to compartment i. To each leak node $i \in Leak$, we associate an independent parameter a_{0i}, the rate of flow from compartment i leaving the system.

To such a graph G and set of leaks $Leak$ we associate the matrix $A(G)$ in the following way:

$$A(G)_{ij} = \begin{cases} -a_{0i} - \sum_{k:i\to k\in E} a_{ki} & \text{if } i = j \text{ and } i \in Leak \\ -\sum_{k:i\to k\in E} a_{ki} & \text{if } i = j \text{ and } i \notin Leak \\ a_{ij} & \text{if } j \to i \text{ is an edge of } G \\ 0 & \text{otherwise} \end{cases}$$

For brevity, we will often use A to denote $A(G)$. Then we construct a system of linear ODEs with inputs and outputs associated to the quadruple $(G, In, Out, Leak)$ as follows:

$$x'(t) = Ax(t) + u(t) \qquad y_i(t) = x_i(t) \text{ for } i \in Out \tag{3}$$

where $u_i(t) \equiv 0$ for $i \notin In$. The coordinate functions $x_i(t)$ are the state variables, the functions $y_i(t)$ are the output variables, and the nonzero functions $u_i(t)$ are the inputs. The resulting model is called a *linear compartment model.*

We use the following convention for drawing linear compartment models [22]. Numbered vertices represent compartments, outgoing arrows from the compartments represent leaks, an edge with a circle coming out of a compartment represents an output, and an arrowhead pointing into a compartment represents an input.

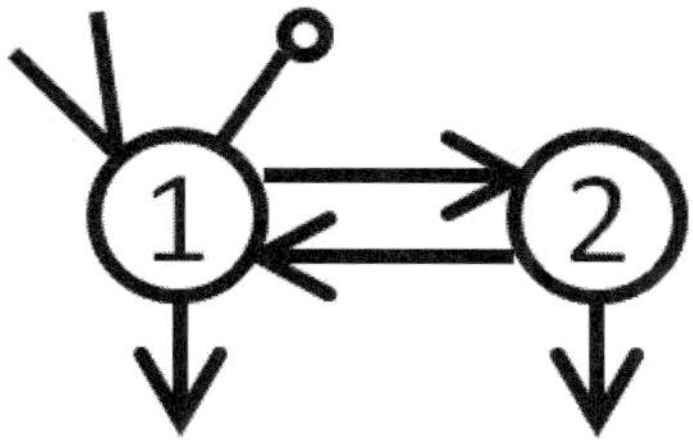

Fig. 1. A 2-compartment model with $In = \{1\}$, $Out = \{1\}$, and $Leak = \{1, 2\}$.

Example 2.2. For the compartment model in Figure 1, the ODE system has the form given in Example 1.1. Since this model has a leak in every compartment, the diagonal entries of $A(G)$ are algebraically independent of the other entries. In this situation, we can re-write the diagonal entries of the matrix A as $a_{11} = -(a_{01}+a_{21})$ and $a_{22} = -(a_{02}+a_{12})$. Thus we have the following ODE system:

$$\begin{pmatrix} x_1' \\ x_2' \end{pmatrix} = \begin{pmatrix} a_{11} & a_{12} \\ a_{21} & a_{22} \end{pmatrix} \begin{pmatrix} x_1 \\ x_2 \end{pmatrix} + \begin{pmatrix} u_1 \\ 0 \end{pmatrix} \qquad y_1 = x_1.$$

§3. Differential Algebra Approach To Identifiability

In this paper we focus on the *structural* versions of identifiability, observability, and indistinguishability (that is, structural identifiability, structural observability, structural indistinguishability). That means we study when these properties hold assuming that we are able to observe trajectories perfectly. *Practical* versions of these problems concern how noise affects the ability to, e.g., infer parameters of the models. Structural answers are important because the structural version of the condition is necessary to insure that the practical version holds. On the other hand, practical versions of these problems depend on the specific data dependent context in which the data might be observed, and might further depend on the particular underlying unknown parameter choices. We will drop "structural" throughout the paper since this will be implicit in the majority of our discussion.

To make the definitions of identifiability, observability, and indistinguishability precise we will use tools from differential algebra. In this approach, we must form the *input-output equations* associated to our model by performing *differential elimination.* We carry out operations in the differential ring

$$\mathbb{Q}(p)[x, y, u, x', y', u', \ldots]$$

with the derivation $\frac{d}{dt}$ with respect to time such that the parameters p are constants with respect to the derivation, and $\frac{d}{dt}x = x'$, etc. Differential algebra was developed by Ritt [60] and Kolchin [41] and has its most well-known applications to the study of the algebraic solution to systems of differential equations [64].

The goal of this differential elimination process for state space models is to eliminate the state variables $x(t)$ and their derivatives, so that the resulting equations are purely in terms of the input variables, output variables, and the parameters. The equations that result from applying the differential elimination process are called the *input-output equations.* We obtain input-output equations in the following form:

$$\sum_i c_i(p)\psi_i(u, y) = 0$$

where $c_i(p)$ are rational functions in the parameter vector p and $\psi_i(u, y)$ are differential monomials in $u(t)$ and $y(t)$. Let $c = (c_1(p), ..., c_m(p))$ denote the vector of coefficients of the input-output equations, which are rational functions in the parameter vector p. This coefficient vector induces a map $c : \mathbb{R}^n \to \mathbb{R}^m$ called the coefficient map, that plays an important role in the study of identifiability and indistinguishability.

For general state space models of the form (2) we can also use ordinary Gröbner basis calculations to determine the input-output equation.

Proposition 3.1. *Consider a state space model of the form* (2) *where f and g are polynomial functions and where there are N state-space variables, $M = 1$ output variable, and R input variables. Let P be the ideal*

$$\begin{aligned}\langle x' - f(x, p, u), \ldots, x^{(N)} - \frac{d^{N-1}}{dt^{N-1}} f(x, p, u), \\ y - g(x, p), \ldots, y^{(N)} - \frac{d^N}{dt^N} g(x, p)\rangle \\ \subseteq \mathbb{Q}(p)[x, y, u, x', y', u', \ldots, x^{(N-1)}, y^{(N-1)}, u^{(N-1)}, x^{(N)}, y^{(N)}].\end{aligned}$$

Then $P \cap \mathbb{Q}(p)[y, u, y', u', \ldots, y^{(N-1)}, u^{(N-1)}, y^{(N)}]$ is not the zero ideal and hence contains an input-output equation.

Although Proposition 3.1 is known in the literature [28, 37], we include a proof because it will illustrate some useful ideas that we will use in other new results later on. Note that although this is stated for a single output, one can apply Proposition 3.1 one output at a time to find input-output equations for each output separately and hence obtain Proposition 3.2.

Proof. Note that P is a prime ideal, since, with a carefully chosen lexicographic term order, it has as its initial ideal

$$\langle x', \ldots, x^{(N)}, y, \ldots, y^{(N)} \rangle$$

which is a prime ideal. Since P is prime, we can consider the algebraic matroid associated to this ideal. To say that $P \cap \mathbb{Q}(p)[y, u, y', u', \ldots, y^{(N-1)}, u^{(N-1)}, y^{(N)}]$ is not the zero ideal is equivalent to saying that the set $\{y, u, y', u', \ldots, y^{(N-1)}, u^{(N-1)}, y^{(N)}\}$ is a dependent set in the associated algebraic matroid. The initial ideal also shows that this ideal is a complete intersection, so it is has codimension N^2+N+1 (since this is the number of equations involved). The total number of variables in our polynomial ring is $N(N+1)+N+1+RN$, where $N(N+1)$ counts the $x, x', \ldots$ variables, $N+1$ counts the $y, y', \ldots$ variables, and RN counts the $u, u', \ldots$ variables. Thus P has dimension $N+RN$. Since the total number of variables in the set $\{y, u, y', u', \ldots, y^{(N-1)}, u^{(N-1)}, y^{(N)}\}$ is $N+1+RN$, these variables must be dependent, i.e. there must exist a relation. Q.E.D.

For multiple outputs, one can again take derivatives up to order N and show that there must exist an input-output equation for each output:

Proposition 3.2. *Consider a state space model of the form* (2) *where f and g are polynomial functions and where there are N state-space variables, M output variables, and R input variables. Let P be the ideal*

$$\begin{aligned} \langle x' - f(x,p,u), \ \ldots, \ x^{(N)} - \frac{d^{N-1}}{dt^{N-1}} f(x,p,u), \\ y - g(x,p), \ \ldots, \ y^{(N)} - \frac{d^N}{dt^N} g(x,p) \rangle \\ \subseteq \mathbb{Q}(p)[x, y, u, x', y', u', \ldots, x^{(N-1)}, y^{(N-1)}, u^{(N-1)}, x^{(N)}, y^{(N)}]. \end{aligned}$$

Then $P \cap \mathbb{Q}(p)[y_i, u, y_i', u', \ldots, y_i^{(N-1)}, u^{(N-1)}, y_i^{(N)}]$ is not the zero ideal and hence contains an input-output equation for each y_i.

Proof. We follow the proof of Proposition 3.1. The number of equations involved is $N^2 + M(N+1)$. The total number of variables in our polynomial ring is $N(N+1)+M(N+1)+RN$. Here $N(N+1)$ counts the $x, x', \ldots$ variables, $M(N+1)$ counts the $y, y', \ldots$ variables, and RN counts the $u, u', \ldots$ variables. Thus P has dimension $N+RN$. For each y_i, the total number of variables in the set $\{y_i, u, y_i', u', \ldots, y_i^{(N-1)}, u^{(N-1)}, y_i^{(N)}\}$ is $N+1+RN$. Thus these variables must be dependent, i.e. there must exist a relation for each y_i. Q.E.D.

Note that one could also work with smaller ideals than P with only up to $k \leq N$ derivatives, as in [48]. In some instances this might produce an input-output equation, but the dimension guarantee that ensures the existence of an input-output equation only occurs when $k = N$.

Example 3.3. Consider the SIR model from Example 2.1. The ideal P in this example is:

$$\begin{aligned}\langle S' - \mu S - \beta S\frac{I}{N} + \mu N,\ & S'' - \mu S' - \beta S\frac{I'}{N} - \beta S'\frac{I}{N}, \\ & S''' - \mu S'' - \beta S\frac{I''}{N} - 2\beta S'\frac{I'}{N} - \beta S''\frac{I}{N}, \\ & I' - (\mu+\gamma)I + \beta S\frac{I}{N},\ I'' - (\mu+\gamma)I' + \beta S'\frac{I}{N} - \beta S\frac{I'}{N}, \\ & I''' - (\mu+\gamma)I'' + \beta S''\frac{I}{N} - 2\beta S'\frac{I'}{N} - \beta S\frac{I''}{N}, \\ & R' - \mu R + \gamma I,\ R'' - \mu R' + \gamma I',\ R''' - \mu R'' + \gamma I'', \\ & y - kI,\ y' - kI',\ y'' - kI'',\ y''' - kI'''\rangle.\end{aligned}$$

This model has no input, so in this case we get a single output equation in the output variable y and the parameters μ, β, γ, N, and k. The output equation is:

$$\begin{aligned}(-\beta kN\mu + kN\mu^2 + kN\mu\gamma)y^2 + (\beta\mu + \beta\gamma)y^3 \\ + kN\mu yy' + \beta y^2 y' - kNy'^2 + kNyy'' = 0.\end{aligned}$$

This differential equation has 6 differential monomials $y^2, y^3, yy', y^2y', y'^2, yy''$, so the coefficient vector c gives a function from $\mathbb{R}^5$ to $\mathbb{R}^6$, given by

$$\begin{aligned}c : \mathbb{R}^5 \to \mathbb{R}^6, \quad & (\mu, \beta, \gamma, N, k) \mapsto \\ & (-\beta kN\mu + kN\mu^2 + kN\mu\gamma, \beta\mu + \beta\gamma, kN\mu, \beta, -kN, kN).\end{aligned}$$

The dynamics of the input and output will only depend on the input-output equation up to a nonzero constant multiple. Hence, the coefficient map is only truly well-defined up to scalar multiplication. There are two natural ways to deal with this issue. The most appealing for an algebraist is to treat the coefficient map as a map into projective space: $c : \mathbb{R}^n \to \mathbb{RP}^{m-1}$. The second approach is to force the equation to have a fixed form that will avoid this issue, by forcing the equation to be monic by dividing through by one of the coefficients. We will take the second approach in this paper. In the output equation in Example 3.3, one possible normalization yields the coefficient map

$$c : \mathbb{R}^5 \to \mathbb{R}^6, \quad (\mu, \beta, \gamma, N, k) \mapsto (-\beta\mu + \mu^2 + \mu\gamma, \frac{\beta\mu + \beta\gamma}{kN}, \mu, \frac{\beta}{kN}, -1, 1).$$

In the standard differential algebra approach to identifiability, we assume that the coefficients $c_i(p)$ of the input-output equations can be recovered uniquely from the input-output data, and thus are assumed to be known quantities. This is a reasonable assumption when the input u is a general enough function and the parameters are generic: in this case the dynamics will yield a unique differential equation. The identifiability question is then: can the parameters of the model be recovered from the coefficients of the input-output equations?

Definition 3.4. Let $c = (c_1(p), ..., c_m(p))$ denote the vector of coefficients of the input-output equations, which are rational functions in the parameter vector p, which we assume to be normalized so that the input-output equations are monic. We consider c as a function from some natural open biologically relevant parameter space $\Theta \subseteq \mathbb{R}^n$.

- The model is *globally identifiable* if $c : \Theta \to \mathbb{R}^m$ is a one-to-one function.
- The model is *generically globally identifiable* if there is a dense open subset $\Theta' \subseteq \Theta$ such that $c : \Theta' \to \mathbb{R}^m$ is one-to-one.
- The model is *locally identifiable* if around any point $p \in \Theta$ there is an open neighborhood $U_p \subseteq \Theta$ such that $c : U_p \to \mathbb{R}^m$ is a one-to-one function.
- The model is *generically locally identifiable* if there is a dense open subset $\Theta' \subseteq \Theta$ such that for all $p \in \Theta'$ there is an open neighborhood $U_p \subseteq \Theta'$ such that $c : U_p \to \mathbb{R}^m$ is a one-to-one function.
- The model is *unidentifiable* if there is a $p \in \Theta$ such that $c^{-1}(c(p))$ is infinite.
- The model is *generically unidentifiable* if there is a dense subset $\Theta' \subseteq \Theta$ such that for all $p \in \Theta'$, $c^{-1}(c(p))$ is infinite.

As can be seen, there are many different variations on the notions of identifiability. Because of problems that might arise on sets of measure zero that can ruin the strongest form of global identifiability, one usually needs to add the generic conditions to get meaningful results. In this paper, we will consider state space models (2) where f and g are polynomial (or rational) functions. This ensures, via the differential elimination procedure, that the coefficient function $c(p)$ is a rational function of the parameters. For linear compartment models this can always be taken to be a polynomial function.

In this paper, we will also focus almost exclusively on generic local identifiability and generic nonidentifiability and will use the following result to determine which of these conditions the model satisfies.

Proposition 3.5. *The model is generically locally identifiable if and only if the rank of the Jacobian of c is equal to n when evaluated at a generic point. Conversely, if the rank of the Jacobian of c is less than n for all choices of the parameters then the model is generically unidentifiable.*

Proof. Since the coefficients in c are all polynomial or rational functions of the parameters, the model is generically locally identifiable if and only if the image of c has dimension equal to the number of parameters, i.e. n. The dimension of the image of a map is equal to the evaluation of the Jacobian at a generic point. Q.E.D.

Example 3.6. SIR Model From Example 3.3, we have the following coefficient map:

$$c : \mathbb{R}^5 \to \mathbb{R}^6, \quad (\mu, \beta, \gamma, N, k) \mapsto (-\beta\mu + \mu^2 + \mu\gamma, \frac{\beta\mu + \beta\gamma}{kN}, \mu, \frac{\beta}{kN}, -1, 1).$$

The Jacobian with respect to the parameter ordering $(k, N, \mu, \gamma, \beta)$ is:

$$\begin{pmatrix} 0 & 0 & -\beta + 2\mu + \gamma & \mu & \mu \\ \frac{-(\mu+\gamma)\beta}{k^2 N} & \frac{-(\mu+\gamma)\beta}{kN^2} & \frac{\beta}{kN} & \frac{\beta}{kN} & \frac{(\mu+\gamma)}{kN} \\ 0 & 0 & 1 & 0 & 0 \\ \frac{-\beta}{k^2 N} & \frac{-\beta}{kN^2} & 0 & 0 & \frac{1}{kN} \\ 0 & 0 & 0 & 0 & 0 \\ 0 & 0 & 0 & 0 & 0 \end{pmatrix}.$$

Since the rank of the Jacobian at a generic point is 4, not $n = 5$, the model is generically unidentifiable.

3.1. Input-output equations for linear models

There have been several methods proposed to find the input-output equations of nonlinear ODE models [3, 5, 25, 26, 43, 48, 54], but for linear models the problem is much simpler. We use Cramer's rule in the following theorem, whose proof can be found in [50]:

Theorem 3.7. *Let ∂ be the differential operator d/dt and let A_{ji} be the submatrix of $\partial I - A$ obtained by deleting the jth row and the ith column of $\partial I - A$. Then the input-output equations are of the form:*

$$\frac{\det(\partial I - A)}{g_i} y_i = \sum_{j \in In} (-1)^{i+j} \frac{\det(A_{ji})}{g_i} u_j$$

where g_i is the greatest common divisor of $\det(\partial I - A)$, $\det(A_{ji})$ such that $j \in In$ for a given $i \in Out$.

Example 3.8. Linear Compartment Model. For the linear 2-compartment model from Example 2.2, we obtain the following input-output equation:

$$y_1'' - (a_{11} + a_{22})y_1' + (a_{11}a_{22} - a_{12}a_{21})y_1 = u_1' - a_{22}u_1.$$

Thus we have the following coefficient map:

$$c : \mathbb{R}^4 \to \mathbb{R}^5,\ (a_{11}, a_{22}, a_{12}, a_{21}) \mapsto (1, -a_{11} - a_{22}, a_{11}a_{22} - a_{12}a_{21}, 1, -a_{22}).$$

We then obtain the Jacobian with respect to the parameter ordering $(a_{11}, a_{22}, a_{12}, a_{21})$:

$$\begin{pmatrix} 0 & 0 & 0 & 0 \\ -1 & -1 & 0 & 0 \\ a_{22} & a_{11} & -a_{21} & -a_{12} \\ 0 & 0 & 0 & 0 \\ 0 & -1 & 0 & 0 \end{pmatrix}.$$

Since the rank of the Jacobian at a generic point is 3, not $n = 4$, the model is generically unidentifiable.

§4. Identifiable functions

One issue that arises in identifiability analysis of state space models is figuring out what to do with a model that is generically unidentifiable. In some circumstances, the natural approach is to develop a new model that has fewer parameters that is identifiable. In other circumstances, the given model is forced upon us by the biology, and we cannot change it. When working with such a generically unidentifiable model, we would still like to determine what functions of the parameters can be determined from given input and output data.

Definition 4.1. Let $c : \Theta \to \mathbb{R}^m$ be the coefficient map, and let $f : \Theta \to \mathbb{R}$ be another function. We say the function f is

- *identifiable* from c if for all $p, p' \in \Theta$, $c(p) = c(p')$ implies $f(p) = f(p')$;
- *generically identifiable* from c if there is an open dense subset $U \subseteq \Theta$ such that f is identifiable from c on U;
- *rationally identifiable* from c if there is a rational function ϕ such that $\phi \circ c(p) = f(p)$ on a dense open subset $U \subseteq \Theta$;
- *locally identifiable* from c if there is an open dense subset $U \subseteq \Theta$ such that for all $p \in U$, there is an open neighborhood U_p such that f is identifiable from c on U_p;
- *non-identifiable* from c if there exists $p, p' \in \Theta$ such that $c(p) = c(p')$ but $f(p) \neq f(p')$; and
- *generically non-identifiable* from c if there is a subset $U \subseteq \Theta$ of nonzero measure such that for all $p \in U$ the set $\{f(p') : p' \in U \text{ and } c(p) = c(p')\}$ is infinite.

Example 4.2. From the linear 2-compartment model in Example 3.8, let $p = (a_{11}, a_{22}, a_{12}, a_{21})$ and let $c_1(p) = -a_{11} - a_{22}, c_2(p) = a_{11}a_{22} - a_{12}a_{21}, c_3(p) = -a_{22}$. Then the functions $a_{11}, a_{22}, a_{12}a_{21}$ are rationally identifiable since

$$a_{11} = -c_1 + c_3, \qquad a_{22} = -c_3 \qquad a_{12}a_{21} = -c_2 - (-c_1 + c_3)c_3.$$

Because we work with polynomial and rational maps c and f in this work, the majority of these conditions can be phrased in algebraic language, and checked using computer algebra.

Proposition 4.3.

(1) *The function $f(p)$ is rationally identifiable from $c(p) = (c_1(p), ..., c_m(p))$ if and only if $\mathbb{R}(f(p), c_1(p), ..., c_m(p)) = \mathbb{R}(c_1(p), ..., c_m(p))$ as field extensions.*

(2) *The function $f(p)$ is locally identifiable from $c(p)$ if and only if $f(p)$ is algebraic over $\mathbb{R}(c_1(p), ..., c_m(p))$.*

(3) *The function $f(p)$ is generically non-identifiable from $c(p)$ if and only if $f(p)$ is transcendental over $\mathbb{R}(c_1(p), ..., c_m(p))$.*

To explain how to use Proposition 4.3 to check the various identifiability conditions we need to introduce some terminology. Associated to a set $S \subseteq \mathbb{R}^m$ we have the *vanishing ideal* $\mathcal{I}(S) \subseteq \mathbb{R}[z_1, \ldots, z_m]$ defined by

$$\mathcal{I}(S) = \langle g \in \mathbb{R}[z_1, \ldots, z_m] : g(s) = 0 \text{ for all } s \in S \rangle.$$

When $S = \mathrm{im}(c)$ for c a rational map, the vanishing ideal can be computed using Gröbner bases and elimination [16]. Associated to the pair of coefficient map c and function f that we want to test identifiability of, we have the augmented map $\tilde{c} : \mathbb{R}^n \to \mathbb{R}^{m+1}$, $p \mapsto (f(p), c(p))$, and the augmented vanishing ideal $\mathcal{I}(\mathrm{im}(\tilde{c})) \subseteq \mathbb{R}[z_0, z_1, \ldots, z_m]$.

Proposition 4.4. [30, Proposition 3] *Suppose that $g(z_0, z) \in \mathcal{I}(\mathrm{im}(\tilde{c}))$ is a polynomial such that z_0 appears in g and that we may write $g(z_0, z) = \sum_{i=0}^{d} g_i(z) z_0^i$ so that $g_d(z)$ is not in $\mathcal{I}(\mathrm{im}(c))$.*

(1) *If g is linear, $g = g_1(z)z_0 - g_0(z)$ then f is rationally identifiable from c by the formula $f = \frac{g_0(c)}{g_1(c)}$. If in addition $g_1(z) \neq 0$ for all $z \in \mathrm{im}(c)$ then f is globally identifiable.*

(2) *If g has higher degree $d > 1$ in z_0, then f is locally identifiable, and there are generically at most d possible values for $f(p')$ among all p' with $c(p) = c(p')$.*

(3) *If no such polynomial g exists then f is generically non-identifiable from c.*

For local identifiability of a function, it is also possible to check using a Jacobian calculation, a result that follows easily from Proposition 3.5.

Proposition 4.5. *Let $c : \mathbb{R}^n \to \mathbb{R}^m$ be the coefficient map. A function $f : \mathbb{R}^n \to \mathbb{R}$ is locally identifiable from c if ∇f is in the span of the rows of the Jacobian $J(c)$. Equivalently, consider the augmented map $\tilde{c} : \mathbb{R}^n \to \mathbb{R}^{m+1}$. Then f is locally identifiable from c if and only if the dimension of the image of $\tilde{c}$ equals the dimension of the image of c.*

§5. Finding identifiable functions

The previous section showed how to check, given the coefficient function c and another function of the parameters f, whether f is identifiable from c (under various variations on the definition of identifiability). In some circumstances, there are natural functions to check for their identifiability (e.g. the individual underlying parameters, or certain specific functions with biological interpretations). However, when these fail to be identifiable, one would like tools to discover new functions that are identifiable in a given state space model. In practice the goal is to find a simple set of functions that generates the field $\mathbb{R}(c_1(p), ..., c_m(p))$ (for globally identifiable functions), or a set of functions $f_1, \ldots, f_k$ that are algebraic over $\mathbb{R}(c_1(p), ..., c_m(p))$ and such that $\mathbb{R}(c_1(p), ..., c_m(p)) \subseteq$

$\mathbb{R}(f_1(p), ..., f_k(p))$ (for locally identifiable functions). The notion "simple" is intentionally left vague; typically, we mean functions of low degree that involve few parameters. While there is no general purpose method guaranteed to solve these problems, there are some useful heuristic approaches that seem to work well in practice. We highlight some of these methods in the present section.

One approach to find identifiable functions is to use Gröbner bases. Specifically, one can find a Gröbner basis of the ideal $\langle c_1(p)-c_1(p^*), c_2(p) - c_2(p^*), ..., c_m(p) - c_m(p^*)\rangle \subseteq \mathbb{R}(p^*)[p]$. We state the main result from [49].

Proposition 5.1. [49, Theorem 1] *If $f(p) - f(p^*)$ is an element of a Gröbner basis of $\langle c_1(p) - c_1(p^*), c_2(p) - c_2(p^*), ..., c_m(p) - c_m(p^*)\rangle$ for some elimination ordering of the parameter vector p, then $f(p)$ is globally identifiable. If instead $f(p) - f(p^*)$ is a factor of an element in the Gröbner basis of $\langle c_1(p) - c_1(p^*), c_2(p) - c_2(p^*), ..., c_m(p) - c_m(p^*)\rangle$ for some elimination ordering of the parameter vector p, then $f(p)$ is locally identifiable.*

In practice, the Gröbner basis computations can be performed by picking a random point p^* and computing a Gröbner basis in the ring $\mathbb{R}[p]$. This certifies identifiability with high probability. The elimination ordering is used since elements in the Gröbner basis at the end of the order are likely to be sparse.

The main issue with the Gröbner basis approach to finding identifiable functions is that it is unclear *a priori* how many Gröbner bases one needs to find in order to generate a full set of algebraically independent identifiable functions. Since Gröbner basis computations can become computationally expensive, we provide another approach to find identifiable functions in this paper, using linear algebra with the Jacobian matrix $J(c)$. Specifically, we describe a sort of converse of Proposition 4.5, which allows us to take appropriate elements in the row span of $J(c)$ and deduce that they came from an identifiable function. We first prove a result in the homogeneous case and then extend to arbitrary coefficient maps via homogenization.

Theorem 5.2. *Let c_i be a homogeneous function of degree d_i, corresponding to a coefficient of the input-output equations. Let $v = f_1(c)\nabla c_1 + f_2(c)\nabla c_2 + ... + f_m(c)\nabla c_m$ be a vector in the span of $J(c)$ over the field $\mathbb{R}(c_1(p), ..., c_m(p))$ (that is, each $f_i \in \mathbb{R}(c_1(p), ..., c_m(p))$). Then the dot product $v \cdot p$ is a rationally identifiable function. If each f_i is locally identifiable then $v \cdot p$ is locally identifiable.*

To prove Theorem 5.2 we make use of the Euler homogeneous function theorem.

Proposition 5.3 (Euler's Homogeneous Function Theorem). *Let f be a homogeneous function of degree d. Then $f = \frac{1}{d}\sum_i p_i \frac{\partial f}{\partial p_i}$.*

Proof of Theorem 5.2. Let $f = (f_1, \ldots, f_m)$ be the row vector of the f_i's. The function $v \cdot p$ has the form

$$v \cdot p = fJ(c)p.$$

The rows of $J(c)$ are the gradients of the c_i's. Since these functions are homogeneous, we have that $J(c)p = (d_1c_1(p), \ldots, d_mc_m(p))^T$ by Euler's homogeneous function theorem. But then

$$v \cdot p = fJ(c)p = f(d_1c_1(p), \ldots, d_mc_m(p))^T = \sum_{i=1}^{m} f_i(p)d_ic_i(p)$$

which expresses $v \cdot p$ as a polynomial function in elements of $\mathbb{R}(c_1(p), ..., c_m(p))$, so $v \cdot p$ is rationally identifiable. If each f_i were locally identifiable, $v \cdot p$ would belong to an algebraic extension of $\mathbb{R}(c_1(p), ..., c_m(p))$ and hence be locally identifiable. Q.E.D.

Theorem 5.2 must be used in conjunction with Gaussian elimination and Proposition 4.4 or 4.5. Indeed, our strategy in implementations is to attempt Gaussian elimination cancellations starting with the Jacobian matrix $J(c)$. At each step when we want to perform an elementary operation, we use Proposition 4.4 or 4.5 to check whether the corresponding multiplier is rationally identifiable or locally identifiable. An approach based completely on linear algebra would only make use of Proposition 4.5 in which case we only deduce local identifiability.

Example 5.4. Let c be the map $p \mapsto (c_1(p), c_2(p), c_3(p))$ from the linear 2-compartment model in Example 4.2. Then the Jacobian $J(c)$ is given by

$$\begin{pmatrix} -1 & -1 & 0 & 0 \\ a_{22} & a_{11} & -a_{21} & -a_{12} \\ 0 & -1 & 0 & 0 \end{pmatrix}.$$

Then applying Gaussian elimination over $\mathbb{R}(c_1(p), c_2(p), c_3(p))$, we obtain:

$$\begin{pmatrix} -1 & 0 & 0 & 0 \\ 0 & -1 & 0 & 0 \\ 0 & 0 & -a_{21} & -a_{12} \end{pmatrix}.$$

This implies that $-a_{11}, -a_{22}$ and $-2a_{12}a_{21}$ are all locally identifiable. Thus, a_{11}, a_{22} and $a_{12}a_{21}$ are locally identifiable.

Remark. Note that in Example 4.2, we obtained that the functions a_{11}, a_{22} and $a_{12}a_{21}$ are rationally identifiable, whereas in Example 5.4, we only obtained that the functions a_{11}, a_{22} and $a_{12}a_{21}$ are locally identifiable. This is the cost of not using a Gröbner basis.

Remark. The identifiable functions obtained using linear algebra on the Jacobian matrix depend heavily on the specific column ordering of the Jacobian matrix chosen. Thus, for a given column ordering (corresponding to a given parameter ordering), we may not generate the "simplest" locally identifiable functions. We do, however, always generate identifiable functions, as opposed to the Gröbner basis approach, in which there is no guarantee of generating elements/factors of elements of the form $f(p) - f(p^*)$ for a given elimination ordering p.

Example 5.5. From the SIR Model in Example 3.3, we can form the following coefficient map, ignoring constant coefficients:

$$c(k, N, \mu, \gamma, \beta) = (-\beta\mu + \mu^2 + \mu\gamma, \frac{(\mu+\gamma)\beta}{kN}, \mu, \frac{\beta}{kN})$$

thus we obtain the following Jacobian with respect to the parameter ordering $(k, N, \mu, \gamma, \beta)$:

$$\begin{pmatrix} 0 & 0 & -\beta + 2\mu + \gamma & \mu & \mu \\ \frac{-(\mu+\gamma)\beta}{k^2N} & \frac{-(\mu+\gamma)\beta}{kN^2} & \frac{\beta}{kN} & \frac{\beta}{kN} & \frac{(\mu+\gamma)}{kN} \\ 0 & 0 & 1 & 0 & 0 \\ \frac{-\beta}{k^2N} & \frac{-\beta}{kN^2} & 0 & 0 & \frac{1}{kN} \end{pmatrix}$$

from this we get the row-reduced Jacobian:

$$\begin{pmatrix} N & k & 0 & 0 & 0 \\ 0 & 0 & 1 & 0 & 0 \\ 0 & 0 & 0 & 1 & 0 \\ 0 & 0 & 0 & 0 & 1 \end{pmatrix}.$$

Thus, dotting each row vector with p and dividing each polynomial by their respective degrees, we find that kN, μ, γ, β are locally identifiable.

When the coefficient functions $c_i(p)$ are not homogeneous functions, we can homogenize the functions by some variable z and add z to the list c of identifiable functions. This results in a similar identifiability result.

Theorem 5.6. *Let $\tilde{c}_i$ be the homogenization of the coefficient function c_i and suppose it has degree d_i. Let $v = f_1(\tilde{c}, z)\nabla\tilde{c}_1 + f_2(\tilde{c}, z)\nabla\tilde{c}_2 + \cdots + f_m(\tilde{c}, z)\nabla\tilde{c_m}$ be a vector in the span of $J(\tilde{c}, z)$ over the field $\mathbb{R}(\tilde{c_1}(p, z),$*

..., $\tilde{c_m}(p,z), z)$. *Then the dot product* $v \cdot (p,z)|_{z=1}$ *is a rationally identifiable function. If* $f_1, \dots, f_m$ *are locally identifiable given* c *then* $v \cdot (p,z)|_{z=1}$ *is locally identifiable.*

Proof. Clearly $v \cdot (p,z)$ is rationally identifiable over the field $\mathbb{R}(\tilde{c_1}(p,z), ..., \tilde{c_m}(p,z), z)$ by Theorem 5.2. We need to show that setting $z = 1$ preserves identifiability. Since $f(p,z) = v \cdot (p,z)$ is algebraic over the field $\mathbb{R}(\tilde{c_1}(p,z), ..., \tilde{c_m}(p,z), z)$, then clearly $f(p,z)|_{z=1}$ is algebraic over the field $\mathbb{R}(\tilde{c_1}(p,z)|_{z=1}, ..., \tilde{c_m}(p,z)|_{z=1}, z|_{z=1})$. Since $\tilde{c_i}(p,z)|_{z=1}$ is precisely $c_i(p)$, then $f(p,z)|_{z=1}$ is in the field $\mathbb{R}(c_1(p), ..., c_m(p))$. If $f_1, \dots, f_m$ are algebraic over $\mathbb{R}(\tilde{c_1}(p,z), ..., \tilde{c_m}(p,z), z)$ then $v \cdot (p,z)|_{z=1}$ is algebraic over $\mathbb{R}(c_1(p), ..., c_m(p))$. Q.E.D.

Example 5.7. Let c be the map $(p_1, p_2, p_3) \mapsto (p_1^2, p_1^2 + p_1p_3 + p_1p_2^2p_3)$. Then the homogenized map $\tilde{c}$ is the map $(p_1, p_2, p_3, z) \mapsto (p_1^2, p_1^2z^2 + p_1p_3z^2 + p_1p_2^2p_3)$. Then the Jacobian $J(\tilde{c}, z)$ is given by

$$\begin{pmatrix} 2p_1 & 0 & 0 & 0 \\ 2p_1z^2 + p_3z^2 + p_2^2p_3 & 2p_1p_2p_3 & p_1z^2 + p_1p_2^2 & 2p_1^2z + 2p_1p_3z \\ 0 & 0 & 0 & 1 \end{pmatrix}.$$

Then applying Gaussian elimination over $\mathbb{R}(\tilde{c_1}(p,z), \tilde{c_2}(p,z), z)$, we obtain:

$$\begin{pmatrix} 1 & 0 & 0 & 0 \\ 0 & 2p_2p_3 & z^2 + p_2^2 & 2(p_1 + p_3)z \\ 0 & 0 & 0 & 1 \end{pmatrix}.$$

Thus, dotting each row vector with (p, z), we obtain $p_1, 3p_2^2p_3 + 2p_1z^2 + 3p_3z^2$, and z are locally identifiable. Dividing by the degree and setting $z = 1$, we obtain that p_1 and $p_2^2p_3 + 2p_1/3 + p_3$ are locally identifiable.

§6. Observability

In this section we explore how algebraic and combinatorial tools can be used to determine whether or not the state variables are observable. Roughly speaking, the state variable x_i is *observable* if it can be recovered from observation of the input and output alone. We will use algebraic language to make this precise and explain how Gröbner bases and matroids can be used to check this condition.

Definition 6.1. Consider a state space model of form (2).

- The state variable x_i is *generically observable* given the input and output trajectories and generic parameter value p if there is a unique trajectory for x_i compatible with the given input-output trajectory.

- The state variable x_i is *rationally observable* given input and output trajectories and generic parameter value p if there is a rational function F such that the trajectory $x_i(t)$ satisfies $x_i(t) = F(y, y', \ldots, u, u', \ldots, p)$.
- The state variable x_i is *generically locally observable* if given a generic parameter vector, there is an open neighborhood U_{x_i} of the trajectory $x_i(t)$ such that there is no other trajectory $\tilde{x}_i(t) \subseteq U_{x_i}$ that is compatible with input-output data.
- The state variable x_i is *generically unobservable* if given the input and output trajectories and a generic parameter value p there are infinitely many trajectories for x_i compatible with the given input-output trajectory.

As usual, when f and g are polynomial functions, we can give equivalent definitions to many of these conditions, and algebraic methods for checking them.

The following proposition gives algebraic conditions for observability. More details on the differential algebra involved can be found in [31].

Proposition 6.2. *Consider a state space model of form* (2). *Let* Π *be the differential ideal generated the polynomials* $x' - f(x, p, u), y - g(x, p)$. *Let* $h \in \Pi \cap \mathbb{Q}(p)[x_i, y, y', \ldots, u, u', \ldots]$ *be a polynomial and write this as* $h = \sum_{j=0}^{k} h_j x_i^j$ *where each* $h_j \in \mathbb{Q}(p)[y, y', \ldots, u, u', \ldots]$, $k \geq 1$, *and* $h_k \notin \Pi$. *Then*

- *If* $k = 1$, *then* x_i *is rationally observable.*
- *If* $k > 1$, *then* x_i *is locally observable.*
- *If there is no polynomial* $h \in \Pi \cap \mathbb{Q}(p)[x_i, y, y', \ldots, u, u', \ldots]$ *satisfying the three conditions then* x_i *is generically unobservable.*

As with computations for finding the input-output equations, one does not need to explicitly use the differential algebra to check the conditions of Proposition 6.2, and it is possible to do this directly via Gröbner bases and properties of the Jacobian matrix.

Proposition 6.3. *Consider a state space model of the form* (2) *where* f *and* g *are polynomial functions and where there are* N *state-space variables,* $M = 1$ *output variable, and* R *input variables. Let* P *be the ideal*

$$\langle x' - f(x, p, u), \ldots, x^{(N-1)} - \frac{d^{N-2}}{dt^{N-2}} f(x, p, u), \quad y - g(x, p), \ldots, y^{(N-1)} - \frac{d^{N-1}}{dt^{N-1}} g(x, p)\rangle$$
$$\subseteq \mathbb{Q}(p)[x, y, u, x', y', u', \ldots, x^{(N-2)}, y^{(N-2)}, u^{(N-2)}, x^{(N-1)}, y^{(N-1)}].$$

Consider an elimination ordering $<$ on

$$\mathbb{Q}(p)[x, y, u, x', y', u', \ldots, x^{(N-2)}, y^{(N-2)}, u^{(N-2)}, x^{(N-1)}, y^{(N-1)}]$$

with three blocks of variables

$$\begin{aligned}\{x, x', \ldots, x^{(N-1)}\} \setminus \{x_i\} &> \{x_i\} \\ &> \{y, u, y', u', \ldots, y^{(N-2)}, u^{(N-2)}, y^{(N-1)}\}.\end{aligned}$$

Then a Gröbner basis for P with respect to $<$ will contain a polynomial of the type indicated in Proposition 6.2 if it exists. Otherwise no such polynomial exists.

The proof of Proposition 6.3 is a combination of the ideas of Propositions 3.1 and 4.4.

Proof. First we need to show that we can find such a polynomial, if it exists, only looking up to derivatives of order $N - 1$. This follows a similar argument as the proof of Proposition 3.1 by dimension counting. The codimension of P is $N(N - 1) + N$, the total number of variables in our polynomial ring is $N^2 + N + R(N - 1)$, and thus P has dimension $N + R(N - 1)$. Since the total number of variables in the set $\{x_i, y, u, y', u', \ldots, y^{(N-2)}, u^{(N-2)}, y^{(N-1)}\}$ is $1 + N + R(N - 1)$, these variables must be dependent, i.e. there must exist a relation. If all the relations that exist do not involve x_i in a nontrivial way, there will not exist such relations if we add more derivatives. Indeed, adding one more set of derivatives then there must exist an input-output equation involving the variable $y^{(N)}$ and lower order terms in y, by the proof of Proposition 3.1. Hence these could be used to eliminate any appearance of $y^{(N)}$ or higher in any putative constraint involving x_i. Since the only equation in our system that involves $y^{(N)}$ was the equation $y^{(N)} - \frac{d^N}{dt^N} g(x, p)$, this means we need not have added it to our system since it cannot be eliminated by interacting with other equations. However, without this equation, there is only a single appearance of $x^{(N)}$, so there is no way to eliminate those variables that involves using those equations, and hence we are reduced to our system just up to order $N - 1$.

Now we will show that the Gröbner basis computation produces the desired equation. Suppose there is an equation h of the desired type in the ideal P. If the Gröbner basis of P did not contain a polynomial of the desired type, then the Gröbner basis of P does not contain a polynomial in the variables $\{x_i, y, u, y', u', \ldots, y^{(N-2)}, u^{(N-2)}, y^{(N-1)}\}$ that involves the variable x_i. Then reducing h by the Gröbner basis cannot produce the zero polynomial, contradicting that we had a Gröbner basis. Q.E.D.

Proposition 6.3 can be generalized to situations where there is more than one output variable. Indeed, from Proposition 3.2, we can obtain input-output equations for each y_i. Following a similar dimension counting argument, we obtain that P has dimension $N+R(N-1)$ and the total number of variables in the set $\{x_i, y, u, y', u', \ldots, y^{(N-2)}, u^{(N-2)}, y^{(N-1)}\}$ is $1+MN+R(N-1)$, thus these variables must be dependent, i.e. there must exist a relation. In this case, one might be able to get away with looking at derivatives of lower orders in some of the variables (i.e. not all the way to $N-1$) however this will depend on the structure of the underlying system. Making this precise depends on terminology from differential algebra that we would like to avoid. See [31] for details. One typical corollary is the following.

Corollary 6.4. *Consider a state space model of the form* (2) *where* f *and* g *are polynomial functions and where there are* N *state-space variables,* $M = 1$ *output variable, and* R *input variables. If the input-output equation has order* N*, then all the state space variables are locally observable.*

Proof. The proof of Proposition 6.3 shows that after adding the $N-1$ derivatives, there must exist a relation among the set $\{x_i, y, u, y', u', \ldots, y^{(N-2)}, u^{(N-2)}, y^{(N-1)}\}$. However, this could not be just among the set of variables $\{y, u, y', u', \ldots, y^{(N-2)}, u^{(N-2)}, y^{(N-1)}\}$ since this would be an input-output equation of order $< N$. Q.E.D.

Example 6.5. From Example 2.2, let our model be of the form:

$$\begin{pmatrix} x_1' \\ x_2' \end{pmatrix} = \begin{pmatrix} a_{11} & a_{12} \\ a_{21} & a_{22} \end{pmatrix} \begin{pmatrix} x_1 \\ x_2 \end{pmatrix} + \begin{pmatrix} u_1 \\ 0 \end{pmatrix}, \qquad y = x_1.$$

Taking derivatives, we have the system of equations:

$$\langle a_{11}x_1 + a_{12}x_2 + u_1 - x_1',\ a_{21}x_1 + a_{22}x_2 - x_2',\ x_1 - y,\ x_1' - y' \rangle.$$

These are polynomials in the polynomial ring $\mathbb{R}(p)[x_1, x_2, x_1', x_2', u_1, y, y']$.

Using the elimination order specified to calculate a Gröbner basis, we see that $a_{11}y_1 + a_{12}x_2 + u_1 - y'$ and $x_1 - y$ are two polynomials of the desired form. Thus the model is rationally observable. Alternatively, the input-output equation for this model is of differential order 2, which equals the number of state variables, so the model is locally observable by Corollary 6.4.

The main problem with this definition of observability is that it appears to require explicit computation of the desired polynomials. However, instead of applying a Gröbner basis to find the desired polynomials, we can examine the algebraic matroid associated to this system.

The algebraic matroid is equivalent to the linear matroid of differentials, for which computations are much simpler. Because the definition of observability distinguishes between a variable and its derivatives, we also treat them separately in our discussion. The ground set of the matroid for an observability computation is

$$E = \left\{ \begin{array}{l|r} x_i, x_i', x_i'', \ldots, x_i^{(N-1)}; & \forall i = 1, \ldots, N \\ y_j, y_j', \ldots, y_j^{(N-1)}; & \forall j = 1, \ldots, M \\ u_k, u_k', \ldots, u_k^{(N-2)}; & \forall k = 1, \ldots, R \end{array} \right\}$$

We can treat the system of ODEs as an ideal of algebraic relations among a set of indeterminates. We use these relations to define the associated Jacobian matrix. This matrix has $N^2 + MN + R(N-1)$ columns, one for each "variable" in the ground set, and $(N-1)N + (N-1)M$ rows, one for each relation. The entries in the matrix are polynomials in $\mathbb{R}[x, x', \ldots, x^{(N-1)}, y, y', \ldots, y^{(N-1)}, u, u', \ldots, u^{(N-2)}]$. The final step is Gaussian elimination in the Jacobian matrix. Unlike the strategy in Section 5, any rational function is permitted here.

Example 6.6. We approach observability of Example 2.2 using the algebraic matroid. The resulting matroid has rank three, with 23 bases and 14 circuits. We can sort this list to find the circuits including x_1 and x_2 while excluding x_1' and x_2'; we find the following circuits:

$$\{x_1, y_1\}, \quad \{x_2, u_1, y_1, y_1'\}, \quad \text{and } \{x_1, x_2, u_1, y_1'\}.$$

The third circuit contains both variables, so is not useful for proving observability; but the first two circuits constitute a proof of observability.

Example 6.7. From the SIR Model in Example 2.1, let our ODE system be of the form:

$$S' = \mu N - \beta S \frac{I}{N} - \mu S$$

$$I' = \beta S \frac{I}{N} - (\mu + \gamma) I$$

$$R' = \gamma I - \mu R$$

$$y = kI.$$

Taking derivatives, we have the system of equations:

$$\langle S' + \mu S + \beta S\frac{I}{N} - \mu N,\ S'' + \mu S' + \beta S\frac{I'}{N} + \beta S'\frac{I}{N},$$
$$I' + (\mu+\gamma)I - \beta S\frac{I}{N},\ I'' + (\mu+\gamma)I' - \beta S'\frac{I}{N} - \beta S\frac{I'}{N},$$
$$R' + \mu R - \gamma I,\ R'' + \mu R' - \gamma I',\ y - kI,\ y' - kI',\ y'' - kI''\rangle.$$

These are polynomials in the polynomial ring

$$\mathbb{R}(p)[S, I, R, S', I', R', S'', I'', R'', y, y', y''].$$

Using the elimination order specified to calculate a Gröbner basis, we find that there are no polynomials in y, y', y'' and R only, so the model is generically unobservable. More precisely, we find the polynomial $-kR' - k\mu R + \gamma y$, but no polynomial involving y, y', y'' and R only.

We use a similar strategy to compute the matroid for the SIR Model. The ground set of the algebraic matroid for the observability computation is

$$E = \{\ S, S', S'',\quad I, I', I'',\quad R, R', R'',\quad y, y', y''\ \}$$

The matroid has rank three, with 123 bases and 146 circuits. We can sort this list to find the circuits including S, I, and R while excluding their derivatives; we find the following circuits for each variable:

$$\begin{array}{lll} \{S, y, y'\} & \{S, y, y''\} & \{S, y', y''\} \\ \{I, y\} & & \{I, y', y''\} \end{array}$$

Any relation in the first row proves that S is observable; similarly, any relation in the second row proves that I is observable. No relation from R exists; an elimination of the original ideal proves that R has no relations that do not also involve its derivatives.

In the linear matroid of differentials this is made more pronounced. In `Macaulay2` [34], the command `kernel(transpose(jacobian(I)))` yields a matrix whose row vectors correspond to variables. The vectors corresponding to R, R', and R'' are nonzero in a coordinate where all other variables are zero. Therefore, any relation including *one* of $\{R, R', R''\}$ must include at least two.

§7. Indistinguishability

Recall two state space models are *indistinguishable* if for any choice of parameters in the first model, there is a choice of parameters in the

The vanishing ideal for model 1 in the polynomial ring $\mathbb{Q}[c_1, c_2, c_3, c_4, c_5]$ is:

$$\langle c_2 - c_5,\ \ c_1c_4 - c_4^2 - c_5\rangle$$

whereas the vanishing ideal for model 2 is

$$\langle c_3 - c_5,\ \ c_2c_4^2 - c_1c_4c_5 + c_5^2\rangle.$$

Since the two vanishing ideals are not equal, the models are generically distinguishable.

Example 7.3. Consider the following variation on the previous example, where we have simply moved an input from compartment 2 to compartment 3.

$$\begin{pmatrix} x_1' \\ x_2' \\ x_3' \end{pmatrix} = \begin{pmatrix} -a_{01} - a_{21} & 0 & 0 \\ a_{21} & -a_{32} & 0 \\ 0 & a_{32} & 0 \end{pmatrix} \begin{pmatrix} x_1 \\ x_2 \\ x_3 \end{pmatrix} + \begin{pmatrix} u_1 \\ 0 \\ u_3 \end{pmatrix} \qquad y_1 = x_3$$

$$\begin{pmatrix} x_1' \\ x_2' \\ x_3' \end{pmatrix} = \begin{pmatrix} -b_{21} & 0 & 0 \\ b_{21} & -b_{02} - b_{32} & 0 \\ 0 & b_{32} & 0 \end{pmatrix} \begin{pmatrix} x_1 \\ x_2 \\ x_3 \end{pmatrix} + \begin{pmatrix} u_1 \\ 0 \\ u_3 \end{pmatrix} \qquad y_1 = x_3.$$

The input-output equations for these models are:

$$\begin{aligned} &y_1''' + (a_{32} + a_{01} + a_{21})y_1'' + (a_{01}a_{32} + a_{21}a_{32})y_1' = \\ &\qquad a_{21}a_{32}u_1 + u_3'' + (a_{01} + a_{21} + a_{32})u_3' + (a_{01}a_{32} + a_{21}a_{32})u_3 \\ &y_1''' + (b_{21} + b_{02} + b_{32})y_1'' + (b_{21}b_{02} + b_{21}b_{32})y_1' = \\ &\qquad\qquad b_{21}b_{32}u_1 + u_3'' + (b_{02} + b_{32} + b_{21})u_3' + (b_{02}b_{21} + b_{32}b_{21})u_3, \end{aligned}$$

respectively. In both cases, the vanishing ideal of the model coefficients is the ideal

$$\langle c_2 - c_5, c_1 - c_4\rangle,$$

which suggests that the two models might be indistinguishable. A simple Jacobian calculation shows that the models are locally identifiable, and hence we can attempt to solve the system $c(p) = c'(p')$ to test for indistinguishability. Solving the system of equations:

$$a_{32} + a_{01} + a_{21} = b_{21} + b_{02} + b_{32}$$

$$a_{01}a_{32} + a_{21}a_{32} = b_{21}b_{02} + b_{21}b_{32}$$

$$a_{21}a_{32} = b_{21}b_{32}$$

we obtain the solutions:

$$\{a_{21} = b_{32}, a_{01} = b_{02}, a_{32} = b_{21}\}$$

$$\{a_{21} = (b_{21}b_{32})/(b_{02} + b_{32}), a_{01} = (b_{02}b_{21})/(b_{02} + b_{32}), a_{32} = b_{02} + b_{32}\}$$

Likewise, one can obtain the solutions:

$$\{b_{21} = a_{32}, b_{02} = a_{01}, b_{32} = a_{21}\}$$

$$\{b_{32} = (a_{21}a_{32})/(a_{01} + a_{21}), b_{02} = (a_{01}a_{32})/(a_{01} + a_{21}), b_{21} = a_{01} + a_{21}\}$$

The parameter spaces for these models have all parameters positive. It is easy to see that for any choice of parameters in the first model, there is a choice of parameters in the second model that gives the same input-output equation, and vice versa. Thus these models are indistinguishable. Note that there are two solutions because the models are locally but not globally identifiable.

Example 7.4. Now consider the following variation of the previous models, where we have added an extra leak parameter to each model and removed the inputs:

$$\begin{pmatrix} x_1' \\ x_2' \\ x_3' \end{pmatrix} = \begin{pmatrix} -a_{01} - a_{21} & 0 & 0 \\ a_{21} & -a_{32} & 0 \\ 0 & a_{32} & -a_{03} \end{pmatrix} \begin{pmatrix} x_1 \\ x_2 \\ x_3 \end{pmatrix} \qquad y_1 = x_3$$

$$\begin{pmatrix} x_1' \\ x_2' \\ x_3' \end{pmatrix} = \begin{pmatrix} -b_{21} & 0 & 0 \\ b_{21} & -b_{02} - b_{32} & 0 \\ 0 & b_{32} & -b_{03} \end{pmatrix} \begin{pmatrix} x_1 \\ x_2 \\ x_3 \end{pmatrix} \qquad y_1 = x_3.$$

The input-output equations for these models is:

$$\begin{aligned} y_1''' + (a_{01} + a_{21} + a_{32} + a_{03})y_1'' + (a_{01}a_{32} + a_{21}a_{32} + a_{32}a_{03} + \\ a_{01}a_{03} + a_{21}a_{03})y_1' + (a_{01}a_{03}a_{32} + a_{03}a_{21}a_{32})y_1 = 0, \\ y_1''' + (b_{21} + b_{02} + b_{32} + b_{03})y_1'' + (b_{21}b_{02} + b_{21}b_{32} + b_{02}b_{03} + \\ b_{03}b_{21} + b_{03}b_{32})y_1' + (b_{02}b_{03}b_{21} + b_{03}b_{21}b_{32})y_1 = 0, \end{aligned}$$

In both cases, the vanishing ideal of the model coefficients is the zero ideal which suggests that the two models might be indistinguishable. These models are clearly unidentifiable since there are 3 coefficients in 4 unknown parameters. Solving the system $c(p) = c'(p')$, we get the following 6 solutions:

$$\{a_{03} = b_{03}, a_{21} = -a_{01} + b_{02} + b_{32}, a_{32} = b_{21}\}$$

$$\{a_{03} = b_{03}, a_{21} = -a_{01} + b_{21}, a_{32} = b_{02} + b_{32}\}$$

$$\{a_{03} = b_{21}, a_{21} = -a_{01} + b_{02} + b_{32}, a_{32} = b_{03}\}$$

$$\{a_{03} = b_{21}, a_{21} = -a_{01} + b_{03}, a_{32} = b_{02} + b_{32}\}$$

$$\{a_{03} = b_{02} + b_{32}, a_{21} = -a_{01} + b_{21}, a_{32} = b_{03}\}$$

$$\{a_{03} = b_{02} + b_{32}, a_{21} = -a_{01} + b_{03}, a_{32} = b_{21}\}$$

when solving for $\{a_{01}, a_{21}, a_{32}, a_{03}\}$. A similar result follows when solving for $\{b_{21}, b_{02}, b_{32}, b_{03}\}$. Thus the models are indistinguishable.

Remark. Note that the vanishing ideals being equal is only a necessary condition for indistinguishability but not in general sufficient. For example, suppose that we restrict to the parameter space consisting of positive parameters and consider the coefficient maps $c(p_1, p_2) = (p_1, p_1 + p_2)$ and $c'(p'_1, p'_2) = (p'_1 + p'_2, p'_2)$. The images in both cases have zero vanishing ideal. However, the models are distinguishable since the image of the first coefficient map is $\{(c_1, c_2) \in \mathbb{R}^2 : c_2 > c_1 > 0\}$ whereas the image of the second coefficient map is $\{(c_1, c_2) \in \mathbb{R}^2 : c_1 > c_2 > 0\}$.

Remark. Some authors also consider a one-sided notion of indistinguishability. In this definition, Model 1 is indistinguishable from Model 2 if every for every choice of parameters in Model 1, there is a choice of parameters in Model 2 that can produce the same dynamics. So Model 2 is a more expressive class of models. It is more difficult to check for this type of indistinguishability because it need not be the case that the input-output equations have the same structure, and so we cannot simply check that the image of the coefficient map of Model 1 is contained in the image of the coefficient map of Model 1. As a simple example, if Model 1 has input-output equation $y' + a_1 y = 0$, and Model 2 has input-output equation $y'' + b_1 y' + b_2 y = 0$, clearly Model 1 is indistinguishable from Model 2, but this is not detectable by comparing the image of the coefficient maps.

§8. Further Reading

We have demonstrated some techniques to test identifiability, observability, and indistinguishability using a differential algebraic approach. There are several other approaches to investigate these concepts, so we outline a few of these other methods now for the interested reader.

For linear models, the global identifiability problem can be solved with the transfer function approach [6] and the similarity transformation approach [69, 72]. For nonlinear models, the differential algebra method has been a powerful technique to test for identifiability [43, 51, 61]. The main advantage of the differential algebra method is that *global* identifiability can be determined. On the other hand, there are many approaches to test *local* identifiability, including the Taylor series method [54], the generating series method [71] (with implementations involving identifiability *tableaus* [4] and exact arithmetic rank [39]), a method based on the implicit function theorem [74, 75], a test for reaction networks [17, 18], and a profile likelihood approach [56]. There are special cases where global identifiability can be determined using a nonlinear variation of the similarity transformation approach [12, 19, 67] and the direct test [20, 21]. These approaches for global and local identifiability are outlined in greater detail and tested on several models in [15] and [57].

For linear models, the concept of observability can be tested using a linear algebra test [38]. These conditions can be translated to conditions on the graph of the linear compartmental models [32, 76]. For nonlinear models, observability can be tested with differential algebra [31, 45]. Alternatively, the nonlinear problem has been approached analytically in [36]. To test local algebraic observability, one can use a probabilistic seminumerical method that solves the problem in polynomial time [62].

For linear models, the indistinguishability problem has been analyzed using geometrical rules [32] and a linear algebra test [78]. For nonlinear models, indistinguishability was introduced in [65]. The problem has been extensively studied for certain classes of nonlinear compartmental models in [13, 14, 33, 70] and more generally in [24].

This paper is concerned with state space models, but identifiability and related concepts are also explored heavily in other contexts. Beltrametti and Robbiano [7] consider the ideal $\langle c_1(p) - c_1(p^*), c_2(p) - c_2(p^*), ..., c_m(p) - c_m(p^*)\rangle$ for detecting identifiability in the context of the Hough transform. Other areas include study of identifiability of graphical models [2, 23, 29, 30, 66] and identifability of phylogenetic models [1, 44, 46, 58].

§9. Appendix: Algebraic Matroids

We review the basics of general matroid theory here and especially main results on algebraic matroids.

Definition 9.1. Let E be a finite set and let $\mathcal{I}$ be a collection of subsets of E satisfying the following three conditions:

(1) $\emptyset \in \mathcal{I}$
(2) If $X \in \mathcal{I}$ and $Y \subseteq X$ then $Y \in \mathcal{I}$, and
(3) If $X, Y \in \mathcal{I}$ with $|X| < |Y|$ then there exists $y \in Y \setminus X$ such that $X \cup \{y\} \in \mathcal{I}$.

The pair $(E, \mathcal{I})$ is called a *matroid* and the elements of $\mathcal{I}$ are called *independent sets.*

A special instance of matroids are sets E of vectors in a vector space where the set $\mathcal{I}$ consists of all linearly independent subsets of E. A matroid that arises in this way is called a *representable* or a *linear matroid.* Various other terminology from linear algebra is also applied in matroid theory. A maximal cardinality subset of $\mathcal{I}$ is called a *basis.* The subsets of E that are not in $\mathcal{I}$ are called *dependent sets.* A minimal dependent set is called a *circuit.* Oxley's text [52] is a standard reference for background on matroids.

The most important type of matroid for us will be the algebraic matroids whose properties we review here.

Proposition 9.2. [52, Thm 6.7.1] *Suppose $\mathbb{K}$ is an extension field of a field $\mathbb{F}$ and E is a finite subset of $\mathbb{K}$. Then the collection $\mathcal{I}$ of subsets of E that are algebraically independent over $\mathbb{F}$ is the set of independent sets of a matroid on E. The resulting matroid is called an* algebraic matroid.

Example 9.3. Let $E = \{a_{11}, a_{22}, a_{12}, a_{21}\}$ and $\mathbb{F} = \mathbb{R}(c_1(p), c_2(p), c_3(p))$, where $c_1(p) = -a_{11} - a_{22}, c_2(p) = a_{11}a_{22} - a_{12}a_{21}, c_3(p) = -a_{22}$ from the linear 2-compartment model in Example 4.2. Then $I = \{\emptyset, \{a_{12}\}, \{a_{21}\}\}$ and $C = \{\{a_{11}\}, \{a_{22}\}, \{a_{12}, a_{21}\}\}$.

In our problem, we have the mapping $p \mapsto (c_1(p), c_2(p), c_3(p))$ and the variety V of interest is the pre-image of a point $\hat{c} = (\hat{c_1}, \hat{c_2}, \hat{c_3})$ under the map c. Note that the map c has a trivial vanishing ideal; the image of this map is the full $\mathbb{R}^3$. The point $\hat{c}$ can therefore be taken to be a generic point of $\mathbb{R}^3$ by setting $\{\hat{c_1}, \hat{c_2}, \hat{c_3}\}$ to be algebraically independent over $\mathbb{R}$. This means that the only algebraic constraints on the p-variables come from the equations $\{c(p) = \hat{c}\}$.

Our associated ideal is $P = \langle c_1(p) - \hat{c_1}, c_2(p) - \hat{c_2}, c_3(p) - \hat{c_3} \rangle$, which contains polynomials in $\mathbb{R}(\hat{c})[p] = \mathbb{R}(\hat{c_1}, \hat{c_2}, \hat{c_3})[a_{11}, a_{22}, a_{12}, a_{21}]$. The ideal P is prime, as confirmed by a Gröbner basis computation at a randomly chosen point; therefore, computation of the algebraic matroid modulo P is well-defined.

Proposition 9.4. [52, Prop 6.7.11] *If a matroid M is algebraic over a field $\mathbb{F}$ of characteristic zero, then M is linearly representable over $\mathbb{F}(T)$ for some finite set T of transcendentals over $\mathbb{F}$.*

The following proposition follows from [42, Proposition 2.14] together with the observation that the tangent space of a variety is the kernel of its Jacobian matrix:

Proposition 9.5. *Let $P = \langle f_1, ..., f_m \rangle$ be a prime ideal contained in $\mathbb{F}[x_1, ..., x_n]$. Define the Jacobian matrix $J(P)$ as:*

$$\left(\frac{\partial f_i}{\partial x_j} : 1 \leq i \leq m, 1 \leq j \leq n \right).$$

This matrix, when considered as a matroid with columns as the ground set and linear independence over $\mathrm{Frac}(\mathbb{F}[x]/P)$ *defining independent set I represents the dual matroid to $M(P)$. The transpose of the matrix spanning the kernel gives the matroid $M(P)$.*

Example 9.6. Let c be the map $p \mapsto (c_1(p), c_2(p), c_3(p))$ from the linear 2-compartment model in Example 4.2. Then the Jacobian $J(c)$ is given by

$$\begin{pmatrix} -1 & -1 & 0 & 0 \\ a_{22} & a_{11} & -a_{21} & -a_{12} \\ 0 & -1 & 0 & 0 \end{pmatrix}$$

A basis for the kernel of this matrix is given by $(0, 0, a_{12}, -a_{21})^T$. Here, linear independence is taken over $\mathrm{Frac}(\mathbb{R}(\hat{c})[p]/P) \cong \mathbb{R}(\hat{c})(a_{12}, a_{21})$. Thus, a vector matroid is given by:

$$\begin{pmatrix} 0 & 0 & a_{12} & -a_{21} \end{pmatrix}$$

where the ground set $E = \{1, 2, 3, 4\}$ and a set of circuits is given by $C = \{\{1\}, \{2\}, \{3, 4\}\}$. This implies that a_{11} and a_{22} are each algebraic over $\mathbb{R}(\hat{c})$, which implies that a_{11} and a_{22} are each locally identifiable. This also implies that $\{a_{12}, a_{21}\}$ is algebraically dependent over $\mathbb{R}(\hat{c})$.

§ Acknowledgments

Nicolette Meshkat was partially supported by the Clare Boothe Luce Program from the Luce Foundation and by the David and Lucille Packard Foundation. Zvi Rosen was partially supported by a Math+X Research Grant from the Simons Foundation. Seth Sullivant was partially supported by the David and Lucille Packard Foundation and by the US National Science Foundation (DMS 0954865 and 1615660).

References

[1] E. S. Allman, J. H. Degnan, and J. A. Rhodes. Determining species tree topologies from clade probabilities under the coalescent. *Journal of Theoretical Biology*, **289** (2011) 96–106.

[2] E. S. Allman, J. A. Rhodes, E. Stanghellini, and M. Valtorta. Parameter identifiability of discrete Bayesian networks with hidden variables. *Journal of Causal Inference*, **3** no. 2 (2015) 189-205.

[3] S. Audoly, G. Bellu, L. D'Angio, M. P. Saccomani, and C. Cobelli, Global identifiability of nonlinear models of biological systems, *IEEE Transactions on Biomedical Engineering* **48** (2001) 55-65.

[4] E. Balsa-Canto, A. Alonso, and J. Banga, An iterative identification procedure for dynamic modeling of biochemical networks, *BMC Systems Biology* **4** (2010) 11.

[5] D. J. Bearup, N. D. Evans, and M. J. Chappell, The input-output relationship approach to structural identifiability analysis, *Computer Methods and Programs in Biomedicine* **109** (2013) 171-181.

[6] R. Bellman and K. J. Astrom, On structural identifiability, *Mathematical Biosciences* **7:3-4** (1970) 329-339.

[7] M. Beltrametti and L. Robbiano. An algebraic approach to Hough transforms. *J. Algebra* **371** (2012), 669–681.

[8] E. Beretta and Y. Takeuchi, Global stability of an SIR epidemic model with time delays, *Journal of Mathematical Biology* **33(3)** (1995) 250-260.

[9] O. N. Bjornstad, B. F. Finkenstadt, and B. T. Grenfell, Dynamics of measles epidemics: estimating scaling of transmission rates using a time series SIR model, *Ecological Monographs* **72:2** (2002) 169-184.

[10] V. Capasso, Mathematical structures of epidemic systems, Springer-Verlag, New York (1993).

[11] C. K. Carter and R. Kohn, On Gibbs sampling for state space models, *Biometrika* **81(3)** (1994) 541-553.

[12] M. Chappell, K. Godfrey, and S. Vajda, Global identifiability of the parameters of nonlinear systems with specific input: A comparison of methods, *Mathematical Biosciences* **102** (1990) 41-73.

[13] M. J. Chapman, K. R. Godfrey, and s. Vajda, Indistinguishability for a class of nonlinear compartmental models, *Mathematical Biosciences* **119(1)** (1994) 77-95.

[14] M. J. Chapman and K. R. Godfrey, Nonlinear compartmental model indistinguishability, *Automatica* **32(3)** (1996) 419-422.

[15] O-T Chis, J. R. Banga, and E. Balsa-Canto, Structural Identifiability of Systems Biology Models: A Critical Comparison of Methods, *PLoS ONE* 6(11):e27755 (2011).

[16] D. A. Cox, J. Little, and D. O'Shea, Ideals, Varieties, and Algorithms: An Introduction to Computational Algebraic Geometry and Commutative Algebra, Springer, New York, 2007.

[17] G. Craciun and C. Pantea, Identifiability of chemical reaction networks, *Journal of Mathematical Chemistry* **44** (2008) 244-259.

[18] F. Davidescu and S. Jorgensen, Structural parameter identifiaiblity analyss for dynamic reaction networks, *Chemical Engineering Science* **63** (2008) 4754-4762.

[19] L. Denis-Vidal and G. Joly-Blanchard, Identifiability of some nonlinear kinetics, *Proceedings of the Third Workshop on Modelling of Chemical Reaction Systems*, Heidelberg (1996).

[20] L. Denis-Vidal and G. Joly-Blanchard, An easy to check criterion for (un)identifaibility of uncontrolled systems and its applications, *IEEE Transactions on Automatic Control* **45** (2000) 768-771.

[21] L. Denis-Vidal, G. Joly-Blanchard, and C. Noiret, Some effective approaches to check the identifiability of untrolled nonlinear systems, *Mathematics in Computers and Simulation* **57** (2001) 35-44.

[22] J. J. DiStefano III, Dynamic Systems Biology Modeling and Simulation. Elsevier, London (2014).

[23] M. Drton, R. Foygel, S. Sullivant, Global identifiability of linear structural equation models. *Ann. Statist.* **39** (2011), no. 2, 865-886.

[24] N. D. Evans, M. J. Chappell, M. J. Chapman, and K. R. Godfrey, Structural indistinguishability between uncontrolled (autonomous) nonlinear analytic systems, *Automatica* **40(11)** (2004) 1947-1953.

[25] N. D. Evans, H. Moyse, D. Lowe, D. Briggs, R. Higgins, D. Mitchell, D. Zehnder, and M. J. Chappell, Structural identifiability of surface binding reactions involving heterogenous analyte: application to surface plasmon resonance experiments, *Automatica* **49** (2012) 48-57.

[26] N. D. Evans and M. J. Chappell, Extensions to a procedure for generating locally identifiable reparameterisations of unidentifiable systems, *Mathematical Biosciences* **168** (2000) 137-159.

[27] E. Fornasini and G. Marchesini, Doubly-indexed dynamical systems: State-space models and structural properties, *Mathematical systems theory* **12(1)** (1978) 59-72.

[28] K. Forsman, Constructive Commutative Algebra in Nonliner Control Theory, Ph.D. thesis, Dept. of Electrical Engineering, Linkoping University, S-581 83 Linkoping, Sweden, 1991.

[29] R. Foygel, J. Draisma, M. Drton, Half-trek criterion for generic identifiability of linear structural equation models. *Ann. Statist.* **40** (2012), no. 3, 1682-1713.

[30] L. Garcia-Puente, S. Spielvogel, and S. Sullivant. Identifying causal effects with computer algebra. *Uncertainty in Artificial Intelligence*, Proceedings of the 26th Conferences, AUAI Press, 2010.

[31] S. T. Glad, Differential Algebraic Modelling of Nonlinear Systems, Realization and Modelling in Systems Theory, Eds: M. A. Kaashoek et al, 1990.

[32] K. R. Godfrey and M. J. Chapman, Identifiability and Indistinguishability of Linear Compartmental Models, *Mathematics and Computers in Simulation* **32** (1990) 273-295.

[33] K. R. Godfrey, M. J. Chapman, and S. Vajda, Identifiability and indistinguishability of nonlinear pharmacokinetic models, *Journal of Pharmacokinetics and Biopharmaceutics* **22(3)** (1994) 229-251.

[34] D.R. Grayson and M. E. Stillman, Macaulay2, a software system for research in algebraic geometry, Available at
`http://www.math.uiuc.edu/Macaulay2/`

[35] J. D. Hamilton, State-space models, in *Handbook of Econometrics* **4** (1994) 3039-3080.

[36] R. Hermann and A. J. Krener, Nonlinear Controllability and Observability, *IEEE Transactions on Automatic Control* **AC-22** (1977) 728-740.

[37] M. Jirstrand, Algebraic Methods for Modeling and Design in Control, Ph.D. thesis, Dept. of Electrical Engineering, Linkping University, S-581 83 Linkoping, Sweden, 1996.

[38] R. E. Kalman, Mathematical description of linear dynamical systems, *SIAM Journal of Control* **1(2)** (1963) 152-192.

[39] J. Karlsson, M. Anguelova, and M. Jirstrand, An efficient method for structural identifiability analysis of large dynamical systems. In: *Proceedings of the 16th IFAC Symposium on System Identification*, 2012.

[40] G. Kitagawa, Monte Carlo Filter and Smoother for Non-Gaussian Nonlinear State Space Models, *Journal of Computational and Graphical Statistics* **5(1)** (1996) 1-25.

[41] E. R. Kolchin, *Differential algebra and algebraic groups*. Pure and Applied Mathematics, Vol. 54. Academic Press, New York-London, 1973.

[42] F. J. Király, Z. Rosen and L. Theran. Algebraic matroids with graph symmetry. arXiv preprint *arXiv*: 1312.3777. 2013.

[43] L. Ljung and T. Glad, On global identifiability for arbitrary model parameterization, *Automatica* **30** (1994) 265-276.

[44] C. Long and S. Sullivant. Identifiability of Jukes-Cantor 3-tree mixtures. *Adv. in Appl. Math.***64** (2015), 89–110.

[45] X. Y. Lu and S. P. Banks, Some New Results in Differential Algebraic Control Theory, IMA Journal of Mathematical Control and Information, 2003.

[46] F. A. Matsen, E. Mossel, M. Steel, Mixed-up trees: the structure of phylogenetic mixtures. *Bull. Math. Biol.* **70** (2008), no. 4, 1115–1139.

[47] C. C. McCluskey, Complete global stability for an SIR epidemic model with delay - Distributed or discrete, *Nonlinear Analysis: Real World Applications* **11(1)** (2010) 55-59.

[48] N. Meshkat, C. Anderson, and J. J. DiStefano III, Alternative to Ritt's Pseudodivision for finding the input-output equations of multi-output models, *Mathematical Biosciences* **239** (2012) 117-123.

[49] N. Meshkat, M. Eisenberg, and J. J. DiStefano III, An algorithm for finding globally identifiable parameter combinations of nonlinear ODE models using Gröbner Bases, *Mathematical Biosciences* **222** (2009) 61-72.

[50] N. Meshkat and S. Sullivant, Identifiable reparametrizations of linear compartment models, *Journal of Symbolic Computation* **63** (2014) 46-67.

[51] F. Ollivier, Le probleme de l'identifaibilite structurelle globale: etude theoretique, methodes effectives et bornes de complexite. Paris, France: These de Doctorat en Science, Ecole Polytechnique (1990).
[52] J. Oxley, Matroid Theory, Oxford Texts in Graduate Mathematics, 1992.
[53] L. Pernebo and L. Silverman, Model reduction via balanced state space representations, *IEEE Transactions on Autoamtic Control* **27(2)** 382-387.
[54] H. Pohjanpalo, System identifiability based on the power series expansion of the solution, *Mathematical Biosciences* **41** (1978) 21-33.
[55] A. Raksanyi, Y. Lecourtier, E. Walter, and A. Venot, Identifiability and Distinguishability Testing Via Computer Algebra, *Mathematical Biosciences* **77** (1985) 245-266.
[56] A. Raue, C. Kreutz, T. Maiwald, J. Bachmann, M. Schilling, U. Klingmuller, and J. Timmer, Structural and practical identifiability analysis of partially observed dynamical models by exploiting the profile likelihood, *Bioinformatics* **25** (2009) 1923-1929.
[57] A. Raue, J. Karlsson, M. P. Saccomani, M. Jirstrand, and J. Timmer, Comparison of approaches for parameter identifiability analysis of biological systems, *Bioinformatics* (2014) 1-9.
[58] J. Rhodes and S. Sullivant. Identifiability of large phylogenetic mixture models. *Bull. Math. Biol.* **74** (2012), no. 1, 212–231.
[59] R. Roesser, A discrete state-space model for linear image processing, *IEEE Transactions on Automatic Control* **20:1** (1975) 1-10.
[60] J. F. Ritt, *Differential Algebra.* American Mathematical Society Colloquium Publications, Vol. XXXIII, American Mathematical Society, New York, N. Y., 1950.
[61] M. P. Saccomani, S. Audoly, and L. D'Angio, Parameter identifiability of nonlinear systems: the role of initial conditions, *Automatica* **39** (2003) 619-632.
[62] A. Sedoglavic, A probabilitistic algorithm to test local algebraic observability in polynomial time, *Journal of Symbolic Computation* **33** (2002) 735-755.
[63] B. Shulgin, L. Stone, and Z. Agur, Pulse vaccination strategy in the SIR epidemic model, *Bulletin of Mathematical Biology* **60** (1998) 1123-1148.
[64] M. van der Put, M. F. Singer, *Galois theory of linear differential equations.* Fundamental Principles of Mathematical Sciences, **328**. Springer-Verlag, Berlin, 2003.
[65] H. Sussmann, Existence and uniqueness of minimal realizations of nonlinear systems, *Mathematical systems theory* **10(1)** (1976) 263-284.
[66] J. Tian, and I. Shpitser, On Identifying Causal Effects, In R. Dechter, H. Geffner, and J. Halpern (Eds.), *Heuristics, Probability and Causality: A Tribute to Judea Pearl*, College Publications, 2010.
[67] S. Vajda, K. Godfrey, and H, Rabitz, Similarity transformation approach to identifiability analysis of nonlinear compartmental models, *Mathematical Biosciences* **93** (1989) 217-248.

[68] M. Verhaegen, Identification of the deterministic part of MIMO state space models given in innovations form from input-output data, Special issue on statistical signal processing and control, *Automatica* **30:1** (1994) 61-74.
[69] E. Walter, Identifiability of State Space Models, *Springer Lecture Notes in Biomathematics* **46**, Springer, Berlin, New York, 1982.
[70] E. Walter and L. Pronzato, On the identifiability and distinguishability of nonlinear parametric models, *Mathematics and Computers in Simulation* **42:2-3** (1996) 125-134.
[71] E. Walter and Y. Lecourtier, Global approaches to identifiability testing for linear and nonlinear state space models, *Mathematics and Computers in Simulation* **24** (1982) 472-482.
[72] E. Walter and Y. Lecourtier, Unidentifiable compartmental models: what to do?, *Mathematical Biosciences* **56** (1981) 1-25.
[73] E. Walter, Y. Lecourtier, and J. Happel, On the Structural Output Distinguishability of Parametric Models and Its Relations with Structurual Identifiability, *IEEE Transactions on Automatic Control* **AC-29** (1984) 56-57.
[74] H. Wu, H. Zhu, H. Miao, and A. S. Perelson, Parameter identifiability and estimation of HIV/AIDS dynamic models, *Bulletin of Mathematical Biology* **70(3)** (2008) 785-799.
[75] X. Xia and C. H. Moog, Identifiability of nonlinear systems with applications to HIV/AIDS models, *IEEE Trans Aut Cont* **48** (2003) 330-336.
[76] R. M. Zazworsky and H. K. Knudsen, Controllability and observability of linear time invariant compartmental models, *IEEE Trans. Automat. Control* **AC-23** (1978) 872-877.
[77] Y. Zeng and S. Wu, State-Space Models: Applications in Economics and Finance, Springer, New York 2013.
[78] L. Zhang, J. Collins, and P. King, Indistinguishability and Identifiability Analysis of Linear Compartmental Models, *Mathematical Biosciences* **103** (1991) 77-95.

(Nicolette Meshkat) *Department of Mathematics and Computer Science, Santa Clara University, Santa Clara, CA*
E-mail address: nmeshkat@scu.edu

(Zvi Rosen) *Department of Mathematics, University of Pennsylvania, Philadelphia, PA*
E-mail address: zvihr@math.upenn.edu

(Seth Sullivant) *Department of Mathematics, North Carolina State University, Raleigh, NC*
E-mail address: smsulli2@ncsu.edu

Advanced Studies in Pure Mathematics 77, 2018
The 50th Anniversary of Gröbner Bases
pp. 207–252

Selected topics on toric varieties

Mateusz Michałek

Abstract.

This article is based on a series of lectures on toric varieties given at RIMS, Kyoto. We start by introducing toric varieties, their basic properties and later pass to more advanced topics relating mostly to combinatorics.

§1. Introduction

There are now many great texts about or strongly related to toric varieties, both classical or new, compact or detailed - just to mention a few: [25, 32, 68, 76, 13]. This is not surprising - toric geometry is a beautiful topic. No matter if you are a student or a professor, working in algebra, geometry or combinatorics, pure or applied mathematics you can always find in it some new theorems, useful methods, astonishing relations. Still, it is absolutely impossible to compete with the texts above neither in scope, level of exposition nor accuracy. We present a review on toric geometry based on ten lectures given at Kyoto University, divided into two parts. The first part is the classical, basic introduction to toric varieties. Our point of view on toric varieties here, is as images of monomial maps. Thus, we relax the normality assumption, but consider varieties as embedded. We hope that this part is completely self-contained - proofs are at least sketched and a motivated reader should be able to reconstruct all details. Further, such an approach should allow the reader not familiar with toric geometry to realize that he encountered toric varieties before.

The second part deals with slightly more advanced topics. These were chosen very subjectively according to interests of the author. We start by recalling the theory of divisors on toric varieties in Section 6

Received December 24, 2016.
2010 *Mathematics Subject Classification.* 14M25.
Key words and phrases. toric variety, algebraic combinatorics, matroid, cut polytope, algebraic phylogenetics.

and their cohomology in Section 7. Then we present basic results on Gröbner degenerations and relations to triangulations of polytopes in Section 8, based on beautiful theory by Bernd Sturmfels [76]. In these three sections our aim completely changes. We do not present proofs, but focus on methods and examples.

In Section 9 we present toric varieties coming from cuts of a graph. Here, our aim was to prove that a conjecture of Sturmfels and Sullivant implies the famous four color theorem, cf. Proposition 9.4. This fact is known to experts and the original proof is due to David Speyer - however so far was not published.

In Section 10 we present relations of matroids, toric varieties and orbits in Grassmannian. Our focus in on famous White's conjecture and finiteness results related to it.

Mathematical biology provides a source of interesting toric varieties in the case of the so-called phylogenetic statistical models. Such a model can be studied as an algebraic variety by solving the polynomial equations which hold among its marginal probability functions in the appropriate field (e.g. $\mathbb{R}$ or $\mathbb{C}$). An interesting subclass of these models are the group-based models discussed in Section 11; for each choice of finite graph Γ (viewed as the underlying topological structure of the phylogeny) and finite group G there is a well-defined toric ideal $I_{\Gamma,G}$ and affine semigroup $M_{\Gamma,G}$. We describe basic toric constructions and known results in this area.

In section 12 we very briefly recall the construction of Cox rings and present results of Brown, Buczyński and Kędzierski on their relations to rational maps of varieties.

In the last section 13 we present some examples related to questions about depth and inner projections of toric varieties.

Throughout the text the reader may find various open problems and conjectures, explicit computations in Macaulay2 [36], relying on Normaliz [16] and 4ti2 [84]. Another great platform for toric computations is Polymake [33]. Further algorithms (e.g. for toric Gröbner basis) are present in CoCoA [17, 50, 5].

There are a lot of very interesting, important topics that we do not address and it is impossible even to list all of those. Let us just mention a few (that we regret most to omit): higher complexity T-varieties [1], many relations to combinatorics, such as e.g. binomial edge ideals [70], toric vector bundles [49, 72], relations to tropical geometry [59, Chapter 6] and secant varieties [67].

§ Acknowledgements

I would like to thank very much Takayuki Hibi and Hiraku Nakajima for providing great working conditions at RIMS, Kyoto. I am grateful to Seth Sullivant and David Speyer for sharing the proof of Proposition 9.4 (the idea is originally due to David Speyer). I express my gratitude to Jarosław Buczyński (section 12 is based on his advice), Michał Lasoń (for contributions to section 10.2), Christopher Manon (for contributions to section 11.2), Karol Palka (for pointing out interesting open problems) and Matteo Varbaro (for contributions to section 10.4). Further I thank Greg Blekherman, Sijong Kwak, Hyunsuk Moon and Rainer Sinn for discoussions on depth and projections.

Part 1. Introduction to toric varieties - basic definitions

§2. The Torus

In algebraic geometry we study an (affine) algebraic variety X, using (locally) rational functions on it - these form a ring R_X.

Example 2.1 (Polynomials and Monomials). *Consider the affine space $\mathbb{C}^n$. The associated ring of polynomial functions is $\mathbb{C}[x_1, \ldots, x_n]$. We represent a polynomial as:*

$$P = \sum_{(a_1,\ldots,a_n)\in\mathbb{N}^n} \lambda_{(a_1,\ldots,a_n)} x_1^{a_1} \cdots x_n^{a_n},$$

with only finitely many $\lambda_{(a_1,\ldots,a_n)} \neq 0$. When $a = (a_1, \ldots, a_n) \in \mathbb{N}^n$ we use a multi-index notation $x^a := x_1^{a_1} \cdots x_n^{a_n}$. Such an expression is called a monomial.

The main object of these lectures is the complex torus $T = (\mathbb{C}^*)^n$, with the structure of the group given by coordinatewise multiplication. On T we have more functions than on the affine space: for $a \in \mathbb{N}^n$ we allow x^a in the denominator.

Definition 2.2 (Laurent polynomial). *We define the ring $\mathbb{C}[x_1, x_1^{-1}, \ldots, x_n, x_n^{-1}]$ consisting of* Laurent polynomials:

$$P = \sum_{(a_1,\ldots,a_n)\in\mathbb{Z}^n} \lambda_{(a_1,\ldots,a_n)} x_1^{a_1} \cdots x_n^{a_n},$$

with finitely many $\lambda_{(a_1,\ldots,a_n)} \neq 0$. We use the same multi-index notation as in Example 2.1.

This is an example of a more general construction of *localization* that is an algebraic analogue of removing closed, codimension one sets from an affine algebraic variety [2].

Given a map of algebraic varieties $f : X \to Y$ we obtain an associated map f^* from functions on Y to functions of X (by composition with $Y \to \mathbb{C}$) [37]. Our first aim is to study algebraic maps $(\mathbb{C}^*)^n \to \mathbb{C}^*$. These correspond to maps of rings $\mathbb{C}[x, x^{-1}] \to \mathbb{C}[x_1, x_1^{-1}, \ldots, x_n, x_n^{-1}]$.

Proposition 2.3. *Every algebraic map $(\mathbb{C}^*)^n \to \mathbb{C}^*$ is given by λx^a for $\lambda \in \mathbb{C}^*$ and $a \in \mathbb{Z}^n$.*

Proof. Fix a lexicographic order on Laurent monomials. For any Laurent polynomial P we denote $LT(P)$ its leading term and $ST(P)$ its smallest term. Consider the ring morphism associated to the given map. Suppose $x \to Q$ and $x^{-1} \to S$. As $xx^{-1} = 1 = QS$ we see that $LT(S)LT(Q) = 1 = ST(S)ST(Q)$. Hence, $LT(Q) = ST(Q)$. Q.E.D.

Corollary 2.4. *Any algebraic map $(\mathbb{C}^*)^n \to (\mathbb{C}^*)^m$ is given by $(\lambda_1 x^{a_1}, \ldots, \lambda_m x^{a_m})$ for $\lambda_1, \ldots, \lambda_m \in \mathbb{C}^*$ and $a_1, \ldots, a_m \in \mathbb{Z}^n$. If in addition the map is a group morphism then it is given by Laurent monomials.*

Definition 2.5 (Characters, Lattice M)**.** *Algebraic group homomorphisms $T \to \mathbb{C}^*$ are called* characters. *They form* a lattice[1] $M_T \simeq \mathbb{Z}^n$, *with the addition induced from the group structure on $\mathbb{C}^*$. Explicitly, given $\chi_1 : T \to \mathbb{C}^*$ and $\chi_2 : T \to \mathbb{C}^*$ we define*

$$\chi_1 + \chi_2 : T \ni t \to \chi_1(t)\chi_2(t) \in \mathbb{C}^*.$$

Example 2.6. *Consider two characters $(\mathbb{C}^*)^2 \to \mathbb{C}^*$:*

$$(x, y) \to x^2 y, \quad (x, y) \to y^{-3}.$$

The first one is identified with $(2, 1) \in \mathbb{Z}^2 \simeq M_T$ and the second one with $(0, -3)$. Their sum is the character $(x, y) \to x^2 y^{-2}$ corresponding to $(2, -2) \in \mathbb{Z}^2$.

Definition 2.7 (One-parameter subgroups, Lattice N)**.** *Algebraic group homomorphisms $\mathbb{C}^* \to T$ are called* one-parameter subgroups. *They form* a lattice $N_T \simeq \mathbb{Z}^n$, *with the addition induced from the group structure on T.*

[1]Throughout the text by a *lattice* we mean a finitely generated free abelian group, i.e. a group isomorphic to some $(\mathbb{Z}^n, +)$.

The similarities in Definitions 2.5 and 2.7 are not accidental. The two lattices are dual, i.e. $M = \mathrm{Hom}_{\mathbb{Z}}(N, \mathbb{Z})$ and $N = \mathrm{Hom}_{\mathbb{Z}}(M, \mathbb{Z})$. In other words, there is a natural pairing $M \times N \to \mathbb{Z}$. Indeed, given a map $m : T \to \mathbb{C}^*$ and $n : \mathbb{C}^* \to T$ we may compose them obtaining $\mathbb{C}^* \to \mathbb{C}^*$ that, by Proposition 2.3 is represented by an integer.

Exercise 2.8. *Check that using the identifications $M \simeq \mathbb{Z}^n$ and $N \simeq \mathbb{Z}^n$ the pairing is given by the usual scalar product.*

The construction below associates to any monoid $\mathfrak{M}$ a ring $\mathbb{C}[\mathfrak{M}]$ known as a monoid algebra. As a vector space over $\mathbb{C}$ the basis of $\mathbb{C}[\mathfrak{M}]$ is given by the elements of $\mathfrak{M}$. The multiplication in $\mathbb{C}[\mathfrak{M}]$ is induced from the monoid action.

Example 2.9. *Consider two monoids*: $\mathfrak{M}_1 = (\mathbb{Z}^n_+, +), \mathfrak{M}_2 = (\mathbb{Z}^n, +)$. *By identifying $x_1^{a_1} \cdots x_n^{a_n}$ with $(a_1, \ldots, a_n)$ we see that:*

$$\mathbb{C}[\mathfrak{M}_1] = \mathbb{C}[x_1, \ldots, x_n], \qquad \mathbb{C}[\mathfrak{M}_2] = \mathbb{C}[x_1, x_1^{-1}, \ldots, x_n, x_n^{-1}].$$

Hence, $\mathfrak{M}_1$ corresponds to the affine space and $\mathfrak{M}_2$ to the torus T. More canonically, noticing that elements of the monoid induce functions on the associated variety, we get that the ring of functions on T equals $\mathbb{C}[M_T]$.

A map of two tori $f : T_1 \to T_2$ corresponds to a map $\mathbb{C}[M_{T_2}] \to \mathbb{C}[M_{T_1}]$. By Corollary 2.4 group morphisms f between two tori correspond to *lattice* maps $\hat{f} : M_{T_2} \to M_{T_1}$ or equivalently $N_{T_1} \to N_{T_2}$.

Definition 2.10 (Saturated sublattice, saturation)**.** *A sublattice $M' \subset M$ is called* saturated *if for every $m \in M$ if $km \in M'$ for some positive integer k, then $m \in M'$.*

For any sublattice $M' \subset M$ we define its saturation *by*:

$$\{m : \text{there is a positive integer } k \text{ such that } km \in M'\}.$$

Example 2.11. *A sublattice $\{(a, a) : a \in \mathbb{Z}\} \subset \mathbb{Z}^2$ is saturated. A sublattice $\{2a : a \in \mathbb{Z}\} \subset \mathbb{Z}$ is not saturated.*

Exercise 2.12.

(i) *Prove that $M' \subset M$ is saturated if and only if there exists a lattice M_1 and a lattice map $M \to M_1$ with kernel M'.*

(ii) *Show that a saturation of a sublattice is also a sublattice.*

In algebraic geometry, just as we study varieties through function on them, we understand a point of a variety by evaluating functions on it. Hence, to $x \in X$ we associate a map $R_X \to \mathbb{C}$ from the ring of functions on X, that sends $f \to f(x)$. One of the fundamental theorems of algebraic geometry, Hilbert's Nullstellensatz asserts that we can go the other way round: given a (nonzero) ring morphism $R_X \to \mathbb{C}$ we can find the (unique) corresponding point $x \in X$.

Example 2.13. *Given a point $(a_1, \ldots, a_n) \in \mathbb{C}^n$ the corresponding morphism $\mathbb{C}[x_1, \ldots, x_n] \to \mathbb{C}$ sends $x_i \to a_i$.*

For a point $t \in T$ we have a morphism $f_t : \mathbb{C}[M] \to \mathbb{C}$. Note that each element of M is not in the kernel (as $f_t(-m)$ is the inverse of $f_t(m)$). Hence, points $t \in T$ correspond to group *morphisms $M \to \mathbb{C}^*$. In coordinates, a point $(t_1, \ldots, t_n)$ corresponds to a map that assigns $\mathbb{Z}^n \ni (a_1, \ldots, a_n) \to \prod_{i=1}^n t_i^{a_i} \in \mathbb{C}^*$.*

Proposition 2.14. *Given a group morphism of tori $f : T_1 \to T_2$ the image equals a subtorus $T' \subset T_2$. We have a canonical isomorphism $M_{T'} = M_{T_2}/\ker \hat{f}$.*

Proof. Consider a subtorus $T' \subset T_2$ with the embedding given by the map $M_{T_2} \to M_{T_2}/\ker \hat{f}$. Our aim is to prove that $T' = \operatorname{Im} f$.

Consider a point of $t \in T_2$ represented by a map $f_t : M_{T_2} \to \mathbb{C}^*$. We have to show that f_t factors through $\hat{f}$ if and only if $t \in T'$, i.e. if and only if $\ker \hat{f} \subset f_t^{-1}(1)$. The implication $\Rightarrow$ is straightforward.

For the other implication consider the injective morphism $i : M_{T'} = M_{T_2}/\ker \hat{f} \to M_{T_1}$. It is enough to show that any morphism $M_{T'} \to \mathbb{C}^*$ factors through i. This follows from the fact that $\mathbb{C}^*$ is a *divisible* group, i.e. an injective $\mathbb{Z}$ module. More directly, in this case, we can extend the morphism to the saturation of $i(M_{T'})$, one element by one (i.e. the basis) and then extend to the whole $M_{T'}$. Q.E.D.

Exercise 2.15. *Show that the ideal (or even vector space) of equations vanishing on the image of T_1 is generated by $\{\chi - 1 : \chi \in \ker \hat{f}\}$.*

Dictionary about torus:

geometry	algebra	combinatorics
torus T	algebra $\mathbb{C}[M_T]$	lattice M_T
point of T	surjective ring morphism $\mathbb{C}[M_T] \to \mathbb{C}$	group morphism $M_T \to \mathbb{C}^*$
algebraic group morphism $T_1 \to T_2$	special ring map $\mathbb{C}[M_{T_2}] \to \mathbb{C}[M_{T_1}]$	lattice map $M_{T_2} \to M_{T_1}$
image of such morphism	kernel of the map	encoded by kernel

The following theorem is the cornerstone of toric geometry (and representation theory). Recall that a representation of a group G on a vector space V is a group morphism $G \to GL(V)$. In other words, each element of G provides a linear transformation of V (in a compatible way). In this lectures we additionally assume that G and the morphism $G \to GL(V)$ are algebraic.

Theorem 2.16. *Each representation of a torus T acting on V induces a decomposition $V = \bigoplus_i V_i$, where for each V_i there exists $m \in M$ such that for any $t \in T$ and $v \in V_i$ we have $t(v) = m(t)v$, i.e. T acts on V_i by rescaling the vectors, with different weight for different i.*

Proof. Consider a map $T \to GL(V)$. It is represented by:

$$f : t \to \sum_{m \in M} m(t) A_m,$$

where A_m is just a square matrix of scalars. By $f(t_1 t_2) = f(t_1) f(t_2)$ we obtain $A_{m_1} A_{m_2} = 0$ for $m_1 \neq m_2$ and $A_m^2 = A_m$. As $f(1) = id$, we obtain that $x = \sum A_m x$. Hence $V = \bigoplus \operatorname{Im} A_m$ and on the image of A_m the torus acts by scaling by the character m. Q.E.D.

§3. Affine Toric Varieties and Cones

Definition 3.1 (Affine Toric Variety). *An affine toric variety is the closure in $\mathbb{C}^m$ of the map $T \to (\mathbb{C}^*)^m \subset \mathbb{C}^m$, where the first one is given by group morphism. Equivalently: it is a closure of a subtorus of $(\mathbb{C}^*)^m$ or a closure of an image of a Laurent monomial map. In particular, the affine toric variety can be identified with a set of m points $S \subset M_T$.*

Example 3.2. *Consider a map $\mathbb{C}^* \to \mathbb{C}^2$ given by $t \to (t^2, t^3)$. The associated toric variety is a* cusp *– the zero locus of $x^3 - y^2$. The two points representing this toric variety are $\{2, 3\} \subset \mathbb{Z}$.*

Remark 3.3. *The definition of (affine) toric variety is often different in other sources and requires the variety to be* normal.

Theorem 3.4. *Let $\tilde{S}$ be the monoid generated by S in M. The toric variety X associated to S is isomorphic to $\operatorname{Spec} \mathbb{C}[\tilde{S}]$. The ideal of $X \subset \mathbb{C}^m$ is linearly spanned by such binomials $y_1^{b_1} \dots y_m^{b_m} - y_1^{c_1} \dots y_m^{c_m}$ that $\sum b_i s_i = \sum c_i s_i$ in M, for $b_i, c_i \in \mathbb{N}$.*

Proof. It is enough to provide the stated description of the ideal.

It is straightforward that binomials of the given form belong to the ideal.

Consider any $f(y)$ in the ideal. We will prove that $f(y)$ is a linear combination of binomials of the given form, inductively on the number of monomials appearing (with nonzero coefficients) in f. If there are no monomials (i.e. $f = 0$), the statement is obvious.

Choose a monomial m appearing in f. If we substitute $y_i \to x^{a_i}$, where a_i represent characters from S, we know that f is zero, as it vanishes on the image. In particular, after the substitution for m (that will remain a monomial, but now in x), the obtained monomial must cancel with some other monomial. This other monomial must come from m' appearing in f. But the fact that after the substitution they cancel, is equivalent to $m - m'$ being the binomial of the given form. Hence, we may subtract this binomial (with appropriate coefficient), reducing the number of monomials appearing in f. Q.E.D.

Remark 3.5. *The proof of the Theorem 3.4 does not depend on the field. In fact, the binomials that generate the ideal of X do not depend on the field, cf. Lemma 10.25.*

Theorem 3.6. *An affine variety on which the torus T acts and has a dense orbit is an affine toric variety.*

Proof. The algebra R_X of the variety embeds into $\mathbb{C}[M]$ as the morphism $t \to tx$ is dominant for (general) $x \in X$. We claim that R_X is linearly spanned by elements of M. Indeed, consider any $g \in R_X$. The torus acts on R_X, in particular on g. Consider the (finite dimensional) vector space spanned by all Tg. By Theorem 2.16 all characters χ_i appearing in $g = \sum_i c_i \chi_i$ (with nonzero coefficients) must belong to R_X.

As R_X is finitely generated, the monoid of characters $\tilde{S}$ in R_X is finitely generated, with generators providing the embedding in the affine space.

It is worth noticing that the dense torus orbit must be in fact also a torus. Q.E.D.

The following definition extends Definition 2.10 to monoids.

Definition 3.7 (Saturated monoid)**.** *A monoid $\tilde{S} \subset M$ is saturated (in M) if and only if $km \in \tilde{S}$ for some $k \in \mathbb{N}_+$, $m \in M$ implies $m \in \tilde{S}$.*

We now introduce a quite subtle notion of *normality*.

Definition 3.8 (Integrally closed, Normal)**.** *We say that a ring $A \subset B$ is integrally closed in B if for any monic (i.e. with the leading coefficient equal to 1) polynomial $f \in A[x]$ if for some $b \in B$ we have $f(b) = 0$ then $b \in A$.*

We say that an integral ring A is normal if it is integrally closed in its ring of fractions.

At this point the definition of normality may look artificial. It turns out that when a ring R_X is normal then the variety X is not 'too singular'. In particular, if X is smooth then R_X is normal. Normality turns out to play a crucial role in many branches of mathematics. Below we will see it appears naturally in toric geometry. Later, we will point out connections to the properties of divisors and finally we will show relations e.g. to matroid theory.

Definition 3.9 (Cone). *By a cone C in a lattice M (resp. vector space V over $\mathbb{R}$ or $\mathbb{Q}$) we mean a subset containing 0 and closed under any nonnegative linear combinations*:

If $\sum \lambda_i c_i \in M$ for $\lambda_i \in \mathbb{R}_+$, $c_i \in C$ then $\sum \lambda_i c_i \in C$.

A cone is called polyhedral *if it is finitely generated (using non-negative linear combinations) and* rational *if its generators are lattice points.*

Note that positive even integers do not form a cone in $\mathbb{Z}$ but they do form a cone in $2\mathbb{Z}$.

Theorem 3.10. *The affine toric variety X is normal if and only if the associated monoid $\tilde{S}$ is saturated in the lattice that it spans.*

A saturated monoid is a cone. Every finitely generated cone is finitely generated as a monoid.

Proof. First let us prove that if X is normal then $\tilde{S}$ is saturated. Consider any point $kc \in \tilde{S}$. We want to prove that $c \in \tilde{S}$. Let M be the lattice spanned by $\tilde{S}$. To improve notation, for $m \in M$ let χ_m be a corresponding character. Consider a polynomial $f(X) = X^k - \chi_{kc}$ with coefficients in the algebra of X. Due to the normality of X we know that χ_c is also in the algebra. Hence $c \in \tilde{S}$.

It remains to prove that if $\tilde{S}$ is saturated, then $\mathbb{C}[\tilde{S}]$ is normal. First note that the quotient field of $\mathbb{C}[\tilde{S}]$ is equal to the quotient field of $\mathbb{C}[M]$. As the torus is smooth, its algebra is normal. (One can also prove it by noticing that its algebra is a UFD - as it is a localization of the polynomial ring.) Consider any monic polynomial $f \in \mathbb{C}[\tilde{S}][x]$. Suppose that g is in the quotient field and satisfies the equation $f(g) = 0$. From the normality of $\mathbb{C}[M]$ we know that $g \in \mathbb{C}[M]$. We can act on the equation $f(g)$ by any point t of the torus T. The action of t on f gives a monic polynomial with coefficients in $\mathbb{C}[\tilde{C}]$. Hence the action of T on g gives polynomials that are in the normalization of $\mathbb{C}[\tilde{S}]$. Considering the action of T on the space of such polynomials, by Theorem 2.16, we conclude that all the characters-monomials appearing in g with nonzero coefficient must be in the normalization of $\mathbb{C}[\tilde{S}]$. Thus we can assume that $g \in M$. Suppose that f is of degree d. Notice that $f(g) = 0$ implies that $dg = d'g + c_0$ for some integer $0 \leq d' < d$ and $c_0 \in \tilde{S}$, as the character χ_{dg} must reduce with some other character. Thus $(d-d')g \in \tilde{S}$ and by normality $g \in \tilde{S}$.

For the cone: generators of the monoid belong to the $\{\sum \lambda_i v_i : 0 \leq \lambda_i \leq 1\}$, where v_i are generators of the cone. Q.E.D.

Hence, we have an equivalence of affine normal toric varieties with the torus $\mathbb{C}[M]$ action and (finitely generated, full dimensional) cones in M.

Example 3.11. *If we consider all lattice points $(a, b) \in \mathbb{N}^2$ such that $a \leq b\sqrt{2}$ then we obtain a monoid that is not finitely generated.*

Example 3.12.

(i) *Looking at Example 3.2 we see that the monoid spanned by $\{2, 3\}$:*
- *as a group spans $\mathbb{Z}$,*
- *is not saturated as $2 \cdot 1 = 2$.*

Indeed, a normal variety is smooth in codimension one (i.e. the singular locus must have codimension at least two). Our variety is a curve, hence is normal if and only if it is smooth. Note that 0 is the singular point of the cusp.

(ii) *We already know that the positive orthant gives rise to the affine space and whole lattice M to the torus.*

(iii) *Consider the cone generated (as a cone) by $(1, 0), (1, 2)$. There is one more monoid generator: $(1, 1)$. Hence, our variety is realized as (the closure of) the monomial map:*

$$(t_1, t_2) \to (t_1, t_1 t_2, t_1 t_2^2).$$

We have an integral linear relation: $(1, 0) + (1, 2) = 2 \cdot (1, 1)$ (all other are generated by this one). Hence, the ring of our variety is: $\mathbb{C}[x, y, z]/(xz - y^2)$.

Exercise 3.13. *Prove that each finitely generated cone has a unique, finite minimal set of generators.*

Hint: consider all elements that do not have a (nontrivial) presentation $c = c_1 + c_2$.

Note that we do not have to choose the minimal set of generators of the cone. Continuing Example 3.12 point iii) we may consider a fourth generator, e.g. $(2, 3) = (1, 2) + (1, 1)$. This provides an embedding in a four dimensional affine space and an isomorphic ring with a different presentation:

$$\mathbb{C}[x, y, z, t]/(xz - y^2, t - yz).$$

Example 3.14.

(i) *The map given by all monomials of degree r (in n variables) is called the r-th Veronese. The associated monoid consists of all points in the positive quadrant with sum divisible by r. Question: Is this toric variety normal?*

(ii) *Consider k groups of (distinct) variables, the i-th group consisting of a_i variables. The map given by all monomials (of degree k) that are of degree one with respect to each group is called the Segre map.*

Both of the above examples are usually considered in projective setting - we will be coming back to them.

Exercise 3.15. *Consider the map $(x, y) \to (x^2, y^2)$. What is the associated toric variety? Is it normal?*

Dictionary about affine toric varieties:

geometry	algebra	combinatorics
affine toric variety	prime binomial ideal	f.g. submonoid in a lattice
point	surjective morphism to $\mathbb{C}$	semigroup map $S \to (\mathbb{C}, \cdot)$
normal	normal ring	cone
toric embedding	special monomial map	choice of generators of the monoid
hypersurfaces that contain	defining ideal	integral relations among generators

§4. Projective Toric Varieties

Definition 4.1 (Projective Toric Variety). *A projective toric variety is the closure in $\mathbb{P}^m$ of the map $T \to (\mathbb{C}^*)^m \subset \mathbb{P}^m$, where the first one is given by a group morphism and the inclusion can be regarded as the locus of points with nonzero coordinates.*

Equivalently: it is a closure of a subtorus of $(\mathbb{C}^*)^m$ or a closure of an image of a Laurent monomial map in a projective space.

Remark 4.2. *As the points in $\mathbb{P}^m$ are regarded up to scalar as $m+1$-tuples of complex numbers, the projective toric variety can be described as a closure of the map given by $m+1$ characters.*

Note that if $m_1, \ldots, m_{m+1}$ are monomials parameterizing the projective toric variety X, it is easy, using an additional variable x_0, to parameterize the affine cone over X:

$$(x_0, \ldots, x_n) \to (x_0 m_1, \ldots, x_0 m_{m+1}) \in \mathbb{A}^{m+1}.$$

Hence, it is 'better' to represent m_i not in the lattice $\mathbb{Z}^n$, but $\mathbb{Z}^{n+1}$, by the inclusion $\mathbb{Z}^n \ni m \to (m, 1) \in \mathbb{Z}^{n+1}$.

Definition 4.3 (Normal polytope). *A lattice polytope (i.e. a polytope whose vertices are lattice points) $P \subset M_{\mathbb{R}}$ is called* normal *(in the lattice M) if and only if for any $k \in \mathbb{N}$ all lattice points in kP are sums of k (not necessary distinct) lattice points of P.*

Definition 4.4 (Projective normality). *A projective algebraic variety is called projectively normal if and only if the affine cone over it is normal.*

Thus, by Theorem 3.10 we obtain the following.

Theorem 4.5. *A projective toric variety is projectively normal if and only if the associated parameterizing monomials $S \in \mathbb{Z}^{n+1} =< S >$ (at height one) generate (as a monoid) all lattice points in the cone generated by them.*

In the situation of the theorem above the lattice points must be all lattice points of a convex polytope, but this is not enough!

Exercise 4.6. *Show that a projective toric variety is projectively normal if and only if the associated polytope is normal.*

A set of points (usually integral points in a polytope) $S \subset \mathbb{Z}^m$ define a projective toric variety, with a projective embedding in a projective space with a distinguished torus T'. The dense torus orbit is the intersection of T' with the variety. Notice that coordinates of the ambient projective space correspond to the points in S. Hence, to choose an affine open chart we have to choose one point and 'set it equal to one'. This corresponds to dividing parameterizing monomials by the chosen one and division corresponds to shifting (subtracting the chosen point). In such a way we obtain the open affine toric variety.

Definition 4.7 (Very ample polytope). *A lattice polytope $P \subset M_{\mathbb{R}}$ is called* very ample *(in the lattice M) if and only if all lattice points in kP are sums of k (not necessary distinct) lattice points of P for k large enough.*

Theorem 4.8. *A set of lattice points in a polytope defines a normal projective toric variety if and only if the polytope is very ample (in the lattice it spans).*

The proof follows from the exercise below.

Exercise 4.9. *Suppose that P is a lattice polytope that spans a lattice M. Show that the following are equivalent:*

- *P is very ample,*
- *for any lattice point $m \in P$ the monoid spanned by $P - m$ is saturated,*
- *for any vertex m of P the monoid spanned by $P - m$ is saturated.*

Exercise 4.10.

- *Show that the unit r-dimensional simplex corresponds to $\mathbb{P}^r$.*
- *The map $\mathbb{P}^r \to \mathbb{P}^{\binom{d+r}{r-1}}$ given by all monomials of degree d is called the* Veronese *embedding. What is the corresponding polytope?*

- *The map* $\mathbb{P}(V_1) \times \cdots \times \mathbb{P}(V_n) \to \mathbb{P}(V_1 \otimes \cdots \otimes V_n)$ *defined by*
$$[v_1] \times \cdots \times [v_n] \to [v_1 \otimes \cdots \otimes v_n]$$
is called the Segre *embedding. What is the corresponding polytope?*
- *What are the defining equations in the two examples above? Are the varieties projectively normal?*

Example 4.11.

- *The set* $\{0, 1, 3, 4\} \subset \mathbb{Z}$ *defines a smooth projective toric variety that is not projectively normal.*
- *The* 3*-dimensional polytope with vertices*

$$(0,0,0), (0,0,-1), (0,1,0), (0,1,-1), (1,0,0), (1,0,-1), (1,1,3), (1,1,4)$$

is very ample, but not normal.

Definition 4.12 (Smooth polytope)**.** *An n-dimensional lattice polytope P is* smooth *if*

- *for any vertex v, there are exactly n edges adjacent to v with lattice points nearest to v denoted by $v_1, \ldots, v_n$ and*
- *the vectors $v - v_1, \ldots, v - v_n$ form a basis of the lattice spanned by P.*

Remark 4.13. *A lattice polytope P defines a smooth projective toric variety if and only if P is smooth.*

Below we present a famous, central conjecture in toric geometry. The first part is due to Oda and the second to Bogvad.

Conjecture 4.14. *A smooth polytope is normal. The associated toric variety is defined by quadrics.*

Proposition 4.15. *The degree of a projective toric variety X represented by a very ample polytope P spanning a lattice M equals the (normalized) volume of P.*

Remark 4.16. *The function $n \to |nP \cap M|$ is a polynomial known as the Ehrhart polynomial.*

Proof. By definition, the degree of X is (up to $\frac{1}{(\dim P)!}$) the coefficient of the leading term of the Hilbert polynomial (that to n associates the dimension of degree n part of R_X). However, this space is spanned by lattice points in nP.

To estimate the number of lattice points in nP and relate it to a volume we cover/inscribe in P small cubes C_i. The statement relating the volume of a cube C_i with the estimate of the degree of the function $n \to |nC_i \cap M|$ easily reduces to one dimension and is an easy exercise. Q.E.D.

Let us return to an affine covering of the toric variety defined by P.

Exercise 4.17. *Suppose a set of characters S defines a projective toric variety X_S. Show that, under this embedding, X_S is covered by principal affine subsets corresponding to vertices of the convex hull (in $M_{\mathbb{R}}$) of S.*

The previous exercise shows that the monoids generated by $S-v$ are of particular interest. When S is formed by lattice points of a polytope these monoids are cones precisely when the polytope is very ample.

4.1. General Toric Varieties

We briefly describe general toric varieties, that do not have to be affine or projective. This is a classical topic, central in toric geometry, well-described in many books and articles. However, in this review, general toric varieties are a side topic - we focus our interest on projective ones.

Definition 4.18 (Fan). *A (finite) collection of (rational, polyhedral) cones in (a vector space over) a lattice N is called a fan if it is closed under taking faces and intersections.*

Definition 4.19 (Dual cone). *Fix a cone $\sigma \subset M$ (resp. $M_{\mathbb{R}}$). We define the dual cone $\sigma^\vee \subset N = M^*$ (resp. $N_{\mathbb{R}}$) consisting of those elements n such that for any $m \in \sigma$ we have $(n, m) \geq 0$.*

If we look at the family of cones $P - v$ (where v runs over vertices of P) in M we do not see any structure.

Exercise 4.20. *Show that $(P-v)^\vee$ (together with their faces) form a fan covering whole N.*

Fans covering N are called *complete*. Not every complete fan comes from a polytope P, as we will see in Section 6. However, there is a general construction that to a fan Σ in N associates a normal toric variety $X(\sigma)$. The idea is to glue together the affine toric varieties $\operatorname{Spec}\mathbb{C}[\sigma^\vee]$ for all $\sigma \in \Sigma$. Precisely, consider the cones $\sigma_3 = \sigma_1 \cap \sigma_2$. We have induced open embeddings $\operatorname{Spec}\mathbb{C}[\sigma_3^\vee] \to \operatorname{Spec}\mathbb{C}[\sigma_i^\vee]$ for $i = 1, 2$ that allow the gluing. This gives a normal algebraic variety with an action of the torus $\operatorname{Spec}\mathbb{C}[N^\vee]$ with a dense orbit.

Conversely, given a normal algebraic variety with an action of a torus T and a dense orbit one can prove that it is represented by a fan. The easy part of the proof is to consider torus invariant affine open subvarieties and show that they glue in a good way. However, the nontrivial part of the theorem relies on the fact that torus invariant affine subvarieties cover the whole variety. This is a theorem of Sumihiro [81, 46].

§5. Affine Toric Varieties - combinatorics and algebra

5.1. Orbit-Cone correspondence

For an affine toric variety corresponding to a cone C the faces of C correspond to orbits of the torus acting on it. Let us present this correspondence in details. We fix a finitely generated monoid C in a lattice M and its generators $\chi_1, \dots, \chi_k \in C$. We know that:

- the dense torus orbit of X contains precisely those points that have all coordinates different from zero,
- the character lattice of the torus acting on X is equal to the sublattice of M spanned by C.

We will generalize this to other orbits.

Theorem 5.1. *Assume that C is a cone. Each orbit will be indexed by a face F of the cone. The face F distinguishes a subset I_F of indices from $\{1, \dots, k\}$ such that $i \in I$ if and only if $\chi_i \in F$. The orbit corresponding to F can be characterized as follows:*

1) *the orbit contains precisely those points that have got coordinates corresponding to $i \in I_F$ different from zero and all other equal to zero,*
2) *the orbit is a torus with a character lattice spanned by elements of F,*
3) *the closure of the orbit is a toric variety given by the cone F,*
4) *each point of the orbit is a projection of the dense torus orbit onto the subspace spanned by basis elements indexed by indices from I_F,*
5) *the inclusion of the orbit in the variety is given by a morphism of algebras $\mathbb{C}[C] \to \mathbb{C}[F]$. This morphism is an identity on $F \subset \mathbb{C}[C]$ and zero on $C \setminus F$.*

Note that each orbit will contain a unique distinguished point given by the projection of the point $(1, \dots, 1) \in \mathbb{C}^k$.

Proof. As in case of the torus we can identify the points of X with monoid morphisms $C \to (\mathbb{C}, \cdot)$.

Point 1): Fix any point $x \in X$. Note that for any $c_1, c_2 \in C$ if $(c_1 + c_2)(x) \neq 0$ then $c_1(x), c_2(x) \neq 0$. Hence, the characters $\chi \in C$ such that $\chi(x) \neq 0$ must form a face F of C. Thus x distinguishes a subset of indices $I_F \subset \{1, \dots, k\}$. Of course the set of points with nonzero coordinates indexed by I_F and other coordinates equal to zero in X is invariant with respect to the action of the torus acting on X. So to prove 1) it is enough to prove that all these points are in one orbit. The point x represents a morphism $C \to (\mathbb{C}, \cdot)$ that is nonzero on F and zero on $C \setminus F$. Consider the restriction of this morphism to F. As it is nonzero it can be extended to a morphism $M' \to \mathbb{C}^*$, where M' is a sublattice generated by F. Next, as M' is saturated, we can extend this morphism to the lattice M generated by C. Thus we obtain a morphism $f : M \to \mathbb{C}^*$ that agrees with the one representing x on F. Note that f represents a point p in the dense torus orbit of X. By the action of p^{-1} on x we obtain a point given by a morphism that associates one to elements from F and zero to elements from $C \setminus F$. Thus we have proved 1). Moreover, we showed that each orbit contains the distinguished point.

Point 2) follows, as morphisms that are nonzero on F and zero on $C \setminus F$ are identified with morphisms from M' to $\mathbb{C}^*$.

Point 3): We already know that the orbit is a torus with the lattice generated by F. This torus is the image of the torus $\operatorname{Spec} \mathbb{C}[M]$ in $\mathbb{C}^k$ by characters from I_F and all other coordinates equal to zero. We see that the orbit corresponding to F is contained in the affine space spanned by basis elements indexed by indices in I_F. In fact, by construction it is the image of $\operatorname{Spec} \mathbb{C}[M]$ by characters χ_i, such that $i \in I_F$. The closure of this torus is exactly given by $\operatorname{Spec} \mathbb{C}[F]$, as generators of the monoid C contained in F are generators of F.

Point 4) is obvious, as the point p constructed in the first part of the proof projects to x.

Point 5) is a consequence of the other points. Q.E.D.

Looking at a cone over a polytope we see that orbits correspond to faces of polytope (the whole polytope to the dense torus orbits, facets to torus invariant Weil divisors,..., vertices to torus invariant points - the containment of closures of torus orbits is as you can see on the polytope).

5.2. Toric ideals

We now characterize ideals $I \subset \mathbb{C}[x_1, \dots, x_n]$ that define toric varieties under a toric embedding.

Theorem 5.2. *The ideal I defines an affine variety that is a closure of a subtorus in $\mathbb{C}^n$ if and only if I is a binomial (i.e. generated by binomials) prime ideal.*

Proof. In Theorem 3.4 we proved the forward implication.

Suppose I is prime and binomial. As $(1, \dots, 1) \in V(I)$ we see that the intersection $X_0 := V(I) \cap (\mathbb{C}^*)^n$ is a nonempty variety and hence dense in $V(I)$. Note that a restriction of a binomial to $(\mathbb{C}^*)^n$ is (up to invertible element) of the form $m - 1$, where m is a character. As X_0 is reduced and irreducible the set $M' := \{m : m - 1 \in I(X_0)\}$ is a saturated sublattice of a lattice $M_{(\mathbb{C}^*)^n}$. The map $\mathbb{C}[x_1^{\pm 1}, \dots, x_n^{\pm 1}]/I(X_0) \to \mathbb{C}[M_{(\mathbb{C}^*)^n}/M']$ is an isomorphism. Hence, X_0 is a subtorus and $V(I)$ and its closure is a toric variety. Q.E.D.

Remark 5.3. *In general, it can be very hard to say if I defines a toric variety (with a different embedding). In particular, there exists a smooth hypersurface H in $\mathbb{C}^5$ given by $x + x^2y + z^2 + t^3$. It is a product of $\mathbb{C}$ and a so-called the Koras-Russel cubic. It is an open problem if it is isomorphic with $\mathbb{C}^4$. Further, it is not known if there exists an automorphism of $\mathbb{C}^5$ that would linearize H.*

Part 2. Topics on Toric Varieties

§6. Divisors

6.1. Weil Divisors

Given any variety X one may consider the set of all irreducible codimension one subvarieties and the free abelian group $D(X)$ generated by it. An integral combination of such subvarieties is called a *Weil divisor*. This group is far too large, thus one regards it modulo an equivalence relation. From now on assume X is regular in codimension one, i.e. the codimension of the singular locus is at least two (e.g. X is normal). In such a case to a rational function f on X we can associate a divisor $\operatorname{div}(f) \in D(X)$ - see e.g. [37, Chapter 6].

Definition 6.1 (Class group)**.** *The class group $Cl(X)$ is defined as $D(X)$ modulo the subgroup generated by all* $\operatorname{div}(f)$.

Let X be a normal toric variety defined by a fan Σ (we may think Σ is a normal fan of a polytope - for a general construction we refer to [25]). The beautiful thing is that to compute $Cl(X)$ we do not have to consider all divisors (just those invariant by the torus action) and not all rational functions (just characters of the torus).

Precisely we recall that rays in Σ correspond to codimension 1 orbits of the torus (hence their closures are codimension one torus invariant subvarieties). For a ray $u \in \Sigma_1$ we abuse the notation and denote also by u the first nonzero lattice point on that ray. We note that for $m \in M$ we have a pairing $\langle m, u \rangle$ that is equal to zero if and only if m does not vanishes or have a pole on the variety D_u associated to

u (think e.g. in terms of the coordinates as in Chapter 5.1). Indeed, $\operatorname{div}(m) = \sum_{u\in\Sigma_1} \langle m, u\rangle D_u$.

Theorem 6.2. *We have an exact sequence:*

$$M \to \bigoplus_{u\in\Sigma_1} \mathbb{Z}D_u \to Cl(X) \to 0.$$

Example 6.3.

(i) *As $\mathbb{P}^n$ corresponds to a simplex the rays of the normal fan are $e_1, \dots, e_n, -e_1 - \dots - e_n \in N$. The exact sequence above becomes:*

$$0 \to \mathbb{Z}^n \to \mathbb{Z}^{n+1} \to \mathbb{Z} \to 0.$$

(ii) *As $\mathbb{P}^1 \times \mathbb{P}^1$ corresponds to a square the rays of the normal fan are $e_1, -e_1, e_2, -e_2$. The exact sequence becomes:*

$$0 \to \mathbb{Z}^2 \to \mathbb{Z}^4 \to \mathbb{Z}^2 \to 0.$$

(iii) *(Rational normal scroll, Hirzebruch surface) Consider a polytope that is the convex hull of $0, f_1, f_2, rf_1 + f_2$. It has four facets corresponding to four rays in the dual fan e_1, e_2, $-e_2$, $-e_1 + re_2$. We obtain $Cl(X) = \mathbb{Z}^2$.*

(iv) *Consider the cone generated by $f_1, f_1 + f_2, f_1 + 2f_2$. The dual cone has ray generators $e_2, 2e_1 - e_2$. We obtain:*

$$0 \to \mathbb{Z}^2 \to \mathbb{Z}^2 \to \mathbb{Z}/2\mathbb{Z} \to 0.$$

6.2. Cartier divisors

A *Cartier* divisor is represented by a(n open) covering U_i of X and rational functions f_i such that f_i/f_j is well defined and nonzero on $U_i \cap U_j$. More formally it is a global section of the quotient sheaf of (the sheaf of) invertible rational functions modulo (the sheaf of) invertible regular functions. Again we consider them modulo divisors given by one rational function (and the covering consisting just of X). The quotient is the *Picard group* $\operatorname{Pic}(X)$ of X.

As before, for toric varieties we only have to consider U_i affine toric (hence represented by cones) and rational functions f_i given by characters of the torus.

Theorem 6.4. *Let Σ be a fan in which each cone is contained in a full dimensional cone. Consider T-invariant Cartier divisors $CDiv_T(X)$ given by:*

- *A continuous, locally linear function represented by $m_\sigma \in M$ for each maximal dimensional cone $\sigma \in \Sigma$.*

We have an exact sequence:

$$M \to CDiv_T(X) \to \mathrm{Pic}(X) \to 0.$$

(Without the assumption that maximal cones are full dimensional the exact sequence also holds, but we must define $CDiv_T$ taking into account that m_σ is a class in $M/(\sigma^\perp)$.)

Exercise 6.5. *Compute the Picard group in cases 1)-3) from Example 6.3. Can you see a general theorem here?*

What happens in case 4)?

In general on a normal variety there is an injection $\mathrm{Pic}(X) \to Cl(X)$. In toric case, we just evaluate the continuous linear function on the ray generators.

Let us now determine the global sections of a line bundle associated to a Cartier divisor D represented by a piecewise linear function f on a fan Σ. As everything is torus equivariant, the sections will be spanned by characters. Let us fix a character $m \in M$. We ask when $\chi_m \in \mathcal{O}(D)$. Let us fix a maximal cone $\sigma \in \Sigma$. Here D is represented by a rational function $m_\sigma \in M$. Hence, we ask when the product (that is the sum in the lattice) $m + m_\sigma$ is well-defined on $\operatorname{Spec} \mathbb{C}[\sigma^\vee]$. This is clearly if and only if $m + m_\sigma \in \sigma^\vee$.

Caution: there are different conventions about signs of f, m_σ etc.

Theorem 6.6. *The global sections of $\mathcal{O}(D)$ are spanned by such m that $m + f$ is nonnegative of $|\Sigma|$. Given a Weil divisor $\sum_{u\in\Sigma_1} a_u D_u$ we define a polyhedron:*

$$P_D := \{m \in M_{\mathbb{Q}} : \langle m, u\rangle \geq -a_u \textit{ for all } u \in \Sigma_1\}$$

From now on suppose Σ is a fan with all maximal cones full dimensional.

Definition 6.7 (Convex, strictly convex)**.** *We say that a piecewise linear function h represented by m_σ on cone $\sigma \in \Sigma$ is* convex *if each $m_\sigma \leq h$ (as functions on $|\Sigma|$). We say it is* strictly convex *if the inequality is strict outside σ.*

Theorem 6.8. *A Cartier divisor D represented by a function f is:*

(i) *globally generated if and only if h is convex,*
(ii) *ample if and only if h is strictly convex,*
(iii) *very ample if and only if h is strictly convex and P_D is very ample.*

Proof. 1) and 3) follow by local affine analysis. 2) is a consequence of 3). Q.E.D.

§7. Cohomology

We start with general remarks on cohomology of line bundles on toric varieties. Let Σ be a fan in $N_{\mathbb{R}} = \mathbb{R}^n$ with rays $r_1, \dots, r_s$. For a Weil divisor $D = \sum_{i=1}^{s} a_i D_i$ and $m \in M$ we define

$$C_{D,m} := \cup_{\sigma \in \Sigma} Conv(r_i : r_i \in \sigma, \langle m, r_i \rangle < -a_i).$$

For a Cartier divisor D_h represented by a (piece-wise linear) function h and $m \in M$ we define

$$C_{D_h,m} := \{u \in |\Sigma| : \langle m, u \rangle < h(u)\}.$$

Proposition 7.1 ([25] Section 9.1). *Note that for a Weil or Cartier torus invariant divisor D, each cohomology is M graded. We have:*

$$H^p(X_\Sigma, \mathcal{O}(D))_m \simeq \tilde{H}^{p-1}(C_{D,m}) \text{ for a Weil divisor } D,$$

$$H^p(X_\Sigma, D_h)_m = \tilde{H}^{p-1}(C_{D_h,m}) \text{ for a Cartier divisor } D_h,$$

where $\tilde{H}$ is the reduced cohomology.

There is an 'Alexander dual' method of computing the cohomology. Assume Σ is complete and smooth. For $I \subset \{1, \dots, m\}$ let C_I be the simplicial complex generated by sets $J \subset I$ such that $\{r_i : i \in J\}$ form a cone in Σ. For $a = (a_i : i = 1, \dots, m)$ let us define $\mathrm{Supp}(a) := C_{\{i : a_i \geq 0\}}$.

Proposition 7.2 ([9]). *The cohomology $H^j(X_\Sigma, L)$ is isomorphic to the direct sum over all $a = (a_i : i = 1, \dots, s)$ such that $\mathcal{O}(\sum_{i=1}^{s} a_i D_{r_i}) = L$ of the $(n-1-j)$-th reduced homology of the simplicial complex* $\mathrm{Supp}(a)$.

To compare Proposition 7.1 and 7.2, let us first look at $H^0(X_\Sigma, \mathcal{O}_D)$. We assume that Σ is a smooth, complete fan and $D = \sum_{i=1}^{s} a_i D_i$. We recall that in this case the basis vectors of $H^0(X_\Sigma, \mathcal{O}_D)$ correspond to lattice points inside

$$P_D := \{m \in M_{\mathbb{Q}} : \langle m, u \rangle \geq -a_u \text{ for all } u \in \Sigma_1\}.$$

In particular,

$$\dim H^0(X_\Sigma, \mathcal{O}_D)_m = \begin{cases} 1 \text{ if } m \in P_D \\ 0 \text{ if } m \notin P_D. \end{cases}$$

Referring to Proposition 7.1 we have:

$$m \in P_D \Leftrightarrow \langle m, u\rangle \geq -a_u \text{ for all } u \in \Sigma_1 \Leftrightarrow \langle m, u\rangle < -a_u \text{ for none } u \in \Sigma_1$$

$$\Leftrightarrow C_{D,m} \text{ is empty} \Leftrightarrow \tilde{H}^{-1}(C_{D,m}) = 1 \Leftrightarrow \tilde{H}^{-1}(C_{D,m}) \neq 0.$$

Referring to Proposition 7.1 we have:

$$m \in P_D \Leftrightarrow \langle m, u\rangle + a_u \geq 0 \text{ for all } u \in \Sigma_1 \Leftrightarrow$$

$$\mathrm{Supp}((\langle m, u\rangle + a_u)_u) = C_{\{1,\dots,m\}} \Leftrightarrow$$

$$H^n(\mathrm{Supp}((\langle m, u\rangle + a_u)_u)) \neq 0 \Leftrightarrow D + \mathrm{div}(m) \text{ contributes 1 to}$$

$$\text{the sum in Propostion 7.2}$$

Example 7.3. *Let $X = \mathbb{P}^n$, i.e. the fan Σ has rays $e_1, \dots, e_n, -e_1 - \dots - e_n$. Let us compute $H^p(X, \mathcal{O}(kD_{e_1}))$. (As all divisors D_{e_i} are equivalent this covers all the cases). We need to consider all decompositions: $\sum a_i D_i \simeq kD_{e_1}$, which is equivalent to $\sum a_i = k$. Note that if there exists $a_{j_1} \geq 0$ and $a_{j_2} < 0$, then* Supp(a) *is a nonempty simplex. Hence, all its reduced homology vanish.*

Suppose $k \geq 0$. Then we cannot have all a_i negative, thus we can assume they are all greater or equal to 0. For each such a, the simplicial complex Supp(a) *is an n-sphere. Hence, the only nonvanishing homology is the n-th one (equal to one), which corresponds to the 0-th cohomology of the divisor. Thus $H^p(\mathbb{P}^n, \mathcal{O}(kD_{e_1})) = 0$ for $k \geq 0$ and $p \neq 0$. For $p = 0$ we need to count the possible decompositions $\sum_{i=1}^{n+1} a_i = k$ for $a_i \geq 0$. Clearly there are $\binom{k+n}{n}$ of those. It is worth noting that they naturally correspond to monomials of degree k in $n+1$ variables - the usual description of the basis of sections.*

Suppose $k < 0$. Now, we only have to consider decompositions with $a_i < 0$. In particular, if $-n-1 < k$ then all cohomology vanish. As the empty simplicial complex has only -1-st reduced homology nonzero (its dimension is equal to 1 in that case) we see that $H^p(\mathbb{P}^n, \mathcal{O}(kD_{e_1})) = 0$ for $k < 0$ and $p \neq n$. For $p = n$ we need to count the number of decompositions $\sum_{i=1}^{n+1} a_i = k$ with $a_i < 0$. Clearly there are $\binom{-k-1}{n}$ such possibilities.

Example 7.4. *Consider the Hirzebruch surface X with the fan given by $e_1, e_2, -e_1 + 2e_2, -e_2$. As an example let us compute $H^2(X, \mathcal{O}(-3D_{e_1} - 5D_{e_2}))$. As before, (and always) nonvanishing top cohomology corresponds to nonvanishing -1-st reduced homology, i.e. empty* *simplex. In other words the dimension of the cohomology equals the number of solutions to the diophantine system of equations*:

$$\begin{aligned} c_1 + c_3 + 2c_4 &= -5 \\ c_2 + c_4 &= -3 \end{aligned}$$

with $c_i < 0$. (*This system corresponds to* $c_1 D_{e_1} + c_2 D_{e_2} + c_3 D_{-e_1+2e_2} + c_4 D_{-e_4} \simeq -5D_{e_1} - 3D_{e_2}$.) *There are two such solutions.*

For more relations among simplicial complexes, techniques of counting homology and cohomology of line bundles we refer to [54, Section 3.2].

§8. Gröbner basis and triangulations of polytopes

Definition 8.1 (Term order). *An order* $<$ *on monomials or equivalently on* $\mathbb{N}^n$ *is called a term order if* 0 *is the unique minimal element and* $a < b$ *implies* $a + c < b + c$ *for any* $a, b, c \in \mathbb{N}^n$.

Definition 8.2 (Initial ideal, Gröbner basis). *For a fixed term order* $<$ *we define* $in_<(f)$ *to be the unique, largest monomial appearing in* f (*with nonzero coefficient*). *For an ideal* I *we define*:

$$in_<(I) :=< in_<(f) : f \in I > .$$

Caution: *it is not enough to take initial forms of* (*any*) *generators of* I *to obtain all generators of* $in_<(I)$.

A set of generators G *of* I *is called* a Gröbner basis (*with respect to* $<$) *if and only if*

$$in_<(I) :=< in_<(f) : f \in G > .$$

A Gröbner basis is called minimal if G *is* minimal *with respect to inclusion* (*among sets satisfying equality above*). *It is called* reduced *if for any* $g' \in G$ *no monomial in* g' *is divisible by* $in_<(g)$ *for some* $g \in G$.

For any ideal I *and order* $<$ *finite Gröbner basis exists. Further minimal, reduced Gröbner basis is unique* (*up to scaling*).

Most important examples of term orders are *weight orders* $<_\omega$ associated to $\omega \in \mathbb{R}^n_{\geq 0}$. Namely we say that $a \leq b$ if and only if $a \cdot w \leq b \cdot w$.

(to make it a total order either we need ω to be irrational or we fix another order $<$ as a tie breaker)

Caution: not every term order is a weight order.

Proposition 8.3 (Proposition 1.11 [76]). *For any term order* $<$ *and any ideal* I *there exists a non-negative integer vector* $\omega \in \mathbb{N}^n$ *such that* $in_{<_\omega}(I) = in_<(I)$.

Remark 8.4. *Geometrically the variety defined by $in_<(I)$ is a degeneration of the variety defined by I. In particular, the most important algebraic invariants (like degree or dimension) remain the same. This fact is used in algebraic software in order to compute such invariants – notice that for generic ω the initial ideal will be monomial.*

For $I \subset \mathbb{C}[x_1, \dots, x_n]$ and a fixed term order $<$ we consider a simplicial complex $\Delta_<(I)$ on $\{1, \dots, n\}$ such that $f = \{i_1, \dots, i_k\}$ is a face of $\Delta_<(I)$ if and only if $\prod_{j=1}^k x_{i_j}$ is not in the radical of $in_<(I)$. Recall that toric ideals are in variables that have natural interpretation as lattice points. Suppose that our toric variety is defined by a lattice polytope P. Our plan is to realize $\Delta_<(I)$ as a 'subdivision' of P. Let P be a $d-1$ dimensional lattice polytope in $\mathbb{R}^{d-1} \times \{1\} \subset \mathbb{R}^d$.

Definition 8.5 (Regular triangulation)**.** *Given a subset of lattice points $P = \{p_1, \dots, p_n\}$ and $\omega \in \mathbb{R}^n$ we define a subdivision Δ_ω by:*

$f = \{p_{i_1}, \dots, p_{i_k}\}$ is a face of Δ_ω if any only if there exists $c \in \mathbb{R}^d$ such that:

$$p_j \cdot c = \omega_j \text{ if } j \in \{i_1, \dots, i_k\} \text{ and}$$

$$p_j \cdot c < \omega_j \text{ if } j \notin \{i_1, \dots, i_k\}.$$

A generic ω defines a triangulation Δ_ω and any such triangulation is called regular.

Remark 8.6. *One should imagine a two dimensional polytope P and the heights w_i above p_i as points in the third dimension. The linear forms c are obtained by putting a sheet of paper from below and stopping at some heights. (If ω is generic such a sheet of paper will touch exactly three points. These are the triangles (after projecting back to P) of the triangulation Δ_ω.)*

Remark 8.7. *Not any triangulation is regular* [76, Example 8.2].

Definition 8.8 (Secondary fan)**.** *The regular triangulations in Definition 8.5 depend on ω. However, the set of $\omega \in \mathbb{R}^n$ that give rise to the same regular triangulation is a cone. These cone form a complete fan in $\mathbb{R}^n$ called the* secondary fan.

Theorem 8.9 (Theorem 8.3 [76])**.** $\Delta_{<_\omega}(I_P) = \Delta_\omega$

Corollary 8.10.

$$\begin{aligned} Rad(in_\omega(I_P)) &= < x^b : b \text{ is a minimal nonface of } \Delta_\omega > \\ &= \bigcap_{\sigma \in \Delta_\omega} < x_i : i \notin \sigma > \end{aligned}$$

What is the multiplicity of a prime ideal in $in_\omega(I_P)$?

Theorem 8.11 (Theorem 8.8 [76]). *For $\sigma \in \Delta_\omega$ of dimension $d-1$ the normalized volume of σ equals the multiplicity of the prime ideal $< x_i : i \notin \sigma >$ in $in_\omega(I_P)$.*

Corollary 8.12 (Corollary 8.9 [76]). *The initial ideal $in_\omega(I_P)$ is square-free if and only if the corresponding regular triangulation is unimodular.*

§9. Cuts and Splits

Let $G = (V, E)$ be a graph. Following [78] we consider two polynomial rings:

$$\mathbb{C}[q] := \mathbb{C}[q_{A|B} : A \cup B = V, A \cap B = \emptyset],$$
$$\mathbb{C}[s,t] := \mathbb{C}[s_{ij}, t_{ij} : \{i,j\} \in E].$$

In the first ring the variables correspond to partitions of V, in the second to each edge we associate two variables. For a given partition $A|B$ we define a subset $\mathrm{Cut}(A|B) \subset E$ of cut edges by:

$$\mathrm{Cut}(A|B) := \{\{i,j\} \in E : i \in A, j \in B \text{ or } j \in A, i \in B\}.$$

There is a natural monomial map:

$$\mathbb{C}[q] \to \mathbb{C}[s,t], \quad q_{A|B} \to \prod_{\{i,j\} \in \mathrm{Cut}(A|B)} s_{ij} \prod_{\{i,j\} \in E \setminus \mathrm{Cut}(A|B)} t_{ij}.$$

As the map is monomial its kernel I_G is a prime binomial ideal, and as the monomials are homogeneous, so is I_G.

Sturmfels and Sullivant posed several conjectures how combinatorics of G relates to algebraic properties of $\mathbb{C}[q]/I_G$. In spite of progress on this topic [29] many still remain open.

Conjecture 9.1 ([78], Conjecture 3.7). *$\mathbb{C}[q]/I_G$ is normal if and only if G is free of K_5 minors.*

By [69] we know that the class of graphs for which $\mathbb{C}[q]/I_G$ is normal is minor closed. Moreover, $\mathbb{C}[q]/I_{K_5}$ is not normal. We will now show how Conjecture 9.1 implies the famous four color theorem. Originally the idea is due to David Speyer. First we need some results on the polytope representing the toric variety. It is called the cut polytope. Its vertices can be described as indicator vectors on the set of edges of G, corresponding to cut edges, for any partition of vertices. From now one we assume that G is a graph without K_5 minor. We have the following theorem due to Seymour.

Theorem 9.2 ([74], [4], Corollary 3.10). *Let $P_G \subset \{1\} \times \mathbb{Q}^{|E|} \subset \mathbb{Q}^{|E|+1}$ be the cut polytope. For any edge $e \in E$ we denote by x_e the corresponding coordinate in $\mathbb{Q}^{|E|}$ and by x_0 the first coordinate. The cone over the polytope P_G is defined by the following inequalities:*

$$0 \leq x_e \leq x_0, \text{ for each edge } e \text{ that does not belong to a triangle,}$$

$$\sum_{e\in F} x_e \leq (|F|-1)x_0 + \sum_{e\in C\setminus F} x_e, \text{ for any cordless cycle } C, F \subset C, |F| \text{ odd.}$$

Lemma 9.3. *The point $p = (3, 2, \ldots, 2) \in \mathbb{Z}^{|E|+1}$ belongs to the lattice spanned by P_G and to the cone over this polytope.*

Proof. Summing up all points corresponding to partitions $\{v\}|V \setminus \{v\}$ we obtain the point $(|V|, 2, \ldots, 2)$. As the partition $\emptyset|V$ corresponds to $(1, 0, \ldots, 0)$ we see that p indeed is in the lattice spanned by P_G. It is straightforward to check that p satisfies the inequalities in Theorem 9.2. Q.E.D.

Proposition 9.4. *Conjecture 9.1 implies the four color theorem.*

Proof. Consider p from Lemma 9.3. If P_G is normal, then $p = p_1 + p_2 + p_3$, where p_i corresponds to a partition $A_i|B_i$. Hence, we have three partitions, such that any edge belongs to precisely two of them. We define four subsets of V:

(i) $(A_1 \cap A_2 \cap A_3) \cup (B_1 \cap B_2 \cap B_3)$,
(ii) $(A_1 \cap A_2 \cap B_3) \cup (B_1 \cap B_2 \cap A_3)$,
(iii) $(A_1 \cap B_2 \cap A_3) \cup (B_1 \cap A_2 \cap B_3)$,
(iv) $(B_1 \cap A_2 \cap A_3) \cup (A_1 \cap B_2 \cap B_3)$.

Clearly, these subsets define a partition of V, i.e. are disjoint and every vertex belongs to one of them. We now prove that this is a proper coloring. As the choice between A_i and B_i was arbitrary, it is enough to prove there are no edges among vertices in $(A_1 \cap A_2 \cap A_3) \cup (B_1 \cap B_2 \cap B_3)$. Indeed, if both vertices of such edge e belonged to $A_1 \cap A_2 \cap A_3$ or $B_1 \cap B_2 \cap B_3$ then $x_e(p) = 0$. However, if one vertex belongs to $A_1 \cap A_2 \cap A_3$ and the other to $B_1 \cap B_2 \cap B_3$ then $x_e(p) = 3$. This finishes the proof. Q.E.D.

Remark 9.5. *We showed that Conjecture 9.1 implies four colorability of any graph without K_5 minor. This apparently stronger statement was in fact classically known to be equivalent to the four color theorem and is a special case of a more general Hadwiger conjecture.*

Remark 9.6. *In the same way one can show that a 4-coloring of G induces a decomposition of $p = p_1 + p_2 + p_3$. The three partitions come from dividing the four colors into two groups of two colors.*

§10. Toric varieties and matroids

Let M be a matroid on a ground set E with the set of bases $\mathfrak{B} \subset \mathcal{P}(E)$ (the reader is referred to [71] for background of matroid theory). Let $S_M := \mathbb{C}[y_B : B \in \mathfrak{B}]$ be a polynomial ring. Let φ_M be the $\mathbb{C}$-homomorphism:

$$\varphi_M : S_M \ni y_B \to \prod_{e \in B} x_e \in \mathbb{C}[x_e : e \in E].$$

The *toric ideal of a matroid* M, denoted by I_M, is the kernel of the map φ_M.

Theorem 10.1 ([82]). *The ideal I_M defines a* normal *toric variety.*

We recall the following theorem.

Theorem 10.2 ([76] Theorem 13.14). *The toric ideal I_P associated to a normal polytope P is generated in degree at most* $\dim P$.

As a corollary of Theorem 10.1 and 10.2 we obtain the following.

Corollary 10.3. *For any matroid M on the ground set E the toric ideal I_E is generated in degree at most $|E|$.*

A result of Gijswijt and Regts strengthens Theorem 10.1.

Theorem 10.4 ([35]). *(Poly)Matroid base polytope (associated to the monomials defining ϕ_M) has an Integer Caratheodory Property. That is, if P_M is a matroid base polytope, then every integer vector in kP_M is a positive, integral sum of affinely independent integer vectors from P_M with coefficients summing up to k. In particular, Caratheodory rank of P_M is as low as possible, it is equal to the dimension of P_M plus 1.*

10.1. Representable matroid

This subsection is not needed in what follows. However, it provides additional motivation to study $V(I_M)$ from the point of view of algebraic geometry. Let M be a representable matroid realized by vectors $v_1, \dots, v_k$ spanning a d-dimensional vector space V. We have a natural map $\mathbb{C}^k \ni e_i \to v_i \in V$ with kernel $K \in G(k-d, k)$. Note, that on $\mathbb{C}^k$ acts a k-dimensional torus T inducing an action on the Grassmannian $G(k-d, k)$.

Theorem 10.5. [34] *The toric variety $V(I_{M^*})$ is isomorphic to the closure of the orbit $G \cdot K$ in $G(k-d, k)$.*

Proof. Let $f_1, \dots, f_{k-d}$ be the basis of K. Let M_k be a $(k-d) \times k$ matrix with i-th row corresponding to f_i. A given Pluücker coordinate

$(e_{a_1} \wedge \cdots \wedge e_{a_{k-d}})^*(f_1 \wedge \cdots \wedge f_{k-d})$ of K equals the maximal minor of M_k distinguished by the columns $\{a_1, \ldots, a_{k-d}\}$. Further, the torus T acts on $e_{a_1} \wedge \cdots \wedge e_{a_{k-d}}$ with weight corresponding to a lattice point in $\mathbb{Z}^k$ that has coordinates indexed by a_i equal to 1 and all other equal to 0. Hence:

- the polytope P_{M^*} has lattice points corresponding to indicator vectors of complements of bases of M,
- the polytope of the toric variety $\overline{T \cdot K}$ has lattice points corresponding to indicator vectors of subsets of $k-d$ columns of M_k giving a nonzero minor.

Thus, it remains to show that a given minor of M_k is nonzero if an only if the corresponding vectors v_i form a complement of a basis of V. Fix a set $S = \{a_1, \ldots, a_{k-d}\} \subset \{1, \ldots, k\}$ and denote its complement by S'. We have following equivalences:

$\{v_i\}_{i \in S}$ is a complement of a basis of $V \Leftrightarrow \{v_i\}_{i \in S'}$ is a basis of $V \Leftrightarrow \{f_j\}_{j=1,\ldots,k-d} \cup \{e_i\}_{i \in S'}$ is a basis of $\mathbb{C}^k \Leftrightarrow$ the minor of M_k distinguished by S is nonzero. Q.E.D.

Remark 10.6. *If we fix a basis of the space V we obtain a matrix representation of v_i which distinguishes a point in $G(d,k)$. The closure of the torus orbit of this point is isomorphic to $V(I_M)$.*

10.2. Open problems

Let us present the major open problems concerning toric ideals associated to matroids.

Conjecture 10.7 (Weak White's conjecture [83]). *For any matroid M the ideal I_M is generated by quadrics.*

We say that $y_{B_1}y_{B_2} - y_{B_3}y_{B_4}$ for $B_i \in \mathfrak{B}$ is a *symmetric exchange* if $B_3 = (B_1 \setminus \{b_1\}) \cup \{b_2\}$ and $B_4 = (B_2 \setminus \{b_2\}) \cup \{b_1\}$ for some $b_1 \in B_1 \setminus B_2$, $b_2 \in B_2 \setminus B_1$ (see [56] for other exchange properties).

Conjecture 10.8 (White's conjecture [83]). *For any matroid M the ideal I_M is generated by symmetric exchanges.*

However, even the following is open.

Conjecture 10.9. *Every quadric in I_M is a linear combination of symmetric exchanges.*

In view of Theorem 10.5 the following turns out to be a very important open problem, that is weaker than White's conjecture.

Conjecture 10.10. *For any representable matroid M, the ideal I_M is generated by quadrics.*

Herzog and Hibi write, that they 'cannot escape from the temptation' to ask the following, stronger questions.

Question 10.11 ([38]). *Let M be a discrete (poly)matroid.*

(i) Does the toric ideal I_M possess a quadratic Gröbner basis?
(ii) Is the base ring $\mathbb{C}[y_B : B \in \mathfrak{B}]/I_M$ Koszul?

A positive answer to the first question would imply the following conjecture, that is a strengthening of Theorem 10.4.

Conjecture 10.12. *For any matroid M the basis polytope P_M has a unimodular triangulation.*

A *k-matroid* is a matroid whose ground set can be partitioned into k pairwise disjoint bases. We call a basis of a k-matroid *complementary* if its complement can be partitioned into $k-1$ pairwise disjoint bases. The *basis graph* of a matroid is a graph with vertices corresponding to bases and an edge between two bases that differ by a pair of elements. The *complementary basis graph* of a k-matroid is the restriction of its basis graph to complementary bases.

Conjecture 10.13. [57] *Complementary basis graph of a k-matroid is connected.*

Notice that Conjecture 10.13 for $k = 2$ coincides with Conjecture 10.9 in a non-commutative setting.

Remark 10.14. *For a 2-matroid it is not known even if some antipodal bases B_1, B_2 (that is, bases such that $B_1 \sqcup B_2 = E$) are always connected in the complementary basis graph. Positive answer would imply that commutative and non-commutative settings of Conjecture 10.9 are equivalent.*

Conjecture 10.15. [57] *Let $k \geq 2$, and let M be a matroid of rank r on the ground set E of size $kr+1$. Suppose $x, y \in E$ are two elements such that both sets $E \setminus x$ and $E \setminus y$ can be partitioned into k pairwise disjoint bases. Then there exist partitions of $E \setminus x$ and $E \setminus y$ into k pairwise disjoint bases which share a common basis.*

If $k \geq 2^{r-1}+1$, then the above Conjecture 10.15 holds [57].

Proposition 10.16. [57] *White's Conjecture 10.8 implies Conjecture 10.13.*
Conjunction of Conjectures 10.13 and 10.15 implies White's Conjecture 10.8.

10.3. Known facts

Several special cases of Conjecture 10.8 are known.

Theorem 10.17. *Conjecture 10.8 holds for graphic* [6], *sparse paving* [8], *strongly base orderable* [55], *and rank* ≤ 3 [45] *matroids.*

Further, the classes of matroids for which the toric ideal is generated by quadrics and that has quadratic Gröbner bases, is closed under series and parallel extensions, series and parallel connections, and 2-sums [75].

Further, partial results are known in general. Let $J_M \subset I_M$ be the ideal generated by symmetric exchanges. Let $m \subset S_M$ be the irrelevant ideal, i.e. the maximal ideal generated by all the variables.

Theorem 10.18. [55] *For any matroid M, we have $J_M : m^\infty = I_M$. That is,* $\operatorname{Proj} S_M/J_M$ *is equal to* $\operatorname{Proj} S_M/I_M$. *In particular, Conjecture 10.8 holds on set-theoretic level.*

We can rephrase the above by saying that homogeneous components of the ideals I_M and J_M of a matroid M, are equal starting from some degree $f(M)$. Even more is known.

Theorem 10.19. [57] *Homogeneous components of ideals I_M, J_M of a matroid M of rank r, are equal starting from degree $f(r)$ depending only on r.*

It follows that checking if White's Conjecture 10.8 is true for matroids of a fixed rank is a decidable (finite) problem.

The first question of Herzog and Hibi 10.11 is already difficult for special classes of matroids. For uniform matroids, it was proved by Sturmfels [76]. Later, it was extended to base-sortable matroids by Blum [7] and further by Herzog, Hibi and Vladoiu [39]. Schweig [73] proved it for lattice path matroids. The only general result concerning Gröbner bases of toric ideals of matroids is the following.

Theorem 10.20. [58] *The toric ideal of a matroid of rank r posesses a Gröbner basis of degree at most $(r + 4)!$.*

Conca [23] proved that the answer to the second Question 10.11 is positive for transversal polymatroids.

10.4. Matroids and finite characteristics

The original idea to look at implications of Theorem 10.18 in case of finite characteristics and pureness is due to Matteo Varbaro.

Definition 10.21 (Pure, F-pure [43]). *A morphism of rings $R \to S$ is called* pure *if for any R-module M the map $M \ni m \to m \otimes 1 \in M \otimes_R S$ is injective.*

If R is an algebra over a field k of characteristic p we say that R is F-pure if the Frobenius morphism $R \ni r \to r^p \in R$ is pure.

Lemma 10.22. *If a k-algebra R is F-pure, then it is reduced.*

Proof. Take $M = R$ in Definition 10.21. For any $a \in \mathbb{N}$ and for any $r \in R$ we have:

$$r^a \to r^a \otimes 1 = 1 \otimes r^{pa}.$$

Thus, we have $r^a \neq 0 \Rightarrow r^{pa} \neq 0$, which proves the lemma. Q.E.D.

Proposition 10.23. *Let C be a cone in a lattice M. Then $k[C]$ is F-pure.*

Proof. The proof is a combination of the following three facts:

(i) A (Lauarent) polynomial ring over k is F-pure (more generally a Noetherian regular k-algebra) [43, Proposition 5.14].

(ii) If S if F-pure and $S = R \oplus M$ as R modules for some R module M then R is F pure [31, Proposition 1.3].

(iii) One can realize $C = \mathbb{Z}^r \times C'$, where $C' = \mathbb{Z}_+^m \cap H$ and H is a linear subspace. In particular, the ring $\mathbb{C}[C']$ as a module is a factor of $\mathbb{C}[\mathbb{Z}_+^m]$ [13, p. 63].

Q.E.D.

Theorem 10.24. *For any matroid M the following are equivalent:*

(i) I_M is generated by symmetric exchanges (White's conjecture holds),

(ii) J_M is saturated with respect to m,

(iii) J_M is prime,

(iv) J_M is primary,

(v) S_M/J_M is an integral domain,

(vi) S_M/J_M is reduced,

(vii) The localisation $(S_M/J_M)_m$ is reduced,

(viii) S_M/J_M is F-pure over some field k of finite characteristic,

(ix) S_M/J_M is F-pure over any field k of finite characteristic,

(x) The localisation $(S_M/J_M)_m$ is F-pure over some/any field.

Proof. Equivalences $i) - vii)$ are direct consequences of Theorem 10.18.

For a fixed field k point $i)$ implies $viii)$ by a combination of Theorem 4.5 and Proposition 10.23 (recall that a toric algebra is normal if the corresponding monoid is a cone - Theorem 3.10).

The equivalence of local and global case follows from [31, Proposition 1.3].

Point $viii)$ implies $vi)$ by Lemma 10.22.

It remains to show that *viii*) is equivalent to *ix*) or in other words that White's conjecture does not depend on the underlying field. More general statement is explained in Lemma 10.25. Q.E.D.

Lemma 10.25. *Consider a finite set of characters $S \subset M \simeq \mathbb{Z}^n$ generating a monoid $\tilde{S}$. Suppose for some field k the kernel I_k of the map:*

$$k[x_1, \ldots, x_n] \to k[\tilde{S}]$$

is generated by binomials $B = \{m_1 - m'_1, \ldots, m_l - m'_l\}$. Then the same binomials generate the toric ideal $I_{k'}$ associated to S over any other field k'.

Proof. Reasoning as in Theorem 3.4 we see that the binomials in the toric ideal do not depend on the field and simply correspond to integral relations among lattice points in S. Further, the ideal is always spanned by such binomials. Thus we only need to show the following:

Any binomial $m - m' \in I_{k'}$ is equal to $\sum_j n_j(m_{i_j} - m'_{i_j})$, where n_j are monomials and $m_{i_j} - m'_{i_j} \in B$.

We know that $m - m' = \sum \lambda_j m_j(m_{i_j} - m'_{i_j})$ for some $\lambda_j \in k^*$. By definition, say that the monomial $m_j m_{i_j}$ is equivalent to $m_j m'_{i_j}$. We generate the equivalence relation and we want to prove that m is equivalent to m'. We may subdivide the terms in $\sum \lambda_j m_j(m_{i_j} - m'_{i_j})$ according to the equivalence class they belong to. Consider all terms in the same class as m. This is a sum of monomials with some coefficients, one of which is m with coefficient one. However, the sum of these coefficients needs to be 0 and the only other monomial with nonzero coefficient that may appear is m'. Q.E.D.

Let us restate a criterion of F-pureness due to Fedder. For an ideal I we let $I^{[p]}$ be the ideal generated by $\{i^p : i \in I\}$.

Theorem 10.26 (Proposition 1.3 and 1.7 [31]). *Let J be a homogeneous ideal in a polynomial ring S over k of characteristic p. Then S/J is F-pure if and only if $J^{[p]} : J \not\subset m^{[p]}$.*

Corollary 10.27. *White's conjecture 10.8 holds if and only if there exists $f \notin m^{[p]}$ such that for any symmetric exchange g we have $gf \in J_M^{[p]}$. Further, we may (but do not have to) assume that we work over the field $\mathbb{Z}_2$, i.e. $p = 2$ and f is homogeneous of degree equal to the number of bases of the matroid. In such a case $f \notin m^{[p]}$ translates to f being a sum of monomials one of which is a (squarefree) product of all variables in S_M.*

We note that in case $k = \mathbb{Z}_2$ the polynomial f distinguishes some multisets of bases - it would be great to understand combinatorial meaning of those. In each particular case of M we can compute (examples of) f.

Example 10.28. *The grading below encodes a uniform rank two matroid on a ground set with four elements. First we compute the toric ideal in Macaulay2.*

```
loadPackage,"Normaliz"
loadPackage("MonomialAlgebras",
Configuration=>{"Use4ti2"=>true})
L={{1,1,0,0},{1,0,1,0},{1,0,0,1},
{0,1,1,0},{0,1,0,1},{0,0,1,1}};
d=#L;
R=ZZ/2[a_1..a_d,Degrees=>L]
J=binomialIdeal R; m=ideal(gens R);
```

We now check Fedder's criterion.

```
Jp=0; mp=0;
for i from 0 to ((numgens J)-1) do
{Jp=Jp+ideal( ((gens J)_i_0)^2)}
for i from 0 to ((numgens m)-1) do
{mp=mp+ideal( ((gens m)_i_0)^2)}
isSubset(Jp:J,mp)
Q=R/mp;
Red=sub(Jp:J,Q)
```

The answer the program gives is:

$$f = a_2a_3a_4a_5 + a_1a_3a_4a_6 + a_1a_2a_5a_6.$$

Below we present the example of a graphic matroid corresponding to a square with one diagonal.

```
L={{1,1,1,0,0},{1,1,0,1,0},{1,0,1,1,0},{0,1,1,1,0},
{1,0,1,0,1},{0,1,0,1,1},{1,0,0,1,1},{0,1,1,0,1}};
d=#L;
R=ZZ/2[a_1..a_d,Degrees=>L]
J=binomialIdeal R; m=ideal(gens R);
Jp=0; mp=0;
for i from 0 to ((numgens J)-1) do
{Jp=Jp+ideal( ((gens J)_i_0)^2)}
for i from 0 to ((numgens m)-1) do
{mp=mp+ideal( ((gens m)_i_0)^2)}
isSubset(Jp:J,mp)
```

```
Q=R/mp;
Red=sub(Jp:J,Q)
```

Here:

$$f = a_1a_4a_5a_6a_7 + a_2a_3a_5a_6a_8 + a_2a_4a_5a_7a_8 + a_1a_3a_6a_7a_8.$$

One of the problems that we encountered is that f may be not uniquely specified (e.g. in Example 10.28 one can multiply f by a variable). Further, it is hard to see a general pattern for f.

§11. Toric varieties and phylogenetics

In this section we present a construction of a family of toric varieties. It is inspired by phylogenetics - a science that aims at reconstruction of the history of evolution. The basic statistical picture is as follows. We start from a (usually large) family of parameters that correspond to various probabilities of mutations (and probability distribution of the common ancestor). These parameters are unknown. However, if we knew them then we could answer questions of type:

'(On a given position in DNA string) what is the probability that a human has C, a gorilla has C and a guenon has A?'

Here, we represent DNA as strings of characters: A, C, G, T. Thus, we obtain a map m from the parameter space to the space of joint probability distributions of states of species that we consider. The latter is a huge space! The states are indexed by a choice of a letter for each species we consider - the dimension is 4 to the power equal to the number of species. Note that there is a distinguished point P in this space: going through whole DNA sequences (that are very long) biologists and statisticians can count the number of times they encounter C for human, C for gorilla and A for guenon etc. In other words, we are getting a(n approximation) of the probability distribution on the (joint) states of the species.

What are the main questions that we would like to answer?

- Does the point P belong to the image of m? (If the answer is no, then it means that some of our assumptions are wrong in the statistical model we have chosen.)
- Can we identify the parameters $m^{-1}(P)$?

Let us focus on the first question. First, we notice that in most interesting examples the map m is algebraic. The approach algebraic geometry proposes is to consider the defining equations of the Zariski closure X of the image of m.

The algebraic variety X we obtain depends on the statistical model we choose, i.e. what we assume on the parameters. The most 'universal' is the General Markov Model that leads to secant varieties of Segre products. Let us discus a different class of so-called group-based models. Their biological motivation comes from the observation that there are certain symmetries among probabilities of mutations. These symmetries can be encoded by the group action. For example the famous Kimura 3-parameter model [48] relies on the fact that $\{A, C, G, T\}$ is naturally divided into two subsets: purines $\{A, G\}$ and pyrimidines $\{C, T\}$. In mathematical language the group $\mathbb{Z}_2 \times \mathbb{Z}_2$ naturally acts on $\{A, C, G, T\}$ and this influences the geometry of the variety X for the 3-Kimura model.

11.1. Construction for group-based models

In general to determine the variety X we need a tree T (that determines how species mutated) and a model (in our case determined by a finite abelian group G). Here, we do not present the general construction referring to [77, 30, 64]. On the other hand, we describe in detail the toric structure of the variety X in case when it is a so-called star or claw-tree $K_{1,n}$, i.e. a tree with one inner vertex and n leaves. The general case may be also obtained by *toric fiber product* [80]. As we will see many conjectures address the case of $K_{1,n}$.

Definition 11.1 (**Flow** [63], [19]). *Let G be a finite abelian group and $n \in \mathbb{N}$. A flow is a sequence of n elements of G summing up to $0 \in G$, the neutral element of G. The set of flows is equipped with a group structure via the coordinatewise action. The group of flows $\mathcal{G}$ is (non-canonically) isomorphic to G^{n-1}.*

Definition 11.2 (**Polytope** $P_{G,n}$, [66], [77]). *Consider the lattice $M \cong \mathbb{Z}^{|G|}$ with a basis corresponding to elements of G. Consider M^n with the basis $e_{(i,g)}$ indexed by pairs $(i, g) \in [n] \times G$. We define an injective map of sets: $\mathcal{G} \to M^n$, by $(g_1, \dots, g_n) \longmapsto \sum_{i=1}^n e_{(i,g_i)}$. The image of this map defines the vertices of the polytope $P_{G,n}$.*

Example 11.3 ([65]). *For $G = (\mathbb{Z}_2, +)$ and $n = 3$, we have four flows:*

$$(0,0,0), (0,1,1), (1,0,1), (1,1,0) \in \mathbb{Z}_2 \times \mathbb{Z}_2 \times \mathbb{Z}_2.$$

Hence, the polytope $P_{\mathbb{Z}_2,3}$ has the following four vertices corresponding to the flows above:

$$(1,0,1,0,1,0), (1,0,0,1,0,1), (0,1,1,0,0,1), (0,1,0,1,1,0) \in \mathbb{Z}^2 \times \mathbb{Z}^2 \times \mathbb{Z}^2,$$

where $(1,0) \in \mathbb{Z}^2$ corresponds to $0 \in \mathbb{Z}_2$ and $(0,1) \in \mathbb{Z}^2$ corresponds to $1 \in \mathbb{Z}_2$.

A more sophisticated example is presented in [61, Example 4.1]. It turns out that the phylogenetic variety X - for group based models - is toric and corresponds to the polytope $P_{G,n}$. We already know by Theorem 3.4 that binomials in the toric ideal correspond to integral relations among lattice points. However, for group based models it is easier to work with flows. Binomials may be identified with a pair of tables of the same size T_0 and T_1 of elements of G, regarded up to row permutation. Each row of such tables has to be a flow. The identification is as follows. Every binomial is a pair of monomials; the variables in such monomials correspond to flows, given by a collection of n elements in G. Every monomial is viewed as a table, whose rows are the variables appearing in the monomial; the number of rows of the corresponding table is the degree of the monomial. Consequently, a binomial is identified with the pair of tables encoding the two monomials respectively.

For a finite abelian group G and the graph $K_{1,n}$ the associated toric variety (represented by the polytope $P_{G,n}$) will be denoted by $X(G, K_{1,n})$. A binomial belongs to $I(X(G, K_{1,n}))$ if and only if the two tables are *compatible*, i.e. for each i, the i-th column of T_0 and the i-th column of T_1 are equal as multisets.

In order to generate a binomial – represented by a pair of tables T_0, T_1 – by binomials of degree at most d we are allowed to select a subset of rows in T_0 of cardinality at most d and replace it with a compatible set of rows, repeating this procedure until both tables are equal.

Example 11.4 ([65]). *For $G = (\mathbb{Z}_2, +)$ and $n = 6$ consider the following two compatible tables:*

$$T_0 = \begin{bmatrix} 1 & 1 & 1 & 1 & 1 & 1 \\ 0 & 0 & 0 & 0 & 0 & 0 \\ 1 & 1 & 0 & 0 & 0 & 0 \end{bmatrix} \text{ and } T_1 = \begin{bmatrix} 0 & 1 & 0 & 1 & 0 & 0 \\ 1 & 0 & 1 & 0 & 0 & 0 \\ 1 & 1 & 0 & 0 & 1 & 1 \end{bmatrix}.$$

Note that the red subtable of T_0 is compatible with the table

$$T' = \begin{bmatrix} 0 & 1 & 0 & 1 & 0 & 0 \\ 1 & 0 & 1 & 0 & 1 & 1 \end{bmatrix}.$$

Hence, we may exchange *them obtaining*:

$$\tilde{T}_0 = \begin{bmatrix} 0 & 1 & 0 & 1 & 0 & 0 \\ 1 & 0 & 1 & 0 & 1 & 1 \\ 1 & 1 & 0 & 0 & 0 & 0 \end{bmatrix}.$$

Note that T_0 *and* $\tilde{T}_0$ *are* compatible. *Now, the brown subtable of* $\tilde{T}_0$ *is compatible with the table*

$$T'' = \begin{bmatrix} 1 & 0 & 1 & 0 & 0 & 0 \\ 1 & 1 & 0 & 0 & 1 & 1 \end{bmatrix}.$$

Finally, we exchange them obtaining T_1. *Hence we have a sequence of tables* $T_0 \rightsquigarrow \tilde{T}_0 \rightsquigarrow T_1$. *More specifically, we started from a degree three binomial given by the pair* T_0, T_1 *and we generated it using degree two binomials of degree two.*

Definition 11.5 (Phylogenetic complexity [77]**).** *Let* $K_{1,n}$ *be the star with* n *leaves, and let* $\phi(G, n)$ *be the maximal degree of a generator in a minimal generating set of* $I(X(G, K_{1,n}))$. *We define the phylogenetic complexity* $\phi(G)$ *of* G *to be* $\sup_{n\in\mathbb{N}} \phi(G, n)$.

A new package to deal with phylogenetic group-based models appeared recently [3] for Macaulay2. The software to generate polytopes $P_{G,n}$ is presented in [27].

11.2. Further properties of group-based models

In studying group-based models, Buczyńska and Wieśniewksi [19], [18] made the startling observation that in the case $G = \mathbb{Z}/2\mathbb{Z}$ the Hilbert function of the affine semigroup algebra $\mathbb{C}[M_{\Gamma,G}]$ associated to a graph Γ (with respect to an appropriate grading) only depends on the number of leaves n and the first Betti number g of Γ. The explanation for this phenomenon was provided by Sturmfels and Xu [79] and Manon [60], where it was shown that the phylogenetic statistical models $M_{\Gamma,\mathbb{Z}/2\mathbb{Z}}$ are closely related to the Wess-Zumino-Witten (WZW) model of conformal field theory, and the moduli space $\mathcal{M}_{C,\vec{p}}(SL_2(k))$ of rank 2 vector bundles on an n-marked algebraic curve $(C, \vec{p})$ of genus g. In particular, the total coordinate ring $\mathcal{V}_{C,\vec{p}}(SL_2(k))$ of this space, which is known to be a direct sum of the so-called conformal blocks of the WZW model, is shown to carry a flat degeneration to each affine semigroup algebra $k[M_{\Gamma,\mathbb{Z}/2\mathbb{Z}}]$. Flat degeneration preserves Hilbert polynomials, explaining the coincidence among the $k[M_{\Gamma,\mathbb{Z}/2\mathbb{Z}}]$.

Kubjas [51] and Donten-Bury [27] showed that Hilbert functions no longer agree for various other finite abelian groups G, so the existence of a common flat deformation cannot hold for the phylogenetic group-based models in general. However, Kubjas and Manon [52] have shown that a generalization of the relationship to the WZW model of conformal field theory and the moduli of vector bundles holds for the cyclic groups $\mathbb{Z}/m\mathbb{Z}$. In particular, these group-based models are related to the corresponding moduli spaces for the algebraic group $SL_m(k)$.

To sum up, group-based models:

(i) can be regarded as basic combinatorial objects encoding a structure of a finite abelian group,
(ii) first appeared in phylogenetics,
(iii) are important also in other fields, such as conformal field theory.

Below we present a table with known facts about generators of ideals for phylogenetic group-based models.

	Group-based Models			
Polynomials defining:	$\mathbb{Z}_2$	$\mathbb{Z}_3$	$\mathbb{Z}_2 \times \mathbb{Z}_2$	G
Generators of the ideal	Degree 2 [77]	Degree 3 [65]	Conjecture [77, Conjecture 30]	Finite [66], Degree $\leq |G|$ [77, Conjecture 29]
Projective scheme		Degree 3 [26]	Degree 4 [62]	Finite [62]
Set-theoretically				Finite [28]
On a Zariski open subset			Degree 4 [63]	Degree $\leq |G|$ [22, 21]

11.3. Open problems

Here we present the main open problems concerning group-based models. Everything is stated in purely toric/combinatorial language. We start from the central conjecture in this context.

Conjecture 11.6 ([77, Conjecture 29]). *For any finite abelian group G, $\phi(G) \leq |G|$.*

It seems crucial to first understand the simplest tree $K_{1,3}$.

Conjecture 11.7. *For any finite abelian group G, $\phi(G,3) \leq |G|$.*

The results of [66] imply that for finite abelian group G the function $\phi(G,\cdot)$ is eventually constant. The ensuing results would be a desired strengthening.

Conjecture 11.8 ([62, Conjecture 9.3]). *We have $\phi(G,n+1) = \max(2,\phi(G,n))$.*

We are grateful to Seth Sullivant for noticing that this is equivalent to $\phi(G,\cdot)$ being constant, apart from the case when $G = \mathbb{Z}_2$ and $n = 3$, when the associated variety is the whole projective space.

Conjecture 11.9. *For any finite (not necessarily abelian) group G, $\phi(G)$ is finite.*

Conjecture 11.8 also implies the following.

Conjecture 11.10 ([77, Conjecture 30]). *The phylogenetic complexity of $G = \mathbb{Z}_2 \times \mathbb{Z}_2$ is 4.*

§12. Maps of Toric Varieties and Cox Rings

Throughout the review we did not mention many important topic, among those Cox rings. In this section we very briefly present the general construction and relations to morphisms.

Local coordinate rings are not always very convenient to work with, especially, when we want to investigate the *global* properties of the variety. Consider the *projective space* $\mathbb{P}^n$ of dimension n. It is glued out of $n+1$ affine spaces of dimension n, so to obtain the description of (for example) a coherent sheaf on the projective space one needs the information about $n+1$ modules over polynomial rings, and a care must be taken to *glue* the modules accordingly. Instead, one may view the projective space globally:

$$\mathbb{P}^n = (\mathbb{A}^{n+1} \setminus \{0\})/\mathbb{C}^*.$$

There are three essential ingredients in this *global description*. Firstly, there is $\mathbb{A}^{n+1}$, an affine space. Secondly, we remove a relatively small subset of the affine space, in this case just one point $\{0\}$. Thirdly, we divide by an action of an algebraic group $\mathbb{C}^*$, the multiplicative group of the base field $\mathbb{C}$. The *homogeneous coordinate ring* of the projective space incorporates all these three ingredients. We just take the polynomial coordinate ring of $\mathbb{A}^{n+1}$; all objects (for example modules or ideals) that are supported in $\{0\}$ are *irrelevant*, and, in particular, if two object differ only at $\{0\}$, then they correspond to the same object on the projective space; all objects must be invariant with respect to the group action, in other words *homogeneous*.

The Cox rings have been first introduced for toric varieties [24], and then generalised to normal varieties with finitely generated divisor class group $Cl(X)$:

$$S[X] := \bigoplus_{[D]\in Cl(X)} H^0(\mathcal{O}_X(D)).$$

A careful choice of the representatives D in each element of $Cl(X)$ must be made in order to obtain a well defined ring structure on $S[X]$. Varieties, for which the Cox ring is finitely generated are called *Mori Dream Spaces (MDS)* [44]. The same varieties arise naturally in Mori theory and Minimal Model Program, as particularly elegant examples illustrating the theory. $S[X]$ is always graded by $Cl(X)$. The main point is that for a MDS there exists a codimension at least 2 subvariety Z of $\operatorname{Spec} S[X]$, such that:

$$X = (\operatorname{Spec} S[X] \setminus Z)/G_X$$

where $G_X = \operatorname{Hom}(Cl(X), \mathbb{C}^*)$ is the group acting on $\operatorname{Spec} S[X]$, corresponding to the grading by $Cl(X)$. This naturally corresponds to the three ingredients of the homogeneous coordinate ring of the projective space. Analogously to the projective case, many global objects on X can be expressed in terms of the Cox ring and vice versa, taking in account the homogeneity and relevance. The Cox ring (as well as its grading and the irrelevant ideal $B = I(Z)$ defining Z) are defined intrinsically, so they do not depend on any embedding or any other choices. Thus it is very convenient to study *global and intrinsic properties* of X.

Affine and projective spaces and normal toric varieties are Mori Dream Spaces. In these cases, the Cox ring is always a polynomial ring, but the grading vary. In fact, the property that $S[X]$ is a polynomial ring characterises toric varieties, see [47] for a recent treatment of this characterisation.

By definition, an algebraic morphism of two affine varieties $\varphi\colon X \to Y$ is a geometric interpretation of an algebra morphism $\varphi^*\colon B \to A$ of their affine coordinate rings. Here $X = \operatorname{Spec} A$ and $Y = \operatorname{Spec} B$. If $X = \mathbb{P}^m$ and $Y = \mathbb{P}^n$ instead, and $A \simeq \mathbb{C}[x_0, \ldots, x_m]$ and $B \simeq \mathbb{C}[y_0, \ldots, y_n]$ are their homogeneous coordinate rings, then any algebraic morphism $\varphi\colon \mathbb{P}^m \to \mathbb{P}^n$ is determined a morphism $B \to A$ satisfying the usual homogeneity and base point freeness conditions. Rational maps between affine varieties or projective spaces have similar interpretations in terms of the fields of fractions of coordinate rings.

Theorem 12.1 ([10], [20]). *Suppose X and Y are Mori Dream Spaces, and $\varphi\colon X \dashrightarrow Y$ is a rational map. Then there exists a description of φ in terms of Cox coordinates, that is a multi-valued map*

$$\Phi\colon \operatorname{Spec} S[X] \longrightarrow \operatorname{Spec} S[Y]$$

such that for all points $x \in X$ and ξ such that $\pi_X(\xi) = x$ and φ is regular at x, the composition $\pi_Y(\Phi(\xi))$ is a single point $\varphi(x) \in Y$.

The notion of *multi-valued map* is modeled on the case of projective space, but may involve roots of homogeneous functions if the target is singular. Just as in the case of projective space, the map must satisfy homogeneity, and relevance condition. The theorem is effective in the sense, that the proof shows how to construct the description.

Similar statement for regular maps between $\mathbb{Q}$-factorial Mori Dream Spaces was obtained by Andreas Hochenegger and Elena Martinengo [41]. Their approach is to use the language of Mori Dream stacks [40]. They use the technique of root constructions, which is parallel to the multi-valued maps.

§13. Examples

Toric varieties provide very fruitful examples. This section is motivated by questions of Sijong Kwak: what happens to the depth under inner projection of a projective variety X? Here, inner projection means a projection from a point $x \in X$. We will denote the (closure of) the image of the projection by X_x. We start by recalling the following general result.

Theorem 13.1 ([53] Theorem 4.1). *If X is defined by quadrics and x is a smooth point of X then depths of X and X_x are equal.*

The following observation was pointed out by Greg Blekherman.

Example 13.2. *In general, the depth may go up under projections from general (in particular, smooth) points. Indeed, if we consider any non aCM variety X we may project it, until it becomes a hypersurfece. In particular, it becomes a complete intersection, hence aCM, hence of maximal depth.*

Before we pass to constructing toric examples we note that inner projections in toric geometry were investigated for many years. The seminal work of Bruns and Gubeladze [12, 15] lead to many interesting examples, disproving important conjectures on characterizations of normal polytopes. From a combinatorial point of view projecting from a torus invariant point corresponding to a vertex v of a lattice polytope P corresponds to considering a toric variety given by lattice points in P distinct from P. The study when such polytopes remain normal, i.e. when the projected variety is projectively normal, were crucial in [12]. Further examples of projectively normal toric varieties that do *not* come from projections of projectively normal toric varieties were found in [14]. Projective normality of toric varieties is related to depth as follows.

Theorem 13.3 (Hochster [42]). *A projectively normal toric variety is aCM.*

Example 13.4. *Consider a toric hypersurface X corresponding to lattice points* $(0, 0, 0)$, $(0, 1, 0)$, $(0, 0, 1)$, $(3, 1, 1)$, $(4, 1, 1)$. *The Macaulay2 code below verifies that it is aCM and not normal.*

```
loadPackage "Depth"
loadPackage "Normaliz"
loadPackage("MonomialAlgebras",
Configuration=>{"Use4ti2"=>true})
L={{1,0,0,0},{1,0,1,0},{1,0,0,1},{1,4,1,1},{1,3,1,1}}
```

```
R=QQ[a_1..a_5,Degrees=>L]
J1=binomialIdeal R
depth (R/J1)==(dim J1)
isNormal (R/J1)
```

The reason is that the singular locus is of codimension one.

Example 13.5 (The depth may go down under projection from a generic (in particular smooth) point). *We start with a normal (aCM) projective toric variety defined below.*

```
L={{1,0,0,0},{1,0,1,0},{1,0,0,1},{1,4,1,1},
{1,3,1,1},{1,2,1,1},{1,-1,0,0}}
R=QQ[a_1..a_7,Degrees=>L]
J2=binomialIdeal R
isNormal (R/J2)
depth (R/J2)==(dim J2)
```

We now project from the point $(1,\dots,1)$ *that belongs to the dense torus orbit.*

```
JJ=sub(J2,{a_2=>a_2+a_1,a_3=>a_3+a_1,a_4=>a_4+a_1,
a_5=>a_5+a_1,a_6=>a_6+a_1,a_7=>a_7+a_1})
JS=eliminate(JJ,a_1);
W=QQ[a_1..a_7]
M=sub(JS,W)+ideal(a_1)
depth (W/M)==(dim M)
```

In a similar way one can construct examples projecting from singular points.

Remark 13.6. *It is* not *possible to project a nonprojectively normal toric variety from a (torus invariant) smooth point and obtain a projectively normal variety (as union of normal polytopes is normal).*

We note that the discussion on projections nicely ties with the conjectures of Bogvad and Oda.

Conjecture 13.7. *A smooth polytope is normal. The associated toric variety is defined by quadrics.*

Proposition 13.8. *Conjecture 13.7 implies that for a smooth polytope P any projection from a torus invariant point remains projectively normal. Further, if we know for a smooth polytope P that there exists a projection from a torus invariant point, that is (projectively) normal, then P is normal.*

Proof. The second statement follows by Remark 13.7. The first statement is based on [11, Theorem 5.1] and [64, Section 11]. Let Q

be the convex hull of lattice points in P distinct from a vertex v. Let $q \in kQ$. We know that $q = \sum_{i=1}^{k} p_i$ for lattice points $p_i \in P$. Let $v_1, \ldots, v_n$ be the first lattice points on edges of P adjacent to v. The only problem is, if some $p_i = v$. But then there must exist $p_j \neq v, v_1, \ldots, v_n$. Note that as $v - v_i$ is a lattice basis we may always find $m \in \mathbb{Z}_+$ such that $mv + p_j = \sum v_i$, where the sum is over any indices (with possible repetitions) $1 \leq i \leq n$. The only way the above relation is generated by quadrics is, if $v + p_j = a + b$ for some lattice points $a, b \in P$. Thus as long as in the decomposition of q the vertex v appears we may change the decomposition in such a way that its multiplicity goes down. This proves normality of Q. Q.E.D.

The previous proposition ties with the general Theorem 13.1. Indeed, if P is defined by quadrics and normal, then it is aCM and we know that the projection is also aCM.

References

[1] K. Altmann, N. Ilten, L. Petersen, H. Süss, R. Vollmert, The geometry of t-varieties, Contributions to Algebraic Geometry, Impanga Lect. Notes (2011).

[2] M.F. Atiyah and I.G. Macdonald, Introduction to commutative algebra, vol. 2, Addison-Wesley Reading, 1969.

[3] H. Baños, N. Bushek, R. Davidson, E. Gross, P. Harris, R. Krone, C. Long, A. Stewart, R. Walker, Phylogenetic trees, arXiv preprint arXiv:1611.05805 (2016).

[4] F. Barahona, A.R. Mahjoub, On the cut polytope, Mathematical programming 36 (1986), no. 2, 157-173.

[5] A.M. Bigatti, R. La Scala, L. Robbiano, Computing toric ideals. J. Symbolic Comput. 27 (1999), no. 4, 351-365.

[6] J. Blasiak, The toric ideal of a graphic matroid is generated by quadrics, Combinatorica 28 (2008), 283-297.

[7] S. Blum, Base-sortable Matroids and Koszulness of Semigroup Rings, European J. Combin. 22 (2001), 937-951.

[8] J. Bonin, Basis-exchange properties of sparse paving matroids, Adv. in Appl. Math. 50 (2013), no. 1, 6-15.

[9] L. Borisov, Z. Hua, On the conjecture of King for smooth toric Deligne-Mumford stacks, Advances in Mathematics 221 (2009), no. 1, 277-301.

[10] G. Brown, Jarosław Buczyński, Maps of toric varieties in Cox coordinates, Fund. Math. 222 (2013), 213-267.

[11] W. Bruns, The quest for counterexamples in toric geometry, Commutative algebra and algebraic geometry (CAAG-2010), Ramanujan Math. Soc. Lect. Notes Ser., vol. 17, Ramanujan Math. Soc., Mysore, 2013, 45-61.

[12] W. Bruns, J. Gubeladze, Normality and covering properties of affine semigroups, J. Reine Angew. Math. 510 (1999), 161-178.

[13] W. Bruns, J. Gubeladze, Polytopes, rings, and k-theory, Springer Science & Business Media, 2009.

[14] W. Bruns, J. Gubeladze, M. Michałek, Quantum jumps of normal polytopes, Discrete Comput. Geom. 56 (2016), no. 1, 181-215.

[15] W. Bruns, J. Gubeladze, N.V. Trung, Normal polytopes, triangulations, and Koszul algebras, J. Reine Angew. Math. 485 (1997), 123-160.

[16] W. Bruns, B. Ichim, C. Söger, Normaliz software, http://www.mathematik.uniosnabrueck.de/normaliz.

[17] A. Capani, G. Niesi, L. Robbiano, (1995). CoCoA, a system for doing computations in commutative algebra. Available via anonymous ftp from cocoa.dima.unige.it

[18] Buczyńska, Phylogenetic toric varieties on graphs, J. Algebraic Combin. 35 (2012), no. 3, 421-460.

[19] W. Buczyńska, J. Wiśniewski, On geometry of binary symmetric models of phylogenetic trees, J. Eur. Math. Soc. 9(3) (2007), 609-635.

[20] J. Buczyński, O. Kędzierski, Maps of mori dream spaces in cox coordinates. Part I: existence of descriptions, arXiv: 1605.06828, 2016.

[21] M. Casanellas, J. Fernández-Sánchez, M. Michałek, Complete intersection for equivariant models, Preprint arXiv:1512.07174 (2015).

[22] M. Casanellas, J. Fernández-Sánchez, M. Michałek, Low degree equations for phylogenetic group-based models, Collect. Math. 66 (2015), no. 2, 203-225.

[23] A. Conca, Linear spaces, transversal polymatroids and ASL domains, J. Algebraic Combin. 25 (2007), 25-41.

[24] D. Cox, The homogeneous coordinate ring of a toric variety, J. Algebraic Geom. 4 (1995), no. 1, 17-50.

[25] D. Cox, J. Little, H. Schenck, Toric varieties. Graduate Studies in Mathematics, 124. American Mathematical Society, Providence, RI, (2011).

[26] M. Donten-Bury, Phylogenetic invariants for $\mathbb{Z}_3$ scheme-theoretically, Ann. Comb. 20 (2016), no. 3, 549-568.

[27] M. Donten-Bury, M. Michałek, Phylogenetic invariants for group-based models, J. Algebr. Stat. 3 (2012), no. 1, 44-63.

[28] J. Draisma, R. Eggermont, Finiteness results for Abelian tree models, J. Eur. Math. Soc. (JEMS) 17 (2015), no. 4, 711-738.

[29] A. Engström, Cut ideals of K4-minor free graphs are generated by quadrics, Michigan Math. J. 60 (2011), no. 3, 705-714.

[30] N. Eriksson, K. Ranestad, B. Sturmfels, S. Sullivant, Phylogenetic algebraic geometry, Projective varieties with unexpected properties (2005), 237-255.

[31] R. Fedder, F-purity and rational singularity, Transactions of the American Mathematical Society 278 (1983), no. 2, 461-480.

[32] W. Fulton, Introduction to Toric Varieties, Annals of Mathematics Studies, vol. 131, Princeton University Press, 1993.

[33] E. Gawrilow, M. Joswig, Polymake: a Framework for Analyzing Convex Polytopes, Polytopes - Combinatorics and Computation (Gil Kalai and Günter M. Ziegler, eds.), Birkhäuser, 2000, pp. 43-74.
[34] I. Gelfand, R. Goresky, R. MacPherson, V. Serganova, Combinatorial geometries, convex polyhedra, and Schubert cells, Advances in Mathematics 63 (1987), no. 3, 301-316.
[35] D. Gijswijt, G. Regts, Polyhedra with the integer caratheodory property, Journal of Combinatorial Theory, Series B 102 (2012), no. 1, 62-70.
[36] D.R. Grayson, M.E. Stillman, Macaulay2, a software system for research in algebraic geometry, Available at http://www.math.uiuc.edu/Macaulay2/.
[37] R. Hartshorne, Algebraic geometry, vol. 52, Springer Science & Business Media, 2013.
[38] J. Herzog, T. Hibi, Discrete Polymatroids, J. Algebraic Combin. 16 (2002), 239-268.
[39] J. Herzog, T. Hibi, M. Vladoiu, et al., Ideals of fiber type and polymatroids, Osaka Journal of Mathematics 42 (2005), no. 4, 807-829.
[40] A. Hochenegger, E. Martinengo, Mori dream stacks, Math. Z. 280 (2015), no. 3-4, 1185-1202.
[41] A. Hochenegger, E. Martinengo, Maps of Mori Dream Spaces, arXiv:1605.06789, 2016.
[42] M. Hochster, Rings of invariants of tori, Cohen-Macaulay rings generated by monomials, and polytopes, Ann. of Math. (2) 96 (1972), 318-337.
[43] M. Hochster, J. Roberts, The purity of the frobenius and local cohomology, Advances in Mathematics 21 (1976), no. 2, 117-172.
[44] Y. Hu, S. Keel, Mori dream spaces and GIT, Michigan Math. J. 48 (2000), 331-348, Dedicated to William Fulton on the occasion of his 60th birthday.
[45] K. Kashiwabara, The toric ideal of a matroid of rank 3 is generated by quadrics, Electron. J. Combin. 17 (2010).
[46] G. Kempf, F. Knudson, D. Mumford, B. Saint-Donat, Toroidal embedding 1, Lecture Notes in Math. 339.
[47] O. Kędzierski, J.Wiśniewski, Differentials of Cox rings: Jaczewski's theorem revisited, J. Math. Soc. Japan 67 (2015), no. 2, 595-608.
[48] M. Kimura, Estimation of evolutionary distances between homologous nucleotide sequences, Proceedings of the National Academy of Sciences 78 (1981), no. 1, 454-458.
[49] A. Klyachko, Equivariant bundles on toral varieties, Mathematics of the USSR-Izvestiya 35 (1990), no. 2, 337.
[50] M. Kreuzer, Martin, L. Robbiano, Computational commutative algebra. 2. Springer-Verlag, Berlin (2005).
[51] K. Kubjas, Hilbert polynomial of the Kimura 3-parameter model, Journal of Algebraic Statistics 3 (2012), no. 1.
[52] K. Kubjas, C. Manon, Conformal blocks, Berenstein-Zelevinsky triangles, and group-based models, J. Algebr. Comb. 40 (2014), no. 3, 861-886.

[53] K. Han, S. Kwak, Analysis on some infinite modules, inner projection, and applications, Transactions of the American Mathematical Society 364 (2012), no. 11, 5791-5812.

[54] M. Lasoń, M. Michałek, On the full, strongly exceptional collections on toric varieties with picard number three, Collectanea mathematica 62 (2011), no. 3, 275-296.

[55] M. Lasoń, M. Michałek, On the toric ideal of a matroid, Adv. Math. 259 (2014), 1-12.

[56] M. Lasoń, List coloring of matroids and base exchange properties, European J. Combin. 49 (2015), 265-268.

[57] M. Lasoń, Degree bounds for the toric ideal of a matroid, arXiv:1601.08199v2.

[58] M. Lasoń, On Gröbner bases of the toric ideal of a matroid, in preparation.

[59] D. Maclagan, B. Sturmfels, Introduction to tropical geometry, vol. 161, American Mathematical Soc., 2015.

[60] C. Manon, Coordinate rings for the moduli stack of $SL_2(\mathbb{C})$ quasi-parabolic principal bundles on a curve and toric fiber products, J. Algebra 365 (2012), 163-183.

[61] M. Michałek, Geometry of phylogenetic group-based models, J. Algebra 339 (2011), no. 1, 339-356.

[62] M. Michałek, Constructive degree bounds for group-based models, J. Combin. Theory, Series A 120 (2013), no. 7, 1672-1694.

[63] M. Michałek, Toric geometry of the 3-Kimura model for any tree, Adv. Geom. 14 (2014), no. 1, 11-30.

[64] M. Michałek, Toric varieties in phylogenetics, Diss. Math. (2015), no. 511, 3-86.

[65] M. Michałek, Finite phylogenetic complexity of $\mathbb{Z}_p$ and invariants for $\mathbb{Z}_3$, European J. Combin. 59 (2017), 169-186.

[66] M. Michałek, E. Ventura, Finite phylogenetic complexity and combinatorics of tables, Algebra and Number Theory 11 (2017), no. 1.

[67] M. Michałek, L. Oeding, P. Zwiernik, Secant cumulants and toric geometry, International Mathematics Research Notices 2015 no. 12 (2015) 4019-4063.

[68] T. Oda, Convex bodies and algebraic geometry, Ergebnisse der Mathematik und ihrer Grenzgebiete (3) [Results in Mathematics and Related Areas (3)], vol. 15, Springer-Verlag, Berlin, 1988, An introduction to the theory of toric varieties, Translated from the Japanese.

[69] H. Ohsugi, Normality of cut polytopes of graphs is a minor closed property, Discrete Mathematics 310 (2010), no. 6, 1160-1166.

[70] H. Ohsugi, T. Hibi, Toric ideals generated by quadratic binomials, Journal of Algebra 218 (1999), no. 2, 509-527.

[71] J. Oxley, Matroid theory, vol. 3, Oxford University Press, USA, 2006.

[72] S. Payne, Toric vector bundles, branched covers of fans, and the resolution property, Journal of Algebraic Geometry 18 (2009), no. 1, 1-36.

[73] J. Schweig, Toric ideals of lattice path matroids and polymatroids, J. Pure Appl. Algebra 215 (2011), 2660-2665.
[74] P. Seymour, Matroids and multicommodity flows, European Journal of Combinatorics 2 (1981), no. 3, 257-290.
[75] K. Shibata, Toric ideals of series and parallel connections of matroids, J. Algebra Appl. 15 (2016), no. 6.
[76] B. Sturmfels, Gröbner Bases and Convex Polytopes, Univ. Lecture Series 8, American Mathematical Society, Providence, 1995.
[77] B. Sturmfels, S. Sullivant, Toric ideals of phylogenetic invariants, J. Comput. Biology 12 (2005), 204-228.
[78] B. Sturmfels, S. Sullivant, Toric geometry of cuts and splits, The Michigan Mathematical Journal 57 (2008), 689-709.
[79] B. Sturmfels, Z. Xu, Sagbi bases of Cox-Nagata rings, J. Eur. Math. Soc. (JEMS) 12 (2010), no. 2, 429-459.
[80] S. Sullivant, Toric fiber products, J. Algebra 316 (2007), no. 2, 560-577.
[81] H. Sumihiro, Equivariant completion, Journal of Mathematics of Kyoto University 14 (1974), no. 1, 1-28.
[82] N. White, The basis monomial ring of a matroid, Adv. Math. 24 (1977), 292-297.
[83] N. White, A unique exchange property for bases, Linear Algebra and its App. 31 (1980), 81-91.
[84] 4ti2 team, 4ti2-a software package for algebraic, geometric and combinatorial problems on linear spaces, www.4ti2.de.

Research Institute for Mathematical Sciences, Kyoto, Japan
Max Planck Institute for Mathematics in the Sciences, Inselstr. 22, Leipzig, Germany
Polish Academy of Sciences, Sniadeckich 8, 00-656 Warsaw, Poland
E-mail address: `wajcha2@poczta.onet.pl`

Advanced Studies in Pure Mathematics 77, 2018
The 50th Anniversary of Gröbner Bases
pp. 253–352

Algorithms for D-modules, integration, and generalized functions with applications to statistics

Toshinori Oaku

Abstract.

This is an enlarged and revised version of the slides presented in a series of survey lectures given by the present author at MSJ SI 2015 in Osaka. The goal is to introduce an algorithm for computing a holonomic system of linear (ordinary or partial) differential equations for the integral of a holonomic function over the domain defined by polynomial inequalities. It applies to the cumulative function of a polynomial of several independent random variables with e.g., a normal distribution or a gamma distribution. Our method consists in Gröbner basis computation in the Weyl algebra, i.e., the ring of differential operators with polynomial coefficients. In the algorithm, generalized functions are inevitably involved even if the integrand is a usual function. Hence we need to make sure to what extent purely algebraic method of Gröbner basis applies to generalized functions which are based on real analysis.

Contents

Received August 30, 2016.
2010 *Mathematics Subject Classification.* 13N10, 13P10, 46F10, 62E15.
Key words and phrases. D-module, Gröbner basis, generalized function, probability density function.

§1. Introduction: aim and an example from statistics

A univariate function is called holonomic if it satisfies a (non-trivial) linear ordinary differential equation. Special functions such as the hypergeometric function or the Bessel function are holonomic, as well as rational functions and their exponential and logarithm. As is well-known, the solutions of a linear ordinary differential equation constitute a finite dimensional vector space.

A D-module is a system of linear (partial or ordinary) differential equations with polynomial (or analytic function) coefficients. There is a special class of D-modules which are called holonomic, the solution spaces of which are finite dimensional vector spaces. This notion was introduced by Mikio Sato and J. Bernstein independently. Bernstein [2], [3] introduced a special class of linear partial differential equations with polynomial coefficients which was called the Bernstein class in [4]. On the other hand, Sato and his collaborators M. Kashiwara, T. Kawai [31] introduced the notion of a holonomic system, which was called at first a maximally overdetermined system, in the category of differential operators with analytic coefficients.

A holonomic function is a differentiable or a generalized function which is a solution of a holonomic system. For example, $\exp(f) = e^f$ is a holonomic function for any polynomial $f = f(x_1, \ldots, x_n)$. In statistics, most of important probability density functions, such as those of the multivariate normal distribution and the gamma distribution are holonomic. Our aim is to find a holonomic system which is satisfied by the integral of a holonomic function over the domain defined by polynomial inequalities.

As an example, let us consider the integral

$$F(t) = \frac{1}{2\pi}\int_{D(t)} \exp\Big(-\frac{1}{2}(x^2+y^2)\Big)\,dxdy,\ \ D(t) = \{(x,y) \in \mathbb{R}^2 \mid xy \leq t\}.$$

It can be regarded as the cumulative distribution function of xy with (x, y) being a random vector with the two dimensional standard normal (Gaussian) distribution. Let us introduce the Heaviside function $Y(t)$ such that $Y(t) = 1$ for $t > 0$ and $Y(t) = 0$ for $t < 0$. (One does not need to mind the value at $t = 0$.) $Y(t)$ is discontinuous at $t = 0$ and its derivative $Y'(t)$ as a generalized function coincides with Dirac's delta function $\delta(t)$. As a generalized function, $\delta(t)$ vanishes outside of $t = 0$ and $t\delta(t) = 0$ holds everywhere in $\mathbb{R}$.

By using the Heaviside function, we rewrite $F(t)$ as

$$F(t) = \frac{1}{2\pi}\int_{\mathbb{R}^2} \exp\Big(-\frac{1}{2}(x^2+y^2)\Big)Y(t-xy)\,dxdy.$$

Differentiation under the integral sign yields

$$v(t) := F'(t) = \frac{1}{2\pi}\int_{\mathbb{R}^2} \exp\Big(-\frac{1}{2}(x^2+y^2)\Big)\delta(t-xy)\,dxdy.$$

The integrand $u(x,y,t) := \exp\Big(-\frac{1}{2}(x^2+y^2)\Big)\delta(t-xy)$ satisfies a holonomic system

$$(\partial_y + x\partial_t + y)u = (\partial_x + y\partial_t + x)u = (t-xy)u = 0$$

with $\partial_x = \partial/\partial x$, $\partial_y = \partial/\partial y$, $\partial_t = \partial/\partial t$ as is easily checked. The integration algorithm for D-modules (see 5.2) outputs an answer

$$(t\partial_t^2 + \partial_t - t)v(t) = 0. \tag{1}$$

In fact, we have an equality

$$y\partial_t(\partial_y + x\partial_t + y) - y(\partial_x + y\partial_t + x) + (\partial_t^2 - 1)(t-xy)$$
$$= -\partial_x y + \partial_y y\partial_t + t\partial_t^2 + \partial_t - t$$

in the ring of differential operators. Since the differential operator on the left-hand side annihilates $u(x,y,t)$, we get

$$(t\partial_t^2 + \partial_t - t)v(t) = \frac{1}{2\pi}\int_{\mathbb{R}^2}(t\partial_t^2 + \partial_t - t)u(x,y,t)\,dxdy$$
$$= \frac{1}{2\pi}\int_{\mathbb{R}^2}\partial_x(yu(x,y,t))\,dxdy - \frac{1}{2\pi}\int_{\mathbb{R}^2}\partial_y(y\partial_t u(x,y,t))\,dxdy = 0.$$

The integrals on the last line vanish since $yu(x,y,t)$ and $y\partial_t u(x,y,t)$ are 'rapidly decreasing' in x, y; this reasoning shall be made precise in 4.3.

It follows that $w(z) := v(-iz)$ satisfies the Bessel differential equation

$$z^2\frac{d^2w}{dz^2} + z\frac{dw}{dz} + z^2 w = 0.$$

Together with the property that $v(t) \to 0$ as $t \to \pm\infty$ and $v(-t) = v(t)$, this implies

$$v(t) = CH_0^{(1)}(i|t|) \qquad (t \neq 0)$$

with some constant C, where $H_0^{(1)}(z)$ is a Hankel function. This fact was observed, for example, by Wishart and Bartlett [36] as a special case. Note that $v(t)$ is discontinuous at $t = 0$ but is integrable and satisfies (1) in the sense of generalized functions on the whole real line $\mathbb{R}$.

It also follows from (1) that the characteristic function, i.e., the Fourier transform

$$\hat{v}(\tau) = \int_{-\infty}^{\infty} e^{it\tau} v(t)\, dt = \int_{\mathbb{R}^2} \exp\Bigl(i\tau xy - \frac{1}{2}(x^2+y^2)\Bigr)\, dxdy$$

satisfies a differential equation

$$(\tau^2+1)\frac{d}{d\tau}\hat{v}(\tau) + \tau\hat{v}(\tau) = 0.$$

Together with $\hat{v}(0) = 1$, this implies $\hat{v}(\tau) = (\tau^2+1)^{-1/2}$. Thus we get an alternative expression

$$v(t) = V_+(t+i0) + V_-(t-i0) = \lim_{\varepsilon\to+0}(V_+(t+i\varepsilon) + V_-(t-i\varepsilon))$$

as a hyperfunction of Mikio Sato ([30]) with

$$V_+(t+is) = \frac{1}{2\pi}\int_{-\infty}^{0} \frac{\exp(-i(t+is)\tau)}{\sqrt{\tau^2+1}}\, d\tau,$$
$$V_-(t+is) = \frac{1}{2\pi}\int_{0}^{\infty} \frac{\exp(-i(t+is)\tau)}{\sqrt{\tau^2+1}}\, d\tau,$$

where $V_+(t+is)$ and $V_-(t+is)$ are holomorphic functions of $t+is$ on the upper half plane $s > 0$ and on the lower half plane $s < 0$ respectively.

In general, for a holonomic function $u(x,y)$ with $x = (x_1,\dots,x_n)$ and $y = (y_1,\dots,y_d)$, let us consider the integral

$$v(y) = \int_{D(y)} u(x,y)\, dx_1\cdots dx_n,$$
$$D(y) = \{x\in\mathbb{R}^n \mid f_j(x,y) \geq 0 \quad (1\leq j\leq m)\}$$

with real polynomials $f_1,\dots,f_m$ in (x,y). We rewrite it as

$$v(y) = \int_{\mathbb{R}^n} u(x,y)Y(f_1(x,y))\cdots Y(f_m(x,y))\, dx_1\cdots dx_n$$

and apply the D-module theoretic integration algorithm to obtain a holonomic system for $v(y)$, assuming that the integrand and its derivatives are 'rapidly decreasing' with respect to the integration variables x. In the process, we also need an algorithm to compute a holonomic system for the product $uY(f_1)\cdots Y(f_m)$ as a generalized function. Then the D-module theory assures us that the obtained system of differential equations for $v(y)$ is holonomic.

Finally, let us remark that we cannot use differential operators with rational function coefficients since generalized functions are involved in the computation. For example, $x\partial_x Y(x) = 0$ does not imply $\partial_x Y(x) = \delta(x) = 0$; we cannot factor out x.

The organization of this article is as follows:

Section 2 is a hopefully concise exposition on the very beginning of the D-module theory; the central subject is holonomic D-modules. More advanced topics such as D-modules with regular singularities are not treated. The presentation is almost self-contained with some arguments and examples supplied in the next section after armed with Gröbner bases.

In Section 3, we introduce Gröbner bases over the ring of differential operators. One point is that we can compute Gröbner bases with respect to arbitrary monomial orders that are not necessarily well-orders, which will be needed in the integration algorithm. We also describe first applications of Gröbner bases to D-module theory: computation of the characteristic variety, and a proof of the equivalence of the two definitions of holonomicity introduced in the previous section.

In Section 4, we briefly review the theory of distributions in the sense of generalized functions from our viewpoint, with mention of the relation with statistical distributions. Especially, we introduce some classes of distributions which are adapted to our integration algorithm developed in the following sections.

Section 5 is a review on the integration of D-modules both from theoretical and algorithmic viewpoints; the material should be more or less standard by now.

In the first subsection of Section 6, we give some examples of integrals which correspond to random variables with respect to the multivariate standard normal distribution such as the example above. In a somewhat technical subsection 6.2, we introduce an algorithm to compute a holonomic system for the product of complex powers of polynomials and a holonomic function. This enables us to compute, in 6.3, a holonomic system for the integral of a holonomic function over the domain defined by arbitrary polynomial inequalities. Finally in 6.4, we treat the integral of a function with some auxiliary parameters which satisfies a holonomic difference-differential system.

The author would like to express his deepest gratitude to the organizers of MSJ SI 2015, especially to Takayuki Hibi, for the invitation and the encouragement. At the same time, the author is grateful to Akimichi Takemura and Nobuki Takayama also for drawing his attention to statistics; their influence is reflected in the appended last phrase of the title.

This work was supported by JSPS Grant-in-Aid for Scientific Research (C) 26400123.

§2. Basics of D-module theory

We review the theory of D-modules, more precisely, of modules over the Weyl algebra, which was initiated by J. Bernstein [2], [3]. A standard reference is the first chapter of [4]. A D-module corresponds to a system of linear (ordinary or partial) differential equations with polynomial coefficients. The notion of holonomic modules, also called the Bernstein class of modules, and its characterizations are most essential. We remark that the notion of holonomic modules over the ring of differential operators with complex analytic coefficients was independently introduced by M. Sato, T. Kawai, and M. Kashiwara [31].

2.1. The ring of differential operators

Let $\mathbb{K}$ be an arbitrary field of characteristic zero. We denote by $\mathbb{K}[x] := \mathbb{K}[x_1, \ldots, x_n]$ the ring of polynomials in indeterminates $x = (x_1, \ldots, x_n)$ with coefficients in $\mathbb{K}$. A *derivation* $\theta : \mathbb{K}[x] \to \mathbb{K}[x]$ is a $\mathbb{K}$-linear map that satisfies

$$\theta(fg) = \theta(f)g + f\theta(g) \qquad (\forall f, g \in \mathbb{K}[x]).$$

The set $\mathrm{Der}_{\mathbb{K}}\mathbb{K}[x]$ of the derivations constitutes a $\mathbb{K}[x]$-module. For $i = 1, \ldots, n$, define a derivation $\partial_i = \partial_{x_i}$ by the partial derivative

$$\partial_i \ : \ \mathbb{K}[x] \ni f \longmapsto \frac{\partial f}{\partial x_i} \in \mathbb{K}[x].$$

Then $\partial_1, \ldots, \partial_n$ are a $\mathbb{K}[x]$-basis of $\mathrm{Der}_{\mathbb{K}}\mathbb{K}[x]$. In fact, if $\theta \in \mathrm{Der}_{\mathbb{K}}\mathbb{K}[x]$, then it is easy to see that

$$\theta = \theta(x_1)\partial_1 + \cdots + \theta(x_n)\partial_n.$$

Let $\mathrm{End}_{\mathbb{K}}\mathbb{K}[x]$ be the $\mathbb{K}$-algebra consisting of the $\mathbb{K}$-linear endomorphisms of $\mathbb{K}[x]$. The ring D_n is defined to be the $\mathbb{K}$-subalgebra of $\mathrm{End}_{\mathbb{K}}\mathbb{K}[x]$ that is generated by $\mathbb{K}[x]$ and $\mathrm{Der}_{\mathbb{K}}\mathbb{K}[x]$, or equivalently, by $x_1, \ldots, x_n$ and $\partial_1, \ldots, \partial_n$. This ring D_n is called the *ring of differential operators* in the variables $x = (x_1, \ldots, x_n)$ with polynomial coefficients, or, more simply, the n-th *Weyl algebra* over $\mathbb{K}$.

An element $a = a(x)$ of $\mathbb{K}[x]$ is regarded as an element of D_n as the multiplication operator $f \mapsto af$ for $f \in \mathbb{K}[x]$. With this identification, D_n contains $\mathbb{K}[x]$ as a subring. The ring D_n is a non-commutative $\mathbb{K}$-algebra. In fact, for $a \in \mathbb{K}[x]$ regarded as an element of D_n, the product in D_n satisfies

$$\partial_i a = a\partial_i + \partial_i(a) = a\partial_i + \frac{\partial a}{\partial x_i} \qquad (i = 1, \dots, n).$$

For a multi-index $\alpha = (\alpha_1, \dots, \alpha_n) \in \mathbb{N}^n$ with $\mathbb{N} = \{0, 1, 2, \dots\}$, we use the notation $x^\alpha = x_1^{\alpha_1} \cdots x_n^{\alpha_n}$, $\partial^\alpha = \partial_x^\alpha = \partial_1^{\alpha_1} \cdots \partial_n^{\alpha_n}$, and $|\alpha| = \alpha_1 + \cdots + \alpha_n$. Then an element P of D_n is uniquely written in a finite sum

$$P = P(x, \partial) = \sum_{\alpha, \beta \in \mathbb{N}^n} a_{\alpha,\beta} x^\alpha \partial^\beta = \sum_{\beta \in \mathbb{N}^n} a_\beta(x) \partial^\beta$$

with $a_{\alpha,\beta} \in \mathbb{K}$ and $a_\beta(x) = \sum_\alpha a_{\alpha,\beta} x^\alpha$, which is called the *normal form* of P. In fact, $a_\beta(x)$ are uniquely determined by the action of P on $\mathbb{K}[x]$ as follows: First we have $a_{(0,\dots,0)}(x) = P1$. Next, we have

$$a_{(1,0,\dots,0)}(x) = Px_1 - a_{(0,\dots,0)}(x)x_1,$$

and so on. Here we need the assumption that the characteristic of $\mathbb{K}$ is zero.

Introducing commutative indeterminates $\xi = (\xi_1, \dots, \xi_n)$ which corresponds to ∂, we associate with this P a polynomial

$$P(x, \xi) := \sum_{\alpha, \beta \in \mathbb{N}^n} a_{\alpha,\beta} x^\alpha \xi^\beta \in \mathbb{K}[x, \xi] = \mathbb{K}[x_1, \dots, x_n, \xi_1, \dots, \xi_n]$$

and call it the *total symbol* of P. Note that P must be in the normal form when ξ is substituted for ∂. By this correspondence, D_n is isomorphic to $\mathbb{K}[x, \xi]$ as a $\mathbb{K}$-vector space but not as a ring.

The product $R = PQ$ in D_n can be effectively computed by using the *Leibniz formula*

$$R(x, \xi) = \sum_{\nu \in \mathbb{N}^n} \frac{1}{\nu!} \left(\frac{\partial}{\partial \xi}\right)^\nu P(x, \xi) \cdot \left(\frac{\partial}{\partial x}\right)^\nu Q(x, \xi) \tag{2}$$

in terms of total symbols, where we use the notation $\nu! = \nu_1! \cdots \nu_n!$ for $\nu = (\nu_1, \dots, \nu_n) \in \mathbb{N}^n$.

Example 2.1. Set $n = 1$ and write $x = x_1$ and $\partial = \partial_1$. Consider the product $R := \partial^m x^m$ with a non-negative integer m. Since the total symbols of ∂^m and x^m are ξ^m and x^m respectively, the Leibniz formula

(2) gives the total symbol $R(x, \xi)$ as

$$R(x,\xi) = \sum_{\nu=0}^{\infty} \frac{1}{\nu!} \left(\frac{\partial}{\partial \xi}\right)^{\nu} \xi^m \cdot \left(\frac{\partial}{\partial x}\right)^{\nu} x^m$$
$$= \sum_{\nu=0}^{m} \frac{1}{\nu!} \{m(m-1)\cdots(m-\nu+1)\}^2 \xi^{m-\nu} x^{m-\nu}.$$

This implies

$$\partial^m x^m = \sum_{\nu=0}^{m} \frac{1}{\nu!} \{m(m-1)\cdots(m-\nu+1)\}^2 x^{m-\nu} \partial^{m-\nu}.$$

Exercise 1. Show that an element $P = \sum_{\beta\in\mathbb{N}} a_\beta(x)\partial^\beta$ of D_n with $a_\beta(x) \in \mathbb{K}[x]$ defines the zero endomorphism of $\mathbb{K}[x]$ if and only if $a_\beta(x) = 0$ for any β.

Exercise 2. Prove the Leibniz formula (2).

Exercise 3. Set $n = 1$ and $x = x_1$, $\partial = \partial_1$. For a positive integer m, prove the formulae

$$x^m\partial^m = x\partial(x\partial-1)\cdots(x\partial-m+1), \quad \partial^m x^m = \partial x(\partial x+1)\cdots(\partial x+m-1).$$

2.2. The D-module formalism

Given $P_1, \ldots, P_r \in D_n$, let us consider a system of linear (partial or ordinary) differential equations

$$P_1 u = \cdots = P_r u = 0 \tag{3}$$

for an unknown function u. Let $I := D_n P_1 + \cdots + D_n P_r$ be the left ideal of D_n generated by $P_1, \ldots, P_r$. Then (3) is equivalent to

$$Pu = 0 \qquad (\forall P \in I).$$

Here we suppose that the unknown function u belongs to some 'function space' $\mathcal{F}$ which is a left D_n-module.

For $\mathcal{F}$ to be a left D_n-module, it is necessary that any function f belonging to $\mathcal{F}$ be infinitely differentiable and multiplication ah by an arbitrary polynomial $a \in \mathbb{K}[x]$ make sense. Here are examples of 'function spaces':

Example 2.2. By the definition, $\mathbb{K}[x]$ has a natural structure of left D_n-module since D_n is a subalgebra of $\mathrm{End}_{\mathbb{K}}\mathbb{K}[x]$. So $\mathbb{K}[x]$ has two structures: a subring of D_n and a left D_n-module. Hence for $f \in \mathbb{K}[x]$ and $P \in D_n$, Pf has two meanings:

- Pf as the product in D_n with f regarded as an element of the subring $\mathbb{K}[x]$ of D_n.
- Pf as the action of P on the element f of the left D_n-module $\mathbb{K}[x]$. In other words, we regard f as a function subject to derivations.

This might cause some confusion. In [29], the action of P on an element f of a left D_n-module is conspicuously denoted $P \bullet f$ for distinction. We shall denote, if needed, $Pf = P(f)$ to clarify the action of P on f, and $Pf = P \cdot f$ to emphasize that it is the product in D_n, following the traditional notation in D-module theory.

Example 2.3. The field $\mathbb{K}(x) = \mathbb{K}(x_1, \dots, x_n)$ of rational functions has a natural structure of left D_n-module. For a point $a = (a_1, \dots, a_n)$ of the affine space $\mathbb{K}^n$, the set $\mathbb{K}[x]_a$ of regular functions at a, i.e., the elements of $\mathbb{K}(x)$ whose denominators do not vanish at a, also has a natural structure of D_n-module. More generally, the localization $\mathbb{K}[x][S^{-1}]$ by a multiplicative subset S of $\mathbb{K}[x]$ is also a left D_n-module.

Example 2.4. Set $\mathbb{K} = \mathbb{C}$. Let $C^\infty(U)$ be the set of the complex-valued C^∞ functions on an open set U of the n-dimensional real Euclidean space $\mathbb{R}^n$. Then each ∂_i acts on $C^\infty(U)$ as differentiation and x_i as multiplication. This makes $C^\infty(U)$ a left D_n-module. Let $C_0^\infty(U)$ be the set of C^∞ functions on U with compact support. More precisely, $f \in C^\infty(U)$ belongs to $C_0^\infty(U)$ if and only if there is a compact subset K of U such that $f(x) = 0$ for any $x \in U \setminus K$. Then $C_0^\infty(U)$ is a left D_n-submodule of $C^\infty(U)$.

Other examples of such $\mathcal{F}$ with $\mathbb{K} = \mathbb{C}$ are the set $\mathcal{O}(\Omega)$ of holomorphic functions on an open subset Ω of $\mathbb{C}^n$, the set $\mathcal{D}'(U)$ of the Schwartz distributions on an open subset U of $\mathbb{R}^n$, and the set $S'(\mathbb{R}^n)$ of tempered distributions, which shall be introduced later, as well as the set $\mathcal{B}(U)$ of the hyperfunctions (of Mikio Sato) on an open subset U of $\mathbb{R}^n$.

Now for a left ideal I of D_n, consider the residue module $M := D_n/I$, which is a left D_n-module generated by the residue class $\overline{1}$ of $1 \in \mathbb{K}[x] \subset D_n$. Fix a left D_n-module $\mathcal{F}$ as your favorite function space. A map $\varphi : M \to \mathcal{F}$ is D_n-linear, or a D_n-homomorphism, if

$$\varphi(u + v) = \varphi(u) + \varphi(v), \quad \varphi(Pu) = P\varphi(u) \quad (\forall u, v \in M, \ \forall P \in D_n).$$

Let $\mathrm{Hom}_{D_n}(M, \mathcal{F})$ be the set of the D_n-homomorphisms of M to $\mathcal{F}$, which is a $\mathbb{K}$-vector space. Since M is generated by $\overline{1}$ as left D_n-module, $\varphi \in \mathrm{Hom}_{D_n}(M, \mathcal{F})$ is uniquely determined by $\varphi(\overline{1}) \in \mathcal{F}$. On the other hand, for φ to be well-defined, it is necessary and sufficient that $\varphi(\overline{1})$ be annihilated by I, i.e., $P\varphi(\overline{1}) = 0$ for any $P \in I$. In conclusion, we have a $\mathbb{K}$-isomorphism

$$\mathrm{Hom}_{D_n}(M, \mathcal{F}) \ni \varphi \overset{\sim}{\longmapsto} \varphi(\overline{1}) \in \{f \in \mathcal{F} \mid Pf = 0 \quad (\forall P \in I)\}.$$

For an element u of a left D_n-module $\mathcal{F}$, we define the annihilator of u in D_n to be the left ideal

$$\mathrm{Ann}_{D_n} u = \{P \in D_n \mid Pu = 0\}.$$

Then we have

$$I = \mathrm{Ann}_{D_n} \overline{1} = \{P \in D_n \mid P\overline{1} = 0 \in M\}$$

by the definition.

We started with a left ideal I of D_n generated by given $P_1, \dots, P_r \in D_n$ and considered a left D_n-module $M = D_n/I$. We can argue in the reverse order: Let M be a finitely generated left D_n-module and let $u_1, \dots, u_m \in M$ be generators of M, i.e., assume that for any $u \in M$, there exist $P_1, \dots, P_m \in D_n$ such that $u = P_1 u_1 + \cdots + P_m u_m$. Set

$$N := \{(P_1, \dots, P_m) \in (D_n)^m \mid P_1 u_1 + \cdots + P_m u_m = 0\},$$

which is a left D_n-submodule of the free module $(D_n)^m$.

Since D_n is a left (and right) Noetherian ring (this can be proved by using a Gröbner basis in D_n), N is also finitely generated over D_n. Hence there exist

$$Q_i = (Q_{i1}, \dots, Q_{im}) \in (D_n)^m \qquad (i = 1, \dots, r)$$

which generate N as left D_n-module. Then we have an exact sequence of left D_n-modules

$$(D_n)^r \xrightarrow{\psi} (D_n)^m \xrightarrow{\varphi} M \longrightarrow 0, \tag{4}$$

which is called a *presentation* of M. Here φ and ψ are homomorphisms of left D_n-modules defined by, for $P_i \in D_n$,

$$\varphi((P_1, \dots, P_m)) = P_1 u_1 + \cdots + P_m u_m,$$
$$\psi((P_1, \dots, P_r)) = \begin{pmatrix} P_1 & \cdots & P_r \end{pmatrix} \begin{pmatrix} Q_{11} & \cdots & Q_{1m} \\ \vdots & & \vdots \\ Q_{r1} & \cdots & Q_{rm} \end{pmatrix}$$

and $N = \ker \varphi = \mathrm{im}\, \psi$ holds.

From (4) we get an exact sequence

$$0 \longrightarrow \mathrm{Hom}_{D_n}(M, \mathcal{F}) \xrightarrow{\varphi^*} \mathrm{Hom}_{D_n}((D_n)^m, \mathcal{F}) \xrightarrow{\psi^*} \mathrm{Hom}_{D_n}((D_n)^r, \mathcal{F}).$$

Since $\mathrm{Hom}_{D_n}((D_n)^m, \mathcal{F})$ is isomorphic to $\mathcal{F}^m$, this yields

$$0 \longrightarrow \mathrm{Hom}_{D_n}(M, \mathcal{F}) \xrightarrow{\varphi^*} \mathcal{F}^m \xrightarrow{\psi^*} \mathcal{F}^r.$$

Regarding the elements of $\mathcal{F}^m$ as column vectors, we have, for $h \in \mathrm{Hom}_{D_n}(M, \mathcal{F})$ and $f_1, \ldots, f_m \in \mathcal{F}$,

$$\varphi^*(h) = \begin{pmatrix} h(u_1) \\ \vdots \\ h(u_m) \end{pmatrix}, \quad \psi^*\left(\begin{pmatrix} f_1 \\ \vdots \\ f_m \end{pmatrix}\right) = \begin{pmatrix} Q_{11} & \cdots & Q_{1m} \\ \vdots & & \vdots \\ Q_{r1} & \cdots & Q_{rm} \end{pmatrix} \begin{pmatrix} f_1 \\ \vdots \\ f_m \end{pmatrix}.$$

Hence we have an isomorphism

$$\begin{aligned} &\mathrm{Hom}_{D_n}(M, \mathcal{F}) \\ &\quad \cong \mathrm{Ker}\,\psi^* = \left\{ {}^t(f_1, \ldots, f_m) \in \mathcal{F}^m \mid \sum_{j=1}^m Q_{ij} f_j = 0 \quad (i = 1, \ldots, r) \right\} \end{aligned}$$

as $\mathbb{K}$-vector space. Note that the generators $u_1, \ldots, u_m$ of M also satisfy the same equations

$$\sum_{j=1}^m Q_{ij} u_j = 0 \qquad (i = 1, \ldots, r)$$

in M. In this way, we can regard a finitely generated left D_n-module M as a system of linear differential equations for unknown functions in a function space which correspond to generators of M.

Example 2.5. Let us consider $\mathbb{K}[x]$ as a left D_n-module. Since D_n contains $\mathbb{K}[x]$ as a subring, $\mathbb{K}[x]$ is generated by 1 as a left D_n-module. For $P \in D_n$, there exist $Q_1, \ldots, Q_n \in D_n$ and $r(x) \in \mathbb{K}[x]$ such that

$$P = Q_1 \partial_1 + \cdots + Q_n \partial_n + r(x).$$

Then the action of P on 1 is $P(1) = r(x)$, which vanishes if and only if $r(x) = 0$. This implies $\mathbb{K}[x] \cong D_n/(D_n\partial_1 + \cdots + D_n\partial_n)$ and a presentation of $\mathbb{K}[x]$ is given by

$$(D_n)^n \xrightarrow{\cdot {}^t(\partial_1, \ldots, \partial_n)} D_n \xrightarrow{\varphi} \mathbb{K}[x] \longrightarrow 0$$

with $\varphi(P) = P(1)$. In the same way we can show

$$\mathrm{Hom}_{D_n}(\mathbb{K}[x], \mathcal{F}) \cong \{f \in \mathcal{F} \mid \partial_1 f = \cdots = \partial_n f = 0\} = \mathbb{K}$$

for $\mathcal{F} = \mathbb{K}[x], \mathbb{K}(x), \mathbb{K}[x]_a$, or for $\mathcal{F} = C^\infty(U)$ with an open connected set U of $\mathbb{R}^n$ if $\mathbb{K} = \mathbb{C}$.

Exercise 4. Confirm the formulae above for $\varphi*$ and ψ^*.

Exercise 5. Construct a $\mathbb{C}$-isomorphism $\mathrm{Hom}_{D_n}(\mathbb{C}[x], C^\infty(U)) \cong \mathbb{C}$ for an open connected set U of $\mathbb{R}^n$, where D_n is the n-th Weyl algebra over $\mathbb{C}$. What happens if U is not connected?

2.3. Weight vector and filtration

A *weight vector* w for D_n is an integer vector

$$w = (w_1, \dots, w_n; w_{n+1}, \cdots, w_{2n}) \in \mathbb{Z}^{2n}$$

with the conditions $w_i + w_{n+i} \geq 0$ for $i = 1, \dots, n$, which are necessary in view of the commutation relation $\partial_i x_i = x_i \partial_i + 1$ in D_n. For a nonzero differential operator P of the form $P = \sum_{\alpha,\beta \in \mathbb{N}^n} a_{\alpha,\beta} x^\alpha \partial^\beta$, we define its *w-order* to be

$$\mathrm{ord}_w(P) = \max\{\langle w, (\alpha, \beta)\rangle \mid a_{\alpha,\beta} \neq 0\}$$

with

$$\langle w, (\alpha, \beta)\rangle := w_1\alpha_1 + \cdots + w_n\alpha_n + w_{n+1}\beta_1 + \cdots + w_{2n}\beta_n.$$

We set $\mathrm{ord}_w(0) := -\infty$. A weight vector w induces the *w-filtration*

$$F_k^w(D_n) := \{P \in D_n \mid \mathrm{ord}_w(P) \leq k\} \qquad (k \in \mathbb{Z})$$

on the ring D_n. In general, for two $\mathbb{K}$-subspaces V, W of D_n, we denote by VW the $\mathbb{K}$-subspace of D_n spanned by products PQ with $P \in V$ and $Q \in W$.

The w-filtration satisfies the properties:

$$F_k^w(D_n) \subset F_{k+1}^w(D_n), \quad \bigcup_{k\in\mathbb{Z}} F_k^w(D_n) = D_n,$$

$$1 \in F_0^w(D_n), \quad F_k^w(D_n)F_l^w(D_n) \subset F_{k+l}^w(D_n), \quad \bigcap_{k\in\mathbb{Z}} F_k^w(D_n) = \{0\}.$$

The *w-graded ring* associated with this filtration is defined to be

$$\mathrm{gr}^w(D_n) := \bigoplus_{k\in\mathbb{Z}} \mathrm{gr}_k^w(D_n), \qquad \mathrm{gr}_k^w(D_n) := F_k^w(D_n)/F_{k-1}^w(D_n).$$

Let P be a nonzero element of D_n with $m := \mathrm{ord}_w(P)$. Then we denote by $\sigma^w(P)$ the residue class of P in $\mathrm{gr}^w_k(D_n) \subset \mathrm{gr}^w(D_n)$. We set $\sigma^w(0) = 0$. It is easy to see that $\sigma^w(PQ) = \sigma^w(P)\sigma^w(Q)$ holds for any $P, Q \in D_n$ by using the Leibniz formula.

If $w_i + w_{n+i} > 0$, then the w-order of $\partial_i x_i - x_i \partial_i = 1$ is zero while that of $x_i \partial_i$ is positive. Hence $\sigma^w(x_i)$ and $\sigma^w(\partial_i)$ commute in $\mathrm{gr}^w(D_n)$. In this case, we denote $\sigma^w(x_i)$ and $\sigma^w(\partial_i)$ simply by x_i and ξ_i regarding these as commutative indeterminates.

On the other hand, if $w_i + w_{n+i} = 0$, then we have

$$\sigma^w(\partial_i)\sigma^w(x_i) - \sigma^w(x_i)\sigma^w(\partial_i) = 1$$

in $\mathrm{gr}^w(D_n)$, the same commutation relation as that for x_i and ∂_i in D_n. Hence we will denote $\sigma^w(x_i)$ and $\sigma^w(\partial_i)$ by x_i and ∂_i for simplicity.

Lemma 2.6. *Assume that $w_i \geq 0$ holds for $i = 1, \dots, 2n$, or else $|w_i| \leq 1$ and $w_i + w_{n+i} = 0$ hold for $i = 1, \dots, n$. Then $F^w_k(D_n)$ is a finitely generated left (and right) $F^w_0(D_n)$-module for each integer k.*

Proof. First, suppose $w_i \geq 0$ for all i. Then for any positive integer k, $F^w_k(D_n)$ is generated over $F^w_0(D_n)$ by the finite set

$$\{x^\alpha \partial^\beta \mid \langle w, (\alpha, \beta)\rangle = k,\ \alpha_i = 0 \text{ if } w_i = 0,\ \beta_i = 0 \text{ if } w_{n+i} = 0\}.$$

Now suppose $|w_i| \leq 1$ and $w_i + w_{n+i} = 0$ for $i = 1, \dots, n$. We may assume $w_i \geq 0$ for $1 \leq i \leq n$ by exchanging x_i and ∂_i if necessary. Each element of D_n is expressed as a linear combination of a finite set of 'monomials' of the form $x^\alpha \partial^\beta$. If $\langle (w_1, \dots, w_n), \alpha\rangle > \langle (w_1, \dots, w_n), \beta\rangle$, then there exists $\gamma \in \mathbb{N}^n$ such that $\alpha - \gamma \in \mathbb{N}^n$ and

$$\langle (w_1, \dots, w_n), \alpha - \gamma\rangle = \langle (w_1, \dots, w_n), \beta\rangle.$$

Then the w-order of $x^\alpha \partial^\beta = x^\gamma x^{\alpha-\gamma}\partial^\beta$ is $\langle (w_1, \dots, w_n), \gamma\rangle \geq 0$ and $x^{\alpha-\gamma}\partial^\beta$ belongs to $F^w_0(D_n)$. Hence $F^w_k(D_n)$ is generated by a finite set

$$\{x^\gamma \mid \langle (w_1, \dots, w_n), \gamma\rangle = k,\ \gamma_i = 0 \text{ if } w_i = 0\}$$

over $F^w_0(D_n)$ if $k > 0$. Likewise, $F^w_k(D_n)$ is generated by a finite set

$$\{\partial^\gamma \mid \langle (w_{n+1}, \dots, w_{2n}), \gamma\rangle = k,\ \gamma_i = 0 \text{ if } w_{n+i} = 0\}$$

over $F^w_0(D_n)$ if $k < 0$ since D_n is spanned by $\partial^\beta x^\alpha$. Q.E.D.

Lemma 2.7. *Assume $|w_i| \leq 1$ for $1 \leq i \leq 2n$. Then for integers j, k one has $F^w_j(D_n)F^w_k(D_n) = F^w_{j+k}(D_n)$ if $j \geq 0$, $k \geq 0$ or else $j \leq 0$, $k \leq 0$.*

Proof. The statement is easily shown if each component of w is 1 or 0. We can argue componentwise. Assume $w_i = -1$, and consequently $w_{n+i} = 1$. Suppose the w-order of $x_i^{\alpha_i}\partial_i^{\beta_i}$ is $j+k$ with $j,k \geq 0$. This means $\beta_i - \alpha_i = j + k$ and consequently $k \leq \beta_i$. Then ∂_i^k belongs to $F_k(D_n)$ and $x_i^{\alpha_i}\partial_i^{\beta_i - k}$ to $F_j(D_n)$. The case $j,k \leq 0$ is similar. Q.E.D.

Note that the lemma above does not hold in general without the assumption on w. For example, if $n=1$ and $w=(1;2)$, then ∂_1 belongs to $F_2(D)$ but does not belong to $F_1(D_1)F_1(D_1)$.

The *Rees algebra* $R^w(D_n)$ associated with the w-filtration is defined by

$$R^w(D_n) := \bigoplus_{k\in\mathbb{Z}} F_k^w(D_n)T^k \subset D_n[T]$$

with an indeterminate T. We have isomorphisms

$$(5) \quad R^w(D_n)/(T-1)R^w(D_n) \cong D_n, \quad R^w(D_n)/TR^w(D_n) \cong \mathrm{gr}^w(D_n)$$

as $\mathbb{K}$-algebra. Note that D_n, $\mathrm{gr}^w(D_n)$, and $R^w(D_n)$ are left (and right) Noetherian rings. This can be proved by using Gröbner bases which will be introduced in the next section.

Let M be a left D_n-module. A family $\{F_k(M)\}_{k\in\mathbb{Z}}$ of $\mathbb{K}$-subspaces $F_k(M)$ of M is called a *w-filtration* on M if it satisfies

(1) $F_k(M) \subset F_{k+1}(M)$ for all $k \in \mathbb{Z}$,

(2) $\bigcup_{k\in\mathbb{Z}} F_k(M) = M$,

(3) $F_j^w(D_n)F_k(M) \subset F_{j+k}(M)$ for all $j,k \in \mathbb{Z}$.

For a w-filtration $\{F_k(M)\}$, let

$$\mathrm{gr}(M) := \bigoplus_{k\in\mathbb{Z}} \mathrm{gr}_k(M), \qquad \mathrm{gr}_k(M) := F_k(M)/F_{k-1}(M)$$

be the associated graded module, which is a left $\mathrm{gr}^w(D_n)$-module.

Definition 2.8. A w-filtration $\{F_k(M)\}$ of a left D_n-module M is called *good* if there exist a finite number of elements $u_i \in F_{k_i}(M)$ and $k_i \in \mathbb{Z}$ $(i = 1,\dots,m)$ such that

$$F_k(M) = F_{k-k_1}^w(D_n)u_1 + \cdots + F_{k-k_m}^w(D_n)u_m \qquad (\forall k \in \mathbb{Z}).$$

It follows from the definition that a left D_n-module M has a good w-filtration if and only if M is finitely generated over D_n. The following lemma is also an immediate consequence of the definition:

Lemma 2.9. *Let $\{F_k(M)\}$ be a good w-filtration on a left D_n-module M. Let N be a left D_n-submodule of M. Define a w-filtration on M/N by*

$$F_k(M/N) := F_k(M)/(F_k(M) \cap N) \subset M/N.$$

Then $\{F_k(M/N)\}$ is a good w-filtration on M/N.

Lemma 2.10. *Let $\{F_k(M)\}$ and $\{F'_k(M)\}$ be w-filtrations on a left D_n-module M. Assume that $\{F_k(M)\}$ is good. Then there exists an integer l such that $F_k(M) \subset F'_{k+l}(M)$ for any $k \in \mathbb{Z}$.*

Proof. There exist $u_i \in F_{k_i}(M)$ such that

$$F_k(M) = F^w_{k-k_1}(D_n)u_1 + \cdots + F^w_{k-k_m}(D_n)u_m \qquad (\forall k \in \mathbb{Z}).$$

There exists an integer l such that each u_i belongs to $F'_l(M)$. Then we have

$$F_k(M) \subset F^w_{k-k_1}(D_n)F'_l(M) + \cdots + F^w_{k-k_m}(D_n)F'_l(M) \subset F'_{k-k_0+l}(M)$$

with $k_0 := \min\{k_1, \ldots, k_m\}$. Q.E.D.

Proposition 2.11. *Let $\{F_k(M)\}$ be a good w-filtration on a left D_n-module M. Then*

(1) *The associated graded module $\mathrm{gr}(M)$ is finitely generated over $\mathrm{gr}^w(D_n)$. In particular, each homogeneous component $\mathrm{gr}_k(M)$ is a finitely generated $\mathrm{gr}^w_0(D_n)$-module if w satisfies the assumption of Lemma 2.6.*

(2) *If $w_i \geq 0$ for all i, then $\{F_k(M)\}$ is bounded below; i.e., there exists $k_0 \in \mathbb{Z}$ such that $F_k(M) = \{0\}$ for any $k \leq k_0$.*

Proof. (1) By the assumption, there exist $u_1, \ldots, u_m \in M$ such that

$$F_k(M) = F^w_{k-k_1}(D_n)u_1 + \cdots + F^w_{k-k_m}(D_n)u_m \qquad (\forall k \in \mathbb{Z}). \tag{6}$$

Hence for any $u \in F_k(M) \setminus F_{k-1}(M)$, there exist $P_i \in F^w_{k-k_i}(D_n)$ such that

$$u = P_1u_1 + \cdots + P_mu_m.$$

Let $\overline{u}$ be the residue class of u in $\mathrm{gr}_k(M)$ and $\overline{u}_i$ be that of u_i in $\mathrm{gr}_{k_i}(M)$. Set $\overline{P}_i = \sigma^w(P_i)$ if $\mathrm{ord}_w(P_i) = k - k_i$, and $\overline{P}_i = 0$ otherwise. Then we have

$$\overline{u} = \overline{P}_1\overline{u}_1 + \cdots + \overline{P}_m\overline{u}_m$$

in $\mathrm{gr}(M)$. Hence $\mathrm{gr}(M)$ is generated by $\overline{u}_i$ $(1 \leq i \leq m)$ over $\mathrm{gr}^w(D_n)$.

(2) We have $F_k(M) = 0$ for $k < \min\{k_1, \ldots, k_m\}$ in view of (6) and $F_{-1}(D_n) = \{0\}$. Q.E.D.

Proposition 2.12. *Regard $L = (D_n)^m$ as a free left D_n-module. Fixing integers $l_1, \dots, l_m$, set*

$$F_k(L) := F^w_{k-l_1}(D_n)e_1 + \cdots + F^w_{k-l_m}(D_n)e_m \qquad (\forall k \in \mathbb{Z})$$

with $e_1 = (1, 0, \dots, 0)$, $\dots$, $e_m = (0, \dots, 0, 1) \in \mathbb{Z}^m$. Let N be a left D_n-submodule of L and assume $P_i = (P_{i1}, \dots, P_{im}) \in L$ $(i = 1, \dots, q)$ generate N and, at the same time, their residue classes $\overline{P}_1$, $\dots$, $\overline{P}_m$ in $\mathrm{gr}(L)$ *generate the graded submodule* $\mathrm{gr}(N)$ *of* $\mathrm{gr}(L)$, *which is associated with the induced filtration $\{F_k(L) \cap N\}$. Suppose $P_i \in F_{k_i}(L) \setminus F_{k_i-1}(L)$. Under these conditions,*

$$F_k(L) \cap N = F^w_{k-k_1}(D_n)P_1 + \cdots + F^w_{k-k_m}(D_n)P_m$$

holds for any $k \in \mathbb{Z}$. In particular, $\{F_k(L) \cap N\}$ is a good w-filtration on N. Moreover, if $w_i \geq 0$ for all i, the assumption that P_i generate N is not necessary.

Proof. This is standard in the case $w_i \geq 0$ for all i, which will suffice for the application in the next subsection. In fact, for any element P of $F_k(L) \cap N$, there exist $Q'_i \in F^w_{k-k_i}(D_n)$ such that

$$P - \sum_{i=1}^{q} Q'_i P_i \in F_{k-1}(L) \cap N$$

by the assumption. Then we can conclude by induction on k since $F_k(L) = \{0\}$ for sufficiently small k. For the general case we need the completion with respect to the filtration; see the proof of Theorem 10.6 in [27] for details. Q.E.D.

The following is an analogue of the Artin-Rees lemma in commutative algebra:

Proposition 2.13. *Let M be a finitely generated left D_n-module and $\{F_k(M)\}$ be a good w-filtration on M. Let N be a left D_n-submodule of M. Then the induced filtration $\{N \cap F_k(M)\}$ on N is good.*

Proof. By the assumption, there exist $u_1, \dots, u_m \in M$ such that

$$F_k(M) = F^w_{k-k_1}(D_n)u_1 + \cdots + F^w_{k-k_m}(D_n)u_m \qquad (\forall k \in \mathbb{Z}).$$

Set $L = (D_n)^m$ and define a D_n-homomorphism $\varphi : L \to M$ by

$$\varphi(A_1, \dots, A_m) = A_1u_1 + \cdots + A_mu_m \qquad (A_i \in D_n).$$

Define a w-filtration on L by

$$F_k(L) = \{(A_1, \dots, A_m) \in L \mid A_i \in F^w_{k-k_i}(D_n) \quad (1 \leq i \leq m)\}.$$

Then $\varphi(F_k(L)) = F_k(M)$ holds for any $k \in \mathbb{Z}$ by the construction. Now $N' := \varphi^{-1}(N)$ is a left D_n-submodule of L and finitely generated since D_n is Noetherian. Hence Proposition 2.12 (or else Theorem 3.14) assures the existence of $Q_1, \dots, Q_p \in N'$ and $l_1 \dots, l_p \in \mathbb{Z}$ such that

$$F_k(L) \cap N' = F^w_{k-l_1}(D_n)Q_1 + \cdots + F^w_{k-l_p}(D_n)Q_p \qquad (\forall k \in \mathbb{Z}).$$

Set $L' = (D_n)^p$ and define a D_n-homomorphism $\psi : L' \to N'$ by

$$\psi(B_1, \dots, B_p) = B_1Q_1 + \cdots + B_pQ_p \qquad (B_1, \dots, B_p \in D_n).$$

Define a w-filtration on L' by

$$F_k(L') = \{(B_1, \dots, B_p) \in L' \mid B_i \in F^w_{k-l_i}(D_n) \quad (1 \le i \le p)\}.$$

Then we have

$$(\varphi \circ \psi)(F_k(L')) = \varphi(F_k(L) \cap N') = F_k(M) \cap N.$$

In fact, if u belongs to $F_k(M) \cap N$, then there exists $Q \in F_k(L)$ such that $u = \varphi(Q)$ and consequently Q belongs to $F_k(L) \cap N'$. Thus $\{F_k(M) \cap N\}$ is a good w-filtration. Q.E.D.

Exercise 6. Let $w \in \mathbb{Z}^{2n}$ be a weight vector for D_n and set

$$d = \min\{w_i + w_{n+i} \mid 1 \le i \le n\}.$$

Suppose $P \in F^w_k(D_n)$ and $Q \in F^w_l(D_n)$ and show that the commutator $[P, Q] := PQ - QP$ belongs to $F^w_{k+l-d}(D_n)$.

Exercise 7. Set $n = 1$, $w = (-1; 1)$, and $M = D_1/I$ with the left ideal $I = D_1(x_1^2\partial_1 - 1)$ of D_1. Set

$$F_k(M) = F^w_k(D_1)/(F^w_k(D_1) \cap I) \qquad (k \in \mathbb{Z}).$$

(1) Show that $\{F_k(M)\}$ is a good w-filtration on M.
(2) Show that $F_k(M) = M$ for any $k \in \mathbb{Z}$ and that $\mathrm{gr}(M) = \{0\}$.

Exercise 8. Set $n = 1$ and regard $\mathbb{K}[x]$ as a left D_1-module. Define $F_k(\mathbb{K}[x]) = \{f \in \mathbb{K}[x] \mid \deg f \le 2k\}$ for $k \in \mathbb{Z}$. Then prove the following:

(1) $\{F_k(\mathbb{K}[x])\}$ is a $(1; 1)$-filtration on $\mathbb{K}[x]$.
(2) The associated graded module $\mathrm{gr}(\mathbb{K}[x])$ is not finitely generated over $\mathrm{gr}^{(1;1)}(D_1)$.
(3) $\{F_k(\mathbb{K}[x])\}$ is not a good $(1; 1)$-filtration, but it is a good $(2; 1)$-filtration.

Exercise 9. Prove the $\mathbb{K}$-algebra isomorphisms (5).

2.4. Holonomic D-module and characteristic variety

Following J. Bernstein [2], [3], let us define the notion of holonomic system by using the weight vector $(\mathbf{1};\mathbf{1}) = (1,\dots,1;1,\dots,1) \in \mathbb{Z}^{2n}$.

Let M be a finitely generated left D_n-module and $\{F_k(M)\}$ a good $(\mathbf{1};\mathbf{1})$-filtration on M. Note that $\mathrm{gr}^{(\mathbf{1};\mathbf{1})}(D_n)$ is isomorphic to the polynomial ring $\mathbb{K}[x,\xi]$ as a graded ring in which indeterminates $x_1, \dots, x_n, \xi_1, \dots, \xi_n$ are all of order one. By Proposition 2.11, $\mathrm{gr}(M)$ is a finitely generated graded $\mathbb{K}[x,\xi]$-module and each $\mathrm{gr}_k(M)$ is a finite dimensional $\mathbb{K}$-vector space. Moreover, $\mathrm{gr}^{(\mathbf{1};\mathbf{1})}(D_n)$ is generated by $\mathrm{gr}_1^{(\mathbf{1};\mathbf{1})}(D_n)$ as $\mathbb{K}$-algebra.

In this situation, it is well-known in commutative algebra (see e.g., [5], [8]) that there exist a (Hilbert) polynomial $H(k) = \sum_{j=0}^{d} c_j k^j \in \mathbb{Q}[k]$ and an integer k_0 such that

$$H(k) = \sum_{j \le k} \dim_{\mathbb{K}} \mathrm{gr}_j(M) = \dim_{\mathbb{K}} F_k(M) \qquad (\forall k \ge k_0)$$

and that $d! c_d$ is a positive integer.

Proposition 2.14. *The leading term $c_d k^d$ of $H(k)$ does not depend on the choice of a good $(\mathbf{1};\mathbf{1})$-filtration $\{F_k(M)\}$. Hence it is an invariant of a finitely generated left D_n-module M. The degree d of $H(k)$ is called the* dimension *of M and denoted* $\dim M$. *The* multiplicity *of M is defined to be the positive integer $d! c_d$ and denoted* $\mathrm{mult}\, M$.

Proof. Let $\{F_k(M)\}$ and $\{F'_k(M)\}$ be two good $(\mathbf{1};\mathbf{1})$-filtrations on M. There exist polynomials $H(k)$, $G(k)$, and an integer k_0 such that

$$\dim_{\mathbb{K}} F_k(M) = H(k), \quad \dim_{\mathbb{K}} F'_k(M) = G(x) \qquad (\forall k \ge k_0).$$

On the other hand, by Lemma 2.10, there exists a non-negative integer k_1 such that

$$F'_{k-k_1}(M) \subset F_k(M) \subset F'_{k+k_1}(M) \qquad (\forall k \in \mathbb{Z}).$$

Hence we have $G(k - k_1) \le H(k) \le G(k + k_1)$ for any $k \ge k_0$. This implies that the leading terms of $H(k)$ and of $G(k)$ coincide. Q.E.D.

Example 2.15. Since

$$\dim_{\mathbb{K}} F_k^{(\mathbf{1};\mathbf{1})}(D_n) = \binom{2n+k}{2n} = \frac{1}{(2n)!}k^{2n} + (\text{lower degree terms in } k),$$

the dimension of D_n as a left D_n module equals $2n$, and the multiplicity is one.

Theorem 2.16 (Bernstein's inequality). *If M is a finitely generated nonzero left D_n-module, then* $\dim M$ *is greater than or equal to n.*

Proof. We follow the argument in §30 of [12], which is based on a lemma by A. Joseph. Let $\{F_k(M)\}$ be a good $(\mathbf{1};\mathbf{1})$-filtration on M. We may assume $F_0(M) \neq \{0\}$ and $F_{-1}(M) = \{0\}$ by shifting k if necessary. Let us define a $\mathbb{K}$-homomorphism

$$\Psi_k : F_k^{(\mathbf{1};\mathbf{1})}(D_n) \ni P \longmapsto \Psi_k(P) \in \mathrm{Hom}_{\mathbb{K}}(F_k(M), F_{2k}(M)),$$

where $\Psi_k(P)$ denotes the natural $\mathbb{K}$-homomorphism $P : F_k(M) \to F_{2k}(M)$. Let us show that Ψ_k is injective by induction on k. First, Ψ_0 is injective since $F_0^{(\mathbf{1};\mathbf{1})}(D_n) = \mathbb{K}$. Now assume Ψ_j is injective if $j \leq k-1$. Let P be a nonzero element of $F_k^{(\mathbf{1};\mathbf{1})}(D_n)$. We may assume $P \notin \mathbb{K}$ since $\Psi_k(P) \neq 0$ otherwise. Then, we have $[P, \partial_i] \neq 0$ or $[P, x_i] \neq 0$ with some i. In fact, $[P, \partial_i] = P\partial_i - \partial_i P$ vanishes if and only if P does not contain x_i, and $[P, x_i]$ vanishes if and only if P does not contain ∂_i.

First assume $[P, \partial_i] \neq 0$. Since $[P, \partial_i]$ belongs to $F_{k-1}^{(\mathbf{1};\mathbf{1})}(D_n)$, there exists an element u of $F_{k-1}(M)$ such that $[P, \partial_i]u \neq 0$ by the induction hypothesis. Hence either $P\partial_i u \neq 0$ or $Pu \neq 0$ holds. Since u and $\partial_i u$ belong to $F_k(M)$, this shows $\Psi_k(P) \neq 0$.

The case $[P, x_i] \neq 0$ can be treated similarly with ∂_i replaced by x_i in the argument above. Thus we have proved that Ψ_k is injective for $k \geq 0$. From this we obtain

$$\begin{aligned}\dim_{\mathbb{K}} F_k^{(\mathbf{1};\mathbf{1})}(D_n) &\leq \dim_{\mathbb{K}} \mathrm{Hom}_{\mathbb{K}}(F_k(M), F_{2k}(M)) \\ &= (\dim_{\mathbb{K}} F_k(M))(\dim_{\mathbb{K}} F_{2k}(M)).\end{aligned}$$

There exists a polynomial $H(k)$ such that $H(k) = \dim_{\mathbb{K}} F_k(M)$ for sufficiently large k. Thus we have

$$H(k)H(2k) \geq \dim_{\mathbb{K}} F_k^{(\mathbf{1};\mathbf{1})}(D_n) = \binom{2n+k}{2n} \qquad (\forall k \gg 0).$$

Comparing the degrees in k, we get $2\deg H(k) \geq 2n$, consequently $\dim M = \deg H(k) \geq n$. Q.E.D.

Definition 2.17. A finitely generated left D_n-module M is called *holonomic* or a *holonomic system* if $\dim M \leq n$, that is if $\dim M = n$ or else $M = 0$.

Example 2.18. Let us show that $\mathbb{K}[x]$ is a holonomic left D_n-module. It is easy to see that

$$F_k(\mathbb{K}[x]) = \{f \in \mathbb{K}[x] \mid \deg f \leq k\} \qquad (k \in \mathbb{Z})$$

constitute a $(\mathbf{1};\mathbf{1})$-filtration on $\mathbb{K}[x]$. Moreover, it is a good filtration since

$$F_k(\mathbb{K}[x]) = \sum_{|\alpha|\le k} \mathbb{K}x^\alpha = F_k^{(\mathbf{1};\mathbf{1})}(D_n)1.$$

It follows that $\dim \mathbb{K}[x] = n$ since

$$\dim_{\mathbb{K}} F_k(\mathbb{K}[x]) = \binom{n+k}{n} = \frac{1}{n!}k^n + (\text{lower order terms in } k).$$

Proposition 2.19. *Let*

$$0 \longrightarrow N \xrightarrow{\varphi} M \xrightarrow{\psi} L \longrightarrow 0$$

be an exact sequence of finitely generated left D_n-modules. Then

(1) *M is holonomic if and only if both N and L are holonomic.*
(2) *If M is holonomic, then* $\operatorname{mult} M = \operatorname{mult} N + \operatorname{mult} L$ *holds, where we define the multiplicity of the zero module to be zero.*

Proof. Let $\{F_k(M)\}$ be a good $(\mathbf{1};\mathbf{1})$-filtration on M and set

$$F_k(N) := \varphi^{-1}(F_k(M)), \qquad F_k(L) := \psi(F_k(M)).$$

Then $\{F_k(N)\}$ is a good filtration on N by Proposition 2.13 and $\{F_k(L)\}$ is a good filtration on L by Lemma 2.9. Hence the assertions follow from

$$\dim_{\mathbb{K}} F_k(M) = \dim_{\mathbb{K}} F_k(N) + \dim_{\mathbb{K}} F_k(L).$$

Q.E.D.

Let us recall another characterization of a holonomic system by using the weight vector $w = (\mathbf{0};\mathbf{1}) = (0,\dots,0;1,\dots,1)$. Let M be a finitely generated left D_n-module and $\{F_k(M)\}$ be a good $(\mathbf{0};\mathbf{1})$-filtration on M. Then $\operatorname{gr}(M)$ is a finitely generated $\mathbb{K}[x,\xi]$-module. Let us denote by $\overline{\mathbb{K}}$ the algebraic closure of $\mathbb{K}$. In general, for a finitely generated $\mathbb{K}[x,\xi]$-module M', its *support* is the algebraic set of $\overline{\mathbb{K}}^{2n}$ defined by

$$\operatorname{Supp} M' := \{(a,b) \in \overline{\mathbb{K}}^n \times \overline{\mathbb{K}}^n \mid M'_{(a,b)} := \overline{\mathbb{K}}[x,\xi]_{(a,b)} \otimes_{\mathbb{K}[x,\xi]} M' \neq 0\},$$

where $\overline{\mathbb{K}}[x,\xi]_{(a,b)}$ denotes the localization of $\overline{\mathbb{K}}[x,\xi]$ at (a,b), i.e., the localization at the maximal ideal corresponding to the point (a,b).

Proposition 2.20. *The support of* $\operatorname{gr}(M)$ *does not depend on the choice of a good $(\mathbf{0};\mathbf{1})$-filtration $\{F_k(M)\}$ on M.*

Proof. We follow the argument of Kashiwara [14]. Let $\{F_k(M)\}$ and $\{F'_k(M)\}$ be good $(\mathbf{0};\mathbf{1})$-filtrations on M and $\mathrm{gr}(M)$ and $\mathrm{gr}'(M)$ be the associated graded modules respectively. Then by Lemma 2.10 we may assume that there exists an integer $k_1 \geq 0$ such that

$$F_{k-k_1}(M) \subset F'_k(M) \subset F_k(M) \qquad (\forall k \in \mathbb{Z})$$

by shifting the index of $F'_k(M)$ if necessary.

Let us argue by induction on k_1. The case $k_1 = 0$ is trivial. Suppose $k_1 = 1$ and consider the following two exact sequences

$$0 \longrightarrow F'_k(M)/F_{k-1}(M) \to F_k(M)/F_{k-1}(M) \to F_k(M)/F'_k(M) \to 0,$$
$$0 \to F_{k-1}(M)/F'_{k-1}(M) \to F'_k(M)/F'_{k-1}(M) \to F'_k(M)/F_{k-1}(M) \to 0.$$

It follows that

$$\begin{aligned} \mathrm{Supp}\,\mathrm{gr}(M) &= \mathrm{Supp} \bigoplus_{k\in\mathbb{Z}} F'_k(M)/F_{k-1}(M) \cup \mathrm{Supp} \bigoplus_{k\in\mathbb{Z}} F_k(M)/F'_k(M), \\ \mathrm{Supp}\,\mathrm{gr}'(M) &= \mathrm{Supp} \bigoplus_{k\in\mathbb{Z}} F'_k(M)/F_{k-1}(M) \\ &\quad \cup \mathrm{Supp} \bigoplus_{k\in\mathbb{Z}} F_{k-1}(M)/F'_{k-1}(M) \end{aligned}$$

since $\overline{\mathbb{K}}[x,\xi]_{(a,b)}$ is a flat module over $\mathbb{K}[x,\xi]$. Hence $\mathrm{Supp}\,\mathrm{gr}(M)$ and $\mathrm{Supp}\,\mathrm{gr}'(M)$ coincide.

Now suppose $k_1 \geq 2$ and set

$$F''_k(M) = F_{k-1}(M) + F'_k(M) \qquad (k \in \mathbb{Z}).$$

Let $\mathrm{gr}''(M)$ be the graded $\mathbb{K}[x,\xi]$-module associated with the good filtration $\{F''_k(M)\}$. It follows from the definition

$$F_{k-1}(M) \subset F''_k(M) \subset F_k(M), \quad F''_{k-k_1+1}(M) \subset F'_k(M) \subset F''_k(M)$$

for any $k \in \mathbb{Z}$. By the induction hypothesis, we have

$$\mathrm{Supp}\,\mathrm{gr}(M) = \mathrm{Supp}\,\mathrm{gr}''(M) = \mathrm{Supp}\,\mathrm{gr}'(M).$$

Q.E.D.

Definition 2.21. Let M be a finitely generated left D_n-module and $\{F_k(M)\}$ be a good $(\mathbf{0};\mathbf{1})$-filtration on M. Then the *characteristic variety* $\mathrm{Char}(M)$ of M is defined to be the support $\mathrm{Supp}\,\mathrm{gr}(M)$ of the graded module $\mathrm{gr}(M)$ associated with $\{F_k(M)\}$.

Since $\mathrm{gr}(M)$ is a graded $\mathbb{K}[x,\xi]$-module with $x_1,\dots,x_n$ of order zero, and $\xi_1,\dots,\xi_n$ of order one, $\mathrm{Char}(M)$ is a homogeneous set with respect to ξ; i.e, if (a,b) belongs to $\mathrm{Char}(M)$, then so does (a,cb) for any $c\in\overline{\mathbb{K}}$.

The following theorem is proved, e.g., in Chapter 3 of [4] by using a homological method based on Auslander-Buchsbaum-Serre theorem (cf. [5], [8]). We will give a more elementary proof in 3.4.

Theorem 2.22. *Let M be a finitely generated left D_n-module. Then the dimension* $\dim M$ *defined through the $(\mathbf{1};\mathbf{1})$-filtration coincides with the Krull dimension (not as a graded module) of the $\mathbb{K}[x,\xi]$-module* $\mathrm{gr}(M)$ *associated with a good $(\mathbf{0};\mathbf{1})$-filtration on M.*

Especially, M is holonomic if and only if the dimension of the characteristic variety is n or else $M=0$. More strongly, it is known that the dimension of each irreducible component of the characteristic variety is of dimension $\geq n$. This fact was first proved by Sato-Kawai-Kashiwara [31] in the analytic category, and by Gabber [9] in a purely algebraic setting. See [33] for extension to general weight vectors.

Example 2.23. Let us regard $\mathbb{K}[x]$ as a left D_n-module. Then

$$F_k(\mathbb{K}[x]) = F_k^{(\mathbf{0};\mathbf{1})}(D_n)1 \qquad (k\in\mathbb{Z})$$

constitute a good $(\mathbf{0};\mathbf{1})$-filtration on $\mathbb{K}[x]$. It is easy to see that $F_k(\mathbb{K}[x]) = \mathbb{K}[x]$ if $k\geq 0$, and $F_k(\mathbb{K}[x])=\{0\}$ if $k\leq -1$. Hence the associated graded module is

$$\mathrm{gr}(\mathbb{K}[x]) = \bigoplus_{k\in\mathbb{Z}} F_k(\mathbb{K}[x])/F_{k-1}(\mathbb{K}[x]) = \mathbb{K}[x].$$

As a $\mathbb{K}[x,\xi]$-module, $\mathbb{K}[x]$ is isomorphic to $\mathbb{K}[x,\xi]/\langle\xi_1,\dots,\xi_n\rangle$, where $\langle\xi_1,\dots,\xi_n\rangle$ denotes the ideal of $\mathbb{K}[x,\xi]$ generated by $\xi_1,\dots,\xi_n$. Hence we get

$$\mathrm{Char}(M) = \{(x,\xi)\in\overline{\mathbb{K}}^{2n} \mid \xi_1=\dots=\xi_n=0\}.$$

Exercise 10. Set $M=D_n/D_n\partial_1^m$ with a positive integer m and the coefficient field $\mathbb{K}=\mathbb{C}$.

(1) Give a presentation of the graded module $\mathrm{gr}(M)$ associated with the good $(\mathbf{1};\mathbf{1})$-filtration

$$F_k(M) = F_k^{(\mathbf{1};\mathbf{1})}(D_n)/(F_k^{(\mathbf{1};\mathbf{1})}(D_n)\cap D_n\partial_1^m)$$

and compute $\dim M$.

(2) Give a presentation of the graded module $\mathrm{gr}(M)$ associated with the good $(\mathbf{0};\mathbf{1})$-filtration

$$F_k(M) = F_k^{(\mathbf{0};\mathbf{1})}(D_n)/(F_k^{(\mathbf{0};\mathbf{1})}(D_n) \cap D_n\partial_1^m)$$

and compute $\mathrm{Char}M$.

§3. Gröbner bases in the ring of differential operators

In this section, we quickly review the theory of Gröbner bases over the Weyl algebra. In D-module theory, one often needs a Gröbner basis with respect to a monomial order which is not a well-ordering; for this we need homogenization technique. A good reference is the first chapter of [29]. See also [23], [27].

3.1. Definitions and basic properties

Recall that $\xi = (\xi_1, \ldots, \xi_n)$ are the commutative variables corresponding to derivations $\partial_i = \partial_{x_i}$ $(i = 1, \ldots, n)$. Let

$$M(x, \xi) = \{x^\alpha \xi^\beta \mid \alpha, \beta \in \mathbb{N}^n\}$$

be the set of the monomials in $\mathbb{K}[x, \xi]$. A total order $\prec$ on $M(x, \xi)$ is called a *monomial order for D_n* if it satisfies

(1) $u \prec v \quad \Rightarrow \quad uw \prec vw \quad (\forall u, v, w \in M(x, \xi))$,
(2) $1 \prec x_i\xi_i$ for any $i = 1, \ldots, n$.

A monomial order $\prec$ is called a *term order* if

(3) $1 \prec x^\alpha \xi^\beta$ for any $(\alpha, \beta) \in \mathbb{N}^{2n} \setminus \{(\mathbf{0}, \mathbf{0})\}$.

This is equivalent to the condition that the monomial order $\prec$ be a well-ordering.

Now fix a monomial order $\prec$. For a nonzero element

$$P = \sum_{\alpha, \beta} a_{\alpha\beta} x^\alpha \partial^\beta \qquad (a_{\alpha,\beta} \in \mathbb{K})$$

of D_n, its initial monomial $\mathrm{in}_\prec(P)$ is defined to be the maximum nonzero monomial

$$\mathrm{in}_\prec(P) = \max{}_\prec\{x^\alpha \xi^\beta \mid a_{\alpha\beta} \neq 0\}$$

of $P(x, \xi)$ with respect to $\prec$. Note that $\mathrm{in}_\prec(P)$ belongs to $\mathbb{K}[x, \xi]$ instead of D_n so that monomial ideals make sense.

By using the Leibniz formula and the conditions (1) and (2), we can verify that $\mathrm{in}_\prec(PQ) = \mathrm{in}_\prec(P)\mathrm{in}_\prec(Q) = \mathrm{in}_\prec(QP)$ holds in $\mathbb{K}[x, \xi]$ for nonzero $P, Q \in D_n$.

Definition 3.1. Let I be a left ideal of D_n. A finite subset G of $I \setminus \{0\}$ is called a *Gröbner basis* of I with respect to a monomial order $\prec$ if

(1) G generates I as a left ideal;
(2) $\mathrm{in}_\prec(G) := \{\mathrm{in}_\prec(P) \mid P \in G\}$ generates the monomial ideal $\mathrm{in}_\prec(I)$ in $\mathbb{K}[x, \xi]$ which is generated by the set $\{\mathrm{in}_\prec(P) \mid P \in I,\ P \neq 0\}$.

First, let us recall

Lemma 3.2 (Dickson). *Every monomial ideal (i.e., an ideal generated by monomials) of* $\mathbb{K}[x, \xi]$ *is finitely generated.*

See e.g., 2.4 of [6] for the proof.

Proposition 3.3. *For any left ideal* I *of* D_n*, and any monomial order* $\prec$*, there exists a Gröbner basis* G *of* I *with respect to* $\prec$*. In particular,* D_n *is left Noetherian.*

Proof. Let G be a finite generating set of I. Since $\mathrm{in}_\prec(I)$ is a monomial ideal of $\mathbb{K}[x, \xi]$, there exists a finite set G' of I such that $\{\mathrm{in}_\prec(P) \mid P \in G'\}$ generates $\mathrm{in}_\prec(I)$ by Lemma 3.2. Then $G \cup G'$ is a Gröbner basis of I with respect to $\prec$. Q.E.D.

For a term order, we can compute a Gröbner basis of I by using division and Buchberger's criterion applied to D_n.

Now let $w \in \mathbb{Z}^{2n}$ be a weight vector for D_n (see 2.3). A monomial order $\prec$ on $M(x, \xi)$ is *adapted* to w if

$$x^\alpha \xi^\beta \prec x^{\alpha'} \xi^{\beta'} \ \Rightarrow\ \langle w, (\alpha, \beta) \rangle \leq \langle w, (\alpha', \beta') \rangle.$$

There exists a term order that is adapted to w if and only if $w_i \geq 0$ for any $i = 1, \ldots, n$.

For an arbitrary monomial order $\prec$ for D_n, define another monomial order $\prec_w$ by

$$\begin{aligned} x^\alpha \xi^\beta \prec_w x^{\alpha'} \xi^{\beta'} \ &\Leftrightarrow\ \langle w, (\alpha, \beta) \rangle < \langle w, (\alpha', \beta') \rangle \\ &\quad \text{or } (\langle w, (\alpha, \beta) \rangle = \langle w, (\alpha', \beta') \rangle \text{ and } x^\alpha \xi^\beta \prec x^{\alpha'} \xi^{\beta'}). \end{aligned}$$

Then $\prec_w$ is adapted to w.

Recall that the residue class in $\mathrm{gr}_k^w(D_n)$ of $P \in F_k(D_n) \setminus F_{k-1}(D_n)$ is denoted by $\sigma^w(P)$ (it is denoted by $\mathrm{in}_w(P)$ in [29]). For a nonzero element P of $\mathrm{gr}^w(D_n)$ and a monomial order $\prec$ for D_n, the initial monomial $\mathrm{in}_\prec(P)$ is defined as a monomial in $\mathbb{K}[x, \xi]$.

The following is an immediate consequence of the definitions.

Lemma 3.4. *If a monomial order $\prec$ for D_n is adapted to a weight vector w for D_n, then one has $\mathrm{in}_\prec(\sigma^w(P)) = \mathrm{in}_\prec(P)$ for any nonzero element P of D_n.*

The following proposition enables us to look at w-filtrations of left D_n-modules from a computational viewpoint. Note that the weight vector w may have negative components and hence the monomial order $\prec_w$ may not be a term order.

Proposition 3.5. *Let w be a weight vector for D_n, $\prec$ be a term order, and I be a left ideal of D_n. Let G be a Gröbner basis of I with respect to $\prec_w$. Then $\mathrm{gr}^w(G) := \{\sigma^w(P) \mid P \in G\}$ generates over $\mathrm{gr}^w(D_n)$ the graded ideal*

$$\mathrm{gr}^w(I) := \bigoplus_{k\in\mathbb{Z}} (I \cap F_k^w(D_n))/(I \cap F_{k-1}^w(D_n))$$

associated with the induced filtration $\{F_k^w(D_n) \cap I\}$ on I.

Proof. Set $G = \{P_1, \ldots, P_r\}$. We denote by $\langle \sigma^w(G) \rangle$ the left ideal of $\mathrm{gr}^w(D_n)$ generated by $\sigma^w(P_1), \ldots, \sigma^w(P_r)$. Let P be a nonzero element of I. Let m be the w-order of P. We have only to show that $\sigma^w(P)$ belongs to $\langle \sigma^w(G) \rangle$.

By the assumption, the monomial $\mathrm{in}_\prec(\sigma^w(P)) = \mathrm{in}_{\prec_w}(P)$ belongs to the monomial ideal $\langle \mathrm{in}_{\prec_w}(G) \rangle$ generated by $\mathrm{in}_{\prec_w}(G)$. Hence there exist $Q_1 \in D_n$ whose total symbol is a monomial, and $i_1 \in \{1, \ldots, r\}$ such that

$$\mathrm{in}_\prec(\sigma^w(P)) = \mathrm{in}_{\prec_w}(Q_1)\mathrm{in}_{\prec_w}(P_{i_1}) = \mathrm{in}_{\prec_w}(Q_1 P_{i_1}).$$

In particular, the w-order of $R_1 := P - Q_1 P_{i_1}$ is $\leq m$. If $\mathrm{ord}_w(R_1) < m$, then $\sigma^w(P) = \sigma^w(Q_1)\sigma^w(P_{i_1})$ belongs to $\langle \mathrm{gr}^w(G) \rangle$ and we are done.

Assume $\mathrm{ord}_w(R_1) = m$. Then we have

$$\sigma^w(R_1) = \sigma^w(P) - \sigma^w(Q_1)\sigma^w(P_{i_1}), \qquad \mathrm{in}_{\prec_w}(R_1) \prec \mathrm{in}_{\prec_w}(P)$$

and R_1 belongs to I. Since the order $\prec_w$ restricted to $\{(\alpha, \beta) \in \mathbb{N}^{2n} \mid \langle w, (\alpha, \beta) \rangle = m\}$ coincides with $\prec$, which is a well-order, this process terminates and we obtain finite number of operators $Q_1, \ldots, Q_l$ and $i_1, \ldots, i_l \in \{1, \ldots, r\}$ such that

$$R_l := P - \sum_{j=1}^{l} Q_j P_{i_j} \in F_{m-1}^w(D_n), \qquad \mathrm{ord}_w(Q_j) + \mathrm{ord}_w(P_{i_j}) = m$$

for $1 \leq j \leq l$. This implies $\sigma^w(P) = \sum_{j=1}^{l} \sigma^w(Q_j)\sigma^w(P_{i_j})$ belongs to $\langle \sigma^w(G) \rangle$. Q.E.D.

Definition 3.6. Let I be a left ideal of D_n. A finite subset G of $I \setminus \{0\}$ is called a *w-involutive basis* of I if the following two conditions hold:

(1) G generates I over D_n.
(2) $\sigma^w(G) := \{\sigma^w(P) \mid P \in G\}$ generates $\mathrm{gr}^w(I)$ over $\mathrm{gr}^w(D_n)$.

Theorem 3.7. *Let I be a left ideal of D_n and set $M = D_n/I$. Let $\prec$ be a term order for D_n and w a weight vector for D_n. Suppose that $G = \{P_1, \dots, P_m\}$ is a Gröbner basis of I with respect to $\prec_w$ and set $k_i = \mathrm{ord}_w(P_i)$. Then*

(1) *G is a w-involutive basis of I.*
(2) *Let $\varphi(P)$ be the residue class of $P \in D_n$ in M and $\psi : (D_n)^m \to D_n$ be the D_n-homomorphism defined by*

$$\psi(A_1, \dots, A_m) = A_1 P_1 + \cdots + A_m P_m \qquad (A_1, \dots, A_m \in D_n).$$

Then the exact sequence

$$(D_n)^m \xrightarrow{\psi} D_n \xrightarrow{\varphi} M \longrightarrow 0$$

induces, for each $k \in \mathbb{Z}$, the exact sequence

$$\bigoplus_{i=1}^m F^w_{k-k_i}(D_n) \xrightarrow{\psi_k} F^w_k(D_n) \xrightarrow{\varphi_k} F_k(M) \longrightarrow 0$$

with $F_k(M) = F^w_k(D_n)/(F^w_k(D_n) \cap I)$.

Proof. (1) is an immediate consequence of Proposition 3.5.

(2) It follows from the definition that φ_k is surjective. Applying Proposition 2.12 to I and G, we have

$$F_k(I) := F^w_k(D_n) \cap I = F^w_{k-k_1}(D_n)P_1 + \cdots + F^w_{k-k_m}(D_n)P_m \qquad (\forall k \in \mathbb{Z}).$$

This completes the proof since $\ker \varphi_k = I \cap F^w_k(D_n) = F_k(I)$. Q.E.D.

We can dispense with Proposition 2.12 if G is obtained by the homogenization introduced in the next subsection (see Theorem 3.14).

Exercise 11. Set $I = D_n\partial_1 + \cdots + D_n\partial_n$ and w be an arbitrary weight vector for w. Show that $G := \{\partial_1, \dots, \partial_n\}$ is a w-involutive basis of I.

Exercise 12. Set $n = 2$ and $P_1 = \partial_1$, $P_2 = \partial_1^2 + \partial_2$. Let I be the left ideal of D_2 generated by P_1 and P_2. Let $w = (w_1, w_2; w_3, w_4)$ be a weight vector for D_2. Show that $\{P_1, P_2\}$ is a w-involutive basis of I if and only if $2w_3 < w_4$.

3.2. Homogenization trick

For a monomial order $\prec$ in which 1 is not the smallest element, the division algorithm cannot be performed directly. To bypass this difficulty, we introduce the $(\mathbf{1};\mathbf{1})$-homogenized ring. First, recall the Rees algebra

$$R^{(\mathbf{1};\mathbf{1})}(D_n) = \bigoplus_{k\in\mathbb{Z}} F_k^{(\mathbf{1};\mathbf{1})}(D_n)T^k$$

of D_n with respect to the $(\mathbf{1};\mathbf{1})$-filtration.

Let $D_n^{(h)}$ be the $\mathbb{K}$-vector space with the basis $\{x^\alpha\partial^\beta h^k \mid \alpha,\beta\in\mathbb{N}^n, k\in\mathbb{N}\}$, where h is a new indeterminate. Define a $\mathbb{K}$-isomorphism $\Psi : R^{(\mathbf{1};\mathbf{1})}(D_n)\to D_n^{(h)}$ by

$$\Psi(x^\alpha\partial^\beta T^k) = x^\alpha\partial^\beta h^{k-|\alpha|-|\beta|}.$$

Note that $x^\alpha\partial^\beta T^k\in R^{(\mathbf{1};\mathbf{1})}(D_n)$ means $|\alpha|+|\beta|\le k$.

We can make $D_n^{(h)}$ a graded $\mathbb{K}$-algebra by using the graded $\mathbb{K}$-algebra structure of $R^{(\mathbf{1};\mathbf{1})}(D_n)$ via Ψ. Let us call this $D^{(h)}$ the *homogenized Weyl algebra*, which was introduced, in connection with Gröbner bases, by Takayama and Assi-Castro-Granger [1] independently. In fact, $D^{(h)}$ was implemented by Takayama in his computer algebra system Kan [34] as early as 1994.

The image of $F_k^{(\mathbf{1};\mathbf{1})}(D_n)T^k$ by Ψ consists of the elements of $D_n^{(h)}$ which are homogeneous of degree k in x,∂,h. For an element P of D_n, we set

$$P^{(h)} := \Psi(PT^k) \quad \text{with } k := \mathrm{ord}_{(\mathbf{1};\mathbf{1})}P,$$

which is called the ($(\mathbf{1};\mathbf{1})$-) homogenization of P. For example, since $\partial_i x_j T^2 = (x_i\partial_j+\delta_{ij})T^2$ holds in $R^{(\mathbf{1};\mathbf{1})}(D_n)$, we have

$$\partial_i x_j = \Psi(\partial_i x_j T^2) = \Psi(x_i\partial_j T^2) + \delta_{ij}\Psi(T^2) = x_i\partial_j + \delta_{ij}h^2$$

in $D_n^{(h)}$. Conversely, the dehomogenization (substituting 1 for h) $D_n^{(h)}\ni P\mapsto P|_{h=1}\in D_n$ defines a ring homomorphism so that $(P^{(h)})|_{h=1}=P$ holds for $P\in D_n$.

For elements P,Q of $D_n^{(h)}$, let $P(x,\xi,h)$ and $Q(x,\xi,h)$ be their total symbols defined in a similar manner as in D_n. Then the total symbol of $R:=PQ$ is given by

$$R(x,\xi,h) = \sum_{\nu\in\mathbb{N}^n}\frac{h^{2\nu}}{\nu!}\left(\frac{\partial}{\partial\xi}\right)^\nu P(x,\xi,h)\cdot\left(\frac{\partial}{\partial x}\right)^\nu Q(x,\xi,h).$$

Definition 3.8. An order $\prec$ on $M(x,\xi,h)=\{x^\alpha\xi^\beta h^j \mid \alpha,\beta\in\mathbb{N}^n, j\in\mathbb{N}\}$ is called a *monomial order for* $D_n^{(h)}$ if

(1) $u \prec v \quad \Rightarrow \quad uw \prec vw \quad (\forall u, v, w \in M(x, \xi, h))$
(2) $h^2 \prec x_i \xi_i$ for any $i = 1, \dots, n$.

A monomial order $\prec$ is called a *term order* if $1 \preceq x^\alpha \xi^\beta h^j$ for any $\alpha, \beta \in \mathbb{N}$ and $j \in \mathbb{N}$.

Definition 3.9. Let $P = \sum_{\alpha,\beta,j} c_{\alpha,\beta,j} x^\alpha \partial^\beta h^j$ be a nonzero element of $D_n^{(h)}$ and $\prec$ a monomial order for $D_n^{(h)}$. Then the *initial monomial* $\mathrm{in}_\prec(P)$ of P is the monomial $x^{\alpha_0}\xi^{\beta_0}h^{j_0}$ such that

$$(\alpha_0, \beta_0, j_0) = \max{}_\prec \{(\alpha, \beta, j) \in \mathbb{N}^{2n+1} \mid c_{\alpha,\beta,j} \neq 0\}.$$

The leading coefficient $\mathrm{LC}_\prec(P)$ and the leading term $\mathrm{LT}_\prec(P)$ are defined to be c_{α_0,β_0,j_0} and $c_{\alpha_0,\beta_0,j_0} x^{\alpha_0}\partial^{\beta_0}h^{j_0}$ respectively. Note that $\mathrm{LT}_\prec(P)$ belongs to $D_n^{(h)}$ while $\mathrm{in}_\prec(P)$ belongs to $\mathbb{K}[x, \xi, h]$.

Definition 3.10. Let J be a left ideal of $D_n^{(h)}$. A finite subset G of $J \setminus \{0\}$ is called a *Gröbner basis* of J with respect to a monomial order $\prec$ if

(1) G generates J as a left ideal;
(2) $\mathrm{in}_\prec(G) := \{\mathrm{in}_\prec(P) \mid P \in G\}$ generates the monomial ideal $\mathrm{in}_\prec(J)$ in $\mathbb{K}[x, \xi, h]$ which is generated by the set $\{\mathrm{in}_\prec(P) \mid P \in J,\ P \neq 0\}$.

Proposition 3.11 (division). *Let $G = \{P_1, \dots, P_m\}$ be a finite set of nonzero elements of $D_n^{(h)}$ and $\prec$ be a term order for $D_n^{(h)}$. Then for any $P \in D_n^{(h)}$, there exist $Q_1, \dots, Q_m, R \in D_n^{(h)}$ such that*

$$P = Q_1P_1 + \cdots + Q_mP_m + R, \qquad \mathrm{in}_\prec(Q_jP_j) \preceq \mathrm{in}_\prec(P) \text{ if } Q_j \neq 0$$

and that $\mathrm{in}_\prec(R)$ is not divisible by $\mathrm{in}_\prec(P_j)$ for $1 \leq j \leq m$. Moreover, if G is a Gröbner basis of the left ideal J generated by G with respect to $\prec$, then $R = 0$ if and only if P belongs to J.

Proof. The existence of Q_j and R can be proved by induction in the well-order $\prec$, in the same way as in the polynomial ring. Suppose that G is a Gröbner basis with respect to $\prec$ and that P belongs to J. If $R \neq 0$, then $\mathrm{in}_\prec(R)$ must be divisible by $\mathrm{in}_\prec(P_j)$ for some j since R belongs to J. This contradicts the assumption. Hence we have $R = 0$. Q.E.D.

Definition 3.12. Let $\prec$ be a term order for $D_n^{(h)}$. For nonzero $P, Q \in D_n^{(h)}$, write $\mathrm{LT}_\prec(P) = ax^\alpha\partial^\beta h^j$ and $\mathrm{LT}_\prec(Q) = bx^{\alpha'}\partial^{\beta'}h^k$ with $a, b \in \mathbb{K} \setminus \{0\}$. Set

$$\begin{aligned}(\alpha'', \beta'', l) &= (\alpha, \beta, j) \wedge (\alpha', \beta', k) \\ &:= (\min\{\alpha_1, \alpha_1'\}, \dots, \min\{\beta_n, \beta_n'\}, \min\{j, k\}).\end{aligned}$$

Then the *S-pair* of P and Q with respect to $\prec$ is an element of $D_n^{(h)}$ defined by

$$\mathrm{sp}_\prec(P, Q) = bx^{\alpha'-\alpha''}\partial^{\beta'-\beta''}h^{k-l}P - ax^{\alpha-\alpha''}\partial^{\beta-\beta''}h^{j-l}Q.$$

Theorem 3.13 (Buchberger's criterion in $D_n^{(h)}$). *Let J be a left ideal of $D_n^{(h)}$ and $\prec$ be a term order for $D_n^{(h)}$. Let $G = \{P_1, \dots, P_m\}$ be a finite subset of $J \setminus \{0\}$ which generates J. Then the following two conditions are equivalent:*

(1) *G is a Gröbner basis of J with respect to $\prec$.*

(2) *If $\mathrm{sp}_\prec(P_i, P_j) \neq 0$ for $1 \leq i < j \leq m$, then there exist $Q_{ijk} \in D_n^{(h)}$ such that*

$$\mathrm{sp}_\prec(P_i, P_j) = Q_{ij1}P_1 + \cdots + Q_{ijm}P_m,$$
$$\mathrm{in}_\prec(Q_{ijk}P_k) \preceq \mathrm{in}_\prec(\mathrm{sp}_\prec(P_i, P_j)) \text{ if } Q_{ijk} \neq 0 \quad (1 \leq k \leq m).$$

Proof. We see that (1) implies (2) by division. Assume (2) and let P be a nonzero element of I. We have only to show that $\mathrm{in}_\prec(P)$ belongs to the monomial ideal $\langle \mathrm{in}_\prec(G)\rangle$. Let us consider the expression of the form

$$P = Q_1P_1 + \cdots + Q_mP_m \qquad (Q_1, \dots, Q_m \in D_n^{(h)}). \tag{7}$$

Since $\prec$ is a well-order, we may assume that this expression is minimal in the sense that $a := \max\{\mathrm{in}_\prec(Q_iP_i) \mid Q_i \neq 0\}$ is minimum among such expressions. Then $\mathrm{in}_\prec(P) \preceq a$ holds. If $\mathrm{in}_\prec(P) = a$, then we are done since a belongs to $\langle \mathrm{in}_\prec(G)\rangle$.

Suppose $\mathrm{in}_\prec(P) \prec a$. Let A be the set of $i \in \{1, \dots, m\}$ such that $\mathrm{in}_\prec(P_iQ_i) = a$. We may assume $A = \{1, \dots, l\}$. We may also assume that the leading coefficients of P_i are all one. Set

$$c_k = \mathrm{LC}_\prec(Q_k), \quad S_k := c_k^{-1}\mathrm{LT}_\prec(Q_k), \quad Q'_k = Q_k - \mathrm{LT}_\prec(Q_k) = Q_k - c_kS_k$$

for $1 \leq k \leq l$. Then we have

$$P = \sum_{k=1}^{l} c_kS_kP_k + \sum_{k=1}^{l} Q'_kP_k + \sum_{k=l+1}^{m} Q_kP_k \tag{8}$$

with the property that $\mathrm{in}_\prec(Q'_kP_k) \prec a$ if $1 \leq k \leq l$ and $Q'_k \neq 0$, and $\mathrm{in}_\prec(Q_kP_k) \prec a$ if $l < k \leq m$ and $Q_k \neq 0$. This implies that the initial monomial of $\sum_{k=1}^{l} c_kS_kP_k$ is smaller than a in $\prec$. The first term can be rewritten as

$$(9)\quad \sum_{k=1}^{l} c_k S_k P_k = \sum_{k=1}^{l-1} (c_1 + \cdots + c_k)(S_k P_k - S_{k+1} P_{k+1}) + (c_1 + \cdots + c_l) S_l P_l.$$

Let $\mathrm{sp}_\prec(P_i, P_j)$ be given by $S_{ji}P_i - S_{ij}P_j$. There exist monomials u_k such that

$$\mathrm{in}_\prec(S_k)\mathrm{in}_\prec(P_k) = \mathrm{in}_\prec(S_{k+1})\mathrm{in}_\prec(P_{k+1}) = u_k \mathrm{LCM}(\mathrm{in}_\prec(P_k), \mathrm{in}_\prec(P_{k+1}))$$

and

$$\mathrm{in}_\prec(S_k) = u_k S_{k+1,k}(x, \xi, h), \quad \mathrm{in}_\prec(S_{k+1}) = u_k S_{k,k+1}(x, \xi, h)$$

for $1 \le k \le l-1$. Take $U_k \in D_n^{(h)}$ whose total symbol is u_k and set

$$A_k := S_k - U_k S_{k+1,k}, \quad B_k := S_{k+1} - U_k S_{k,k+1}.$$

Then we have

$$S_k P_k - S_{k+1} P_{k+1} = U_k (S_{k+1,k} P_k - S_{k,k+1} P_{k+1}) + A_k P_k - B_k P_{k+1}.$$

Combined with (9), this yields

$$\begin{aligned}\sum_{k=1}^{l} c_k S_k P_k &= \sum_{k=1}^{l-1} (c_1 + \cdots + c_k) U_k \mathrm{sp}_\prec(P_k, P_{k+1}) + (c_1 + \cdots + c_l) S_l P_l \\ &\quad + \sum_{k=1}^{l-1} (c_1 + \cdots + c_k)(A_k P_k - B_k P_{k+1}) \\ &= \sum_{j=1}^{m} \sum_{k=1}^{l-1} (c_1 + \cdots + c_k) U_k Q_{k,k+1,j} P_j + (c_1 + \cdots + c_l) S_l P_l \\ &\quad + \sum_{k=1}^{l-1} (c_1 + \cdots + c_k)(A_k P_k - B_k P_{k+1}).\end{aligned}$$

Here the initial monomials of $U_k Q_{k,k+1,j} P_j$, $A_k P_k$, and $B_k P_{k+1}$ are smaller than a, as well as the initial monomial of $\sum_{k=1}^{l} c_k S_k P_k$, while the initial monomial of $S_l P_l$ is a. It follows that $c_1 + \cdots + c_l = 0$ and hence

$$\sum_{k=1}^{l} c_k S_k P_k = \sum_{j=1}^{m}\sum_{k=1}^{l-1}(c_1+\cdots+c_k)U_k Q_{k,k+1,j}P_j \\ + \sum_{k=1}^{l-1}(c_1+\cdots+c_k)(A_kP_k - B_kP_{k+1}).$$

Substituting this for the first term of the right-hand side of (8) gives an expression of P which contradicts the minimality of (7). Hence we must have $\mathrm{in}_\prec(P) = a$. This completes the proof. Q.E.D.

This criterion assures that the Buchberger algorithm applies to $D_n^{(h)}$. Note also that this criterion and the proof works in D_n if $\prec$ is a term order for D_n.

Now let $\prec$ be an arbitrary monomial order for D_n. We define a monomial order $\prec_h$ on $M(x,\xi,h)$ by

$$x^\alpha\xi^\beta h^j \prec_h x^{\alpha'}\xi^{\beta'}h^k \;\Leftrightarrow\; |\alpha|+|\beta|+j < |\alpha'|+|\beta'|+k \\ \text{or } (|\alpha|+|\beta|+j = |\alpha'|+|\beta'|+k \text{ and } x^\alpha\xi^\beta \prec x^{\alpha'}\xi^{\beta'}).$$

Then $\prec_h$ is a term order for $D_n^{(h)}$. Hence the division and the Buchberger algorithm works with $\prec_h$ in $D_n^{(h)}$. If P is a nonzero homogeneous element of $D_n^{(h)}$, then $\mathrm{in}_\prec(P|_{h=1}) = \mathrm{in}_{\prec_h}(P)|_{h=1}$ holds.

Theorem 3.14. *Let I be the left ideal of D_n generated by nonzero elements $P_1,\ldots,P_r$ of I, and $\prec$ be an arbitrary monomial order for D_n. Let J be a left ideal of $D_n^{(h)}$ generated by $P_1^{(h)},\ldots,P_r^{(h)}$ and $\{Q'_1,\ldots,Q'_m\}$ be a Gröbner basis of J with respect to $\prec_h$, which can be computed by Buchberger's algorithm.*

Set $Q_i := Q'_i|_{h=1}$. Then $\{Q_1,\ldots,Q_m\}$ is a Gröbner basis of I with respect to $\prec$. Moreover, for any nonzero element P of I, there exist $U_1,\ldots,U_m \in D_n$ such that

$$P = U_1Q_1+\cdots+U_mQ_m, \qquad \mathrm{in}_\prec(U_iQ_i) \preceq \mathrm{in}_\prec(P) \text{ if } U_i \neq 0.$$

In particular, if $\prec$ is adapted to w, then $Q_1,\ldots,Q_m$ are a w-involutive basis of I; more precisely, one has

$$I\cap F_k^w(D_n) = F_{k-k_1}^w(D_n)Q_1+\cdots+F_{k-k_m}^w(D_n)Q_m \qquad (\forall k\in\mathbb{Z})$$

with $k_i := \mathrm{ord}_w(Q_i)$.

Proof. Let P be a nonzero element of I. Then there exist $A_1, \dots, A_r \in D_n$ such that $P = A_1 P_1 + \cdots + A_r P_r$. Homogenizing the both sides of this equation, we obtain

$$h^l P^{(h)} = h^{l_1} A_1^{(h)} P_1^{(h)} + \cdots + h^{l_m} A_m^{(h)} P_m^{(h)}$$

with some $l, l_1, \dots, l_m \in \mathbb{N}$. Hence $h^l P^{(h)}$ belongs to J. Since $Q'_1, \dots, Q'_m$ are a Gröbner basis of J, division algorithm in $D_n^{(h)}$ produces an expression

$$h^l P^{(h)} = \sum_{j=1}^{m} B_j Q'_j$$

with some homogeneous elements B_j of $D_n^{(h)}$ such that $\mathrm{in}_{\prec_h}(B_j Q'_j) \preceq_h \mathrm{in}_{\prec_h}(h^l P^{(h)})$ if $B_j \neq 0$. Dehomogenization yields

$$P = \sum_{j=1}^{m} B_j|_{h=1} Q_j, \qquad \mathrm{in}_{\prec}(B_j|_{h=1} Q_j) \preceq \mathrm{in}_{\prec}(P). \tag{10}$$

In particular, $\mathrm{in}_{\prec}(P)$ is divisible by $\mathrm{in}_{\prec}(Q_j)$ for some j. Hence $Q_1, \dots, Q_m$ are a Gröbner basis of I with respect to $\prec$. The last statement of the theorem also follows from (10) if $\prec$ is adapted to w. Q.E.D.

Example 3.15. Set $n = 2$, $w = (0, 1; 0, -1)$, and $P_1 = x_1 - x_2^2$, $P_2 = 2x_2\partial_1 + \partial_2$. Fix a term order $\prec$ for D_2 which is adapted to the weight vector $(1, 1; 1, 1)$ such that $x_1\xi_1 \succ x_2\xi_2$. Let us compute a Gröbner basis of $I := D_2 P_1 + D_2 P_2$ with respect to the monomial order$\prec_w$. Homogenization gives

$$P_1^{(h)} = x_1 h - \underline{x_2^2}, \qquad P_2^{(h)} = \underline{2x_2\partial_1} + \partial_2 h$$

with the leading terms with respect to $\prec_{wh}$, which is the term order for $D_2^{(h)}$ defined by $\prec_w$, being underlined. Their S-pair is

$$P'_3 := \mathrm{sp}_{\prec_w}(P_1^{(h)}, P_2^{(h)}) = 2\partial_1 P_1^{(h)} + x_2 P_2^{(h)} = \underline{2x_1\partial_1 h} + x_2\partial_2 h + 2h^3.$$

By using the Buchberger criterion, we can check that $P_1^{(h)}$, $P_2^{(h)}$, P'_3 are a Gröbner basis of the left ideal $J := D_2^{(h)} P_1^{(h)} + D_2^{(h)} P_2^{(h)}$ of $D_2^{(h)}$ with respect to $\prec_{wh}$. Hence P_1, P_2, and $P_3 := P'_3|_{h=1} = 2x_1\partial_1 + x_2\partial_2 + 2$ are a Gröbner basis of I with respect to $\prec_w$.

The notion and the algorithm of Gröbner basis can be extended to submodules of free modules over D_n or $D_n^{(h)}$ of finite rank.

Exercise 13. In the example above, confirm that $P_1^{(h)}$, $P_2^{(h)}$, P'_3 are a Gröbner basis of J with respect to $\prec_{wh}$.

3.3. Computation of the characteristic variety and the singular locus

Let I be a left ideal of D_n and consider the left D_n-module $M = D_n/I$. As was seen in 2.4, the most fundamental invariants of M are the dimension and the characteristic variety. Now let us deduce a more concrete description of the characteristic variety. Let P be a nonzero differential operator written in the form

$$P = P(x, \partial) = \sum_{\alpha,\beta\in\mathbb{N}^n} a_{\alpha,\beta} x^\alpha \partial^\beta \qquad (a_{\alpha,\beta} \in \mathbb{K})$$

and set $m := \mathrm{ord}_{(\mathbf{0};\mathbf{1})}(P)$. Then the *principal symbol* of P is the polynomial defined by

$$\sigma(P)(x, \xi) = \sum_{|\beta|=m} \sum_{\alpha} a_{\alpha,\beta} x^\alpha \xi^\beta.$$

It can be identified with the residue class of P in $\mathrm{gr}^{(\mathbf{0};\mathbf{1})}(D_n) \cong \mathbb{K}[x, \xi]$. Note that $\sigma(P)(x, \xi)$ is homogeneous with respect to ξ.

In general, let w be a weight vector for D_n satisfying $w_i \geq 0$ for $i = 1, \ldots, 2n$, and $w_i + w_{n+i} > 0$ for $i = 1, \ldots, n$. Let $\prec$ be an arbitrary term order for D_n. Then the Buchberger algorithm applied to I with the term order $\prec_w$ yields a Gröbner basis $G = \{P_1, \ldots, P_m\}$ of I with respect to $\prec_w$. Proposition 3.5 assures that G is a w-involutive basis of I; that is, $\sigma^w(G)$ generates the graded ideal $\mathrm{gr}^w(I)$ associated with the filtration $\{F_k^w(D_n) \cap I\}$ on I.

Let $\mathrm{gr}^w(M)$ be the graded module associated with the good w-filtration $F_k^w(M) := F_k^w(D_n)/(F_k^w(D_n) \cap I)$. Then there exists a graded exact sequence

$$0 \longrightarrow \mathrm{gr}^w(I) \longrightarrow \mathrm{gr}^w(D_n) \longrightarrow \mathrm{gr}^w(M) \longrightarrow 0.$$

Note that $\mathrm{gr}^w(D_n)$ can be regarded as $\mathbb{K}[x, \xi]$ by the assumption on w. Hence one has an isomorphism

$$\mathrm{gr}^w(M) \cong \mathbb{K}[x, \xi]/\mathrm{gr}^w(I) \cong \mathbb{K}[x, \xi]/(\mathbb{K}[x, \xi]\sigma^w(P_1) + \cdots + \mathbb{K}[x, \xi]\sigma^w(P_m))$$

as $\mathbb{K}[x, \xi]$-module. In particular, setting $w = (\mathbf{0}; \mathbf{1})$, we obtain

$$\mathrm{Char}(M) = \{(x, \xi) \in \overline{\mathbb{K}}^{2n} \mid \sigma(P_1)(x, \xi) = \cdots = \sigma(P_m)(x, \xi) = 0\}.$$

Let $\pi : \overline{\mathbb{K}}^{2n} \ni (x, \xi) \mapsto x \in \overline{\mathbb{K}}^n$ be the projection. Then the *singular locus* of M is defined by

$$\mathrm{Sing}(M) := \pi(\mathrm{Char}(M) \setminus (\overline{\mathbb{K}}^n \times \{0\})).$$

It is an algebraic set of $\overline{\mathbb{K}}^n$ since $\mathrm{gr}(M)$ is homogeneous with respect to ξ. In particular, if M is holonomic, then $\mathrm{Sing}(M)$ is an algebraic set of codimension ≥ 1, or an empty set, since $\mathrm{Char}(M)$ is homogeneous with respect to ξ.

The set $\mathrm{Char}(M) \setminus (\{0\} \times \overline{\mathbb{K}}^n)$ can be regarded as the subset of $\overline{\mathbb{K}}^n \times \mathbb{P}^{n-1}(\overline{\mathbb{K}})$, where $\mathbb{P}^{n-1}(\overline{\mathbb{K}})$ is the $(n-1)$-dimensional projective space over $\overline{\mathbb{K}}$. Thus the problem of finding $\mathrm{Sing}(M)$ from $\mathrm{Char}(M)$ is completely solved by what is called the projective elimination theory, as is described in Chapter 8 of [6] in detail with a complete proof.

Proposition 3.16. *Assume that $\mathbb{K}$ is an algebraically closed field of characteristic zero and set $M = D_n/I$ with a left ideal I of D_n. Let $f_1(x,\xi),\dots,f_m(x,\xi)$ be polynomials homogeneous in ξ which generate $\mathrm{gr}^{(\mathbf{0};\mathbf{1})}(I)$. Let J_i be the ideal of $\mathbb{K}[x,\xi]$ generated by $f_1,\dots,f_m$ with the variable ξ_i replaced by 1. Set $I_i = J_i \cap \mathbb{K}[x]$. Then* $\mathrm{Sing}(M)$ *is the algebraic subset of $\mathbb{K}^n$ defined as the zeros of the ideal $I_1 \cap \cdots \cap I_n$.*

Thus we can compute $\mathrm{Sing}(M)$ from $\mathrm{Char}(M)$ by using appropriate Gröbner bases in $\mathbb{K}[x,\xi]$; this fact was pointed out in [21], where it is also noticed that the characteristic variety as is defined here coincides with the analytic definition using the differential operators with analytic coefficients. Even if M is generated by more than one elements over D_n, we can compute $\mathrm{Char}(M)$ by using a Gröbner basis for a submodule of the free module (see [21] for details).

Example 3.17. Let

$$P = a_m(x)\partial^m + a_{m-1}(x)\partial^{m-1} + \cdots + a_0(x) \qquad (a_i(x) \in \mathbb{K}[x],\ a_m(x) \neq 0)$$

be a linear ordinary differential operator of order $m \geq 1$ with $x = x_1$ and $\partial = \partial_1$. Set $M = D_1/D_1P$. Then we have

$$\begin{aligned}\mathrm{Char}(M) &= \{(x,\xi) \in \overline{\mathbb{K}}^2 \mid \sigma(P)(x,\xi) = a_m(x)\xi^m = 0\} \\ &= \{(x,\xi) \in \overline{\mathbb{K}}^2 \mid a_m(x) = 0\} \cup \{(x,0) \mid x \in \overline{\mathbb{K}}\}.\end{aligned}$$

Hence M is holonomic and $\mathrm{Sing}(M) = \{x \in \mathbb{K} \mid a_m(x) = 0\}$, a point of which is called a *singular point* of P.

Example 3.18. Let f be an arbitrary nonzero polynomial of $x = (x_1,\dots,x_n)$. For each $i = 1,\dots,n$, $\partial_i f = f\partial_i + f_i$ annihilates the rational function $1/f$, with $f_i := \partial f/\partial x_i$. Set $M = D_n/I$ with

$$I := D_n\partial_1 f + \cdots + D_n\partial_n f.$$

This is a 'naive' D-module for the rational function $1/f$.

For example, set $n = 2$ and $f(x) = x_1^3 - x_2^2$ which has a cusp singularity at the origin. We can check that $\partial_1 f$ and $\partial_2 f$ are a $(\mathbf{0};\mathbf{1})$-involutive basis of I. This gives

$$\begin{aligned}\mathrm{Char}(M) &= \{(x,\xi) \in \overline{\mathbb{K}}^4 \mid \xi_1 f(x_1,x_2) = \xi_2 f(x_1,x_2) = 0\}\\ &= \{(x,\xi) \mid x_1^3 - x_2^2\} \cup \{(x,\xi) \mid \xi = 0\},\\ \mathrm{Sing}(M) &= \{(x_1,x_2) \mid x_1^3 - x_2^2 = 0\}.\end{aligned}$$

Hence the dimension of $\mathrm{Char}(M)$ is 3, consequently M is not holonomic. In fact, I is much smaller than $\mathrm{Ann}_{D_2}(1/f)$, which is generated by $3x_1^2\partial_2 + 2x_2\partial_1$ and $2x_1\partial_1 + 3x_2\partial_2 + 6$. There is an algorithm to compute $\mathrm{Ann}_{D_n}(1/f)$ for an arbitrary polynomial f and $D_n/\mathrm{Ann}_{D_n}(1/f)$ is always holonomic (see [22], [29]).

Exercise 14. In the example above with $n = 2$ and $f = x_1^3 - x_2^2$

(1) Verify that $\partial_1 f$ and $\partial_2 f$ are a $(\mathbf{0};\mathbf{1})$-involutive basis of the ideal I which they generate.
(2) Verify that $P_1 := 3x_1^2\partial_2 + 2x_2\partial_1$ and $P_2 := 2x_1\partial_1 + 3x_2\partial_2 + 6$ annihilate $1/f$.
(3) Find a $(\mathbf{0};\mathbf{1})$-involutive basis of $J := D_2P_1 + D_2P_2$ and verify that D_2/J is holonomic.
(4) Find the singular locus of D_2/J.

Exercise 15. Find the characteristic variety and the singular locus of the left D_n-module $M = D_n/(D_nx_1 + \cdots + D_nx_n)$.

Exercise 16. Let $\mathbb{K} = \mathbb{C}$ and let $f(x) \in \mathbb{C}[x] = \mathbb{C}[x_1,\ldots,x_n]$. Consider a C^∞ function $e^{f(x)}$ on $\mathbb{R}^n$. Set $f_i = \partial_i(f)$ and $M := D_n/I$ with

$$I := D_n(\partial_1 - f_1) + \cdots + D_n(\partial_n - f_n).$$

(1) Show that $I = \mathrm{Ann}_{D_n} e^{f(x)} := \{P \in D_n \mid Pe^{f(x)} = 0\}$.
(2) Show that $\mathrm{Hom}_{D_n}(M, C^\infty(\mathbb{R}^n)) \cong \mathbb{C}e^{f(x)}$.
(3) Find the characteristic variety and the singular locus of M.

3.4. Equivalence of two definitions of holonomicity

The purpose of this subsection is to prove Theorem 2.22 by using only basic tools in commutative algebra and Gröbner basis.

Definition 3.19 ([17]). A map φ from $\mathbb{N}$ to $\{t \in \mathbb{R} \mid t \geq 0\}$ is said to be of *polynomial growth* if there exists $\nu \in \mathbb{R}$ such that $\varphi(n) \leq n^\nu$ for $n \gg 0$. Then we define the degree of φ by

$$\deg\varphi = \inf\{\nu \mid \varphi(n) \leq n^\nu \text{ for } n \gg 0\}.$$

We set $\deg\varphi = \infty$ if φ is not of polynomial growth.

Definition 3.20 ([5]). A function $\varphi : \mathbb{Z} \to \mathbb{R}$ is called a *quasi-polynomial* (of period r) if there exist a positive integer r and polynomials H_i ($i = 0, 1, \dots, r-1$) such that

$$\varphi(jr+i) = H_i(jr+i) \qquad (\forall j \in \mathbb{Z},\, 0 \le \forall i \le r-1).$$

Then one has $\deg \varphi = \max_{0 \le i \le r-1} \deg H_i$.

Let w be a weight vector such that $w_i > 0$ for $1 \le i \le 2n$. In this case we call w positive and denote it by $w > 0$.

Proposition 3.21 ([33], Theorem 5.2). *Let $w \in \mathbb{N}^{2n}$ be a positive weight vector for D_n. Let $\{F_k(M)\}$ be a good w-filtration on a left D_n-module M. Then the degree of the function*

$$\varphi(k) = \dim_{\mathbb{K}} F_k(M)$$

of $k \ge 0$ does not depend either on w or on the choice of a good w-filtration $\{F_k(M)\}$.

Proof. For a fixed w, the fact that $\deg \varphi$ does not depend on the choice of a good w-filtration can be proved by using Lemma 2.10.

Let $u_1, \dots, u_m$ be a set of generators of M. Define a good w-filtration and a good $(\mathbf{1};\mathbf{1})$-filtration on M by

$$F_k^w(M) := F_k^w(D_n)u_1 + \cdots + F_k^w(D_n)u_m,$$
$$F_k^{(\mathbf{1};\mathbf{1})}(M) := F_k^{(\mathbf{1};\mathbf{1})}(D_n)u_1 + \cdots + F_k^{(\mathbf{1};\mathbf{1})}(D_n)u_m$$

respectively and $\mathrm{gr}^w(M)$ and $\mathrm{gr}^{(\mathbf{1};\mathbf{1})}(M)$ be associated graded modules. Set $\psi(k) = \dim_{\mathbb{K}} F_k^{(\mathbf{1};\mathbf{1})}(M)$. Then there exists a polynomial G such that $\psi(k) = G(k)$ for all $k \gg 0$. On the other hand, since there exists $P(T) \in \mathbb{Z}[T, T^{-1}]$ such that

$$\sum_{k \in \mathbb{Z}} \dim_{\mathbb{K}} \mathrm{gr}_k^w(M)\, T^k = \frac{P(T)}{\prod_{i=1}^n (1 - T^{w_i})(1 - T^{w_{n+i}})},$$

there exists a quasi-polynomial H of some period r such that $\varphi(k) = H(k)$ for all $k \gg 0$ (see e.g., Proposition 4.4.1 and Theorem 4.4.3 of [5]). Let d be an integer not smaller than $\max\{w_i \mid 1 \le i \le 2n\}$. Then the inclusion

$$F_k^w(D_n) \subset F_k^{(\mathbf{1};\mathbf{1})}(D_n) \subset F_{dk}^w(D_n) \qquad (\forall k \ge 0)$$

yields

$$F_k^w(M) \subset F_k^{(\mathbf{1};\mathbf{1})}(M) \subset F_{dk}^w(M) \qquad (\forall k \ge 0),$$

and hence

$$\varphi(k) \le \psi(k) \le \varphi(dk).$$

This implies $\deg \varphi = \deg \psi$. Q.E.D.

Proposition 3.22. *Let $w \in \mathbb{N}^{2n}$ be a positive weight vector and $M' = \oplus_j M'_j$ be a finitely generated graded module over $\mathbb{K}[x,\xi]$ in which x_i is of order w_i and ξ_i is of order w_{n+i}. Then the degree of the function*

$$\varphi(k) = \sum_{j \le k} \dim_{\mathbb{K}} M'_j$$

coincides with the Krull dimension of M' as a (not graded) $\mathbb{K}[x,\xi]$-module.

Proof. Let $u'_1, \dots, u'_m$ be w-homogeneous generators of M' with $u'_i \in M'_{k_i}$. Then we have

$$M'_k = \mathbb{K}[x,\xi]^w_{k-k_1} u'_1 + \cdots + \mathbb{K}[x,\xi]^w_{k-k_m} u'_m,$$
$$\mathbb{K}[x,\xi]^w_k := \sum_{\langle w, (\alpha,\beta)\rangle = k} \mathbb{K} x^\alpha \xi^\beta.$$

Forgetting the w-graded structure of M', set

$$F_k(M') = F^{(\mathbf{1};\mathbf{1})}_{k-k_1}(\mathbb{K}[x,\xi]) u'_1 + \cdots + F^{(\mathbf{1};\mathbf{1})}_{k-k_m}(\mathbb{K}[x,\xi]) u'_m,$$
$$F^{(\mathbf{1};\mathbf{1})}_k(\mathbb{K}[x,\xi]) := \sum_{|\alpha|+|\beta| \le k} \mathbb{K} x^\alpha \xi^\beta.$$

Let $\mathrm{gr}(M') = \oplus_k F_k(M')/F_{k-1}(M')$ be the associated graded module. It is well-known in commutative algebra (e.g., Corollary 13.6 of [8]) that the Krull dimension of M' coincides with the degree of the function $\psi(k) := \dim_{\mathbb{K}} F_k(M')$, which equals a polynomial for $k \gg 0$. (One can dispense with the Krull dimension by adopting the degree of $\psi(k)$ as the definition of the dimension as in Chapter 9 of [6].) On the other hand, we can show that $\varphi(k)$ is a quasi-polynomial for $k \gg 0$ and its degree coincides with that of ψ by the same argument over $\mathbb{K}[x,\xi]$ instead of D_n as the proof of Proposition 3.21. Q.E.D.

Now let us prove Theorem 2.22. If M is generated by $u_1, \dots, u_m$, then it is easy to see that

$$\dim M = \max_{1 \le i \le m} \dim D_n u_i, \qquad \mathrm{Char}(M) = \bigcup_{1 \le i \le m} \mathrm{Char}(D_n u_i)$$

hold. Hence we may assume that M is generated by a single element, and consequently that $M = D_n/I$ with a left ideal I of D_n.

Let $\prec$ be a term order for D_n adapted to the weight vector $(\mathbf{1};\mathbf{1})$. Let $G=\{P_1,\dots,P_m\}$ be a Gröbner basis of I with respect to the term order $\prec_{(\mathbf{0};\mathbf{1})}$. There exist $Q_{ijk}\in D_n$ such that

$$\mathrm{sp}_{\prec_{(\mathbf{0};\mathbf{1})}}(P_i,P_j)=\sum_{k=1}^{m}Q_{ijk}P_k,$$
$$\mathrm{in}_{\prec_{(\mathbf{0};\mathbf{1})}}(Q_{ijk}P_k)\preceq_{(\mathbf{0};\mathbf{1})}\mathrm{in}_{\prec_{(\mathbf{0};\mathbf{1})}}(\mathrm{sp}_{\prec_{(\mathbf{0};\mathbf{1})}}(P_i,P_j))\quad\text{if } Q_{ijk}\neq 0.$$

Set $w=(1,\dots,1;d,\dots,d)=(d-1)(\mathbf{0};\mathbf{1})+(\mathbf{1};\mathbf{1})$. If we take d large enough, then the initial monomials of P_i and Q_{ijk} with respect to $\prec_{(\mathbf{0};\mathbf{1})}$ stay unchanged with $\prec_{(\mathbf{0};\mathbf{1})}$ replaced by $\prec_w$ since $\prec$ is adapted to $(\mathbf{1};\mathbf{1})$. Hence G is also a Gröbner basis of I with respect to $\prec_w$ in view of the Buchberger criterion in D_n.

Let $\mathrm{gr}^{(\mathbf{0};\mathbf{1})}(M)$ and $\mathrm{gr}^w(M)$ be the graded modules associated with filtrations

$$F_k^{(\mathbf{0};\mathbf{1})}(M)=F_k^{(\mathbf{0};\mathbf{1})}(D_n)/(F_k^{(\mathbf{0};\mathbf{1})}(D_n)\cap I),$$
$$F_k^{w}(M)=F_k^{w}(D_n)/(F_k^{w}(D_n)\cap I)\qquad(k\in\mathbb{Z})$$

respectively. Then from the argument above and Proposition 3.5 we have

$$\mathrm{gr}^{(\mathbf{0};\mathbf{1})}(M)=\mathbb{K}[x,\xi]/\langle\sigma^{(\mathbf{0};\mathbf{1})}(G)\rangle,\qquad \mathrm{gr}^w(M)=\mathbb{K}[x,\xi]/\langle\sigma^w(G)\rangle$$

and

$$\mathrm{in}_\prec(\sigma^{(\mathbf{0};\mathbf{1})}(P_i))=\mathrm{in}_{\prec_{(\mathbf{0};\mathbf{1})}}(P_i)=\mathrm{in}_{\prec_w}(P_i)=\mathrm{in}_\prec(\sigma^w(P_i))\quad(1\le i\le m).$$

Since the Hilbert polynomial of $\mathbb{K}[x,\xi]/\langle\sigma^w(G)\rangle$ coincides with that of $\mathbb{K}[x,\xi]/\langle\mathrm{in}_\prec(\sigma^w(G))\rangle$ (see Chapter 9 of [6]), it follows that $\mathrm{gr}^{(\mathbf{0};\mathbf{1})}(M)$ and $\mathrm{gr}^w(M)$ have the same Krull dimension. Set

$$F_k^{(\mathbf{1};\mathbf{1})}(M)=F_k^{(\mathbf{1};\mathbf{1})}(D_n)/(F_k^{(\mathbf{1};\mathbf{1})}(D_n)\cap I).$$

By Propositions 3.21 and 3.22, the Krull dimension of $\mathrm{gr}^w(M)$ is

$$\deg\Big(\sum_{j\le k}\dim_{\mathbb{K}}\mathrm{gr}_j^w(M)\Big)=\deg\dim_{\mathbb{K}}F_k^w(M)=\deg\dim_{\mathbb{K}}F_k^{(\mathbf{1};\mathbf{1})}(M),$$

where deg denotes the degree as a function of k. It is also equal to $\dim M$ by the definition. This completes the proof of Theorem 2.22.

§4. Distributions as generalized functions

We briefly review the theory of distributions (generalized functions) introduced by L. Schwartz [32]. See also [10] for plenty of interesting examples. The main purpose here is to introduce some classes of distributions which adapt nicely to the integration algorithm based on differentiation under the integral sign. The terminology coined by Schwartz has an origin in probability theory. So we also consider probability and cumulative distribution functions associated with the multivariate normal distribution or the gamma distribution as integrals of generalized functions.

4.1. Definitions and basic properties

Definition 4.1. Let $C_0^\infty(U)$ be the set of the complex-valued C^∞ functions on an open set U of $\mathbb{R}^n$ with compact support. Here the support of a C^∞ function f on U is defined to be the closure in U of the set $\{x \in U \mid f(x) \neq 0\}$ and denoted by $\operatorname{supp} u$. A *distribution* u on U is a linear mapping

$$u : C_0^\infty(U) \ni \varphi \longmapsto \langle u, \varphi \rangle \in \mathbb{C}$$

such that $\lim_{j\to\infty}\langle u, \varphi_j\rangle = 0$ holds for a sequence $\{\varphi_j\}$ of $C_0^\infty(U)$ if there is a compact set $K \subset U$ such that $\operatorname{supp}\varphi_j \subset K$ for any j and

$$\lim_{j\to\infty} \sup_{x\in U} |\partial^\alpha \varphi_j(x)| = 0 \quad \text{for any } \alpha \in \mathbb{N}^n,$$

where $x = (x_1, \ldots, x_n)$ and $\partial^\alpha = \partial_1^{\alpha_1} \cdots \partial_n^{\alpha_n}$ with $\partial_j = \partial/\partial x_j$. If this is the case, we say that $\{\varphi_j\}$ converges to 0 in $C_0^\infty(U)$, which makes $C_0^\infty(U)$ a topological vector space that is also denoted by $\mathcal{D}(U)$. The set of the distributions on U is denoted by $\mathcal{D}'(U)$.

A Lebesgue measurable function $u(x)$ defined on an open set U of $\mathbb{R}^n$ is called locally integrable on U if it is integrable on any compact subset of U. We can regard a locally integrable function $u(x)$ on U as a distribution on U through the pairing

$$\langle u, \varphi \rangle = \int_U u(x)\varphi(x)\, dx \qquad (\forall \varphi \in C_0^\infty(U)).$$

Identifying two locally integrable functions which are equal to each other almost everywhere in U (i.e., outside a set of measure 0), we can regard the set $L^1_{\mathrm{loc}}(U)$ of the locally integrable functions on U as a subspace of $\mathcal{D}'(U)$.

Let u be a distribution on U. The derivative $\partial_k u$ of u with respect to x_k is defined by

$$\langle \partial_k u, \varphi \rangle = -\langle u, \partial_k \varphi \rangle \quad \text{for any } \varphi \in C_0^\infty(U).$$

For a C^∞ function a on U, the product au is defined by

$$\langle au, \varphi \rangle = \langle u, a\varphi \rangle \quad \text{for any } \varphi \in C_0^\infty(U).$$

In particular, by these actions of the derivations and the polynomial multiplications, $\mathcal{D}'(U)$ has a natural structure of left D_n-module.

Example 4.2. Set $n = 1$. The Heaviside function $Y(x)$ is the measurable function on $\mathbb{R}$ such that $Y(x) = 1$ for $x > 0$ and $Y(x) = 0$ for $x < 0$. The Dirac delta function $\delta(x)$ is a distribution on $\mathbb{R}$ defined by

$$\langle \delta(x), \varphi \rangle = \varphi(0) \qquad (\forall \varphi \in C_0^\infty(\mathbb{R})).$$

The derivative $Y'(x)$ of $Y(x)$ as a distribution coincides with $\delta(x)$ since

$$\langle Y'(x), \varphi(x) \rangle = -\langle Y(x), \varphi'(x) \rangle = -\int_0^\infty \varphi'(x)\,dx = \varphi(0) = \langle \delta(x), \varphi \rangle$$

holds for any $\varphi \in C_0^\infty(\mathbb{R})$. The derivative $\delta'(x)$ of $\delta(x)$ is defined by

$$\langle \delta'(x), \varphi(x) \rangle = -\langle \delta(x), \varphi'(x) \rangle = -\varphi'(0).$$

In the same way, the k-th derivative $\delta^{(k)}(x) \in \mathcal{D}'(\mathbb{R})$ is defined by

$$\langle \delta^{(k)}(x), \varphi(x) \rangle = (-1)^k \langle \delta(x), \varphi^{(k)}(x) \rangle = (-1)^k \varphi^{(k)}(0).$$

Example 4.3. The n-dimensional delta function $\delta(x)$ is the distribution defined by $\langle \delta(x), \varphi \rangle = \varphi(0)$ for any $\varphi \in C_0^\infty(\mathbb{R}^n)$.

Let $u \in \mathcal{D}'(U)$ with an open set U of $\mathbb{R}^n$. Let V be an open subset of U. Then there exists a natural inclusion $C_0^\infty(V) \subset C_0^\infty(U)$. The restriction $v := u|_V$ of u to V is defined by

$$\langle v, \varphi \rangle = \langle u, \varphi \rangle \qquad (\forall \varphi \in C_0^\infty(V)).$$

Then $U \longmapsto \mathcal{D}'(U)$, where U are open sets of $\mathbb{R}^n$, constitutes a sheaf on $\mathbb{R}^n$. For $u \in \mathcal{D}'(U)$, the *support* $\operatorname{supp} u$ is defined to be the smallest closed set Z in U such that $u|_{U \setminus Z} = 0$, i.e., $\langle u, \varphi \rangle = 0$ for any $\varphi \in C_0^\infty(U \setminus Z)$. For example, with $x = x_1$ we have $\operatorname{supp} \delta(x) = \{0\}$ and $\operatorname{supp} Y(x) = \{x \in \mathbb{R} \mid x \geq 0\}$.

The set of the distributions on U whose supports are compact sets of U is denoted by $\mathcal{E}'(U)$. ($\mathcal{E}'(U)$ means the dual space of $\mathcal{E}(U) = C^\infty(U)$.)

Let u belong to $\mathcal{E}'(U)$ and $K := \operatorname{supp} u$ be its support. Then for an arbitrary $\varphi \in C^\infty(U)$, the pairing

$$\langle u, \varphi(x)\rangle = \langle u, \chi(x)\varphi(x)\rangle$$

is well-defined with a cut-off function $\chi \in C_0^\infty(U)$ such that $\chi(x) = 1$ on an open set $V \subset U$ such that $K \subset V$. This pairing does not depend on the choice of χ. In fact, assume $\tilde{\chi} \in C_0^\infty(U)$ satisfies the same condition. Then since

$$\operatorname{supp}(\chi - \tilde{\chi}) \cap \operatorname{supp} u = \emptyset,$$

$\langle u, \chi\varphi\rangle = \langle u, \tilde{\chi}\varphi\rangle$ holds for any $\varphi \in C^\infty(U)$.

Definition 4.4. A C^∞ function φ on $\mathbb{R}^n$ is called a *rapidly decreasing function* if $|P\varphi|$ is bounded on $\mathbb{R}^n$ for any differential operator $P \in D_n$. The set of the rapidly decreasing functions on $\mathbb{R}^n$ is denoted by $\mathcal{S}(\mathbb{R}^n)$; it contains $C_0^\infty(\mathbb{R}^n)$ as a subspace.

Definition 4.5. A *tempered distribution* u on $\mathbb{R}^n$ is a $\mathbb{C}$-linear mapping

$$u : \mathcal{S}(\mathbb{R}^n) \ni \varphi \longmapsto \langle u, \varphi\rangle \in \mathbb{C}$$

such that $\lim_{j\to\infty}\langle u, \varphi_j\rangle = 0$ holds for any sequence $\{\varphi_j\}$ of $\mathcal{S}(\mathbb{R}^n)$ which satisfies the condition

$$\lim_{j\to\infty} \sup_{x\in\mathbb{R}^n} |P\varphi_j(x)| = 0 \quad \text{for any } P \in D_n.$$

The sequence $\{\varphi_j\}$ with this condition is said to converge to 0 in $\mathcal{S}(\mathbb{R}^n)$, which makes $\mathcal{S}(\mathbb{R}^n)$ a topological vector space. The set of the tempered distributions on $\mathbb{R}^n$ is denoted by $\mathcal{S}'(\mathbb{R}^n)$, which can be regarded as a subspace of $\mathcal{D}'(\mathbb{R}^n)$ since $C_0^\infty(\mathbb{R}^n)$ is dense in $\mathcal{S}(\mathbb{R}^n)$. Moreover, $\mathcal{E}'(\mathbb{R}^n)$ is a subspace of $\mathcal{S}'(\mathbb{R}^n)$. Any $P \in D_n$ defines a continuous linear endomorphism of $\mathcal{S}(\mathbb{R}^n)$ and of $\mathcal{S}'(\mathbb{R}^n)$. Hence $\mathcal{S}'(\mathbb{R}^n)$ is a left D_n-submodule of $\mathcal{D}'(\mathbb{R}^n)$.

Let $f(x) = f(x_1, \dots, x_n)$ be a complex-valued locally integrable function on $\mathbb{R}^n$ of polynomial growth; i.e., assume that there exists a constant $C > 0$ and a non-negative integer m such that

$$|f(x)| \le C(1 + |x|^2)^m = C(1 + x_1^2 + \cdots + x_n^2)^m \qquad (\forall x \in \mathbb{R}^n).$$

Then $f(x)$ defines a tempered distribution through the pairing

$$\langle f, \varphi\rangle = \int_{\mathbb{R}^n} f(x)\varphi(x)\,dx \qquad (\forall \varphi \in \mathcal{S}(\mathbb{R}^n)).$$

Example 4.6. Let $f_1, \dots, f_p$ be polynomials in $x = (x_1, \dots, x_n)$ with real coefficients. Let $\lambda_1, \dots, \lambda_p$ be complex numbers such that $\mathrm{Re}\, \lambda_j \geq 0$ $(j = 1, \dots, p)$. Set

$$((f_1)_+^{\lambda_1} \cdots (f_p)_+^{\lambda_p})(x) = \begin{cases} f_1(x)^{\lambda_1} \cdots f_p(x)^{\lambda_p} & \text{if } f_1(x) > 0, \dots, f_p(x) > 0 \\ 0 & \text{otherwise} \end{cases},$$

where we use the convention that $t^\lambda = \exp(\lambda \log t)$ with real $\log t$ for $t > 0$ and $\lambda \in \mathbb{C}$. Then it is easy to see that $(f_1)_+^{\lambda_1} \cdots (f_p)_+^{\lambda_p}$ is locally integrable and of polynomial growth, thus can be regarded as a tempered distribution. In particular, $(f_1)_+^0 \cdots (f_p)_+^0$ coincides with $Y(f_1) \cdots Y(f_p)$.

Theorem 4.7 (Sato-Kawai-Kashiwara). *Let M be a finitely generated left D_n-module. For a point a of $\mathbb{R}^n$, let us denote by $\mathcal{D}'_a$ the stalk of the sheaf $\mathcal{D}'$ at a, which is the inductive limit $\mathcal{D}'_a = \varinjlim \mathcal{D}'(U)$ where U are open neighborhoods of a.*

(1) *If M is holonomic, then $\mathrm{Hom}_{D_n}(M, \mathcal{D}'_a)$ is a finite dimensional vector space over $\mathbb{C}$ for any $a \in \mathbb{R}^n$.*

(2) *Let U be an open set of $\mathbb{R}^n$. Then any distribution solution of M is real analytic on $U' := U \setminus \mathrm{Sing}(M)$; i.e., the natural $\mathbb{C}$-linear map*

$$\mathrm{Hom}_{D_n}(M, \mathcal{A}(U')) \longrightarrow \mathrm{Hom}_{D_n}(M, \mathcal{D}'(U'))$$

is an isomorphism, where $\mathcal{A}(U')$ denotes the set of complex-valued real analytic functions on U'.

In fact, this theorem holds in a weaker assumption that M is an analytic D-module and with $\mathcal{D}'(U)$ being replaced by the set $\mathcal{B}(U)$ of hyperfunctions. Under this weaker assumption, the statement (1) is due to Kashiwara (see Theorem 5.1.7 of [14] for a more refined formulation). The statement (2) was first noticed by M. Sato with the introduction of the theory of microfunctions developed together with Kawai and Kashiwara ([31]).

Example 4.8. We have $\mathrm{Hom}_{D_n}(\mathbb{C}[x], \mathcal{D}'(\mathbb{R}^n)) \cong \mathbb{C}$. In fact, $\mathbb{C}[x] = D_n/(D_n\partial_1 + \cdots + D_n\partial_n)$ and we can prove that if $u \in \mathcal{D}'(\mathbb{R}^n)$ satisfies $\partial_1 u = \cdots = \partial_n u = 0$, then u is a constant function (see Exercise 19). Since $\mathrm{Sing}(\mathbb{C}[x]) = \emptyset$, u is real analytic on $\mathbb{R}^n$.

Example 4.9. Set $M := D_n/(D_n x_1 + \cdots + D_n x_n)$ with $\mathbb{K} = \mathbb{C}$. Then $\mathrm{Hom}_{D_n}(M, \mathcal{D}'(\mathbb{R}^n))$ is one dimensional and spanned by the n-dimensional delta function $\delta(x)$. Since $\mathrm{Sing}(M) = \{0\}$, u is real analytic (zero in fact) on $\mathbb{R}^n \setminus \{0\}$.

Exercise 17. Let u be a distribution on $\mathbb{R}^n$ satisfying $x_1 u = \cdots = x_n u = 0$.

(1) Show that the support of u is contained in $\{0\}$ and hence u belongs to $\mathcal{E}'(\mathbb{R}^n)$.

(2) Prove that u is a constant multiple of the n-dimensional delta function $\delta(x)$. Use the fact that for $\varphi \in C^\infty(\mathbb{R}^n)$, there exist $\varphi_i \in C^\infty(\mathbb{R}^n)$ $(i = 1, \ldots, n)$ such that

$$\varphi(x) = \varphi(0) + x_1\varphi_1(x) + \cdots + x_n\varphi_n(x).$$

Exercise 18. Set $n = 1$ and write $x = x_1$, $\partial = \partial_1$. Define a $\mathbb{C}$-linear map $u : \mathcal{S}(\mathbb{R}) \longrightarrow \mathbb{C}$ by

$$\langle u, \varphi \rangle = \lim_{\varepsilon \to +0} \int_\varepsilon^\infty \frac{\varphi(x) - \varphi(-x)}{x}\, dx \qquad (\varphi \in \mathcal{S}(\mathbb{R})).$$

(1) Show that u belongs to $\mathcal{S}'(\mathbb{R})$.

(2) Show that $xu = 1$, and hence $\partial x u = 0$ holds.

4.2. Product of distributions

The product of two distributions cannot be defined in general. There are some cases where the product is well-defined:

(1) Let U be an open set of $\mathbb{R}^n$ and V an open set of $\mathbb{R}^m$. For $u_1 \in \mathcal{D}'(U)$ and $u_2 \in \mathcal{D}'(V)$, their tensor product $u_1 \otimes u_2$, which is also denoted by $u_1(x)u_2(y)$ with $x = (x_1, \ldots, x_n)$ and $y = (y_1, \ldots, y_m)$, is defined as the unique distribution on $U \times V$ such that

$$\langle u_1 \otimes u_2,\, \varphi_1(x)\varphi_2(y) \rangle = \langle u_1,\, \varphi_1 \rangle \langle u_2,\, \varphi_2 \rangle$$

holds for any $\varphi_1 \in C_0^\infty(U)$ and $\varphi_2 \in C_0^\infty(V)$. Then

$$\langle u_1 \otimes u_2,\, \varphi(x, y) \rangle = \langle u_1,\, \langle u_2,\, \varphi(x, y) \rangle_y \rangle$$

holds for $\varphi(x, y) \in C_0^\infty(U \times V)$, where $\langle\rangle_y$ denotes the pairing of $\mathcal{D}'(V)$ and $C_0^\infty(V)$ with x fixed. See Chapter 4 of [32] for details. If $u_1 \in \mathcal{S}'(\mathbb{R}^n)$ and $u_2 \in \mathcal{S}'(\mathbb{R}^m)$, then $u_1 \otimes u_2$ belongs to $\mathcal{S}'(\mathbb{R}^{n+m})$.

(2) Let U be an open set of $\mathbb{R}^n$. For $u_1 \in C^\infty(U)$ and $u_2 \in \mathcal{D}'(U)$, the product $u = u_1 u_2$ is well-defined as an element of $\mathcal{D}'(U)$ and the Leibniz rule $\partial_i(u_1 u_2) = (\partial_i u_1)u_2 + u_1(\partial_1 u_2)$ holds for $i = 1, \ldots, n$. In fact, the pairing

$$\langle u_1 u_2, \varphi \rangle = \langle u_2, u_1 \varphi \rangle$$

is well-defined for $\varphi \in C_0^\infty(U)$, and we have

$$\begin{aligned}\langle \partial_i(u_1u_2),\, \varphi\rangle &= -\langle u_1u_2,\, \partial_i\varphi\rangle = -\langle u_2,\, u_1\partial_i\varphi\rangle\\ &= -\langle u_2,\, \partial_i(u_1\varphi) - (\partial_i u_1)\varphi\rangle\\ &= \langle \partial_i u_2,\, u_1\varphi\rangle + \langle u_2,\, (\partial_i u_1)\varphi\rangle\\ &= \langle u_1(\partial_i u_2) + (\partial_i u_1)u_2,\, \varphi\rangle.\end{aligned}$$

(3) Let u_1 belong to $\mathcal{S}(\mathbb{R}^n)$ and u_2 to $\mathcal{S}'(\mathbb{R}^n)$. Then u_1u_2 is well-defined as an element of $\mathcal{S}'(\mathbb{R}^n)$ and the Leibniz rule holds.

(4) Let u_1 and u_2 be measurable functions on U. Let p, q be positive real numbers or infinity such that $1/p + 1/q = 1$. If u_1 is locally L^p and u_2 is locally L^q (i.e., $|u_1|^p$ and $|u_2|^q$ are locally integrable if $p, q > 1$; the measure of the set $\{x \in U \mid |u_1(x)| > a\}$ is zero for some $a \in \mathbb{R}$ in case $p = \infty$), then the product $u = u_1u_2$ is well-defined as a locally integrable function. But the Leibniz rule does not make sense; in fact, the product $(\partial_1 u_1)u_2$ cannot be defined in general.

However, for example in one variable $x = x_1$, the product $\delta(x)^2$ or $Y(x)\delta(x)$ cannot be defined as distributions. If $u(x)$ is locally integrable, then $Y(x)u(x)$ is also a locally integrable function. But $\delta(x)u(x)$ cannot be defined in general. In particular, the Leibniz rule

$$\partial_x(Y(x)u(x)) = Y(x)u'(x) + \delta(x)u(x)$$

does not make sense in general because the products on the right-hand side cannot be defined unless u is C^∞ while the left-hand side is well-defined as distribution.

Proposition 4.10. *Let $f(x)$ be a real-valued C^∞ function on an open subset U of $\mathbb{R}^n$. If $u(t)$ belongs to $\mathcal{D}'(\mathbb{R})$ and $v(x)$ belongs to $\mathcal{D}'(U)$, then $u(t - f(x))v(x)$ is well-defined as an element of $\mathcal{D}'(\mathbb{R} \times U)$ and the Leibniz formulae*

$$\begin{aligned}&\frac{\partial}{\partial t}(u(t - f(x))v(x)) = u'(t - f(x))v(x),\\ &\frac{\partial}{\partial x_i}(u(t - f(x))v(x)) = u(t - f(x))\frac{\partial}{\partial x_i}v(x) - \frac{\partial f}{\partial x_i}u'(t - f(x))v(x)\end{aligned}$$

hold for $i = 1, \ldots, n$ with $u'(t)$ being the derivative of $u(t)$. Moreover, if $f(x)$ is a polynomial, $u(t)$ belongs to $\mathcal{S}'(\mathbb{R})$ and $v(x)$ belongs to $\mathcal{S}'(\mathbb{R}^n)$, then $u(t - f(x))v(x)$ belongs to $\mathcal{S}'(\mathbb{R}^{n+1})$.

Proof. For $\varphi(t, x) \in C_0^\infty(\mathbb{R} \times U)$, set

$$\langle u(t - f(x))v(x),\, \varphi(t, x)\rangle = \langle u(t)v(x),\, \varphi(t + f(x), x)\rangle.$$

Since $\varphi(t+f(x),x)$ belongs to $C_0^\infty(\mathbb{R}\times U)$, this defines an element of $\mathcal{D}'(\mathbb{R}\times U)$. For $\varphi \in C_0^\infty(\mathbb{R}\times U)$, we have

$$\begin{aligned}
&\left\langle \frac{\partial}{\partial t}(u(t-f(x))v(x)),\ \varphi(t,x)\right\rangle = -\left\langle u(t-f(x))v(x),\ \frac{\partial\varphi}{\partial t}(t,x)\right\rangle\\
&= -\left\langle u(t)v(x),\ \frac{\partial\varphi}{\partial t}(t+f(x),x)\right\rangle\\
&= \langle u'(t)v(x),\ \varphi(t+f(x),x)\rangle = \langle u'(t-f(x))v(x),\ \varphi(t,x)\rangle
\end{aligned}$$

and

$$\begin{aligned}
&\left\langle \frac{\partial}{\partial x_i}(u(t-f(x))v(x)),\ \varphi(t,x)\right\rangle = -\left\langle u(t-f(x))v(x),\ \frac{\partial\varphi}{\partial x_i}(t,x)\right\rangle\\
&= -\left\langle u(t)v(x),\ \frac{\partial\varphi}{\partial x_i}(t+f(x),x)\right\rangle\\
&= -\left\langle u(t)v(x),\ \frac{\partial}{\partial x_i}\varphi(t+f(x),x) - \frac{\partial f}{\partial x_i}\frac{\partial\varphi}{\partial t}(t+f(x),x)\right\rangle\\
&= \langle u(t)\partial_{x_i}v(x),\ \varphi(t+f(x),x)\rangle - \left\langle u'(t)\frac{\partial f}{\partial x_i}v(x),\ \varphi(t+f(x),x)\right\rangle\\
&= \langle u(t-f(x))\partial_{x_i}v(x),\ \varphi(t,x)\rangle - \left\langle u'(t-f(x))\frac{\partial f}{\partial x_i}v(x),\ \varphi(t,x)\right\rangle.
\end{aligned}$$

If $\varphi(t,x)$ belongs to $\mathcal{S}(\mathbb{R}^{n+1})$ and $f(x)$ is a polynomial, then $\varphi(t+f(x),x)$ also belongs to $\mathcal{S}(\mathbb{R}^{n+1})$, as is seen by the inequality $t^2 \le 2((t+f(x))^2 + f(x)^2)$. This implies the last assertion. Q.E.D.

Example 4.11. Let $f(x)$ be a polynomial in $x=(x_1,\dots,x_n)$ with real coefficients and $v(x)$ be an element of $\mathcal{S}'(\mathbb{R}^n)$. Then $Y(t-f(x))v(x)$ and $\delta^{(k)}(t-f(x))v(x)$ with a non-negative integer k are well-defined as elements of $\mathcal{S}'(\mathbb{R}^{n+1})$ and we have

$$\begin{aligned}
&\frac{\partial}{\partial t}(Y(t-f(x))v(x)) = \delta(t-f(x))v(x),\\
&\frac{\partial}{\partial t}(\delta^{(k)}(t-f(x))v(x)) = \delta^{(k+1)}(t-f(x))v(x).
\end{aligned}$$

Exercise 19. Let u be a distribution on $\mathbb{R}^n$ satisfying $\partial_1 u = 0$.

(1) Show that $\langle u,\varphi\rangle = 0$ if $\varphi \in C_0^\infty(\mathbb{R}^n)$ satisfies

$$\int_{-\infty}^{\infty} \varphi(t,x_2,\dots,x_n)\,dt = 0.$$

(2) Show that there exists a distribution v on $\mathbb{R}^{n-1}$ such that $u = 1\otimes v$.

4.3. Integrals of distributions

Let us consider distributions in variables (x, y) with $x = (x_1, \dots, x_n)$ and $y = (y_1, \dots, y_d)$. We regard x as the integration variables and y as parameters. Let $\varpi : \mathbb{R}^{n+d} \ni (x, y) \mapsto y \in \mathbb{R}^d$ be the projection. Let U be an open set of $\mathbb{R}^d$ and let u be a distribution defined on $\varpi^{-1}(U) = \mathbb{R}^n \times U$.

We would like to define the integral

$$\int_{\mathbb{R}^n} u(x, y)\, dx = \int_{\mathbb{R}^n} u(x, y)\, dx_1 \cdots dx_n$$

along the fibers of ϖ (i.e., with respect to x) as a distribution on U. However, we need some 'tameness' of u with respect to x for this integral to be well-defined. Let us introduce the following two sufficient conditions:

(1) Let u be a distribution on $\varpi^{-1}(U)$ such that $\varpi : \operatorname{supp} u \to \mathbb{R}^d$ is proper, i.e., for any compact set K of U, $\varpi^{-1}(K) \cap \operatorname{supp} u$ is compact.

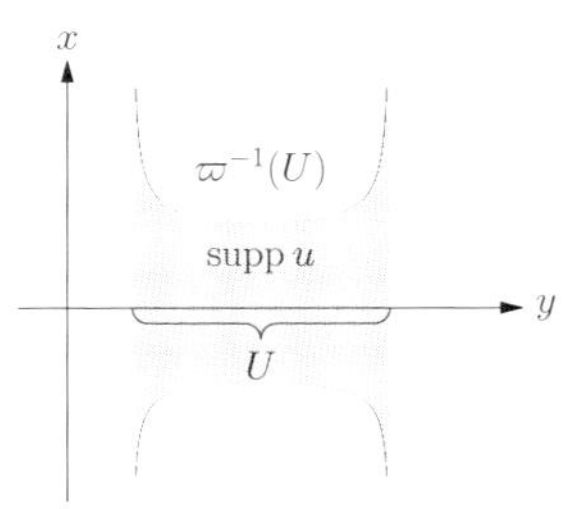

Fig. 1. An example of the support of $u \in \mathcal{E}'\mathcal{D}'(\mathbb{R}_x \times U)$

Let us denote by $\mathcal{E}'\mathcal{D}'(\mathbb{R}^n \times U)$ the set of such distributions, which constitutes a left D_{n+d}-submodule of $\mathcal{D}'(\mathbb{R}^n \times U)$. The integral of $u \in \mathcal{E}'\mathcal{D}'(\mathbb{R}^n \times U)$ with respect to x is defined by

$$\left\langle \int_{\mathbb{R}^n} u(x, y)\, dx,\ \varphi(y) \right\rangle = \langle u(x, y), 1(x)\varphi(y)\rangle \qquad (\forall \varphi(y) \in C_0^\infty(U)),$$

where $1(x)$ denotes the constant function with value 1. This integral belongs to $\mathcal{D}'(U)$. More precisely, the pairing above is defined as follows:

Choose $\chi(x, y) \in C^\infty(\varpi^{-1}(U))$ such that $\chi(x, y) = 1$ on an open set W of $\varpi^{-1}(U)$ which contains $\operatorname{supp} u$ and that $\varpi : \operatorname{supp} \chi \to U$ is proper, by using a partition of unity. Then we define

$$\langle u(x, y),\ 1(x)\varphi(y)\rangle := \langle u(x, y),\ \chi(x, y)\varphi(y)\rangle.$$

The right-hand side does not depend on such $\chi(x,y)$ since $\operatorname{supp}(1-\chi)\cap\operatorname{supp}u=\emptyset$.

(2) Let $\mathcal{SS}'(\mathbb{R}^n\times\mathbb{R}^d)$ be the subspace of $\mathcal{S}'(\mathbb{R}^{n+d})$ consisting of distributions of the form

$$u(x,y)=\sum_{j=1}^{m}u_j(x)v_j(x,y)\qquad(m\in\mathbb{N},\ u_j\in\mathcal{S}(\mathbb{R}^n),\ v_j\in\mathcal{S}'(\mathbb{R}^{n+d})). \tag{11}$$

We also denote $\mathcal{SS}'(\mathbb{R}^n_x\times\mathbb{R}^d_y)$ to clarify the variables. Then $\mathcal{SS}'(\mathbb{R}^n\times\mathbb{R}^d)$ is a left D_{n+d}-submodule of $\mathcal{S}'(\mathbb{R}^{n+d})$. The integral of $u(x,y)$ is naturally defined as an element of $S'(\mathbb{R}^d)$ by the pairing

$$\left\langle\int_{\mathbb{R}^n}u(x,y)\,dx,\ \varphi(y)\right\rangle=\sum_{j=1}^{m}\langle v_j(x,y),\ u_j(x)\varphi(y)\rangle\qquad(\forall\varphi\in\mathcal{S}(\mathbb{R}^d)).$$

This integral does not depend on the choice of expression (11). In fact, assume $u(x,y)=0$ in (11) and take $\chi(x)\in C_0^\infty(\mathbb{R}^n)$ such that $\chi(x)=1$ if $|x|\le 1$. Then for an arbitrary constant $r>0$, we have an equality

$$0=\left\langle\sum_{j=1}^{m}u_j(x)v_j(x,y),\ \chi\Big(\frac{x}{r}\Big)\varphi(y)\right\rangle=\sum_{j=1}^{m}\Big\langle v_j(x,y),\ \chi\Big(\frac{x}{r}\Big)u_j(x)\varphi(y)\Big\rangle$$

for any $\varphi\in\mathcal{S}(\mathbb{R}^d)$. Since $\chi(x/r)u_j(x)\varphi(y)$ converges to $u_j(x)\varphi(y)$ in $\mathcal{S}(\mathbb{R}^{n+d})$ as $r\to\infty$, we get

$$\sum_{j=1}^{m}\langle v_j(x,y),\ u_j(x)\varphi(y)\rangle=0.$$

Proposition 4.12 (differentiation under the integral sign). *Let $u(x,y)$ belong to $\mathcal{E}'\mathcal{D}'(\mathbb{R}^n\times U)$ with an open subset U of $\mathbb{R}^d$, or else to $\mathcal{SS}'(\mathbb{R}^n\times\mathbb{R}^d)$. Then*

$$P(y,\partial_y)\int_{\mathbb{R}^n}u(x,y)\,dx=\int_{\mathbb{R}^n}P(y,\partial_y)u(x,y)\,dx$$

holds for any $P=P(y,\partial_y)\in D_d$.

Proof. Let $u(x,y)=\sum_{j=1}^m u_j(x)v_j(x,y)$ with $u_j\in\mathcal{S}(\mathbb{R}^n)$ and $v_j\in\mathcal{S}'(\mathbb{R}^{n+d})$. Then for any $\varphi\in\mathcal{S}(\mathbb{R}^d)$, we have

$$\begin{aligned}
&\left\langle \partial_{y_i}\int_{\mathbb{R}^n} u(x,y)\,dx,\ \varphi(y)\right\rangle = -\left\langle \int_{\mathbb{R}^n} u(x,y)\,dx,\ \partial_{y_i}\varphi(y)\right\rangle\\
&= -\sum_{j=1}^m \langle v_j(x,y),\ u_j(x)\partial_{y_i}\varphi(y)\rangle = -\sum_{j=1}^m \langle v_j(x,y),\ \partial_{y_i}(u_j(x)\varphi(y))\rangle\\
&= \sum_{j=1}^m \langle \partial_{y_i}v_j(x,y),\ u_j(x)\varphi(y)\rangle = \left\langle \int_{\mathbb{R}^n} \partial_{y_i}u(x,y)\,dx,\ \varphi(y)\right\rangle
\end{aligned}$$

and

$$\begin{aligned}
&\left\langle y_i\int_{\mathbb{R}^n} u(x,y)\,dx,\ \varphi(y)\right\rangle = \sum_{j=1}^m \langle v_j(x,y),\ y_iu_j(x)\varphi(y)\rangle\\
&\quad= \sum_{j=1}^m \langle y_iv_j(x,y),\ u_j(x)\varphi(y)\rangle = \left\langle \int_{\mathbb{R}^n} y_iu(x,y)\,dx,\ \varphi(y)\right\rangle.
\end{aligned}$$

The case $u\in\mathcal{E}'\mathcal{D}'(\mathbb{R}^n\times U)$ can be proved in the same manner. Q.E.D.

Example 4.13. Let $\delta(t)$ be the univariate delta function. We have $\int_{\mathbb{R}}\delta(t)\,dt=1$ since

$$\int_{\mathbb{R}}\delta(t)=\langle\delta(t),1(t)\rangle=1.$$

More generally, let $f(t,x)=f(t,x_1,\dots,x_n)$ be a C^∞ function on $\mathbb{R}\times U$ with an open set U of $\mathbb{R}^n$. Since $\operatorname{supp}f(t,x)\delta(t)\subset\{(t,x)\mid t=0\}$, $f(t,x)\delta(t)$ belongs to $\mathcal{E}'\mathcal{D}'(\mathbb{R}\times U)$. Note that $f(t,x)\delta(t)=f(0,x)\delta(t)$ holds since there exists $g\in C^\infty(\mathbb{R}\times U)$ such that $f(t,x)-f(0,x)=tg(t,x)$. Hence we get

$$\int_{\mathbb{R}} f(t,x)\delta(t)\,dt=\int_{\mathbb{R}} f(0,x)\delta(t)\,dt=f(0,x)\int_{\mathbb{R}}\delta(t)\,dt=f(0,x).$$

Example 4.14. Set $x=(x_1,\dots,x_n)$ and let a be an arbitrary positive constant. Let $f(x)$ be a real polynomial in x. Then $\exp(-a|x|^2)Y(t-f(x))$ belongs to $\mathcal{SS}'(\mathbb{R}^n_x\times\mathbb{R}_t)$. Hence the integral

$$F(t)=\int_{\mathbb{R}^n}\exp(-a|x|^2)Y(t-f(x))\,dx$$

is well-defined as an element of $\mathcal{S}'(\mathbb{R}_t)$. Up to a constant multiple, $F(t)$ is the cumulative distribution function of $f(x)$ with x being the random vector with an n-dimensional normal (Gaussian) distribution. By Proposition 4.12 the derivative $F'(t)$ is given by the integral

$$F'(t) = \int_{\mathbb{R}^n} \exp(-a|x|^2)\delta(t - f(x))\,dx$$

as an element of $\mathcal{S}'(\mathbb{R})$.

Example 4.15. Let $f(x)$ be a real polynomial in $x = (x_1, \ldots, x_n)$, and $a_1, \ldots, a_n, b_1, \ldots, b_n$ be positive real numbers. Let us consider the integral

$$F(t) = \int_{\mathbb{R}^n} e^{-b_1x_1-\cdots-b_nx_n} Y(t - f(x))(x_1)_+^{a_1-1} \cdots (x_n)_+^{a_n-1}\,dx,$$

which can be regarded, up to a constant multiple depending on a_i, b_i, as the cumulative distribution function of $f(x)$ with x being the random vector with a multi-dimensional gamma distribution. Let $\chi(t)$ be a C^∞ function on $\mathbb{R}$ such that $\chi(t) = 1$ for $t \geq -1$ and $\chi(t) = 0$ for $t \leq -2$. Then we have

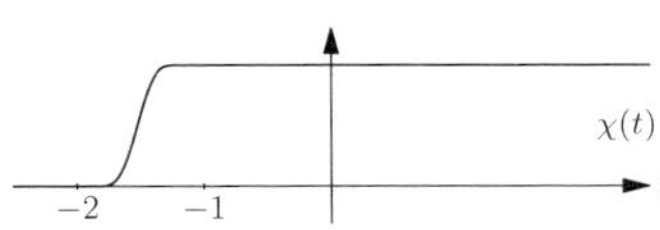

Fig. 2. The graph of $\chi(t)$ in Example 4.15

$$\begin{aligned} &e^{-b_1x_1-\cdots-b_nx_n}(x_1)_+^{a_1-1} \cdots (x_n)_+^{a_n-1} \\ &\qquad = e^{-b_1x_1-\cdots-b_nx_n}\chi(x_1)\cdots\chi(x_n)(x_1)_+^{a_1-1} \cdots (x_n)_+^{a_n-1} \end{aligned}$$

and $e^{-b_1x_1-\cdots-b_nx_n}\chi(x_1)\cdots\chi(x_n)$ belongs to $\mathcal{S}(\mathbb{R}^n_x)$. Hence the integrand belongs to $\mathcal{SS}'(\mathbb{R}^n_x \times \mathbb{R}_t)$ and consequently $F(t)$ is well-defined as an element of $\mathcal{S}'(\mathbb{R})$. Its derivative is given by

$$F'(t) = \int_{\mathbb{R}^n} e^{-b_1x_1-\cdots-b_nx_n}\delta(t - f(x))(x_1)_+^{a_1-1} \cdots (x_n)_+^{a_n-1}\,dx.$$

Example 4.16. Set $x = (x_1, \ldots, x_n)$ and let $a_1, \ldots, a_n, b_1, \ldots, b_n$ be positive real numbers. For a real polynomial $f(x)$, set

$$\begin{aligned} &F(t) \\ &= \int_{\mathbb{R}^n} (x_1)_+^{a_1-1} \cdots (x_n)_+^{a_n-1}(1-x_1)_+^{b_1-1} \cdots (1-x_n)_+^{b_n-1} Y(t-f(x))\,dx. \end{aligned}$$

The integrand is integrable and belongs to $\mathcal{E}'\mathcal{D}'(\mathbb{R}^n_x \times \mathbb{R}_t)$ since its support is contained in the n-cube $[0,1] \times \cdots \times [0,1]$. Hence $F(t)$ and its derivatives are well-defined as elements of $\mathcal{D}'(\mathbb{R})$. (In fact, $F(t)$ belongs to $\mathcal{S}'(\mathbb{R})$ since the integrand belongs also to $\mathcal{SS}'(\mathbb{R}^n_x \times \mathbb{R}_t)$ multiplied by a suitable cut-off function.) Up to a constant multiple, $F(t)$ is the cumulative distribution function of $f(x)$ with x regarded as the random vector with a multivariate beta distribution.

The following proposition will play a crucial role in the integration algorithm for holonomic distributions, which will be introduced in the following sections.

Proposition 4.17. *Let $u(x,y)$ belong to $\mathcal{E}'\mathcal{D}'(\mathbb{R}^n \times U)$ with an open subset U of $\mathbb{R}^d$, or else to $\mathcal{SS}'(\mathbb{R}^n \times \mathbb{R}^d)$. Then one has*

$$\int_{\mathbb{R}^n} \partial_{x_i} u(x,y)\,dx = 0 \qquad (i = 1, \dots, n).$$

Proof. First, let us assume that $u(x,y)$ belongs to $\mathcal{E}'\mathcal{D}'(\mathbb{R}^n \times U)$. Let $\chi(x,y)$ be an element of $C^\infty(\mathbb{R}^n \times U)$ which takes the value 1 on an open subset of $\mathbb{R}^n \times U$ containing $\operatorname{supp} u$ such that the projection map $\operatorname{supp}\chi \to U$ is proper. Then we have, by the definition of the integral,

$$\begin{aligned}\left\langle \int_{\mathbb{R}^n} \partial_{x_i} u(x,y)\,dx,\ \varphi(y) \right\rangle &= \langle \partial_{x_i} u(x,y),\ \chi(x,y)\varphi(y)\rangle \\ &= -\langle u(x,y),\ \partial_{x_i}\chi(x,y)\varphi(y)\rangle = 0\end{aligned}$$

for any $\varphi \in C_0^\infty(U)$ since $\partial_{x_i}\chi(x,y)$ vanishes on an open set containing $\operatorname{supp} u$.

Next, let us assume that $u(x,y)$ belongs to $\mathcal{SS}'(\mathbb{R}^n \times \mathbb{R}^d)$. We may assume, without loss of generality, that $u(x,y) = v(x)w(x,y)$ with $v \in \mathcal{S}(\mathbb{R}^n)$ and $w \in \mathcal{S}'(\mathbb{R}^{n+d})$. Then it follows from the definition of the integral that

$$\begin{aligned}&\left\langle \int_{\mathbb{R}^n} \partial_{x_i}(v(x)w(x,y))\,dx,\ \varphi(y) \right\rangle \\ &= \left\langle \int_{\mathbb{R}^n} (\partial_{x_i} v(x))w(x,y)\,dx,\ \varphi(y) \right\rangle + \left\langle \int_{\mathbb{R}^n} v(x)(\partial_{x_i} w(x,y))\,dx,\ \varphi(y) \right\rangle \\ &= \langle w(x,y),\, \partial_{x_i} v(x)\varphi(y)\rangle + \langle \partial_{x_i} w(x,y),\, v(x)\varphi(y)\rangle \\ &= \langle w(x,y),\, \partial_{x_i} v(x)\varphi(y)\rangle - \langle w(x,y),\, \partial_{x_i}(v(x)\varphi(y))\rangle = 0\end{aligned}$$

holds for any $\varphi \in \mathcal{S}(\mathbb{R}^n)$. Q.E.D.

4.4. Holonomic distributions

We assume $\mathbb{K} = \mathbb{C}$. A distribution $u(x) \in \mathcal{D}(U)$, with an open set U of $\mathbb{R}^n$, is called *holonomic* if there exist a finite set of differential operators $P_1, \ldots, P_m$ which annihilate u, i.e., $P_i u = 0$ holds in $\mathcal{D}'(U)$ for $i = 1, \ldots, m$ such that the left D_n-module $D_n/(D_n P_1 + \cdots + D_n P_m)$ is holonomic. In other words, $u(x)$ is holonomic if and only if $D_n/\mathrm{Ann}_{D_n} u$ is a holonomic D_n-module, where

$$\mathrm{Ann}_{D_n} u = \{P \in D_n \mid Pu = 0\}$$

is the annihilator (ideal) of u.

For example, the univariate delta function $\delta(x)$ and the Heaviside function $Y(x)$ are holonomic since $x\delta(x) = x\partial_x Y(x) = 0$.

Proposition 4.18. *If elements u and v of $\mathcal{D}'(U)$ are holonomic, then $C_1 u + C_2 v$ and Pu are also holonomic for any $C_1, C_2 \in \mathbb{C}$ and $P \in D_n$.*

Proof. Set $I = \mathrm{Ann}_{D_n} u$ and $J = \mathrm{Ann}_{D_n} v$. Then D_n/I and D_n/J are holonomic. Since the annihilator of $C_1 u + C_2 v$ contains $I \cap J$, we have only to show that $D_n/(I \cap J)$ is holonomic. The left D_n-homomorphism of D_n to $(D_n)^2$ which sends $Q \in D_n$ to $(Q, -Q)$ induces an injective homomorphism

$$D_n/(I \cap J) \longrightarrow (D_n/I) \oplus (D_n/J).$$

This implies that $D_n/(I \cap J)$ is holonomic since $(D_n/I) \oplus (D_n/J)$ is holonomic.

The left ideal $I : P = \{Q \in D_n \mid QP \in I\}$ coincides with $\mathrm{Ann}_{D_n} Pu$. The left D_n-endomorphism of D_n which sends $Q \in D_n$ to QP induces an injective homomorphism $D_n/(I : P) \to D_n/I$. Hence $D_n/(I : P)$ is holonomic. Q.E.D.

Definition 4.19. Let $f(x)$ be a real polynomial in $x = (x_1, \ldots, x_n)$. Set $\partial_x = (\partial_1, \ldots, \partial_n)$ with $\partial_i = \partial/\partial x_i$. Let t be a single variable and set $\partial_t = \partial/\partial t$. For $P = P(x, \partial_x) \in D_n$, define $\tau(P, f) \in D_{n+1}$ by

$$\tau(P, f) = P(x, \partial_1 + f_1 \partial_t, \ldots, \partial_n + f_n \partial_t)$$

with $f_i = \partial f/\partial x_i$. This substitution is well-defined since $x_1, \ldots, x_n$ and $\partial_1 + f_1\partial_t, \ldots, \partial_n + f_n\partial_t$ satisfy the same commutation relations as $x_1, \ldots, x_n$ and $\partial_1, \ldots, \partial_n$.

Proposition 4.20. *Let $f(x)$ be a real polynomial in $x = (x_1, \ldots, x_n)$ and suppose that $v \in \mathcal{D}'(U)$ with an open set U of $\mathbb{R}^n$ is holonomic. Then $Y(t - f(x))v(x)$ is holonomic. More concretely, let I be a left ideal of*

D_n which is contained in $\mathrm{Ann}_{D_n} v$ such that D_n/I is holonomic. Let J be the left ideal of D_{n+1} which is generated by $\{\tau(P,f) \mid P \in I\}$. Then left ideals

$$J_0 := J + D_{n+1}(t - f(x))\partial_t, \qquad J_1 := J + D_{n+1}(t - f(x))$$

of D_{n+1} annihilate $Y(t - f(x))v(x)$ and $\delta(t - f(x))v(x)$ respectively and both D_{n+1}/J_0 and D_{n+1}/J_1 are holonomic.

Proof. We have

$$(\partial_i + f_i\partial_t)(Y(t - f(x))v(x)) = Y(t - f(x))\partial_i v(x),$$
$$(\partial_i + f_i\partial_t)(\delta(t - f(x))v(x)) = \delta(t - f(x))\partial_i v(x)$$

for $i = 1, \ldots, n$ by Proposition 4.10. Hence

$$\tau(P,f)(Y(t - f(x))v(x)) = Y(t - f(x))Pv(x),$$
$$\tau(P,f)(\delta(t - f(x))v(x)) = \delta(t - f(x))Pv(x)$$

hold for any $P \in D_n$. It follows that J_0 and J_1 annihilate $Y(t-f(x))v(x)$ and $\delta(t - f(x))v(x)$ respectively.

Let us show that D_{n+1}/J_0 is holonomic. Since D_n/I is holonomic, its characteristic variety $\mathrm{Char}(D_n/I)$ is an n-dimensional algebraic set of $\mathbb{C}^{2n}$. By the definition, we have

$$\begin{aligned}
&\mathrm{Char}(D_{n+1}/J_0)\\
&\subset \Big\{(x,t,\xi,\tau) \in \mathbb{C}^{2(n+1)} \mid \sigma(P)(x, \xi_1 + f_1\tau, \ldots, \xi_n + f_n\tau) = 0\\
&\qquad (\forall P \in I),\ (t - f(x))\tau = 0\Big\}\\
&= \Big\{(x,t,\xi,\tau) \mid (x, \xi_1 + f_1\tau, \ldots, \xi_n + f_n\tau) \in \mathrm{Char}(D_n/I),\ t = f(x)\Big\}\\
&\quad \cup \Big\{(x,t,\xi,\tau) \mid (x, \xi_1, \ldots, \xi_n) \in \mathrm{Char}(D_n/I),\ \tau = 0\Big\}.
\end{aligned}$$

Since the last two sets are in one-to-one correspondence with the set $\mathrm{Char}(D_n/I) \times \mathbb{C}$, the dimension of $\mathrm{Char}(D_{n+1}/J)$ is $n+1$, which implies that D_{n+1}/J_0 is a holonomic module. Similarly, D_{n+1}/J_1 is also holonomic. Q.E.D.

Example 4.21. Let $f(x)$ be a real polynomial in $x = (x_1, \ldots, x_n)$ and $a_1, \ldots, a_n$ be positive real numbers. Set

$$u(x,t) = \exp(-a_1x_1^2 - \cdots - a_nx_n^2)\delta(t - f(x)).$$

Then $u = u(x,t)$ satisfies a holonomic system

$$(t - f(x))u = (\partial_1 + f_1\partial_t + 2a_1x_1)u = \cdots = (\partial_n + f_n\partial_t + 2a_nx_n)u = 0.$$

Lemma 4.22. *Let a be a positive real number. Then the univariate locally integrable function t_+^{a-1} in t satisfies $(t\partial_t - a + 1)t_+^{a-1}$ in $\mathcal{S}'(\mathbb{R})$.*

Proof. Let $\varphi(t)$ belong to $\mathcal{S}(\mathbb{R})$. Then we have

$$\begin{aligned}\langle t\partial_t t_+^{a-1}, \varphi(t)\rangle &= -\langle t_+^{a-1}, \partial_t(t\varphi(t))\rangle = -\langle t_+^{a-1}, \varphi(t)\rangle - \langle t_+^{a-1}, t\varphi'(t)\rangle \\ &= -\int_0^\infty t^{a-1}\varphi(t)\,dt - \int_0^\infty t^a\varphi'(t)\,dt \\ &= -\int_0^\infty t^{a-1}\varphi(t)\,dt + a\int_0^\infty t^{a-1}\varphi(t)\,dt \\ &= \langle (a-1)t_+^{a-1}, \varphi(t)\rangle\end{aligned}$$

by integration by parts. Q.E.D.

Example 4.23. Let $f(x)$ be a real polynomial in $x = (x_1, \dots, x_n)$ and $a_1, \dots, a_n, b_1, \dots, b_n$ be positive real numbers. Set

$$u(x,t) = \exp(-b_1x_1 - \cdots - b_nx_n)\delta(t - f(x))(x_1)_+^{a_1-1}\cdots(x_n)_+^{a_n-1}.$$

Then $u = u(x,t)$ satisfies a holonomic system

$$(t - f(x))u = (x_i(\partial_i + f_i\partial_t + b_i) - a_i + 1)u = 0 \quad (i = 1, \dots, n).$$

Exercise 20. Set $n = 1$, $x = x_1$, $\partial = \partial_1$ and $M := D_1/D_1x\partial$. Show that $\mathrm{Hom}_{D_1}(M, \mathcal{D}'(\mathbb{R}))$ is two dimensional and spanned by $Y(x)$ and $Y(-x)$.

Exercise 21. Let $\lambda_1, \dots, \lambda_n$ be complex numbers such that $\mathrm{Re}\,\lambda_i > -1$ for $i = 1, \dots, n$. Set $f(x) = (x_1)_+^{\lambda_1}\cdots(x_n)_+^{\lambda_n}$.

(1) Show that $f(x)$ is locally integrable on $\mathbb{R}^n$ and belongs to $\mathcal{S}'(\mathbb{R}^n)$.

(2) Show that $f(x)$ satisfies linear differential equations

$$(x_1\partial_1 - \lambda_1)f(x) = \cdots = (x_n\partial_n - \lambda_n)f(x) = 0.$$

(3) Find the characteristic variety and the singular locus of the left D_n-module

$$M := D_n/(D_n(x_1\partial_1 - \lambda_1) + \cdots + D_n(x_n\partial_n - \lambda_n)).$$

4.5. Distributions with smooth parameters

Let Ω be an open set of $\mathbb{R}^p$. Introducing parameters $a = (a_1, \dots, a_p)$, let us define the space $\mathcal{ES}(\Omega \times \mathbb{R}^n)$ of rapidly decreasing functions with smooth parameters as the set of $\varphi(a,x) \in C^\infty(\Omega \times \mathbb{R}_x^n)$ such that

$$\sup_{K\times\mathbb{R}^n} |x^\alpha\partial_x^\beta\partial_a^\gamma\varphi(a,x)| < \infty$$

for any $\alpha, \beta \in \mathbb{N}^n$, $\gamma \in \mathbb{N}^p$, and any compact subset K of Ω. It is easy to see that $\mathcal{ES}(\Omega \times \mathbb{R}^n)$ is a left D_{p+n}-submodule of $C^\infty(\Omega \times \mathbb{R}^n)$. For example, e^{-ax^2} belongs to $\mathcal{ES}(\Omega \times \mathbb{R})$ with $\Omega = \{a \in \mathbb{R} \mid a > 0\}$.

Let $\mathcal{ESS}'(\Omega \times \mathbb{R}^n \times \mathbb{R}^d)$ be the set of $u \in \mathcal{D}'(\Omega \times \mathbb{R}^n_x \times \mathbb{R}^d_y)$ which can be written as

$$u(a, x, y) = \sum_{j=1}^{m} u_j(a, x) v_j(x, y)$$

with $u_j \in \mathcal{ES}(\Omega \times \mathbb{R}^n)$, $v_j \in \mathcal{S}'(\mathbb{R}^n \times \mathbb{R}^d)$, and $m \in \mathbb{N}$. Then the integral $\int_{\mathbb{R}^n} u(a, x, y)\, dx$ of u with respect to x is defined by

$$\left\langle \int_{\mathbb{R}^n} u(a, x, y)\, dx,\ \varphi(a, y) \right\rangle = \sum_{j=1}^{m} \int_{\Omega} \langle v_j(x, y),\, u_j(a, x)\varphi(a, y)\rangle_{(x,y)}\, da$$

as an element of $\mathcal{D}'(\Omega \times \mathbb{R}^d)$ for $\varphi \in C_0^\infty(\Omega \times \mathbb{R}^d)$, where $\langle , \rangle_{(x,y)}$ denotes the pairing of $\mathcal{D}'(\mathbb{R}^n_x \times \mathbb{R}^d_y)$ and $C_0^\infty(\mathbb{R}^n_x \times \mathbb{R}^d_y)$ with a fixed, which is C^∞ with respect to a. The well-definedness of the integal can be proved by using a suitable cut-off function in the same way as in the case without parameters. For example, $e^{-ax^2}\delta(y - x^3)$ belongs to $\mathcal{ESS}'(\Omega \times \mathbb{R}_x \times \mathbb{R}_y)$ with $\Omega = \{a \in \mathbb{R} \mid a > 0\}$.

We can also consider the case $d = 0$, which we denote $\mathcal{ESS}'(\Omega \times \mathbb{R}^n)$. For example, $e^{ax - x^3}Y(x) = e^{ax - x^3}\chi(x)Y(x)$ belongs to $\mathcal{ESS}'(\mathbb{R}_a \times \mathbb{R}_x)$, where $\chi(x)$ is the same cut-off function as in Example 4.15. If $u(a, x)$ belongs to $\mathcal{ESS}'(\Omega \times \mathbb{R}^n)$, then $\int_{\mathbb{R}^n} u(a, x)\, dx$ belongs to $C^\infty(\Omega)$.

Proposition 4.24. *Let Ω be an open set of $\mathbb{R}^p_a$ and let u belong to $\mathcal{ESS}'(\Omega \times \mathbb{R}^n_x \times \mathbb{R}^d_y)$.*

(1) *$P \int_{\mathbb{R}^n} u(a, x, y)\, dx = \int_{\mathbb{R}^n} Pu(a, x, y)\, dx$ holds for any differential operator $P \in D_{p+d}$ in the variables (a, y).*

(2) *$\int_{\mathbb{R}^n} \partial_{x_i} u(a, x, y)\, dx = 0$ holds for any $i = 1, \dots, n$.*

The proof is similar to the case without parameters.

Example 4.25. Set $x = (x_1, \dots, x_n)$ and let $a > 0$ and $b = (b_1, \dots, b_n) \in \mathbb{R}^n$. Let $f(x)$ be a real polynomial in x and set

$$u(x, t, a, b) = \left(\frac{a}{\pi}\right)^{\frac{n}{2}} \exp(-a(x - b)^2)Y(t - f(x))$$

with $(x - b)^2 = \sum_{j=1}^{n}(x_j - b_j)^2$. Then $u(x, t, a, b)$ belongs to $\mathcal{ESS}'(\Omega \times \mathbb{R}^n_x \times \mathbb{R}_t)$ with $\Omega = \{(a, b) \in \mathbb{R} \times \mathbb{R}^n \mid a > 0\}$. Moreover, $u(x, t, a, b)$

satisfies a holonomic system in the whole variables (x, t, a, b). Hence the integral

$$F(t, a, b) = \int_{\mathbb{R}^n} u(x, t, a, b)\, dx$$

is well-defined and holonomic as an element of $\mathcal{D}'(\mathbb{R}_t \times \Omega)$.

Example 4.26. Set $x = (x_1, \dots, x_n)$ and let a_{ij} $(1 \le i, j \le n)$ and b_j $(1 \le j \le n)$ be real parameters such that $a_{ij} = a_{ji}$. Set $A = (a_{ij})$ and $b = (b_1, \dots, b_n)$. Let $f(x)$ be a real polynomial in x. Then

$$u(x, t, A, b) = \exp\left(\sum_{i,j=1}^{n} a_{ij} x_i x_j + \sum_{i=1}^{n} b_i x_i\right) Y(t - f(x))$$

belongs to $\mathcal{ESS}'(\Omega \times \mathbb{R}^n_x \times \mathbb{R}_t)$, where Ω is the set of (A, b) with $b \in \mathbb{R}^n$ and a negative definite n by n symmetric matrix $A = (a_{ij})$, i.e.,

$$(-1)^k \det \begin{pmatrix} a_{11} & \cdots & a_{1k} \\ \vdots & & \vdots \\ a_{k1} & \cdots & a_{kk} \end{pmatrix} > 0 \qquad (1 \le k \le n).$$

The integral

$$F(t, A, b) = \int_{\mathbb{R}^n} u(x, t, A, b)\, dx$$

is well-defined and holonomic as an element of $\mathcal{D}'(\mathbb{R}_t \times \Omega)$ since the integrand $u(x, t, A, b)$ satisfies a holonomic system including the parameters.

§5. D-module theoretic integration algorithm

We first recall the notion of integration of D-modules, which is purely algebraic. The most crucial fact is that the integration preserves holonomicity. Then we recall an algorithm for precisely computing the D-module theoretic integration, which was first introduced in [26] and [27]. See also Chapter 5 of [29].

5.1. Integration as an operation on D-modules

Set $x = (x_1, \dots, x_n)$ and $y = (y_1, \dots, y_d)$. In this section we set $X = \mathbb{K}^{n+d}$ and $Y = \mathbb{K}^d$ to simplify the notation. Let $\varpi : X \ni (x, y) \longmapsto y \in Y$ be the projection. We denote by $D_X = D_{n+d}$ the ring of differential operators in the variables (x, y), and by $D_Y = D_d$ the ring of differential operators in the variables y.

The module

$$D_{Y \leftarrow X} := D_X / (\partial_{x_1} D_X + \cdots + \partial_{x_n} D_X)$$

has a structure of (D_Y, D_X)-bimodule. The *integral* of a left D_X-module M along the fibers of ϖ, or the *direct image* by ϖ is defined to be

$$\varpi_* M := D_{Y\leftarrow X} \otimes_{D_X} M = M/(\partial_{x_1} M + \cdots + \partial_{x_n} M).$$

This is a left D_Y-module since any element of D_Y commutes with ∂_{x_j}. For an element u of M, let $[u]$ be its residue class in $\varpi_* M$. If M is generated by $u_1, \dots, u_r$ over D_X, then $\varpi_* M$ is generated by the set $\{x^\alpha [u_j] \mid 1 \le j \le r,\ \alpha \in \mathbb{N}^n\}$ over D_Y.

Now assume $\mathbb{K} = \mathbb{C}$ and let φ be a D_X-homomorphism from M to $\mathcal{E}'\mathcal{D}'(\mathbb{R}^n \times U)$ with an open set $U \subset \mathbb{R}^d$, or else to $\mathcal{S}\mathcal{S}'(\mathbb{R}^n \times \mathbb{R}^d)$. Let us define a $\mathbb{C}$-linear map φ' from M to $\mathcal{D}'(U)$ or to $\mathcal{S}'(\mathbb{R}^d)$ by

$$\varphi'(u) = \int_{\mathbb{R}^n} \varphi(u)\, dx \qquad (\forall u \in M),$$

which is D_Y-linear by Proposition 4.12. Moreover, Proposition 4.17 implies

$$\partial_{x_1} M + \cdots + \partial_{x_n} M \subset \operatorname{Ker} \varphi'.$$

Hence φ' induces a D_Y-homomorphism

$$\varpi_*(\varphi) : \varpi_* M \longrightarrow \mathcal{D}'(U) \quad \text{or} \quad \varpi_*(\varphi) : \varpi_* M \longrightarrow \mathcal{S}'(\mathbb{R}^d).$$

The generators $x^\alpha [u_j]$ of $\varpi_* M$ with $1 \le j \le r$ and $\alpha \in \mathbb{N}^n$ are sent by $\varpi_*(\varphi)$ to

$$\varpi_*(\varphi)(x^\alpha [u_j]) = \int_{\mathbb{R}^n} x^\alpha \varphi(u_j)\, dx.$$

In conclusion, we have defined $\mathbb{C}$-linear maps

$$\varpi_* : \operatorname{Hom}_{D_X}(M, \mathcal{E}'\mathcal{D}'(\mathbb{R}^n \times U)) \longrightarrow \operatorname{Hom}_{D_Y}(\varpi_* M, \mathcal{D}'(U)),$$
$$\varpi_* : \operatorname{Hom}_{D_X}(M, \mathcal{S}\mathcal{S}'(\mathbb{R}^n \times \mathbb{R}^d)) \longrightarrow \operatorname{Hom}_{D_Y}(\varpi_* M, \mathcal{S}'(\mathbb{R}^d)).$$

Theorem 5.1 (Bernstein, Kashiwara). *If M is a holonomic D_X-module, then $\varpi_* M$ is a holonomic D_Y-module.*

Proof. We follow the argument in Chapter 1 of [4]. By induction on n, we have only to prove the holonomicity of $\varpi_* M$ in case $n = 1$. We may assume $M \neq 0$. First assume $\partial_{x_1} : M \to M$ is injective. Let $\{F_k(M)\}$ be a good $(\mathbf{1};\mathbf{1})$-filtration on M. There exists a polynomial $H(k)$ of degree $d+1$ such that $\dim_{\mathbb{K}} F_k(M) = H(k)$ for any $k \gg 0$. Set

$$F_k(\varpi_* M) := F_k(M)/(F_k(M) \cap \partial_{x_1} M) \qquad (k \in \mathbb{Z}).$$

Since $F_k(M) \cap \partial_{x_1} M$ contains $\partial_{x_1} F_{k-1}(M)$ and $\partial_{x_1} : M \to M$ is injective, we have

$$\begin{aligned}\dim_{\mathbb{K}} F_k(\varpi_* M) &= \dim_{\mathbb{K}} F_k(M) - \dim_{\mathbb{K}}(F_k(M) \cap \partial_{x_1} M) \\ &\leq \dim_{\mathbb{K}} F_k(M) - \dim_{\mathbb{K}} F_{k-1}(M) = H(k) - H(k-1)\end{aligned}$$

for sufficiently large k.

Let N be a finitely generated nonzero left D_Y-submodule of $\varpi_* M$. Then $F_k(N) := F_k(\varpi_* M) \cap N$ constitute a $(\mathbf{1}, \mathbf{1})$-filtration on N. There exists a good $(\mathbf{1}, \mathbf{1})$-filtration $\{F'_k(N)\}$ on N since it is finitely generated. Let $H'(k)$ be the associated Hilbert polynomial. Then by Lemma 2.10 there exists $k_0 \in \mathbb{N}$ such that

$$H'(k) = \dim_{\mathbb{K}} F'_k(N) \leq \dim_{\mathbb{K}} F_{k+k_0}(N) \leq H(k+k_0) - H(k+k_0-1)$$

holds if k is sufficiently large. Let $c_d k^d$ be the leading term of $H(k) - H(k-1)$. Then the inequality above implies that N is holonomic with $\operatorname{mult} N \leq d! c_d$. Hence we are done if $\varpi_* M$ is finitely generated over D_Y.

Otherwise, there exist finitely generated nonzero D_Y-submodules N_j ($j \in \mathbb{N}$) of $\varpi_* M$ such that $N_j \subsetneq N_{j+1}$. This implies that N_j are holonomic and $\operatorname{mult} N_j < \operatorname{mult} N_{j+1}$ holds in view of Proposition 2.19. This contradicts the inequality $\operatorname{mult} N_j \leq d! c_d$. Thus $\varpi_* M$ must be finitely generated over D_Y and hence holonomic.

In general case, set

$$N = \{u \in M \mid \partial_{x_1}^{\nu} u = 0 \text{ for some } \nu \in \mathbb{N}\}.$$

Then N is a left D_X-module since $\partial_{x_1}^{\nu} u = 0$ implies

$$\partial_{x_1}^{\nu+1}(x_1 u) = x_1 \partial_{x_1}^{\nu+1} u + (\nu+1) \partial_{x_1}^{\nu} u = 0.$$

Let us show that $\partial_{x_1} : N \to N$ is surjective. Suppose $u \in N$ satisfies $\partial_{x_1} u = 0$. Then we have $\partial_{x_1} x_1 u = u$. Hence u belongs to $\partial_{x_1} M$. Now assume that for any $v \in N$, v belongs to $\partial_{x_1} M$ if $\partial_{x_1}^{\nu} v = 0$. Suppose $u \in N$ satisfies $\partial_{x_1}^{\nu+1} u = 0$. Then we have

$$\partial_{x_1}^{\nu}(\partial_{x_1} x_1 u - (\nu+1) u) = x_1 \partial_{x_1}^{\nu+1} u = 0.$$

By the induction hypothesis, there exists $v \in N$ such that

$$\partial_{x_1} x_1 u - (\nu+1) u = \partial_{x_1} v.$$

This implies that u belongs to $\partial_{x_1} N$.

From the exact sequence

$$0 \longrightarrow N \longrightarrow M \longrightarrow M/N \longrightarrow 0,$$

we get an exact sequence

$$\varpi_* N \longrightarrow \varpi_* M \longrightarrow \varpi_*(M/N) \longrightarrow 0$$

of left D_Y-modules. Here $\varpi_* N = 0$ holds since $N = \partial_{x_1} N$. Hence $\varpi_* M$ is isomorphic to $\varpi_*(M/N)$, which is holonomic since $\partial_{x_1} : M/N \to M/N$ is injective. Q.E.D.

In particular, if a holonomic D_X-module M is generated by a single element u, then $\varpi_* M$ is generated by a finite number of residue classes $x^\alpha[u]$ with $\alpha \in \mathbb{N}^n$. In general, let M be a left D_X-module generated by u. Setting $I = \mathrm{Ann}_{D_X} u$, we have an isomorphism

$$\varpi_* M \cong D_X/(\partial_{x_1} D_X + \cdots + \partial_{x_n} D_X + I).$$

From a computational viewpoint, we are mainly interested in the submodule $D_Y[u]$ of $\varpi_* M$. The isomorphism above induces

$$D_Y[u] \cong D_Y/(D_Y \cap (\partial_{x_1} D_X + \cdots + \partial_{x_n} D_X + I)).$$

The map ϖ_* and the inclusion $D_Y[u] \to \pi_* M$ induces $\mathbb{C}$-linear maps

$$\begin{aligned} \varpi_* &: \mathrm{Hom}_{D_X}(M, \mathcal{E}'\mathcal{D}'(\mathbb{R}^n \times U)) \longrightarrow \mathrm{Hom}_{D_Y}(D_Y[u], \mathcal{D}'(U)), \\ \varpi_* &: \mathrm{Hom}_{D_X}(M, \mathcal{S}\mathcal{S}'(\mathbb{R}^n \times \mathbb{R}^d)) \longrightarrow \mathrm{Hom}_{D_Y}(D_Y[u], \mathcal{S}'(\mathbb{R}^d)). \end{aligned}$$

This means that for a solution in $\mathcal{E}'\mathcal{D}'(\mathbb{R}^n \times U)$ or in $\mathcal{S}\mathcal{S}'(\mathbb{R}^n \times \mathbb{R}^d)$ of a system M of differential equations, its integral with respect to x is a solution of $D_Y[u]$.

Example 5.2. The integral of the holonomic D_{n+d}-module $\mathbb{K}[x, y]$ along the fibers of the projection $\varpi : \mathbb{K}^{n+d} \ni (x, y) \mapsto y \in \mathbb{K}^d$ is $\{0\}$ as a D_d-module because $\partial_{x_j} : \mathbb{K}[x, y] \to \mathbb{K}[x, y]$ is surjective for $1 \leq j \leq n$.

Example 5.3. Set

$$M = D_{n+d}/(D_{n+d}x_1 + \cdots + D_{n+d}x_n + D_{n+d}y_1 + \cdots + D_{n+d}y_d)$$

and $\varpi : \mathbb{K}^{n+d} \ni (x, y) \mapsto y \in \mathbb{K}^d$ be the projection. Then there exists a natural isomorphism

$$\varpi_* M \cong D_d/(D_d y_1 + \cdots + D_d y_d).$$

In fact, by the definition we can write $\varpi_* M = D_{n+d}/N$ with

$$\begin{aligned} N := D_{n+d}x_1 + \cdots + D_{n+d}x_n + D_{n+d}y_1 + \cdots + D_{n+d}y_d \\ + \partial_{x_1} D_{n+d} + \cdots + \partial_{x_n} D_{n+d}. \end{aligned}$$

The ring extension $D_d \to D_{n+d}$ induces a homomorphism

$$\varphi : D_d/(D_d y_1 + \cdots + D_d y_d) \longrightarrow D_{n+d}/N$$

of left D_d-modules. Every element of $D_d/(D_d y_1 + \cdots + D_d y_d)$ is uniquely written as a linear combination of the residue classes $[\partial_y^\gamma]$ with $\gamma \in \mathbb{N}^d$. It follows that φ is injective. Every element of D_{n+d} is uniquely written as a linear combination of $\partial_x^\alpha x^\beta \partial_y^\gamma y^\delta$ with $\alpha, \beta \in \mathbb{N}^n$ and $\gamma, \delta \in \mathbb{N}^d$. It belongs to N unless $\alpha = \beta = 0$ and $\delta = 0$. This implies that φ is surjective.

Exercise 22. Set $n = d = 1$ and write $x = x_1$, $\partial_x = \partial_{x_1}$, $y = y_1$, $\partial_y = \partial_{y_1}$. Compute the integral of

$$M := D_2/(D_2\partial_y + D_2 x^2)$$

along the fibers of the projection $\varpi : \mathbb{K}^2 \ni (x, y) \mapsto y \in \mathbb{K}$. Note that $\varpi_* M$ is generated by $[1]$ and $[x]$. Deduce a presentation of the submodule $D_1[u]$ of $\varpi_* M$.

Exercise 23. Let $(a, x, y) = (a_1, \ldots, a_p, x_1, \ldots, x_n, y_1, \ldots, y_d) \in \Omega \times \mathbb{R}^n \times \mathbb{R}^d$ with an open set Ω of $\mathbb{R}^p$. Let D_{p+n+d} and D_{p+d} be the ring of differential operators in the variables (a, x, y) and that in the variables (a, y) respectively. Let $\varpi : \mathbb{C}^{p+n+d} \ni (a, x, y) \mapsto (a, y) \in \mathbb{C}^{p+d}$ be the projection. Let M be a finitely generated left D_{p+n+d}-module. Construct a $\mathbb{C}$-linear map

$$\varpi_* : \mathrm{Hom}_{D_{p+n+d}}(M, \mathcal{ESS}'(\Omega \times \mathbb{R}^n \times \mathbb{R}^d)) \to \mathrm{Hom}_{D_{p+d}}(\varpi_* M, \mathcal{D}'(\Omega \times \mathbb{R}^d))$$

in terms of the integration.

5.2. An algorithm for integration

In what follows, we assume that a left module M over $D_X = D_{n+d}$ is generated by a single element u for the sake of simplicity; it is easy to extend the following arguments so as to work in the case where M is generated by several elements and as well to yield the torsion groups associated with the integration, which are nothing but the relative de Rham cohomology groups of M along the fibers of ϖ (see [26]).

Now let us fix the weight vector

$$w := (\underbrace{1, \ldots, 1}_{n}, \underbrace{0, \ldots, 0}_{d}; \underbrace{-1, \ldots, -1}_{n}, \underbrace{0, \ldots, 0}_{d}) \in \mathbb{Z}^{2(n+d)}$$

for D_{n+d} and set

$$\theta := -(\partial_{x_1} x_1 + \cdots + \partial_{x_n} x_n) = -x_1\partial_{x_1} - \cdots - x_n\partial_{x_n} - n.$$

That is, we define the weights of x_i and of ∂_{x_i} to be 1 and -1 respectively, and the weights of y_j and ∂_{y_j} to be 0. In fact, we could work with a more general weight vector

$$w = (w_1, \dots, w_n, 0, \dots, 0; -w_1, \dots, -w_n, 0, \dots, 0)$$

with positive integers $w_1, \dots, w_n$ as in Chapter 5 of [29] by modifying the following arguments accordingly. Set

$$F_k(M) := F_k^w(D_X)u, \qquad \mathrm{gr}_k(M) := F_k(M)/F_{k-1}(M) \qquad (k \in \mathbb{Z}).$$

Then $\{F_k(M)\}$ is a good w-filtration on M.

Theorem 5.4. *If M is a holonomic D_X-module, then there exists a nonzero polynomial $b(s) \in \mathbb{K}[s]$ in s such that $b(\theta)\mathrm{gr}_0(M) = 0$. Such $b(s)$ of minimum degree is called the b-function of M with respect to the weight vector w and the filtration $\{F_k(M)\}$, or the b-function for integration along the fibers of ϖ. Moreover, $b(\theta + k)\mathrm{gr}_k(M) = 0$ holds for any $k \in \mathbb{Z}$.*

Proof. The graded module $\mathrm{gr}(M) = \bigoplus_{k\in\mathbb{Z}} \mathrm{gr}_k(M)$ is also a holonomic module over $\mathrm{gr}^w(D_{n+d}) \cong D_{n+d}$ by Theorem 2.2.1 of [29]. From $(\theta + 1)x_j = x_j\theta$ and $(\theta - 1)\partial_{x_j} = \partial_{x_j}\theta$ it follows that $(\theta + k)P = P\theta$ holds if P is homogeneous of order k with respect to w. Hence the sum of $\theta + k : \mathrm{gr}_k(M) \to \mathrm{gr}_k(M)$ for each $k \in \mathbb{Z}$ defines an endomorphism of the left $\mathrm{gr}^w(D_{n+d})$-module $\mathrm{gr}(M)$. There exists the minimal polynomial $b(s) \in \mathbb{K}[s]$ of this endomorphism since the space of the endomorphisms of a holonomic D-module is finite dimensional as was first shown by Kashiwara. A direct statement of this fact is found e.g., as Theorem 4.45 of [15]; this also follows from Theorem 6.6 in Chapter 1 of [4] combined with Lemma 7.14 and Theorem 7.15 in Chapter 2 of [4]. Q.E.D.

Note that a non-holonomic D_{n+d}-module can have a b-function in the above sense. The following arguments only rely on the existence of the b-function hence applies also to such non-holonomic modules.

Let us begin with an algorithm for computing the intersection of a left ideal I of D_{n+d} with the subring $D_d[x_1\partial_{x_1}, \dots, x_n\partial_{x_n}]$ by using multi-homogenization (Proposition 4.3 of [27]). Let us set $\theta_j = x_j\partial_{x_j}$ for $1 \le j \le n$.

Algorithm 5.5. Input: A set G_0 of generators of a left ideal I of D_{n+d}.
Output: A set G of generators of the left ideal $I \cap D_d[\theta_1, \dots, \theta_n]$ of $D_d[\theta_1, \dots, \theta_n]$.

(1) Introducing new variables u_j, v_j for $j = 1, \dots, n$, let $h(P) \in D_{n+d}[u]$ be the multi-homogenization of $P \in D_{n+d}$; i.e., for each $j = 1, \dots, n$, $h(P)$ is homogeneous with respect to the weight in which x_j and u_j are of order -1, ∂_{x_j} is of order 1, while x_i, ∂_{x_i} for $i \neq j$ and y_i, ∂_{y_i} for all i are of order zero.

(2) Let J be the left ideal of $D_{n+d}[u, v]$ generated by the set

$$\{h(P) \mid P \in G_0\} \cup \{1 - u_j v_j \mid j = 1, \dots, n\}.$$

(3) Compute a set G_1 of generators of the ideal $J \cap D_{n+d}$ by eliminating u, v via an appropriate Gröbner basis.

(4) Since each element P of G_1 is multi-homogeneous and free of u, v, there exist unique $\nu_1, \dots, \nu_n \in \mathbb{Z}$ and $Q(\theta_1, \dots, \theta_n) \in D_d[\theta_1, \dots, \theta_n]$ such that

$$S_{1,\nu_1} \cdots S_{n,\nu_n} P = Q(\theta_1, \dots, \theta_n),$$

where we set $S_{j,\nu_j} = \partial_{x_j}^{\nu_j}$ if $\nu_j \geq 0$ and $S_{j,\nu_j} = x_j^{-\nu_j}$ otherwise. Let G be the set of such $Q(\theta_1, \dots, \theta_n)$ for each $P \in G_1$.

See the proof of Proposition 4.3 of [27] for the correctness of this algorithm. The following algorithm was also presented in [27] (Algorithm 4.6):

Algorithm 5.6 ($b(s)$ with respect to w)**.** Input: $I := \mathrm{Ann}_{D_{n+d}} u$. Output: The b-function $b(s)$ of $M = D_{n+d}u$ with respect to w if it exists. 'None' if it does not.

(1) Compute a Gröbner basis $G = \{P_1, \dots, P_r\}$ of I with respect to a monomial order which is adapted to the weight vector w defined above.

(2) Set $\sigma^w(G) = \{\sigma^w(P_1), \dots, \sigma^w(P_r)\}$ and let $\mathrm{gr}^w(I)$ be the left ideal of D_{n+d} generated by $\sigma^w(G)$.

(3) Compute a set of generators of $\mathrm{gr}^w(I) \cap D_d[\theta_1, \dots, \theta_n]$ by Algorithm 5.5.

(4) Compute the intersection

$$\mathrm{gr}^w(I) \cap \mathbb{K}[\theta_1, \dots, \theta_n] = (\mathrm{gr}^w(I) \cap D_d[\theta_1, \dots, \theta_n]) \cap \mathbb{K}[\theta_1, \dots, \theta_n]$$

by using a Gröbner basis.

(5) Setting $\theta = -\theta_1 - \cdots - \theta_n - n$, compute

$$B := \mathrm{gr}^w(I) \cap \mathbb{K}[\theta_1, \dots, \theta_n] \cap \mathbb{K}[\theta]$$

by using a Gröbner basis. If $B \neq \{0\}$, let $b(\theta)$ be a generator of B. If $B = \{0\}$, then there exists no b-function of M with respect to w.

If M is holonomic, or more generally if we know that there exists a (nonzero) b-function in advance, then we can employ more efficient algorithm by Noro [19] which calculates $b(s)$ directly as the minimal polynomial of θ with modular computation; this algorithm is available as a function named 'generic_bfct' in a computer algebra system Risa/Asir [20].

Proposition 5.7. *Suppose that a left D_X-module $M = D_X u = D_X/I$ has a b-function $b(s)$ with respect to the weight vector w as above and the good w-filtration $F_k(M) := F_k^w(D_X)u$. Let k_1 be the largest integer root, if any, of $b(s)$. Let k_1 be an arbitrary integer if $b(s)$ has no integral root. Then the exact sequence*

$$M^n \overset{(\partial_{x_1},\dots,\partial_{x_n})}{\longrightarrow} M \longrightarrow \varpi_* M \longrightarrow 0$$

induces an exact sequence

$$F_{k_1+1}(M)^n \overset{(\partial_{x_1},\dots,\partial_{x_n})}{\longrightarrow} F_{k_1}(M) \longrightarrow \varpi_* M \longrightarrow 0.$$

To prove this proposition we need a lemma on the Koszul complex. In general, let $L = \oplus_{k\in\mathbb{Z}} L_k$ be a graded module over $\mathrm{gr}^w(D_X) = D_{n+d}$ with L_k being the homogeneous part of order k. For any integer k, let us define the Koszul complex $\mathcal{K}^\bullet(L[k], \partial_{x_1}, \dots, \partial_{x_n})$ to be the complex

$$0 \longrightarrow L_{k+n} \otimes_{\mathbb{Z}} \wedge^0 \mathbb{Z}^n \overset{\delta_n}{\longrightarrow} L_{k+n-1} \otimes_{\mathbb{Z}} \wedge^1 \mathbb{Z}^n \overset{\delta_{n-1}}{\longrightarrow} \cdots \overset{\delta_1}{\longrightarrow} L_k \otimes_{\mathbb{Z}} \wedge^n \mathbb{Z}^n.$$

Here $\wedge^l \mathbb{Z}$ is the free $\mathbb{Z}$-module spanned by $e_{i_1} \wedge \cdots \wedge e_{i_l}$ with the unit vectors $e_1, \dots, e_n$ of $\mathbb{Z}^n$ satisfying $e_i \wedge e_j + e_j \wedge e_i = 0$. The homomorphism δ_l is defined by

$$\delta_l(v \otimes e_{i_1} \wedge \cdots \wedge e_{i_l}) = \sum_{j=1}^{n} (\partial_{x_j} v) e_j \wedge e_{i_1} \wedge \cdots \wedge e_{i_l}.$$

For example, we have $L_k \otimes_{\mathbb{Z}} \wedge^n \mathbb{Z}^n = L_k e_1 \wedge \cdots \wedge e_n \cong L_k$,

$$L_{k+1} \otimes_{\mathbb{Z}} \wedge^{n-1} \mathbb{Z}^n = \bigoplus_{j=1}^{n} L_{k+1} e_{\hat{j}} \cong (L_{k+1})^n,$$
$$L_{k+2} \otimes_{\mathbb{Z}} \wedge^{n-2} \mathbb{Z}^n = \bigoplus_{1 \le i < j \le n} L_{k+2} e_{\hat{i}\hat{j}} \cong (L_{k+1})^{n(n-1)/2}$$

with

$$e_{\hat{j}} := e_1 \wedge \cdots \wedge e_{j-1} \wedge e_{j+1} \wedge \cdots \wedge e_n,$$
$$e_{\hat{i}\hat{j}} := e_1 \wedge \cdots \wedge e_{i-1} \wedge e_{i+1} \wedge \cdots \wedge e_{j-1} \wedge e_{j+1} \wedge \cdots \wedge e_n,$$

and

$$\delta_1\left(\sum_{j=1}^n v_j e_{\hat{j}}\right) = \sum_{j=1}^n (-1)^{j-1}(\partial_{x_j} v_j) e_1 \wedge \cdots \wedge e_n,$$

$$\delta_2\left(\sum_{1\le i<j\le n} v_{ij} e_{\hat{i}\hat{j}}\right) = \sum_{i=1}^n \left(\sum_{j=1}^{i-1}(-1)^{j-1}\partial_{x_j} v_{ji} + \sum_{j=i+1}^n (-1)^j \partial_{x_j} v_{ij}\right) e_{\hat{i}}.$$

Lemma 5.8. *Assume that there exists a nonzero polynomial $b(s) \in \mathbb{K}[s]$ such that $b(\theta + j)L_j = 0$ for any $j \in \mathbb{Z}$. If $b(k) \neq 0$ holds for an integer k, then the Koszul complex $\mathcal{K}^\bullet(L[k], \partial_{x_1}, \ldots, \partial_{x_n})$ is exact.*

Proof. We argue by induction on n. First let us prove the lemma for $n = 1$; i.e., that the homomorphism

$$\delta_1 : L_{k+1} \otimes_{\mathbb{Z}} \wedge^0 \mathbb{Z} \ni v \longmapsto (\partial_{x_1} v)e_1 \in L_k \otimes_{\mathbb{Z}} \wedge^1 \mathbb{Z}$$

is an isomorphism. Let v be an element of L_k. Then $b(\theta+k)v = 0$ holds. There exists a polynomial $c(\theta) \in \mathbb{K}[\theta]$ such that $b(\theta + k) - b(k) = \theta c(\theta)$. This implies

$$b(k)v = -\theta c(\theta)v = \partial_{x_1} x_1 c(-\partial_{x_1} x_1)v.$$

It follows that δ_1 is surjective. Next suppose $v \in L_{k+1}$ satisfies $\partial_{x_1} v = 0$ in L_k. Then we get

$$0 = b(\theta + k + 1)v = b(-x_1\partial_{x_1} + k)v = b(k)v,$$

and consequently $v = 0$ since $b(k) \neq 0$.

Now suppose $n \geq 2$ and that the lemma has been proved with n replaced by $n-1$. The Koszul complex $\mathcal{K}^\bullet(L[k], \partial_{x_1}, \ldots, \partial_{x_n})$ is isomorphic to the total complex associated with the double complex

$$(12)\qquad \begin{array}{ccccccc}
\cdots \longrightarrow & L_{k+2} \otimes_{\mathbb{Z}} \wedge^{n-2}\mathbb{Z}^{n-1} & \xrightarrow{\delta_1'} & L_{k+1} \otimes_{\mathbb{Z}} \wedge^{n-1}\mathbb{Z}^{n-1} & \longrightarrow 0 \\
& \downarrow \partial_{x_n} & & \downarrow \partial_{x_n} & \\
\cdots \longrightarrow & L_{k+1} \otimes_{\mathbb{Z}} \wedge^{n-2}\mathbb{Z}^{n-1} & \xrightarrow{\delta_1'} & L_{k} \otimes_{\mathbb{Z}} \wedge^{n-1}\mathbb{Z}^{n-1} & \longrightarrow 0
\end{array}$$

where the two horizontal sequences are $\mathcal{K}^\bullet(L[k+1], \partial_{x_1}, \ldots, \partial_{x_{n-1}})$ and $\mathcal{K}^\bullet(L[k], \partial_{x_1}, \ldots, \partial_{x_{n-1}})$ respectively. In fact, there is an isomorphism

$$\left(L_{k+l} \otimes_{\mathbb{Z}} \wedge^{n-l}\mathbb{Z}^{n-1}\right) \oplus \left(L_{k+l} \otimes_{\mathbb{Z}} \wedge^{n-l-1}\mathbb{Z}^{n-1}\right) \cong L_{k+l} \otimes_{\mathbb{Z}} \wedge^{n-l}\mathbb{Z}^{n}$$

for $0 \le l \le n$ (with the convention $\wedge^j \mathbb{Z}^n = 0$ if $j < 0$ or $j > n$) defined by the one-to-one correspondence

$$\Big(\sum_{1 \le i_1 < \cdots < i_{n-l} \le n-1} v_{i_1 \ldots i_{n-l}} e_{i_1} \wedge \cdots \wedge e_{i_{n-l}}, \\ \sum_{1 \le i_1 < \cdots < i_{n-l-1} \le n-1} v_{i_1 \ldots i_{n-l-1}} e_{i_1} \wedge \cdots \wedge e_{i_{n-l-1}} \Big) \\ \longleftrightarrow \sum_{1 \le i_1 < \cdots < i_{n-l} \le n-1} v_{i_1 \ldots i_{n-l}} e_{i_1} \wedge \cdots \wedge e_{i_{n-l}} \\ + \sum_{1 \le i_1 < \cdots < i_{n-l-1} \le n-1} v_{i_1 \ldots i_{n-l-1}} e_{i_1} \wedge \cdots \wedge e_{i_{n-l-1}} \wedge e_n,$$

in which the homomorphism ∂_{x_n} is given by

$$\partial_{x_n} \Big(\sum_{1 \le i_1 < \cdots < i_{n-l} \le n-1} v_{i_1 \ldots i_{n-l}} e_{i_1} \wedge \cdots \wedge e_{i_{n-l}} \Big) \\ = \sum_{1 \le i_1 < \cdots < i_{n-l} \le n-1} \partial_{x_n} v_{i_1 \ldots i_{n-l}} e_{i_1} \wedge \cdots \wedge e_{i_{n-l}} \wedge e_n.$$

Hence the homomorphism

$$\big(L_{k+l} \otimes_{\mathbb{Z}} \wedge^{n-l} \mathbb{Z}^{n-1}\big) \oplus \big(L_{k+l} \otimes_{\mathbb{Z}} \wedge^{n-l-1} \mathbb{Z}^{n-1}\big) \\ \longrightarrow \big(L_{k+l-1} \otimes_{\mathbb{Z}} \wedge^{n-l+1} \mathbb{Z}^{n-1}\big) \oplus \big(L_{k+l-1} \otimes_{\mathbb{Z}} \wedge^{n-l} \mathbb{Z}^{n-1}\big)$$

defined by $\big(\delta'_l \oplus (-1)^{n-l} \partial_{x_n}\big) \oplus \delta'_{l+1}$ corresponds to δ_l.

Let L'_j and L''_j be the kernel and the cokernel of $\partial_{x_n} : L_{j+1} \to L_j$ respectively and set $L' := \oplus_{j \in \mathbb{Z}} L'_j$ and $L'' := \oplus_{j \in \mathbb{Z}} L''_j$, which are graded D_{n+d-1}-modules. For any $v \in L'_j$ we have

$$b(-\partial_{x_1} x_1 - \cdots - \partial_{x_{n-1}} x_{n-1} + j) v \\ = b(-\partial_{x_1} x_1 - \cdots - \partial_{x_{n-1}} x_{n-1} - x_n \partial_{x_n} + j) v = b(\theta + j + 1) v = 0$$

by the assumption on L since L'_j is a subset of L_{j+1}.

Now let v be an element of L_j and $\overline{v}$ be its residue class in L''_j. Then we have $b(\theta + j) v = 0$ and

$$b(-\partial_{x_1} x_1 - \cdots - \partial_{x_{n-1}} x_{n-1} + j) v - b(\theta + j) v \in \partial_{x_n} L_{j+1}.$$

This implies

$$b(-\partial_{x_1} x_1 - \cdots - \partial_{x_{n-1}} x_{n-1} + j) \overline{v} = 0.$$

Hence $\mathcal{K}^\bullet(L'[k], \partial_{x_1}, \ldots, \partial_{x_{n-1}})$ and $\mathcal{K}^\bullet(L''[k], \partial_{x_1}, \ldots, \partial_{x_{n-1}})$ are both exact by the induction hypothesis if $b(k) \neq 0$.

We have an exact sequence

$$0 \to \mathcal{K}^\bullet(L'[k], \partial_{x_1}, \ldots, \partial_{x_{n-1}}) \to \mathcal{K}^\bullet(L[k+1], \partial_{x_1}, \ldots, \partial_{x_{n-1}})$$
$$\xrightarrow{\partial_{x_n}} \mathcal{K}^\bullet(L[k], \partial_{x_1}, \ldots, \partial_{x_{n-1}}) \to \mathcal{K}^\bullet(L''[k], \partial_{x_1}, \ldots, \partial_{x_{n-1}}) \to 0$$

of chain maps. Here the central chain map defined by ∂_{x_n} is a quasi-isomorphism, i.e., induces isomorphisms of the homology groups since the leftmost and the rightmost complexes are exact. This implies that the total complex associated with the double complex (12) is exact, which can be verified by diagram chasing. Summing up, we have shown that $\mathcal{K}^\bullet(L[k], \partial_{x_1}, \ldots, \partial_{x_n})$ is exact. Q.E.D.

Now let us prove Proposition 5.7. First let us show that $F_{k_1}(M) \to \varpi_* M$ is surjective. Let v be an element of $F_k(M)$ with $k > k_1$. Applying Lemma 5.8 to $\mathrm{gr}(M)$, we get $v_1, \ldots, v_n \in F_{k+1}(M)$ such that

$$v - \partial_{x_1} v_1 - \cdots - \partial_{x_n} v_n \in F_{k-1}(M)$$

since $b(k) \neq 0$. By induction, we see that there exist $v'_1, \ldots, v'_n \in F_{k+1}(M)$ such that

$$v - \partial_{x_1} v'_1 - \cdots - \partial_{x_n} v'_n \in F_{k_1}(M).$$

Thus $F_{k_1}(M) \to \varpi_* M$ is surjective.

Next, suppose the residue class $[v]$ in $\varpi_*(M)$ of $v \in F_{k_1}(M)$ vanishes. Then there exist $v_1, \ldots, v_n \in F_{k+1}(M)$ with some k such that $v = \partial_{x_1} v_1 + \cdots + \partial_{x_n} v_n$. Assume $k > k_1$. Let $\overline{v}_j$ be the residue class of v_j in $\mathrm{gr}_{k+1}(M)$. Then we have $\partial_{x_1} \overline{v}_1 + \cdots + \partial_{x_n} \overline{v}_n = 0$ in $\mathrm{gr}_k(M)$. By Lemma 5.8, there exist $v_{ij} \in F_{k+2}(M)$ such that their residue classes $\overline{v}_{ij}$ in $\mathrm{gr}_{k+2}(M)$ satisfy

$$\overline{v}_i = \sum_{j=1}^{n} (-1)^{i+j-1} \partial_{x_j} \overline{v}_{ij}, \qquad \overline{v}_{ij} + \overline{v}_{ji} = 0.$$

Hence $v'_i := v_i - \sum_{j=1}^{n} (-1)^{i+j-1} \partial_{x_j} v_{ij}$ belongs to $F_k(M)$ and we get a new expression

$$v = \sum_{i=1}^{n} \partial_{x_i} \left(v'_i + \sum_{j=1}^{n} (-1)^{i+j-1} \partial_{x_j} v_{ij} \right) = \sum_{i=1}^{n} \partial_{x_i} v'_i.$$

Proceeding inductively, we can show that v belongs to $\partial_{x_1} F_{k_1+1}(M) + \cdots + \partial_{x_n} F_{k_1+1}(M)$. This completes the proof of Proposition 5.7.

Now let

$$(D_X)^r \xrightarrow{\psi} D_X \xrightarrow{\varphi} M \longrightarrow 0$$

be a presentation of M with

$$\begin{aligned} &\varphi(P) = Pu \quad (\forall P \in D_X), \\ &\psi((Q_1, \dots, Q_r)) = Q_1 P_1 + \cdots + Q_r P_r \qquad (\forall Q_1, \dots, Q_r \in D_X). \end{aligned}$$

Here we assume that $P_1, \dots, P_r$ are a w-involutive basis of $I = \mathrm{Ann}_{D_X} u$ with $\mathrm{ord}_w(P_i) = m_i$. This implies that the sequence

$$\oplus_{i=1}^r F_{k-m_i}(D_X) \xrightarrow{\psi} F_k(D_X) \xrightarrow{\varphi} F_k(M) \longrightarrow 0$$

is exact for any $k \in \mathbb{Z}$. Set $F_k[\mathbf{m}]((D_{Y \leftarrow X})^r) := \oplus_{i=1}^r F_{k-m_i}(D_{Y \leftarrow X})$ with $\mathbf{m} = (m_1, \dots, m_r)$, and so on, where $\{F_k(D_{Y \leftarrow X})\}$ denotes the filtration induced by $\{F_k^w(D_X)\}$. Then ψ induces homomorphisms

$$\overline{\psi} : (D_{Y \leftarrow X})^r \longrightarrow D_{Y \leftarrow X}, \qquad \overline{\psi} : F_k[\mathbf{m}]((D_{Y \leftarrow X})^r) \longrightarrow F_k(D_{Y \leftarrow X}).$$

Let k_1 be an integer as in Proposition 5.7. Then we have a commutative diagram

$$\begin{array}{ccccccc}
F_{k_1+1}[\mathbf{m}]((D_X)^r)^n & \xrightarrow{(\psi,\dots,\psi)} & F_{k_1+1}(D_X)^n & \xrightarrow{(\varphi,\dots,\varphi)} & F_{k_1+1}(M)^n & \longrightarrow & 0 \\
\downarrow & & \downarrow{\scriptstyle (\partial_{x_1},\dots,\partial_{x_n})} & & \downarrow{\scriptstyle (\partial_{x_1},\dots,\partial_{x_n})} & & \\
F_{k_1}[\mathbf{m}]((D_X)^r) & \xrightarrow{\psi} & F_{k_1}(D_X) & \xrightarrow{\varphi} & F_{k_1}(M) & \longrightarrow & 0 \\
\downarrow & & \downarrow & & \downarrow & & \\
F_{k_1}[\mathbf{m}]((D_{Y \leftarrow X})^r) & \xrightarrow{\overline{\psi}} & F_{k_1}(D_{Y \leftarrow X}) & \longrightarrow & \varpi_* M & \longrightarrow & 0 \\
\downarrow & & \downarrow & & \downarrow & & \\
0 & & 0 & & 0 & &
\end{array}$$

where the upper leftmost homomorphisms send

$$\begin{pmatrix} Q_{11} & \cdots & Q_{1r} \\ \vdots & & \vdots \\ Q_{n1} & \cdots & Q_{nr} \end{pmatrix} \in F_{k_1+1}[\mathbf{m}]((D_X)^r)^n$$

to

$$\begin{pmatrix} Q_{11} & \cdots & Q_{1r} \\ \vdots & & \vdots \\ Q_{n1} & \cdots & Q_{nr} \end{pmatrix} \begin{pmatrix} P_1 \\ \vdots \\ P_r \end{pmatrix} \in F_{k_1+1}(D_X)^n,$$

$$\begin{pmatrix} \partial_{x_1} & \cdots & \partial_{x_n} \end{pmatrix} \begin{pmatrix} Q_{11} & \cdots & Q_{1r} \\ \vdots & & \vdots \\ Q_{n1} & \cdots & Q_{nr} \end{pmatrix} \in F_{k_1}[\mathbf{m}]((D_X)^r)$$

respectively. In the commutative diagram, the three vertical sequences and the two horizontal sequences except the one at the bottom are exact in view of Proposition 5.7. This implies that the horizontal sequence at the bottom is also exact; i.e.,

$$\varpi_* M = \operatorname{coker}(\overline{\psi} : F_{k_1}[\mathbf{m}]((D_{Y\leftarrow X})^r) \longrightarrow F_{k_1}(D_{Y\leftarrow X})).$$

Note that

$$F_{k_1}(D_{Y\leftarrow X}) = \bigoplus_{|\alpha|\le k_1} x^\alpha D_Y, \quad F_{k_1}[\mathbf{m}]((D_{Y\leftarrow X})^r) = \bigoplus_{i=1}^{r} \bigoplus_{|\alpha|\le k_1 - m_i} x^\alpha D_Y$$

as left D_Y-modules. Hence $\overline{\psi}$ is a homomorphism of free left D_Y-modules of finite rank so that $\operatorname{coker}\overline{\psi}$ can be explicitly computed by linear algebra over D_Y. This gives the relations among the generators $\{x^\alpha[u] \mid |\alpha| \le k_1\}$ of $\varpi_* M$. In particular, we get

Proposition 5.9. *One has* $\varpi_* M = 0$ *if* $b(k) \neq 0$ *for any non-negative integer* k.

By elimination, we obtain $\operatorname{Ann}_{D_Y}[u]$ so that $D_Y[u] \cong D_Y/\operatorname{Ann}_{D_Y}[u]$ is a left D_Y-submodule of $\varpi_* M$. It is easy to see by the construction that

$$\operatorname{Ann}_{D_Y}[u] = D_Y \cap (\partial_{x_1} D_X + \cdots + \partial_{x_n} D_X + \operatorname{Ann}_{D_X} u)$$

holds; the right-hand side is called the integration ideal of $I = \operatorname{Ann}_{D_Y} u$. Summing up we have obtained

Algorithm 5.10 (integration ideal). Input: A set G_0 of generators of $I := \operatorname{Ann}_{D_X} u$.
Output: A set G of generators of the integration ideal $\operatorname{Ann}_{D_Y}[u]$ of I.

(1) Compute a Gröbner basis $G_1 = \{P_1, \dots, P_r\}$ of I with respect to a monomial order which is adapted to the weight vector w.

(2) Using G_1, compute the b-function $b(s)$ of $M = D_X/I$ with respect to w by steps (2)–(5) of Algorithm 5.6.
(3) If $b(s)$ has no non-negative integer root, then $[u] = 0$; quit. Otherwise let k_1 be the largest integral root of $b(s)$.
(4) Set $\mathbf{m} = (m_1, \ldots, m_r)$ with $m_j = \mathrm{ord}_w(P_j)$.
(5) Express the homomorphism

$$(D_Y)^{l_1} \cong F_{k_1}[\mathbf{m}]((D_{Y\leftarrow X})^r) \xrightarrow{\overline{\psi}} F_{k_1}(D_{Y\leftarrow X}) \cong (D_Y)^{l_0}$$

of free D_Y-modules, which is induced by ${}^t(P_1, \ldots, P_r)$, as an $l_1 \times l_0$ matrix $A = (A_{ij})$ with

$$l_1 = \sum_{i=1}^{r} \binom{n + k_1 - m_i}{n}, \quad l_0 = \binom{n + k_1}{n}, \quad A_{ij} \in D_Y.$$

(6) Compute a set G_2 of generators of the submodule

$$D_Y e_1 \cap \left\{ \sum_{i=1}^{l_1} Q_i(A_{i1}, \ldots, A_{il_0}) \mid Q_i \in D_Y \right\}$$

of $D_Y^{l_0}$ with $e_1 = (1, 0, \ldots, 0) \in \mathbb{Z}^{l_0}$, which corresponds to $\overline{1} \in F_{k_1}(D_{Y\leftarrow X})$, by a Gröbner basis with respect to what is called a 'position-over-term' order. Let G be the set of the first elements of G_2.

For practical computation of integration, computer algebra systems such as Risa/Asir [20], Macaulay2 [11], and SINGULAR [7] are available. An implementation of the D-module theoretic integration algorithm was first supplied with Kan/sm1 [34] (see [26]). We make use of a Risa/Asir library 'nk_restriction.rr' by Hiromasa Nakayama for computing various examples in the next section. One can also use a Macaulay2 package 'Dmodules' by Anton Leykin and Harrison Tsai, or a SINGULAR library 'dmodapp_lib' by Viktor Levandovskyy and Daniel Andres.

Example 5.11. Set $x = x_1$, $\partial_x = \partial_{x_1}$ and so on with $n = d = 1$ and $w = (1, 0, ; -1, 0)$. Let I be the left ideal of D_2 generated by $P_1 = y - x^2$, $P_2 = 2x\partial_y + \partial_x$ and set $M = D_2/I$. We denote by u the residue class of 1 in M. It is easy to see that P_1 and P_2 annihilate $\delta(y - x^2)$, which belongs to $\mathcal{E}'\mathcal{D}'(\mathbb{R}_x \times \mathbb{R}_y)$.

From Example 3.15, P_1, P_2, and $P_3 = 2y\partial_y + x\partial_x + 2$ are a w-involutive basis of I. The b-function $b(s)$ of $M = D/I$ with respect to w divides $s(s-1)$ since $\sigma^w(P_1) = -x^2$ and hence $\partial_x^2 x^2 = \partial_x x(\partial_x x + 1)$ annihilates $\mathrm{gr}_0(M) = F_0(M)/F_{-1}(M)$ with $F_k(M) := F_k^w(D_2)/(F_k^w(D_2) \cap$

I). Let $\varpi : X = \mathbb{C}^2 \ni (x,y) \mapsto x \in \mathbb{C} = Y$ be the projection. Since the w-orders of P_1, P_2, P_3 are $2, 1, 0$ respectively, we have an exact sequence

$$F_{-1}(D_{Y\leftarrow X}) \oplus F_0(D_{Y\leftarrow X}) \oplus F_1(D_{Y\leftarrow X}) \xrightarrow{\overline{\psi}} F_1(D_{Y\leftarrow X}) \to \varpi_* M \to 0,$$

where $\overline{\psi}$ is induced by ${}^t(P_1, P_2, P_3)$. With $F_{-1}(D_{Y\leftarrow X}) = \{0\}$ omitted, $\overline{\psi}$ is explicitly given by

$$\begin{aligned}\overline{\psi}((Q_1, Q_2 + Q_3 x)) &= [Q_1 P_2 + (Q_2 + Q_3 x) P_3] \\ &= [Q_1(2x\partial_y + \partial_x) + (Q_2 + Q_3 x)(2y\partial_y + \partial_x x + 1)] \\ &= [Q_2(2y\partial_y + 1) + 2(Q_1\partial_y + Q_3 y\partial_y)x]\end{aligned}$$

for $Q_1, Q_2, Q_3 \in D_Y = D_1$, where the bracket denotes the residue class in $D_{Y\leftarrow X}$. Hence the homomorphism $\overline{\psi}$ is represented by the matrix

$$A = \begin{pmatrix} 0 & 2\partial_y \\ 2y\partial_y + 1 & 0 \\ 0 & 2y\partial_y \end{pmatrix}.$$

Thus $\varpi_* M$ is isomorphic to the direct sum

$$\varpi_* M = D_1[u] \oplus D_1[xu] \cong D_1/D_1(2y\partial_y + 1) \,\oplus\, D_1/D_1\partial_y.$$

This implies that $f_0(y) := \int_{-\infty}^{\infty} \delta(y - x^2)\,dx$ and $f_1(y) := \int_{-\infty}^{\infty} x\delta(y - x^2)\,dx$ satisfies

$$(2y\partial_y + 1) f_0(y) = \partial_y f_1(y) = 0.$$

Noticing $f_0(y) = f_1(y) = 0$ for $y < 0$, we get $f_0(y) = C y_+^{-1/2}$ and $f_1(y) = 0$ (naturally!) with a constant C. We can use the formula

$$\delta(1 - x^2) = \delta((x-1)(x+1)) = \frac{1}{2}\delta(x-1) + \frac{1}{2}\delta(x+1)$$

to obtain $C = f_0(1) = 1$. We conclude that the distribution $f_0(y)$ coincides with the locally integrable function $y_+^{-1/2}$ on whole $\mathbb{R}$ because the differential equation $(2y\partial_y + 1) f(y) = 0$ has no distribution solution $f(y)$ whose support is $\{0\}$. See Proposition 6.1 in the next subsection.

Exercise 24. Find a differential equation for

$$g_0(y) := \int_{-\infty}^{\infty} Y(y - x^2)\,dx$$

by using the integration algorithm and determine $g_0(y)$ as a distribution on $\mathbb{R}$ explicitly.

Exercise 25. For a positive integer n, find a differential equation for $v(y) = \int_{-\infty}^{\infty} \delta(y - x^n)\,dx$ and determine $v(y)$ explicitly.

Exercise 26. Set $u(x,t) = e^{tx-x^3}Y(x)$, which belongs to the space $\mathcal{ESS}'(\mathbb{R}_t \times \mathbb{R}_x)$ introduced in 4.5.

(1) Find a holonomic system for $u(x,t)$; confirm that it is holonomic.

(2) Find a holonomic system, i.e., a linear ordinary differential equation for

$$v(t) := \int_0^{\infty} e^{tx-x^3}\,dx = \int_{-\infty}^{\infty} e^{tx-x^3}Y(x)\,dx.$$

§6. Integration of holonomic distributions

We first apply the integration algorithm in 5.2 and classes of distributions introduced in 4.3 to the integral of a holonomic distribution over the whole space. For integrals over domains defined by arbitrary polynomial inequalities, we need more sophisticated method in order to compute a holonomic system for the product of Heaviside functions and the given integrand, which will be introduced in 6.2 with correctness proofs. This method also provides us with an integration algorithm for functions satisfying difference-differential holonomic systems, which will be explained in 6.4. Algorithms in this section complement the ones introduced in [24] with more detailed arguments.

6.1. Integrals of holonomic distributions over the whole space

We assume $\mathbb{K} = \mathbb{C}$. Let $u(x,y) = u(x_1,\dots,x_n,y_1,\dots,y_d)$ be a distribution in $\mathcal{E}'\mathcal{D}'(\mathbb{R}^n \times U)$ with an open set U of $\mathbb{R}^d$, or in $\mathcal{SS}'(\mathbb{R}^n \times \mathbb{R}^d)$. Suppose that $u(x,y)$ is holonomic and that we have a left ideal I of D_{n+d} which annihilates $u(x,y)$ such that D_{n+d}/I is holonomic. Then the integral $\varpi_* M = M/(\partial_{x_1} M + \cdots + \partial_{x_n} M)$ of M gives a holonomic system of linear differential equations for

$$v(y) := \int_{\mathbb{R}^d} u(x,y)\,dx,$$

which belongs to $\mathcal{D}'(U)$ or to $\mathcal{S}'(\mathbb{R}^d)$, as was explained so far.

Let us first consider the standard normal distribution whose density function is given by $(2\pi)^{-n/2}\exp(-|x|^2/2)$. Let $f(x)$ be an arbitrary real polynomial in $x = (x_1,\dots,x_n)$. Then the cumulative function

$$F(t) = (2\pi)^{-\frac{n}{2}} \int_{\mathbb{R}^n} \exp\Bigl(-\frac{1}{2}|x|^2\Bigr) Y(t - f(x))\,dx$$

of the random variable $f(x)$ is well-defined as an element of $\mathcal{S}'(\mathbb{R})$ since the integrand belongs to $\mathcal{SS}'(\mathbb{R}^n_x \times \mathbb{R}_t)$. It is also a continuous function if $f(x)$ is non-constant. The density function $F'(t)$ is given by the integral

$$F'(t) = (2\pi)^{-\frac{n}{2}} \int_{\mathbb{R}^n} \exp\Big(-\frac{1}{2}|x|^2\Big)\delta(t - f(x))\,dx$$

as an element of $\mathcal{S}'(\mathbb{R})$.

Since the integrands of $F(t)$ and of $F'(t)$ are holonomic by Proposition 4.20 and belong to $\mathcal{SS}'(R^n_x \times \mathbb{R}_t)$, we obtain linear ordinary differential equations which $F(t)$ and $F'(t)$ satisfy as elements of $\mathcal{S}'(\mathbb{R})$ by the integration algorithm.

The characteristic function of $v(t) := F'(t)$ is the tempered distribution $\hat{v}$ on $\mathbb{R}$ defined by

$$\langle \hat{v}, \varphi\rangle = \langle v, \hat{\varphi}\rangle, \qquad \hat{\varphi}(\tau) := \int_{-\infty}^{\infty} e^{i\tau t}\varphi(t)\,dt$$

for $\varphi \in \mathcal{S}(\mathbb{R})$ with $i = \sqrt{-1}$. Let us show that $\hat{v}(\tau)$ is given by

$$\hat{v}(\tau) = (2\pi)^{-\frac{n}{2}} \int_{\mathbb{R}^n} \exp\Big(i\tau f(x) - \frac{1}{2}|x|^2\Big)\,dx.$$

Set $\psi(x) = (2\pi)^{-n/2}\exp\Big(-\dfrac{1}{2}|x|^2\Big)$. (Indeed ψ can be an arbitrary element of $\mathcal{S}(\mathbb{R}^n)$.) Then by the definition of the integral of an element of $\mathcal{SS}'(\mathbb{R}^n_x \times \mathbb{R}_t)$ and Proposition 4.10, we get, for any $\chi \in \mathcal{S}(\mathbb{R})$,

$$\begin{aligned}
\langle \hat{v}, \chi\rangle &= \Big\langle \int_{\mathbb{R}^n} \psi(x)\delta(t - f(x))\,dx,\ \hat{\chi}\Big\rangle = \langle \delta(t - f(x)),\ \psi(x)\hat{\chi}(t)\rangle \\
&= \langle 1(x)\delta(t),\ \psi(x)\hat{\chi}(t + f(x))\rangle = \int_{\mathbb{R}^n} \psi(x)\hat{\chi}(f(x))\,dx \\
&= \int_{\mathbb{R}^n} \psi(x)\Big(\int_{-\infty}^{\infty} e^{itf(x)}\chi(t)\,dt\Big)\,dx \\
&= \int_{-\infty}^{\infty}\Big(\int_{\mathbb{R}^n} \psi(x)e^{itf(x)}\,dx\Big)\chi(t)\,dt.
\end{aligned}$$

This means the identity above. Moreover, $\hat{v}(\tau)$ belongs also to $C^\infty(\mathbb{R})$ since $\exp\Big(i\tau f(x) - \dfrac{1}{2}|x|^2\Big)$ belongs to $\mathcal{ES}(\mathbb{R}_\tau \times \mathbb{R}^n_x)$.

If $v(t)$ satisfies a differential equation $Pv = 0$ with $P = P(t, \partial_t) = \sum_{j=0}^m a_j(t)\partial_t^{m-j}$, then $\hat{v}$ satisfies $\hat{P}\hat{v} = 0$, where

$$\hat{P} = P(-i\partial_\tau, -i\tau) = \sum_{j=0}^m a_j(-i\partial_\tau)(-i\tau)^{m-j}.$$

Now let

$$P = a_0(t)\partial_t^m + a_1(t)\partial_t^{m-1} + \cdots + a_m(t) \tag{13}$$

be a linear ordinary differential operator with analytic functions $a_j(x)$ defined on a neighborhood of $t_0 \in \mathbb{R}$. In general, for an analytic function $f(t)$ near t_0, the order $\mathrm{ord}_{t_0} f(t)$ of $f(t)$ at t_0 is defined to be the smallest non-negative integer k such that $f^{(k)}(t_0) \neq 0$. The point t_0 is called a *regular singular point* of P if $\mathrm{ord}_{t_0} a_j(t) \geq \mathrm{ord}_{t_0} a_0(t) - j$ for $j = 1, \ldots, m$. Set $k = \mathrm{ord}_{t_0} a_0(t)$. With P being multiplied by a power of t or of ∂_t on the left, we may assume $k = m$. Then the *indicial polynomial* of P at a regular singular point t_0 is defined to be

$$b(s) := \sum_{j=0}^{m} c_j s(s-1)\cdots(s-m+j+1), \qquad c_j := \lim_{t \to t_0} \frac{a_j(t)}{(t-t_0)^{m-j}}.$$

This is nothing but the b-function with respect to the weight vector $(-1;1)$. The roots of $b(s) = 0$ are called the *characteristic exponents* of P at t_0.

The following well-known facts often provide us with information on the behavior of the distribution solutions near a singular point:

Proposition 6.1. *Let $t_0 \in \mathbb{R}$ be a regular singular point of an ordinary differential operator* (13).

(1) *If P has no negative integer as a characteristic exponent, then P has no distribution (or even hyperfunction) solution whose support is $\{t_0\}$ on a neighborhood of t_0.*

(2) *If the real part of each characteristic exponent of P at t_0 is greater than -1, then any distribution (or even hyperfunction) solution of the differential equation $Pu = 0$ coincides with a Lebesgue integrable function on a neighborhood of t_0.*

The simplest proof of this proposition would be to consider a distribution as a hyperfunction, which is defined as the boundary value of a complex analytic function to the real line (see [30]), and employ the theory of ordinary differential equations with regular singularities in the complex domain.

As a first example, let us deduce the density function of the χ^2 distribution in statistics.

Example 6.2 (χ^2 distribution). Set

$$u(x,t) = (2\pi)^{-\frac{n}{2}} \exp\Bigl(-\frac{1}{2}|x|^2\Bigr)\delta(t - |x|^2), \qquad v(t) = \int_{\mathbb{R}^n} u(x,t)\,dx$$

with $|x|^2 = x_1^2 + \cdots + x_n^2$. Then $u(x,t)$ belongs to $\mathcal{SS}'(\mathbb{R}_x^n \times \mathbb{R}_t)$ and thus $v(t)$ is well-defined as a tempered distribution on $\mathbb{R}$. Note that $v(t)$ is the density function of the χ^2 distribution. By Example 4.21, $u(x,t)$ satisfies a holonomic system

$$(t - |x|^2)u = (\partial_i + 2x_i\partial_t + x_i)u = 0 \qquad (i = 1, \ldots, n).$$

Since

$$\begin{aligned}
&\sum_{i=1}^n x_i(\partial_i + 2x_i\partial_t + x_i) + (1 + 2\partial_t)(t - |x|^2) \\
&= \sum_{i=1}^n x_i\partial_i + 2|x|^2\partial_t + |x|^2 + (1 + 2\partial_t)(t - |x|^2) \\
&= \sum_{i=1}^n x_i\partial_i + 2\partial_t t + t = \sum_{i=1}^n \partial_i x_i + 2t\partial_t + t - n + 2,
\end{aligned}$$

we know that $v(t)$ satisfies

$$(2t\partial_t + t - n + 2)v(t) = 0.$$

This differential equation has 0 as a regular singular point with the characteristic exponent $n/2 - 1$, which is greater than -1. Hence $v(t)$ is integrable on $\mathbb{R}$. Solving this equation by quadrature and noting that $v(t) = 0$ for $t < 0$, we conclude that

$$v(t) = Ce^{-t/2}t_+^{n/2-1}$$

with some constant C, which can be determined by

$$C = \left(\int_0^\infty e^{-t/2}t^{n/2-1}\,dt\right)^{-1} = \frac{1}{2^{n/2}\Gamma\left(\dfrac{n}{2}\right)}.$$

The characteristic function $\hat{v}(\tau) = \int_{-\infty}^\infty e^{i\tau t}v(t)\,dt$ satisfies

$$((2\tau + i)\partial_\tau + n)\hat{v}(\tau) = 0.$$

Together with $\hat{v}(0) = 1$, this implies

$$\hat{v}(\tau) = (1 - 2i\tau)^{-n/2}.$$

The following example was proposed by A. Takemura (see [18]):

Example 6.3 (sum of cubes of standard normal random variables). Set

$$v(t) = (2\pi)^{-\frac{n}{2}} \int_{\mathbb{R}^n} \exp\Big(-\frac{|x|^2}{2}\Big)\delta(t - x_1^3 - \cdots - x_n^3)\, dx.$$

If $n = 2$, $v(t)$ satisfies the ordinary differential equation $Pv(t) = 0$ with

$$P = 729t^3{\partial_t}^6 + 6561t^2{\partial_t}^5 + 12555t{\partial_t}^4 \\ + (81t^2 + 3240){\partial_t}^3 + 243t{\partial_t}^2 + 60\partial_t + 2t.$$

The origin is a regular singular point of P with the indicial polynomial $b(s) = s(s-1)^2(s-2)(3s+1)(3s-7)$ up to a constant multiple.

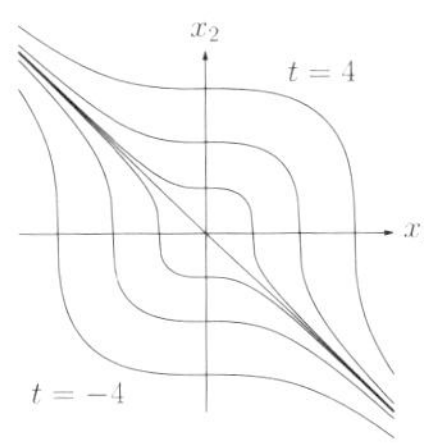

Fig. 3. Curves $x_1^3 + x_2^3 = t$ with $t = -4, -1, -1/8, 0, 1/8, 1, 4$

If $n = 3$, $v(t)$ satisfies the ordinary differential equation $Pv(t) = 0$ with

$$P = 6561t^4{\partial_t}^9 + 118098t^3{\partial_t}^8 + 607257t^2{\partial_t}^7 + (1458t^3 + 944055t){\partial_t}^6 \\ + (13122t^2 + 280665){\partial_t}^5 + 25920t{\partial_t}^4 + (99t^2 + 8100){\partial_t}^3 + 297t{\partial_t}^2 \\ + 90\partial_t + 2t.$$

Its indicial polynomial at 0 is

$$b(s) = s^2(s-1)(s-2)(s-3)(s-4)^2(3s-4)(3s-8)$$

up to a constant multiple. Hence in both cases, $v(t)$ is Lebesgue integrable on $\mathbb{R}$ and real analytic on $\mathbb{R} \setminus \{0\}$. In [18] it is proved that $v(t)$ satisfies a linear differential equation of order $3n$ with a regular singularity at the origin.

Example 6.4. Let us consider

$$v(t) = (2\pi)^{-\frac{n}{2}} \int_{\mathbb{R}^n} \exp\Big(-\frac{|x|^2}{2}\Big)\delta(t - x_1^4 - x_2^4 - \cdots - x_n^4)\, dx.$$

If $n = 2$, then $v(t)$ is annihilated by

$$128t^3{\partial_t}^4 + 768t^2{\partial_t}^3 + (-24t^2 + 864t){\partial_t}^2 + (-48t + 96)\partial_t + t - 6,$$

which has a regular singularity at 0 with the indicial polynomial $b(s) = s^2(2s-1)(2s+1)$ up to a constant multiple. If $n = 3$, $v(t)$ is annihilated by

$$\begin{aligned}&2048t^4{\partial_t}^6 + 24576t^3{\partial_t}^5 + (-768t^3 + 77568t^2){\partial_t}^4\\ &\quad + (-4608t^2 + 64512t){\partial_t}^3 + (88t^2 - 5328t + 7560){\partial_t}^2\\ &\quad + (176t - 720)\partial_t - 3t + 27,\end{aligned}$$

which has a regular singular point at 0 with the indicial polynomial

$$b(s) = s(s-1)(4s+1)(4s-1)(4s-3)(4s-5)$$

up to a constant multiple.

Example 6.5. Let us consider

$$v(t) = (2\pi)^{-\frac{n}{2}} \int_{\mathbb{R}^n} \exp\Big(-\frac{|x|^2}{2}\Big)\delta(t - x_1x_2\cdots x_n)\,dx.$$

If $n = 2$, then $v(t)$ is annihilated by $t{\partial_t}^2 + \partial_t - t$, which has 0 as a regular singular point with the indicial polynomial $b(s) = s^2$. If $n = 3$, then $v(t)$ is annihilated by $t^2{\partial_t}^3 + 3t{\partial_t}^2 + \partial_t + t$, which has 0 as a regular singular point with the indicial polynomial $b(s) = s^3$. If $n = 4$, then $v(t)$ is annihilated by $t^3{\partial_t}^4 + 6t^2{\partial_t}^3 + 7t{\partial_t}^2 + \partial_t - t$ with the indicial polynomial $b(s) = s^4$ at 0.

Example 6.6. Introducing parameters $a = (a_1, a_2)$, let us consider the density function

$$v(t,a) = (2\pi)^{-1} \int_{\mathbb{R}^2} \exp\Big(-\frac{1}{2}(x_1^2 + x_2^2)\Big)\delta(t - (x_1 + a_1)(x_2 + a_2))\,dx_1dx_2.$$

The integrand belongs to $\mathcal{SS}'(\mathbb{R}^2_x \times (\mathbb{R}_t \times \mathbb{R}^2_a))$, hence $v(t,a)$ to $\mathcal{S}'(\mathbb{R}_t \times \mathbb{R}^2_a)$. Integrating the holonomic module over D_5 for the integrand shows that $v(t,a)$ satisfies a holonomic system $P_jv(t,a) = 0$ $(1 \le j \le 6)$ with

$$\begin{aligned}P_1 &= -4t\partial_{a_1} + \partial^3_{a_2} + 4a_2\partial^2_{a_2} + (4{a_2}^2 + 4)\partial_{a_2} + 4a_2,\\ P_2 &= (\partial_{a_2} + 2a_2)\partial_{a_1} + 2a_1\partial_{a_2} - 4t + 4a_2a_1,\\ P_3 &= 2t\partial_t + \partial^2_{a_2} + 2a_2\partial_{a_2} + 2, \qquad P_4 = (\partial_{a_1} + 2a_1)\partial_t + 2\partial_{a_2},\\ P_5 &= (\partial_{a_2} + 2a_2)t\partial_t + 2\partial_{a_1}, \qquad P_6 = \partial^2_{a_1} + 2a_1\partial_{a_1} - \partial^2_{a_2} - 2a_2\partial_{a_2}.\end{aligned}$$

The characteristic variety of $M := D_3/(D_3P_1 + \cdots + D_3P_6)$ is

$$\begin{aligned}&\{(t, a_1, a_2; \tau, \xi_1, \xi_2) \in \mathbb{C}^6 \mid \xi_2^2 = \xi_1\xi_2 = \tau\xi_2 = \xi_1^2 = \tau\xi_1 = t\tau^2 = 0\}\\ &= \{t = \xi_1 = \xi_2 = 0\} \cup \{\tau = \xi_1 = \xi_2 = 0\}.\end{aligned}$$

The singular locus of M is $\{t = 0\}$. By elimination we obtain an operator

$$P = t\partial_t^4 + 3\partial_t^3 + (-2t - a_1a_2)\partial_t^2 + (a_1^2 + a_2^2 - 3)\partial_t + t - a_1a_2$$

in t with parameters a_1, a_2 which annihilates $v(t, a)$. The indicial polynomial of P at $t = 0$ is $s^2(s-1)(s-2)$. The Fourier transform gives us a differential equation

$$\Big((\tau^2+1)^2\frac{d}{d\tau} + \tau^3 + ia_1a_2\tau^2 + (a_1^2 + a_2^2 + 1)\tau - ia_1a_2\Big)\hat{v}(\tau) = 0$$

for the characteristic function $\hat{v}(\tau)$. By quadrature we obtain

$$\hat{v}(\tau) = \frac{1}{\sqrt{\tau^2+1}}\exp\Big(\frac{2ia_1a_2\tau + a_1^2 + a_2^2}{2(\tau^2+1)} - \frac{1}{2}(a_1^2 + a_2^2)\Big).$$

Exercise 27. Compute a differential equation for

$$v(t) = \int_{\mathbb{R}^3} \exp\Big(-\frac{x_1^2 + x_2^2 + x_3^2}{2}\Big)\delta(t - x_1^2 + x_2^2 + x_3^2)\, dx_1dx_2dx_3$$

and one for its characteristic function $\hat{v}(\tau)$. Give an explicit formula for $\hat{v}(\tau)$.

Exercise 28. Set $u(x, t, a, b) = e^{tx}x_+^{a-1}(1-x)_+^{b-1}$ with positive real parameters a, b. We regard it as an element of $\mathcal{E}'\mathcal{D}'(\mathbb{R}_x \times \mathbb{R}_t)$ with a and b fixed. Deduce a linear differential equation (in t) for

$$v(t, a, b) := \int_0^1 e^{tx}x^{a-1}(1-x)^{b-1}\, dx = \int_{-\infty}^{\infty} u(x, t, a, b)\, dx$$

regarding a, b as parameters.

6.2. Powers of polynomials times a holonomic function

Let us begin with the simplest example: For a complex number λ with non-negative real part, the distribution x_+^λ on $\mathbb{R}$ satisfies a holonomic system $(x\partial_x - \lambda)x_+^\lambda = 0$. In particular, we have $x\partial_xY(x) = 0$. This amounts to introducing the differential equation $(x\partial_x - s)x^s = 0$ for a formal function x^s, which corresponds to x_+^λ, and specializing the parameter s to λ. We cannot regard $x^0 = 1$ because x^0 does not correspond to the constant function 1 but to $Y(x) = x_+^0$.

Now let $f_1(x), \ldots, f_p(x)$ be non-constant real polynomials in the variables $x = (x_1, \ldots, x_n)$. Let $v(x)$ be a holonomic locally integrable function on U. Then

$$\tilde{v}(x) = (f_1)_+^{\lambda_1} \cdots (f_p)_+^{\lambda_p} v(x)$$

is also locally integrable on U for complex numbers $\lambda_1, \ldots, \lambda_p$ with non-negative real parts. Especially, we have $\tilde{v}(x) = Y(f_1) \cdots Y(f_p) v(x)$ if $\lambda_j = 0$ $(j = 1, \ldots, p)$. Our purpose is to compute a holonomic system for $\tilde{v}(x)$.

Our strategy is as follows: First we work in a purely algebraic setting and consider the D-module generated by the tensor product $u \otimes f_1^{\lambda_1} \cdots f_p^{\lambda_p}$; we show that this D-module is holonomic and introduce an algorithm to compute its structure. Then we 'realize' these arguments and apply to the corresponding distribution $\tilde{v}(x)$, which lives in the 'real world'.

Let $\mathbb{K}$ be a field of characteristic zero and $f_1, \ldots, f_p \in \mathbb{K}[x] = \mathbb{K}[x_1, \ldots, x_n]$ be non-constant polynomials. Let us consider a 'function' $f_1^{s_1} \cdots f_p^{s_p}$ with indeterminates (as parameters) $s = (s_1, \ldots, s_p)$. More precisely, set

$$\mathcal{L} := \mathbb{K}[x, (f_1 \cdots f_p)^{-1}, s] f_1^{s_1} \cdots f_p^{s_p},$$

which is regarded as a free $\mathbb{K}[x, (f_1 \cdots f_p)^{-1}, s]$-module generated by the 'symbol' $f_1^{s_1} \cdots f_p^{s_p}$. Then $\mathcal{L}$ is a left $D_n[s]$-module with the natural derivations

$$\partial_{x_i}(f_1^{s_1} \cdots f_p^{s_p}) = \sum_{j=1}^{p} s_j \frac{\partial f_j}{\partial x_i} f_j^{-1} f_1^{s_1} \cdots f_p^{s_p} \qquad (i = 1, \ldots, n).$$

In what follows we denote $f^s = f_1^{s_1} \cdots f_p^{s_p}$ for the sake of simplicity when there is no fear of confusion.

Let $M = D_n u = M/I$ be a holonomic left D_n-module generated by an element $u \in M$ with the left ideal $I = \mathrm{Ann}_{D_n} u$. Let us consider the tensor product $M \otimes_{\mathbb{K}[x]} \mathcal{L}$ as $\mathbb{K}[x]$-module, which has also a natural structure of left $D_n[s]$-module induced by the derivations

$$\partial_{x_i}(u' \otimes v) = (\partial_{x_i} u') \otimes v + u' \otimes (\partial_{x_i} v) \quad (u' \in M,\ v \in \mathcal{L},\ i = 1, \ldots, n).$$

Our aim is to compute the annihilator (in $D_n[s]$) of $u \otimes f^s \in M \otimes_{\mathbb{K}[x]} \mathcal{L}$. For this purpose, define shift (difference) operators E_j by

$$E_j : \mathcal{L} \ni a(x, s_1, \ldots, s_p) f^s \longmapsto a(x, s_1, \ldots, s_j + 1, \ldots, s_p) f_j f^s \in \mathcal{L}$$

for $j = 1, \ldots, p$, which are bijective with the inverse shifts $E_j^{-1} : \mathcal{L} \to \mathcal{L}$.

Let $D_n\langle s, E, E^{-1}\rangle$ be the D_n-algebra generated by $s = (s_1, \dots, s_p)$, $E = (E_1, \dots, E_p)$, and $E^{-1} = (E_1^{-1}, \dots, E_p^{-1})$. We introduce new variables $t = (t_1, \dots, t_p)$ and the associated derivations $\partial_t = (\partial_{t_1}, \dots, \partial_{t_p})$. Let D_{n+p} be the ring of differential operators with respect to the variables $(x, t) = (x_1, \dots, x_n, t_1, \dots, t_p)$.

Let $\mu : D_{n+p} \to D_n\langle s, E, E^{-1}\rangle$ be the D_n-algebra homomorphism (Mellin transform) of D_n defined by

$$\mu(t_j) = E_j, \quad \mu(\partial_{t_j}) = -s_j E_j^{-1}.$$

This homomorphism is well-defined since

$$\begin{aligned}\mu(\partial_{t_j} t_j - t_j \partial_{t_j}) &= \mu(\partial_{t_j})\mu(t_j) - \mu(t_j)\mu(\partial_{t_j}) \\ &= -s_j E_j^{-1} E_j - E_{s_j}(-s_i)E_j^{-1} = 1.\end{aligned}$$

It can be extended to an isomorphism

$$\mu : D_{n+p}[(t_1 \cdots t_p)^{-1}] \xrightarrow{\sim} D_n\langle s, E, E^{-1}\rangle$$

such that $\mu(t_j^{-1}) = E_j^{-1}$. Hence we can regard D_{n+p} as a subring of $E\langle s, E, E^{-1}\rangle$ through μ. With this identification, we have

$$t_j = E_j, \qquad \partial_{t_j} = -s_j E_j^{-1}, \qquad s_j = -\partial_{t_j} t_j = -t_j \partial_{t_j} - 1.$$

Thus we have inclusions

$$D_n[s] \subset D_n\langle s, E\rangle \subset D_{n+p} \subset D_n\langle s, E, E^{-1}\rangle = D_{n+p}[(t_1 \cdots t_p)^{-1}]$$

of rings. Note that $M \otimes_{\mathbb{K}[x]} \mathcal{L}$ has a structure of left $D_n\langle s, E, E^{-1}\rangle$-module with s and E acting only on $\mathcal{L}$.

Lemma 6.7. *The submodule $D_{n+p} f^s$ of $\mathcal{L}$ is a free $\mathbb{K}[x]$-module generated by $\partial_t^\nu f^s$ with $\nu \in \mathbb{N}^p$.*

Proof. Let J be the left ideal of D_{n+p} generated by

$$P_i := \partial_{x_i} + \sum_{j=1}^{p} \frac{\partial f_j}{\partial x_i} \partial_{t_j} \quad (i = 1, \dots, n), \quad t_j - f_j \quad (j = 1, \dots, p).$$

It is easy to see that J annihilates f^s. In order to show that J coincides with $\mathrm{Ann}_{D_{n+p}} f^s$, let $\prec$ be a lexicographic term order for D_{n+p} such that $\partial_{t_j} \prec \partial_{x_i}$, $x_j \prec \partial_{x_i}$, and $x_i \prec t_j$ for $1 \le i \le n$ and $1 \le j \le p$. Then by division with respect to $\prec$, every element P of D_{n+p} is written in the form

$$P = \sum_{i=1}^{n} U_i P_i + \sum_{j=1}^{p} V_j (t_j - f_j) + \sum_{\nu \in \mathbb{N}^p} r_\nu(x) \partial_t^\nu$$

with $U_i, V_j \in D_{n+p}$ and $r_\nu(x) \in \mathbb{K}[x]$. This implies

$$Pf^s = \sum_{\nu\in\mathbb{N}^p} r_\nu(x)\partial_t^\nu f^s = \sum_{\nu\in\mathbb{N}^p} (-1)^{|\nu|}[s]_\nu r_\nu(x) f^{s-\nu}$$

with

$$\begin{aligned}[s]_\nu &= [s_1]_{\nu_1}\cdots[s_p]_{\nu_p}\\ &= s_1(s_1-1)\cdots(s_1-\nu_1+1)\cdots s_p(s_p-1)\cdots(s_p-\nu_p+1).\end{aligned}$$

Hence $Pf^s = 0$ holds if and only if $r_\nu(x) = 0$ for each ν since $\mathcal{L}$ is a free $\mathbb{K}[s]$-module. This also implies that an element of $D_{n+p}f^s$ is uniquely written in the form $\sum_{\nu\in\mathbb{N}^p} r_\nu(x)\partial_t^\nu f^s$ with $r_\nu(x) \in \mathbb{K}[x]$. This completes the proof. Q.E.D.

Our primary purpose in the following algebraic arguments is to compute the $D_n[s]$-submodule $D_n[s](u\otimes f^s)$ of $M\otimes_{\mathbb{K}[x]}\mathcal{L}$. With this purpose in mind, we consider the following modules:

- The D_{n+p}-submodule $N := D_{n+p}(u\otimes f^s)$ of $M\otimes_{\mathbb{K}[x]}\mathcal{L}$,
- The $D_n[s]$-submodule $N_s := D_n[s](u\otimes f^s)$ of N,
- The D_{n+p}-submodule $N' := D_{n+p}(u\otimes f^s)$ of $M\otimes_{\mathbb{K}[x]} D_{n+p}f^s$,
- The $D_n[s]$-submodule $N'_s := D_n[s](u\otimes f^s)$ of N'.

We will see that N' coincides with $M\otimes_{\mathbb{K}[x]} D_{n+p}f^s$ in fact. The inclusion $D_{n+p}f^s \subset \mathcal{L}$ induces a natural homomorphism

$$\iota\ :\ M\otimes_{\mathbb{K}[x]} D_{n+p}f^s \longrightarrow M\otimes_{\mathbb{K}[x]}\mathcal{L}$$

such that $\iota(N') = N$ and $\iota(N'_s) = N_s$. Let us first determine the structure of N'.

Algorithm 6.8 ($N' = M\otimes_{\mathbb{K}[x]} D_{n+p}f^s$). Input: A set G_0 of generators of I with $M = D_n/I$ and non-constant polynomials $f_1,\ldots,f_p \in \mathbb{K}[x]$.

For $P = P(x,\partial_{x_1},\ldots,\partial_{x_n}) \in G_0$, set

$$\tau(P,f_1,\ldots,f_p) := P\left(x,\,\partial_{x_1}+\sum_{j=1}^{p}\frac{\partial f_j}{\partial x_1}\partial_{t_j},\ldots,\partial_{x_n}+\sum_{j=1}^{p}\frac{\partial f_j}{\partial x_n}\partial_{t_j}\right)$$

following [35]. This substitution is well-defined in the ring D_{n+p} since the operators which are substituted for $\partial_{x_1},\ldots,\partial_{x_n}$ commute with one another.

Output: $G := \{\tau(P,f_1,\ldots,f_p) \mid P \in G_0\} \cup \{t_j - f_j(x) \mid j = 1,\ldots,p\}$ generates $J := \mathrm{Ann}_{D_{n+p}}(u\otimes f^s)$ and $M\otimes_{\mathbb{K}[x]} D_{n+p}f^s = N' = D_{n+p}/J$ is holonomic.

Proof. In view of the equality

$$\left(\partial_{x_i} + \sum_{j=1}^{p} \frac{\partial f_j}{\partial x_i}\partial_{t_j}\right)(u \otimes f^s)$$
$$= (\partial_{x_i} u) \otimes f^s + u \otimes \left(\partial_{x_i} + \sum_{j=1}^{p} \frac{\partial f_j}{\partial x_i}\partial_{t_j}\right) f^s$$
$$= (\partial_{x_i} u) \otimes f^s + u \otimes \left(\partial_{x_i} + \sum_{j=1}^{p}(-s_j) f_j^{-1} \frac{\partial f_j}{\partial x_i}\right) f^s = (\partial_{x_i} u) \otimes f^s$$

in $M \otimes_{\mathbb{K}[x]} \mathcal{L}$, we have, for any $P \in D_n$ and $j = 1, \ldots, p$,

$$\tau(P, f_1, \ldots, f_p)(u \otimes f^s) = (Pu) \otimes f^s, \quad (t_j - f_j)(u \otimes f^s) = u \otimes (t_j - f_j) f^s.$$

Hence J annihilates $u \otimes f^s$ in $M \otimes_{\mathbb{K}[x]} \mathcal{L}$. These relations also imply that $M \otimes_{\mathbb{K}[x]} D_{n+p} f^s$ is generated by $u \otimes f^s$ over D_{n+p} since $D_{n+p} f^s$ is generated by $\partial_t^\nu f^s$ ($\nu \in \mathbb{N}^p$) over $\mathbb{K}[x]$.

Conversely, suppose that $P \in D_{n+p}$ annihilates $u \otimes f^s$. We can rewrite P in the form

$$P = \sum_{\alpha \in \mathbb{N}^n, \nu \in \mathbb{N}^p} p_{\alpha,\nu}(x) \partial_t^\nu \Big(\partial_{x_1} + \sum_{j=1}^{p} \frac{\partial f_j}{\partial x_1}\partial_{t_j}\Big)^{\alpha_1} \cdots \Big(\partial_{x_n} + \sum_{j=1}^{p} \frac{\partial f_j}{\partial x_n}\partial_{t_j}\Big)^{\alpha_n} + \sum_{j=1}^{p} Q_j \cdot (t_j - f_j(x))$$

with $p_{\alpha,\nu}(x) \in \mathbb{K}[x]$ and $Q_j \in D_{n+p}$. Setting $P_\nu := \sum_{\alpha \in \mathbb{N}^n} p_{\alpha,\nu}(x) \partial_x^\alpha$, we get

$$0 = P(u \otimes f^s) = \sum_{\nu \in \mathbb{N}^p} \partial_t^\nu \tau(P_\nu, f_1, \ldots, f_p)(u \otimes f^s) = \sum_{\nu \in \mathbb{N}^p} P_\nu u \otimes \partial_t^\nu f^s.$$

It follows in view of Lemma 6.7 that each P_ν belongs to I. Hence we have

$$P = \sum_{\nu \in \mathbb{N}^p} \partial_t^\nu \tau(P_\nu, f_1, \ldots, f_p) + \sum_{j=1}^{p} Q_j \cdot (t_j - f_j(x)) \in J.$$

Finally, let us show that D_{n+p}/J is holonomic. Since D_n/I is holonomic, its characteristic variety $\mathrm{Char}(D_n/I)$ is an n-dimensional algebraic set of $\overline{\mathbb{K}}^{2n}$. By the definition, we have

$$\begin{aligned}
&\mathrm{Char}(D_{n+p}/J)\\
&\subset \Big\{(x,t,\xi,\tau) \in \overline{\mathbb{K}}^{2(n+p)} \mid \sigma(P)\Big(x, \xi_1 + \sum_{j=1}^{p} \frac{\partial f_j}{\partial x_1}\tau_j, \dots\Big) = 0 \ (\forall P \in I),\\
&\qquad t_j = f_j(x)\ (1 \le j \le p)\Big\}\\
&= \Big\{(x,t,\xi,\tau) \in \overline{\mathbb{K}}^{2(n+p)} \mid \Big(x, \xi_1 + \sum_{j=1}^{p} \frac{\partial f_j}{\partial x_1}\tau_j, \dots\Big) \in \mathrm{Char}(D_n/I),\\
&\qquad t_j = f_j(x)\ (1 \le j \le p)\Big\}.
\end{aligned}$$

Since the set on the last line is in one-to-one correspondence with the set $\mathrm{Char}(D_n/I) \times \overline{\mathbb{K}}^{p}$, the dimension of $\mathrm{Char}(D_{n+p}/J)$ is $n+p$, which implies that D_{n+p}/J is a holonomic module. This proves the correctness of the algorithm. Q.E.D.

Now that we have a set of generators of $J = \mathrm{Ann}_{D_{n+p}}(u \otimes f^s)$, we can compute $\mathrm{Ann}_{D_n[s]}(u \otimes f^s) = J \cap D_n[s]$ by using Algorithm 5.5. Thus we get an explicit presentation of the $D_n[s]$-submodule $N'_s = D_n[s](u \otimes f^s)$ of $M \otimes_{\mathbb{K}[x]} D_{n+p} f^s$. Finally let us specialize the parameters $s = (s_1, \dots, s_n)$ to $\lambda = (\lambda_1, \dots, \lambda_p) \in \mathbb{K}^p$. Set

$$\begin{aligned}
N'(\lambda) &= N'_s/((s_1 - \lambda_1)N'_s + \dots + (s_p - \lambda_p)N'_s),\\
N(\lambda) &= N_s/((s_1 - \lambda_1)N_s + \dots + (s_p - \lambda_p)N_s),
\end{aligned}$$

which are left D_n-modules.

Proposition 6.9. *Set $f = f_1 \cdots f_p$. The homomorphism*

$$\iota : N' = M \otimes_{\mathbb{K}[x]} D_{n+p} f_1^{s_1} \cdots f_p^{s_p} \longrightarrow M \otimes_{\mathbb{K}[x]} \mathcal{L}$$

is injective if and only if M is f-saturated, or f-torsion free, i.e., $f : M \to M$ is injective. In particular, ι induces isomorphisms $N' \cong N$ and $N'_s \cong N_s$ if M is f-saturated

Proof. By Lemma 6.7, an arbitrary element w of $M \otimes_{\mathbb{K}[x]} D_{n+p} f^s$ is uniquely written in a finite sum

$$w = \sum_{\nu \in \mathbb{N}^p} u_\nu \otimes \partial_t^\nu f_1^{s_1} \cdots f_p^{s_p}$$

with $u_\nu \in M$. Then we have

$$\iota(w) = \sum_{\nu\in\mathbb{N}^p} (-1)^{|\nu|}[s]_\nu u_\nu \otimes f_1^{s_1-\nu_1}\cdots f_p^{s_p-\nu_p}.$$

Since $\mathcal{L}$ is isomorphic to $\mathbb{K}[x,s,f^{-1}]$ with $f = f_1\cdots f_p$ as a $\mathbb{K}[x,s]$-module, $M\otimes_{\mathbb{K}[x]}\mathcal{L}$ is isomorphic to the localization $M[s,f^{-1}]$ of $M[s] := M\otimes_{\mathbb{K}}\mathbb{K}[s]$ with respect to f as a $\mathbb{K}[x,s]$-module. Hence $\iota(w)$ vanishes if and only if

$$\sum_{\nu\in\mathbb{N}^p} (-1)^{|\nu|}[s]_\nu f^{-\nu_1}\cdots f^{-\nu_p}u_\nu$$

vanishes in $M[s,f^{-1}]$. This amounts to the condition $u_\nu = 0$ in $M[f^{-1}]$ for all ν. Finally, u_ν vanishes in $M[f^{-1}]$ if and only if $f^{m_\nu}u_\nu = 0$ holds in M with a non-negative integer m_ν. This implies the assertion. Q.E.D.

The following theorem was proved in case $p = 1$ by Kashiwara [13] in the analytic setting.

Theorem 6.10. *For any $\lambda \in \mathbb{K}^p$, $N(\lambda)$ is a holonomic D_n-module.*

Proof. First, let us show that $N'(\lambda)$ is holonomic if either no component of λ belongs to $\mathbb{N}$ or M is f-saturated with $f = f_1\cdots f_p$.

We may assume $M \neq \{0\}$. Set $F_k(N') = F_k^{(\mathbf{1};\mathbf{1})}(D_{n+p})(u\otimes f^s)$ for $k\in\mathbb{Z}$. Since N' is holonomic, there exists a polynomial $H_0(k)$ of degree $n+p$ such that

$$\dim_{\mathbb{K}} F_k(N') = H_0(k) \qquad (\forall k \gg 0).$$

We define a filtration $F_k(D_n[s])$ on the ring $D_n[s]$ by

$$F_k(D_n[s]) = \Big\{\sum_{\alpha,\beta,\gamma} a_{\alpha,\beta,\gamma}x^\alpha\partial_x^\beta s^\gamma \mid |\alpha|+|\beta|+2|\gamma| \leq k\Big\}.$$

Then $F_k(N'_s) := F_k(D_n[s])(u\otimes f^s)$ defines a good filtration on N'_s as a filtered module over the filtered ring $D_n[s]$. The associated graded module $\mathrm{gr}(N'_s)$ is a graded module over the graded ring $\mathbb{K}[x,\xi,s]$ in which x_i, ξ_i are of order one, and s_j are of order two. Hence (see e.g., [5], [8], [12]) there exists $Q(T)\in\mathbb{Z}[T,T^{-1}]$ such that

$$\begin{aligned}\sum_{k=-\infty}^{\infty}\dim_{\mathbb{K}}\mathrm{gr}_k(N'_s)T^k &= \sum_{k=-\infty}^{\infty}(\dim_{\mathbb{K}}F_k(N'_s) - \dim_{\mathbb{K}}F_{k-1}(N'_s))T^k \\ &= \frac{Q(T)}{(1-T)^{2n}(1-T^2)^p}\end{aligned}$$

holds as a formal Laurent series in an indeterminate T. It follows that there exists a polynomial $H_1(k) \in \mathbb{Q}[k]$ such that

$$\dim_{\mathbb{K}} F_{2k}(N'_s) = H_1(k) \qquad (\forall k \gg 0).$$

The degree of $H_1(k)$ is at most $n+p$ since $F_k(N'_s) \subset F_k(N')$.

Now, let us consider the $D_n[s]$-endomorphism of N'_s defined by $s_p - \lambda_p = -\partial_{t_p} t_p - \lambda_p$. Let v be an element of N'_s such that $(s_p - \lambda_p)v = 0$. As an element of N', v is written uniquely in the form

$$v = \sum_{\nu \in \mathbb{N}^p} v_\nu \otimes \partial_t^\nu f^s$$

with $v_\nu \in M$. Then $(s_p - \lambda_p)v = 0$ is equivalent to

$$(14) \qquad (\nu_p - \lambda_p)v_\nu - f_p v_{\nu - e_p} = 0 \qquad (\forall \nu = (\nu_1, \dots, \nu_p) \in \mathbb{N}^p)$$

with $e_p = (0, \dots, 0, 1) \in \mathbb{N}^p$ since

$$s_p \partial_{t_p}^j = -\partial_{t_p} t_p \partial_{t_p}^j = -\partial_{t_p}^{j+1} t_p + j \partial_{t_p}^j, \qquad t_p f^s = f_p f^s.$$

First let us assume that $\lambda_p \neq 0, 1, 2, \dots$. Then from (14) we deduce $v_\nu = 0$ for any $\nu \in \mathbb{N}^p$ by induction on ν_p. Hence $s_p - \lambda_p : N'_s \to N'_s$ is injective.

Next, assume that $\lambda_p \in \mathbb{N}$ and that M is f-saturated. There exists d such that $v_\nu = 0$ if $\nu_p \geq d$. Then from (14), it follows that $f^j v_{(\nu', d-j)} = 0$ for all $0 \leq j \leq d$ and $\nu' \in \mathbb{N}^{p-1}$ by induction on j. Hence we have $v_\nu = 0$ for all $\nu \in \mathbb{N}^p$ if M is f-saturated.

Hereafter we suppose that $s_p - \lambda_p$ defines an injective endomorphism of N'_s. Then we have an exact sequence

$$0 \longrightarrow N'_s \stackrel{s_p - \lambda_p}{\longrightarrow} N'_s \longrightarrow N'_s/(s_p - \lambda_p)N'_s \longrightarrow 0$$

of left $D_n[s']$-modules with $s' = (s_1, \dots, s_{p-1})$. Let us regard $N'' := N'/(s_p - \lambda_p)N'$ as a filtered module over the filtered ring $D_n[s_1, \dots, s_{p-1}]$ and define a good filtration on N'' by

$$F_k(N'') = F_k(N'_s)/(F_k(N'_s) \cap (s_p - \lambda_p)N'_s).$$

Since $(s_p - \lambda_p)F_{2k-2}(N'_s)$ is contained in $F_{2k}(N'_s) \cap (s_p - \lambda_p)N'_s$, we have

$$\begin{aligned}
\dim_{\mathbb{K}} F_{2k}(N'') &= \dim_{\mathbb{K}} F_{2k}(N'_s) - \dim_{\mathbb{K}}(F_{2k}(N'_s) \cap (s_p - \lambda_p)N'_s) \\
&\leq \dim_{\mathbb{K}} F_{2k}(N'_s) - \dim_{\mathbb{K}} F_{2k-2}(N'_s) = H_1(k) - H_1(k-1)
\end{aligned}$$

for sufficiently large k. Note that $H_1(k) - H_1(k-1)$ is a polynomial of degree at most $n+p-1$. Proceeding inductively, we obtain a polynomial $H_p(k)$ of degree at most n such that

$$\dim_{\mathbb{K}} F_{2^p k}(N'(\lambda)) = H_p(k) \qquad (\forall k \gg 0)$$

with a good $(\mathbf{1};\mathbf{1})$-filtration

$$F_k(N'(\lambda)) := F_k(N'_s)/(F_k(N'_s) \cap ((s_1-\lambda_1)N'_s + \cdots + (s_p-\lambda_p)N'_s))$$

on $N'(\lambda)$. On the other hand, there exists a polynomial $G(k)$ such that

$$\dim_{\mathbb{K}} F_k(N'(\lambda)) = G(k) \qquad (\forall k \gg 0).$$

Hence $H_p(k) = G(2^p k)$ holds for sufficiently large k. This implies

$$\dim N'(\lambda) = \deg G(k) = \deg H_p(k) \leq n.$$

Thus $N'(\lambda)$ is a holonomic D_n-module under the assumption above.

Finally, let us show $N(\lambda)$ is holonomic. The localization $M[f^{-1}]$ has a natural structure of left D_n-module and is holonomic as such ([13], [4]; see also Theorem 2.14 in [25] for a constructive proof). Moreover, $M[f^{-1}]$ is f-saturated by the definition. Let $\rho : M \ni Pu \mapsto Pu \otimes 1 \in M[f^{-1}]$ be the canonical homomorphism. Then by Proposition 6.9, the canonical homomorphism

$$\iota : \rho(M) \otimes_{\mathbb{K}[x]} D_{n+p} f^s \longrightarrow \rho(M) \otimes_{\mathbb{K}[x]} \mathcal{L}$$

is injective. Hence the submodule $\tilde{N}_s := D_n[s](\rho(u) \otimes f^s)$ of $\rho(M) \otimes_{\mathbb{K}[x]} \mathcal{L}$ is isomorphic to the submodule $\tilde{N}'_s := D_n[s](\rho(u) \otimes f^s)$ of $\rho(M) \otimes_{\mathbb{K}[x]} D_{n+p} f^s$. On the other hand, ρ induces an isomorphism $M \otimes_{\mathbb{K}[x]} \mathcal{L} \to \rho(M) \otimes_{\mathbb{K}[x]} \mathcal{L}$ by the definition. This induces an isomorphism $N_s \cong \tilde{N}_s$. Summing up, we obtain an isomorphism

$$N(\lambda) \cong \tilde{N}'(\lambda) := \tilde{N}'_s/((s_1-\lambda_1)\tilde{N}'_s + \cdots + (s_p-\lambda_p)\tilde{N}'_s).$$

The module of the right-hand side is holonomic by the argument above. This completes the proof. Q.E.D.

Note that if $\mathbb{K}$ is algebraically closed then Theorem 6.10 holds under a weaker assumption that M be holonomic on $\{x \in \mathbb{K}^n \mid f_1(x) \cdots f_p(x) \neq 0\}$ since it implies that $M[f^{-1}]$ is a holonomic D_n-module ([13]; see also Theorem 3.14 of [25] for an elementary proof). We do not know if $N'(\lambda)$, which can be computed directly, is always holonomic; the proof of Theorem 4 in [24] is insufficient. In general, $N(\lambda)$ is stronger and more suited to our application below than $N'(\lambda)$. Summing up we have obtained

Algorithm 6.11 ($N(\lambda)$)**.** Input: A set G_0 of generators of I with $M = D_n/I = D_n u$, non-constant polynomials $f_1, \ldots, f_p \in \mathbb{K}[x]$, and $\lambda \in \mathbb{K}^p$.
Output: $N(\lambda) = D_n/J$.

(1) Compute the D_n-submodule $\rho(M) = D_n\rho(u)$ of the localization $M[(f_1 \cdots f_p)^{-1}]$ with $\rho(u) = u \otimes 1$ by using the localization algorithm of [25], which is a modification of the one in [28].
(2) Compute a set G_1 of generators of the annihilator of $\rho(u) \otimes f^s$ in $\rho(M) \otimes_{\mathbb{K}[x]} D_{n+p} f^s$ by Algorithm 6.8.
(3) Compute generators $P_1(s), \ldots, P_r(s)$ of the annihilator

$$\mathrm{Ann}_{D_n[s]}(\rho(u) \otimes f^s) = D_n[s] \cap \mathrm{Ann}_{D_{n+p}}(\rho(u) \otimes f^s)$$

by using G_1 and Algorithm 5.5.
(4) Let J be the left ideal of D_n generated by $P_1(\lambda), \ldots, P_r(\lambda)$.

In practice, we can skip step (1) and proceed to steps (2),(3),(4) with $\rho(M)$ replaced by M. This gives us $N'(\lambda)$ which might be weaker than $N(\lambda)$.

Now let us return to the 'real world'. Assume that $f_1, \ldots, f_p \in \mathbb{R}[x]$ and let $v(x)$ be a locally integrable holonomic function on an open set U of $\mathbb{R}^n$. Then

$$\tilde{v}(x, \lambda) := v(x)(f_1)_+^{\lambda_1} \cdots (f_p)_+^{\lambda_p}$$

is well-defined as a locally integrable function on U if the real parts of $\lambda_1, \ldots, \lambda_p$ are non-negative. More precisely, $\tilde{v}(x, \lambda)$ belongs to the space $\mathcal{O}(\mathbb{C}_+^p, \mathcal{D}'(U))$ of $\mathcal{D}'(U)$-valued holomorphic functions on

$$\mathbb{C}_+^p = \{\lambda = (\lambda_1, \ldots, \lambda_p) \in \mathbb{C}^p \mid \mathrm{Re}\, \lambda_j > 0 \ (1 \le j \le p)\},$$

that is, $\langle \tilde{v}(x, \lambda), \varphi(x) \rangle$ is a holomorphic function of $\lambda = (\lambda_1, \ldots, \lambda_p)$ on $\mathbb{C}_+^p$ for any $\varphi \in C_0^\infty(U)$. Moreover, $\langle \tilde{v}(x, \lambda), \varphi(x) \rangle$ is continuous on the closure $\overline{\mathbb{C}_+^p}$.

Let I be a left ideal of D_n which annihilates $v(x)$ such that $M := D_n/I$ is holonomic. Set $M = D_n/I = D_n u$ with $u = \overline{1}$ and $\mathcal{L} = \mathbb{K}[x, (f_1 \cdots f_p)^{-1}, s] f^s$ with $\mathbb{K} = \mathbb{C}$. In order to apply the algebraic arguments so far to $\tilde{v}(x, \lambda)$, we need the following fact:

Theorem 6.12. *Suppose that $P(s) \in D_n[s]$ annihilates $u \otimes f^s$ in $M \otimes_{\mathbb{C}[x]} \mathcal{L}$ and that the real parts of the components of $\lambda \in \mathbb{C}^p$ are non-negative. Then $P(\lambda)\tilde{v}(x, \lambda)$ vanishes as a distribution on U. Hence $\tilde{v}(x, \lambda)$ is a solution of the holonomic system $N(\lambda)$.*

Before proving this theorem, let us begin with

Lemma 6.13. *Suppose that $P(s) \in D_n[s]$ annihilates $u \otimes f^s$ in $M \otimes_{\mathbb{C}[x]} \mathcal{L}$. Then $P(\lambda)\tilde{v}(x,\lambda)$ vanishes as a distribution on $U_f := \{x \in U \mid f(x) \neq 0\}$ with $f = f_1 \cdots f_p$ for any $\lambda \in \mathbb{C}^p_+$.*

Proof. We denote $f_+^\lambda = (f_1)_+^{\lambda_1} \cdots (f_p)_+^{\lambda_p}$. The $\mathbb{C}$-bilinear homomorphism

$$\begin{aligned}\Phi \,:\, M \times \mathcal{L} \ni (Pu,\, a(x,s,f^{-1})f^s)& \\ \longmapsto a(x,\lambda,f^{-1})f_+^\lambda Pv(x) &\in \mathcal{O}(\mathbb{C}^p_+, \mathcal{D}'(U_f))\end{aligned}$$

with $P \in D_n$, $a \in \mathbb{C}[x,s,t]$ is well-defined since $f_1,\dots,f_p$ do not vanish on U_f. Moreover, Φ is $\mathbb{C}[x]$-balanced in the sense that

$$\Phi(c(x)Pu,\, a(x,s,f^{-1})) = \Phi(Pu,\, c(x)a(x,s,f^{-1})) \qquad (\forall c(x) \in \mathbb{C}[x]).$$

Hence Φ induces a $\mathbb{C}[x]$-homomorphism

$$\Psi \,:\, M \otimes_{\mathbb{C}[x]} \mathcal{L} \longrightarrow \mathcal{O}(\mathbb{C}^p_+, \mathcal{D}'(U_f))$$

such that

$$\Psi((Pu) \otimes a(x,s,f^{-1})f^s) = \Phi(Pu,\, a(x,s,f^{-1})f^s) = a(x,\lambda,f^{-1})f_+^\lambda Pv$$

because of the universality of the tensor product. It is easy to see that

$$\Psi((Pu) \otimes s_j a(x,s,f^{-1})f^s) = \lambda_j \Psi((Pu) \otimes a(x,s,f^{-1})f^s) \quad (1 \le j \le p),$$

and

$$\begin{aligned}\Psi(\partial_i((Pu) \otimes a(x,s,f^{-1})f^s)) &= \Psi((\partial_i Pu) \otimes a(x,s,f^{-1})f^s) \\ &\quad + \Psi((Pu) \otimes (\partial_i a(x,s,f^{-1}))f^s)) \\ &\quad + \Psi\Big((Pu) \otimes \sum_{j=1}^p s_j f_j^{-1} \frac{\partial f_j}{\partial x_i} a(x,s,f^{-1})f^s\Big) \\ &= a(x,\lambda,f^{-1})f_+^\lambda \partial_i Pv(x) + \partial_i a(x,\lambda,f^{-1})f_+^\lambda Pv(x) \\ &\quad + \sum_{j=1}^p \lambda_j f_j^{-1} \frac{\partial f_j}{\partial x_i} a(x,\lambda,f^{-1})f_+^\lambda Pv(x) \\ &= \partial_i \left(a(x,\lambda,f^{-1})f_+^\lambda Pv(x)\right) \qquad (1 \le i \le n)\end{aligned}$$

hold as distributions on U_f since f_+^λ is real analytic in x there. This implies that

$$\Psi(P(s)(u \otimes f^s)) = P(\lambda)\Psi(u \otimes f^s) = P(\lambda)(f_+^\lambda v(x))$$

holds for any $P(s) \in D_n[s]$. Hence the right-hand side vanishes as an element of $\mathcal{O}(\mathbb{C}^p_+, \mathcal{D}'(U_f))$ if $P(s)(u \otimes f^s) = 0$ in $M \otimes_{\mathbb{C}[x]} \mathcal{L}$. Q.E.D.

Next we generalize a lemma by Kashiwara and Kawai [16], which corresponds to the case $p = 1$:

Proposition 6.14. *Let U be an open set of $\mathbb{R}^n$ and let $f_1, \dots, f_p$ be real-valued real analytic functions on U such that $\{x \in U \mid f_1(x) > 0, \dots, f_p(x) > 0\}$ is not empty. Let $v(x)$ be a locally integrable function on U. Set $f = f_1 \cdots f_p$ and $U_f = \{x \in U \mid f(x) \neq 0\}$. Let $s_1, \dots, s_p$ be indeterminates and $\lambda_1, \dots, \lambda_p$ be complex variables. Assume that $P(s_1, \dots, s_p) \in D_n[s_1, \dots, s_p]$ satisfies*

$$P(\lambda_1, \dots, \lambda_p)((f_1)_+^{\lambda_1} \cdots (f_p)_+^{\lambda_p} v) = 0 \quad \textit{in } \mathcal{D}'(U_f)$$

if $\mathrm{Re}\,\lambda_j$ $(j = 1, \dots, p)$ are sufficiently large. Then one has

$$P(\lambda_1, \dots, \lambda_p)((f_1)_+^{\lambda_1} \cdots (f_p)_+^{\lambda_p} v) = 0 \quad \textit{in } \mathcal{D}'(U)$$

for any $\lambda_j \in \mathbb{C}$ with $\mathrm{Re}\,\lambda_j \geq 0$ $(j = 1, \dots, p)$.

Proof. We argue by induction on p. First let us set $p = 1$ and recall the proof of Lemma 2.9 in [16]. We denote $s = s_1$, $f = f_1$, and $\lambda = \lambda_1$. Let φ be an element of $C_0^\infty(U)$. Then its support $K = \mathrm{supp}\,\varphi$ is a compact subset of U. Let $\chi(t)$ be a C^∞ function of a variable t such that $\chi(t) = 1$ for $|t| \leq 1/2$ and $\chi(t) = 0$ for $|t| \geq 1$. Let τ be a real number with $0 < \tau < 1$. Since the support of $\big(1 - \chi(\frac{f}{\tau})\big)\varphi$ is contained in U_f, we have by the assumption that

$$\begin{aligned}\langle P(\lambda)(f_+^\lambda v),\ \varphi\rangle &= \left\langle P(\lambda)(f_+^\lambda v),\ \chi(\frac{f}{\tau})\varphi\right\rangle \\ &= \int_K {}^tP(\lambda)\Big(\chi(\frac{f}{\tau})\varphi\Big)(f_+^\lambda v)\,dx,\end{aligned}$$

where ${}^tP(\lambda)$ denotes the adjoint operator of $P(\lambda)$. Let m be the order of $P(s)$ and d be the degree of $P(s)$ in s. Then there exists a constant $C > 0$ such that

$$\sup_{x \in K} \left| {}^tP(\lambda)\Big(\chi(\frac{f(x)}{\tau})\varphi(x)\Big)\right| \leq C(1 + |\lambda|)^d \tau^{-m} \qquad (0 < \forall\tau < 1).$$

Assume Re $\lambda > m$ and $0 < \tau < 1$. Then we have

$$\begin{aligned}&\Big|\int_K {}^tP(\lambda)\Big(\chi(\frac{f}{\tau})\varphi\Big)(f_+^\lambda v)\,dx\Big|\\&\quad\le C(1+|\lambda|)^d\tau^{-m}\int_{\{x\in K\,|\,0\le f(x)\le\tau\}}|f_+^\lambda v(x)|\,dx\\&\quad\le C(1+|\lambda|)^d\tau^{-m+\mathrm{Re}\,\lambda}\int_K|v(x)|\,dx.\end{aligned}$$

This implies

$$\langle P(\lambda)(f_+^\lambda v),\ \varphi\rangle = \lim_{\tau\to+0}\int_K {}^tP(\lambda)\Big(\chi(\frac{f}{\tau})v\Big)(f_+^\lambda v)\,dx = 0.$$

By the uniqueness of analytic continuation with respect to λ, we know that $P(\lambda)(f_+^\lambda v) = 0$ in $\mathcal{D}'(U)$ if Re $\lambda \ge 0$.

Now suppose that the assertion of the proposition is proved with p replaced by $p-1$. We use the notation $s = (s_1,\dots,s_p)$ and $\lambda = (\lambda_1,\dots,\lambda_p)$. Set

$$V = \{x\in U \mid f_1(x) = \cdots = f_p(x) = 0\}.$$

With the assumption of the proposition, we have

$$P(\lambda)((f_1)_+^{\lambda_1}\cdots(f_p)_+^{\lambda_p}v) = 0$$

on $U\setminus V$ if Re $\lambda_j \ge 0$ for $j = 1,\dots,p$. In fact, for a point x_0 of $U\setminus V$, we may assume $f_p(x_0) > 0$. Then replacing v by $f_p^{\lambda_p}v$ and U by a neighborhood U_{x_0} of x_0, we conclude that $P(\lambda)((f_1)_+^{\lambda_1}\cdots(f_p)_+^{\lambda_p}v)$ vanishes as an element of $\mathcal{D}'(U_{x_0})$ if Re $\lambda_j \ge 0$ $(1\le j\le p)$ by the induction hypothesis.

Since the support of $\big(1-\chi(\frac{f_1}{\tau})\cdots\chi(\frac{f_p}{\tau})\big)\varphi$ is contained in $U\setminus V$, we have

$$\begin{aligned}&\Big\langle P(\lambda)((f_1)_+^{\lambda_1}\cdots(f_p)_+^{\lambda_p}v),\ \varphi\Big\rangle\\&\quad=\Big\langle P(\lambda)(f_1)_+^{\lambda_1}\cdots(f_p)_+^{\lambda_p}v),\ \chi(\frac{f_1}{\tau})\cdots\chi(\frac{f_p}{\tau})\varphi\Big\rangle\\&\quad=\int_U {}^tP(\lambda)\Big(\chi(\frac{f_1}{\tau})\cdots\chi(\frac{f_p}{\tau})\varphi\Big)(f_1)_+^{\lambda_1}\cdots(f_p)_+^{\lambda_p}v\,dx.\end{aligned}$$

Let m be the order of $P(s)$ and d be the total degree of $P(s)$ in s. Then there exists a constant $C > 0$ such that

$$\sup_{x\in K}\Big|{}^tP(\lambda)\Big(\chi(\frac{f_1(x)}{\tau})\cdots\chi(\frac{f_p(x)}{\tau})\varphi(x)\Big)\Big| \le C(1+|\lambda_1|+\cdots+|\lambda_p|)^d\tau^{-m}$$

holds if $0 < \forall\tau < 1$. Set $K(\tau) := \{x \in K \mid 0 \le f_j(x) \le \tau \ (j = 1, \dots, p)\}$ with $K = \operatorname{supp}\varphi$. Then we have

$$\begin{aligned}
&\left|\left\langle P(\lambda)((f_1)_+^{\lambda_1}\cdots(f_p)_+^{\lambda_p}v),\ \varphi\right\rangle\right| \\
&\le C(1+|\lambda_1|+\cdots+|\lambda_p|)^d\tau^{-m}\int_{K(\tau)}|(f_1)_+^{\lambda_1}\cdots(f_p)_+^{\lambda_p}v(x)|\,dx \\
&\le C(1+|\lambda_1|+\cdots+|\lambda_p|)^d\tau^{\mathrm{Re}\ \lambda_1+\cdots+\mathrm{Re}\ \lambda_p-m}\int_K|v(x)|\,dx
\end{aligned}$$

if $0 < \forall\tau < 1$. This implies that $\left\langle P(\lambda)((f_1)_+^{\lambda_1}\cdots(f_p)_+^{\lambda_p}u),\ \varphi\right\rangle$ vanishes if $\mathrm{Re}\ \lambda_j \ge 0$ $(1 \le j \le p)$ and $\mathrm{Re}\ \lambda_1 + \cdots + \mathrm{Re}\ \lambda_p > m$, and hence only if $\mathrm{Re}\ \lambda_j \ge 0$ $(1 \le j \le p)$ by the uniqueness of analytic continuation. Q.E.D.

Proof of Theorem 6.12: The first statement follows from Lemma 6.13 and Proposition 6.14. Let us fix $\lambda^0 = (\lambda_1^0, \dots, \lambda_p^0) \in \overline{\mathbb{C}_+^p}$. Suppose $P \in D_n$ annihilates the residue class $(u \otimes f^s)|_{s=\lambda^0}$ in $N(\lambda^0)$ of $u \otimes f^s \in M \otimes_{\mathbb{K}[x]} \mathcal{L}$. Then there exist $P(s), Q_j(s) \in D_n[s]$ such that

$$P = P(s) + (s_1 - \lambda_1^0)Q_1(s) + \cdots + (s_p - \lambda_p^0)Q_p(s), \qquad P(s)(u \otimes f^s) = 0.$$

It follows from the first statement that

$$P\tilde{v}(x,\lambda) = (\lambda_1 - \lambda_1^0)Q_1(\lambda)\tilde{v}(x,\lambda) + \cdots + (\lambda_p - \lambda_p^0)Q_p(\lambda)\tilde{v}(x,\lambda)$$

for $\lambda \in \overline{\mathbb{C}_+^p}$, and hence $P\tilde{v}(x,\lambda^0) = 0$. This implies that there exists a D_n-linear map from $N(\lambda^0)$ to $\mathcal{D}'(U)$ which sends $(u \otimes f^s)|_{s=\lambda_0}$ to $\tilde{v}(x,\lambda^0)$. This completes the proof of Theorem 6.12.

6.3. Integrals over the domain defined by polynomial inequalities

We assume $\mathbb{K} = \mathbb{C}$. As in 4.3, set $x = (x_1, \dots, x_n)$ and $y = (y_1, \dots, y_d)$. Assume that $f_1, \dots, f_p$ are real polynomials in (x, y) and let $v(x, y)$ be a holonomic locally integrable function on an open set of $\mathbb{R}^{n+d}$ and let $\lambda_1, \dots, \lambda_p$ be complex numbers with non-negative real parts. We assume that

$$\tilde{v}(x,y) := v(x,y)(f_1)_+^{\lambda_1}\cdots(f_p)_+^{\lambda_p}$$

belongs to $\mathcal{E}'\mathcal{D}'(\mathbb{R}_x^n \times U)$ with an open set U of $\mathbb{R}_y^d$, or to $\mathcal{SS}'(\mathbb{R}_x^n \times \mathbb{R}_y^d)$. Let I be a left ideal of D_{n+d} which annihilates $v(x, y)$ such that $M := D_{n+d}/I$ is holonomic. Set $M = M/I = D_{n+d}u$ with $u = \overline{1}$ and $\mathcal{L} = \mathbb{C}[x, y, (f_1 \cdots f_p)^{-1}, s]f^s$. Then Algorithm 6.11 yields a holonomic system for $\tilde{v}(x, y)$ by virtue of Theorem 6.12. Thus by the integration algorithm, we get a holonomic system for

$$w(y) := \int_{\mathbb{R}^n} v(x, y)(f_1)_+^{\lambda_1} \cdots (f_p)_+^{\lambda_p} \, dx.$$

In particular, setting $\lambda_1 = \cdots = \lambda_p = 0$, we obtain a holonomic system for

$$w(x) = \int_{D(y)} v(x, y) \, dx = \int_{\mathbb{R}^n} v(x, y) Y(f_1(x, y)) \cdots Y(f_p(x, y)) \, dx$$

with

$$D(y) = \{x \in \mathbb{R}^n \mid f_j(x, y) \geq 0 \ (1 \leq j \leq p)\}.$$

As examples, let us consider truncated multi-dimensional normal distributions: Let $f_1, \ldots, f_p$ be real polynomials in $x = (x_1, \ldots, x_n)$ and set

$$D = \{x \in \mathbb{R}^n \mid f_j(x) \geq 0 \ (1 \leq j \leq p)\}.$$

Then $\exp\Big(-\dfrac{|x|^2}{2}\Big) Y(f_1) \ldots Y(f_p)$ is, up to a constant multiple, the probability density function of the standard normal distribution truncated by D. Let $f(x)$ be a real polynomial, which we regard as a random variable. Then the cumulative and the density functions of $f(x)$ are given by

$$\begin{aligned} F(t) &= \int_D \exp\Big(-\frac{|x|^2}{2}\Big) Y(t - f(x)) \, dx \\ &= \int_{\mathbb{R}^n} \exp\Big(-\frac{|x|^2}{2}\Big) Y(t - f(x)) Y(f_1(x)) \cdots Y(f_p(x)) \, dx \end{aligned}$$

and

$$F'(t) = \int_{\mathbb{R}^n} \exp\Big(-\frac{|x|^2}{2}\Big) \delta(t - f(x)) Y(f_1(x)) \cdots Y(f_p(x)) \, dx$$

respectively up to constant multiples. The integrands belong to the space $\mathcal{SS}'(\mathbb{R}^n_x \times \mathbb{R}_t)$.

Example 6.15. Setting $f(x) = |x|^2$ and

$$D = \{x = (x_1, \ldots, x_n) \in \mathbb{R}^n \mid x_i \geq 0 \ (1 \leq i \leq n), \ x_1 + \cdots + x_n \leq 1\},$$

let us consider the density function

$$v(t) = \int_{\mathbb{R}^n} \exp\Big(-\frac{|x|^2}{2}\Big) \delta(t - |x|^2) Y(x_1) \cdots Y(x_n) Y(1 - x_1 - \cdots - x_n) \, dx$$

(up to a constant multiple) of the random variable $|x|^2$. If $n = 2$, then $v(t)$ satisfies a differential equation

$$\{4t(t-1)(2t-1){\partial_t}^2+4(-2t^3+6t^2-5t+1)\partial_t+2t^3-9t^2+9t-2\}v(t)=0.$$

Its indicial polynomials at 0, 1, and 1/2 are s^2, $s(s-1)$, and $s(2s-1)$ respectively. It can be verified that 1 is an apparent singular point e.g., by an algorithm described in Chapter 1 of [23]. Hence $v(t)$ belongs to $L^1(\mathbb{R})$ and real analytic on $\mathbb{R} \setminus \{0, 1/2\}$.

If $n = 3$, $v(t)$ is annihilated by

$$8t(t-1)(2t-1)(3t-1){\partial_t}^3 + (-72t^4 + 276t^3 - 308t^2 + 116t - 12){\partial_t}^2 \\ + (36t^4 - 210t^3 + 308t^2 - 162t + 28)\partial_t - 6t^4 + 47t^3 - 83t^2 + 53t - 11.$$

Its indicial polynomials at $0, 1, 1/2, 1/3$ are $s(s-1)(2s-1)$, $s(s-1)(s-2)$, $s(s-1)(2s-3)$, $s(s-1)^2$ respectively up to constant multiples. The point 1 is an apparent singular point.

Example 6.16. Set $n = 2$ and

$$v(t) = \int_{\mathbb{R}^2} \exp\Big(-\frac{|x|^2}{2}\Big)\delta(t - x_1 - x_2)Y(1 - x_1^2 - x_2^2)\,dx.$$

Then $v(t)$ is annihilated by two operators

$$P_1 = 2t(t^2-2){\partial_t}^2 + (-t^4 + 2t^2 + 4)\partial_t - t^3,$$
$$P_2 = 4(t^2-2){\partial_t}^3 + 12t{\partial_t}^2 + (-t^4 - 8t^2 + 12)\partial_t - t^3 - 6t,$$

neither of which is a multiple of the other. The singular locus of the D_1-module $D_1/(D_1P_1 + D_1P_2)$ is $\{t \mid t^2 - 2 = 0\} = \{\sqrt{2}, -\sqrt{2}\}$.

Example 6.17. Set $n = 2$, $D = \{x = (x_1, x_2) \mid x_1^3 - x_2^2 \geq 0\}$ and consider

$$v(t) = \int_{\mathbb{R}^2} \exp\Big(-\frac{|x|^2}{2}\Big)\delta(t - f(x))Y(x_1^3 - x_2^2)\,dx_1dx_2$$

for a real polynomial $f(x)$. If $f(x) = x_1$, then $v(t)$ is annihilated by

$$2t{\partial_t}^2 + (-3t^3 - 4t^2 - 1)\partial_t + 3t^4 + 2t^3 - t;$$

Its indicial polynomial at 0 is $s(2s-3)$.

If $f(x) = x_1^2 + x_2^2$, then $v(t)$ is annihilated by

$$16t^3(27t-4){\partial_t}^4 + (-864t^4 + 3368t^3 - 320t^2){\partial_t}^3 \\ + (648t^4 - 4956t^3 + 5724t^2 - 268t){\partial_t}^2 \\ + (-216t^4 + 2462t^3 - 5484t^2 + 1654t - 12)\partial_t \\ + 27t^4 - 409t^3 + 1351t^2 - 760t + 6.$$

The indicial polynomials at 0 and 27/4 are $s^2(4s-1)(4s-3)$ and $s(s-1)(s-2)(s-3)$ respectively up to constant multiples. The point 27/4 is an apparent singular point.

Example 6.18. Set

$$v(t) = \frac{1}{2\pi}\int_{\mathbb{R}^2} \exp\Big(-\frac{x_1^2+x_2^2}{2}\Big)\delta(t - x_1x_2)\,dx_1dx_2.$$

This is the density function of the random variable x_1x_2 with the standard normal distribution (x_1, x_2) and satisfies $(t\partial_t^2 + \partial_t - t)v(t) = 0$. Consider the density function

$$w(t) = \int_{\mathbb{R}^n} \delta(t - x_1^2 - x_2^2)v(x_1)v(x_2)\,dx_1dx_2$$

of $x_1^2 + x_2^2$, where (x_1, x_2) is the random vector with the probability density function $v(x_1)v(x_2)$. The integrand belongs to $\mathcal{E}'\mathcal{D}'(\mathbb{R}_x^2 \times \mathbb{R}_t)$, a holonomic system for which can be computed by using Algorithm 6.11. The integration algorithm gives

$$(8t^3{\partial_t}^4 + 48t^2{\partial_t}^3 - (6t^2 - 56t){\partial_t}^2 - (12t-8)\partial_t + t - 2)w(t) = 0.$$

The indicial equation at 0 is s^4.

6.4. Integrals with auxiliary difference parameters

Let us take as an example the integral

$$v(t; a, b) = \frac{1}{B(a,b)}\int_{-\infty}^{\infty} \delta(t - x + x^2)x_+^{a-1}(1-x)_+^{b-1}\,dx$$

for positive real numbers a, b. The integrand is holonomic in x and t, but not in (x, t, a, b) if we regard a, b as variables. Let us apply the algorithm of the preceding sections: first,

$$u(x, t; a, b) := \frac{1}{B(a,b)}\delta(t - x + x^2)x_+^{a-1}(1-x)_+^{b-1}$$

is annihilated by two operators

$$(15)\ t-x+x^2, \quad (-x^2+x)\partial_x+(2x^3-3x^2+x)\partial_t+(a+b-2)x-a+1.$$

Since the initial part of $t-x+x^2$ with respect to the weight vector $(1,0;-1,0)$ for $(x,t,\partial_x,\partial_t)$ is x^2 and $\partial_x^2 x^2=\partial_x x(\partial_x x+1)$, we know that the b-function $b(s)$ with respect to this weight vector divides, in fact equals, $s(s-1)$, which does not depend on a,b. Hence the integration algorithm can be safely applied although the integrand is not holonomic in the variables (x,t,a,b) and produces a differential equation

$$(16)\quad (t^2(1-4t){\partial_t}^2+((4a+4b-18)t^2+(-a-b+3)t)\partial_t \\ +(-a^2+(-2b+7)a-b^2+7b-12)t+(b-1)a-b+1)v(t;a,b)=0.$$

It has regular singularities at $t=0$ and $t=1/4$. The characteristic exponents at 0 are $a-1$ and $b-1$; those at $1/4$ are 0 and $-1/2$. Note that $v(t;a,b)$ vanishes if $t<0$ or $t>1/4$ and its explicit formula for $0<t<1/4$ can be obtained directly. The computation above assures us that $v(t;a,b)$ satisfies (16) on the whole $\mathbb{R}$ as a distribution in t.

In what follows we treat the case where the integrand has some auxiliary parameters with respect to which the integrand satisfies difference equations. In general, let D_n be the ring of differential operators defined over $\mathbb{K}=\mathbb{C}$. As in 6.2 define the D_n-algebra homomorphism $\mu: D_{n+p}\to D_n\langle a,E_a,E_a^{-1}\rangle$ by

$$\mu(t_j)=E_{a_j}, \quad \mu(\partial_{t_j})=-a_jE_{a_j}^{-1} \qquad (1\le j\le p),$$

where E_{a_j} denotes the shift operator $a_j\mapsto a_j+1$. Conversely, we define a D_n-algebra homomorhphism $\hat\mu: D_n\langle a,E_a\rangle\to D_{n+p}$ by

$$\hat\mu(a_j)=-\partial_{t_j}t_j, \quad \hat\mu(E_{a_j})=t_j \qquad (1\le j\le p).$$

Then $\mu\circ\hat\mu$ coincides with the inclusion map $D_n\langle a,E_a\rangle\subset D_n\langle a,E_a,E_a^{-1}\rangle$.

Definition 6.19. A left ideal I of $D_n\langle a,E_a\rangle$ is called a *holonomic* $D_n\langle a,E_a\rangle$-ideal if

$$J:=\mu^{-1}(D_n\langle a,E_a,E_a^{-1}\rangle I)=\{P\in D_{n+p}\mid \mu(P)\in D_n\langle a,E_a,E_a^{-1}\rangle I\}$$

is a holonomic ideal of D_{n+p}, i.e., D_{n+p}/J is a holonomic D_{n+p}-module.

Definition 6.20. A subset Ω of $\mathbb{C}^p$ is said to be *shift-invariant* if $a\in\Omega$ implies that $a+(1,0,\dots,0)$, $\dots$, $a+(0,\dots,0,1)$ also belong to Ω.

Definition 6.21. Let Ω be a shift-invariant subset of $\mathbb{C}^p$. We define a pair of classes $(\mathcal{F}_{n,d}(\Omega), \mathcal{F}_{0,d}(\Omega))$ as one which satisfies the following properties:

(1) Any element $u = u(x,y,a)$ of $\mathcal{F}_{n,d}(\Omega)$ is a map from Ω to $\mathcal{D}'(\mathbb{R}^n_x \times \mathbb{R}^d_y)$; any element $v = v(y,a)$ of $\mathcal{F}_{0,d}(\Omega)$ is a map from Ω to $\mathcal{D}'(\mathbb{R}^d_y)$.

(2) $\mathcal{F}_{n,d}(\Omega)$ is a left $D_{n+d}\langle a, E_a\rangle$-module and $\mathcal{F}_{0,d}(\Omega)$ is a left $D_d\langle a, E_a\rangle$-module.

(3) Let $u = u(x,y,a)$ be an arbitrary element of $\mathcal{F}_{n,d}(\Omega)$. Then the integral $\int_{\mathbb{R}^n} u(x,y,a)\,dx$ is well-defined and belongs to $\mathcal{F}_{0,d}(\Omega)$. Moreover,

$$P\int_{\mathbb{R}^n} u(x,y,a)\,dx = \int_{\mathbb{R}^n} Pu(x,y,a)\,dx, \qquad \int_{\mathbb{R}^n} \partial_{x_i} u(x,y,a)\,dx = 0$$

hold for any $u \in \mathcal{F}_{n,d}(\Omega)$, $P \in D_d\langle a, E_a\rangle$, and $i = 1, \dots, n$.

(4) $E_{a_j} : \mathcal{F}_{0,d}(\Omega) \to \mathcal{F}_{0,d}(\Omega)$ defines an injective $\mathbb{C}$-linear map for each $j = 1, \dots, p$.

As the first examle, let Ω be a shift-invariant open subset of $\mathbb{C}^p$ and define the space $\mathcal{O}(\Omega, \mathcal{S}'(\mathbb{R}^d))$ to be the set of $\mathcal{S}'(\mathbb{R}^d)$-valued holomorphic functions:

$$\begin{aligned}&\mathcal{O}(\Omega, \mathcal{S}'(\mathbb{R}^d))\\ &\qquad = \{u : \Omega \to \mathcal{S}'(\mathbb{R}^d) \mid \langle u(y,a), \varphi(y)\rangle_y \in \mathcal{O}(\Omega)\ (\forall \varphi \in \mathcal{S}(\mathbb{R}^d))\}.\end{aligned}$$

Let $\mathcal{O}(\Omega, \mathcal{SS}'(\mathbb{R}^n \times \mathbb{R}^d))$ be the set of functions from Ω to $\mathcal{S}'(\mathbb{R}^n \times \mathbb{R}^d)$ of the form

$$u(x,y,a) = \sum_{j=1}^{m} u_j(x) v_j(x,y,a)$$

with $m \in \mathbb{N}$, $u_j \in \mathcal{S}(\mathbb{R}^n)$, $v_j \in \mathcal{O}(\Omega, \mathcal{S}'(\mathbb{R}^{n+d}))$. Then the integral of $u(x,y,a)$ with respect to x is defined by

$$\left\langle \int_{\mathbb{R}^n} u(x,y,a)\,dx,\ \varphi(y) \right\rangle = \sum_{j=1}^{m} \langle v_j(x,y,a),\ u_j(x)\varphi(y)\rangle_{(x,y)}$$

for any $\varphi \in \mathcal{S}(\mathbb{R}^d)$, which is well-defined as an element of $\mathcal{O}(\Omega, \mathcal{S}'(\mathbb{R}^d))$ independent of the expression of $u(x,y,a)$ above. Thus the pair

$$\mathcal{F}_{n,d}(\Omega) = \mathcal{O}(\Omega, \mathcal{SS}'(\mathbb{R}^n \times \mathbb{R}^d)), \quad \mathcal{F}_{0,d}(\Omega) = \mathcal{O}(\Omega, \mathcal{S}'(\mathbb{R}^d))$$

satisfies the conditions of Definition 6.21.

The second example is the pair

$$\mathcal{F}_{n,d}(\Omega) = \{u : \Omega \to \mathcal{SS}'(\mathbb{R}^n \times \mathbb{R}^d)\}, \quad \mathcal{F}_{0,d}(\Omega) = \{u : \Omega \to \mathcal{S}'(\mathbb{R}^d)\}$$

with a subset Ω of $\mathbb{C}^p$ such that $a \in \Omega$ implies $(a_1, \dots, a_j \pm 1, \dots, a_p) \in \Omega$ for $j = 1, \dots, p$.

Algorithm 6.22 (difference-differential equations for an integral).
Input: A set G_0 of generators of a holonomic ideal I of $D_{n+d}\langle a, E_a\rangle$ which annihilates $u(x, y, a) \in \mathcal{F}_{n,d}(\Omega)$ with a shift-invariant subset Ω of $\mathbb{C}^p$.
Output: A set G of generators of a holonomic ideal of $D_d\langle a, E_a\rangle$ which annihilates $v(y, a) = \int_{\mathbb{R}^n} u(x, y, a)\, dx$.

(1) Let J be the left ideal of D_{n+d+p} generated by $\hat{\mu}(G_0)$.

(2) Compute $\tilde{J} := \{P \in D_{n+d+p} \mid t^\nu P \in J\ (\exists \nu \in \mathbb{N}^p)\}$ as the annihilator of $\overline{1} \otimes 1$ in the localization

$$(D_{n+d+p}/J)[(t_1 \cdots t_p)^{-1}] = (D_{n+d+p}/J) \otimes_{\mathbb{K}[x,y,t]} \mathbb{K}[x, y, t, (t_1 \cdots t_p)^{-1}],$$

where $\overline{1}$ is the residue class of 1 in D_{n+d+p}/J, by using the localization algorithm of [25].

(3) Compute a set G_1 of generators of the integration ideal

$$N := D_{d+p} \cap (\partial_{x_1} D_{n+d+p} + \cdots + \partial_{x_n} D_{n+d+p} + \tilde{J})$$

of $\tilde{J}$ by Algorithm 5.10.

(4) Let P be an element of G_1. Then there exists a (componentwise) minimal $\nu = (\nu_1, \dots, \nu_p) \in \mathbb{Z}^p$ such that $Q := E_a^\nu \mu(P)$ belongs to $D_d\langle a, E_a\rangle$. (Set $\nu_j = 0$ if $\mu(P)$ does not contain E_{a_j}.) Let us denote this Q by $\mathrm{nm}(\mu(P))$. Set $G := \{\mathrm{nm}(\mu(P)) \mid P \in G_1\}$.

Proof. First, let us prove that $D_{n+d+p}/\tilde{J}$ is holonomic. It suffices to show that $\tilde{J}$ contains $\mu^{-1}(D_{n+d}\langle a, E_a, E_a^{-1}\rangle I)$, which is a holonomic ideal by the assumption. Let P be an element of this set. Then there exist $Q \in I$ and $\nu \in \mathbb{N}^p$ such that

$$\mu(t^\nu P) = E_a^\nu \mu(P) = Q = \mu(\hat{\mu}(Q)).$$

Hence $t^\nu P = \hat{\mu}(Q)$ belongs to J. This implies $P \in \tilde{J}$.

Next let us prove that each element P of G annihilates $v(y, a)$. By the definition of G, there exist $Q_i \in D_{n+d+p}$, $R \in \tilde{J}$ and $\nu \in \mathbb{N}^p$ such that

$$E_a^\nu P = \sum_{i=1}^{n} \partial_{x_i} \mu(Q_i) + \mu(R). \tag{17}$$

Taking the components of ν large enough, we may also assume that $\mu(Q_i)$ belong to $D_d\langle a, E_a\rangle$ and $\mu(R)$ belongs to I so that $\mu(Q_i)u \in \mathcal{F}_{n,d}(\Omega)$ and $\mu(R)u = 0$. Thus (17) implies

$$E_a^\nu Pv(y,a) = \int_{\mathbb{R}^n} E_a^\nu Pu(x,y,a)\,dx = \sum_{i=1}^n \int_{\mathbb{R}^n} \partial_{x_i}\mu(Q_i)u(x,y,a)\,dx = 0$$

by (2) and (3) of Definition 6.21, and hence $Pv = 0$ by (4).

Finally, let us show the ideal $\tilde{I}$ of $D_d\langle a, E_a\rangle$ generated by G is holonomic. Theorem 5.1 assures that D_{d+p}/N is holonomic. Hence it suffices to show that $\mu^{-1}(D_d\langle a, E_a, E_a^{-1}\rangle\tilde{I})$ contains N, which is easy to see by the definition of $\tilde{I}$. This completes the correctness proof of the algorithm. Q.E.D.

In practice we can skip step (2) if J of step (1) is already holonomic.

Example 6.23. Let us come back to the example given at the beginning of this subsection. First note that the integrand

$$u(x,t;a,b) = \frac{1}{B(a,b)}\delta(t - x + x^2)x_+^{a-1}(1-x)_+^{b-1},$$

which belongs to $\mathcal{O}(\Omega, \mathcal{E}'\mathcal{D}'(\mathbb{R}_x \times \mathbb{R}_t))$ with $\Omega = \{(a,b) \in \mathbb{C}^2 \mid \mathrm{Re}\, a > 0,\ \mathrm{Re}\, b > 0\}$, satisfies difference equations

$$(aE_a - (a+b)x)u(x,t;a,b) = 0, \quad (bE_b - (a+b)(1-x))u(x,t;a,b) = 0$$

with the shift operators $E_a : a \mapsto a+1$ and $E_b : b \mapsto b+1$, in addition to (15). With these inputs, Algorithm 6.22 returns a set of generators of a holonomic ideal of $D_1\langle a, b, E_a, E_b\rangle$ which annihilates $v(t;a,b) = \int_{-\infty}^{\infty} u(x,t;a,b)\,dx$. For example, $v(t;a,b)$ is annihilated by

$$\begin{aligned}(4E_bE_at^2 - E_bE_at)\partial_t + \{&(b(-2E_b+1) + 2E_b + 1)E_a \\ &+ a(-E_b+1) + b(2E_b^2 - 2E_b + 1) + 2E_b^2 - E_b\}t.\end{aligned}$$

Computing the intersection with the subring $D_1[a,b]$, we get (16) again. If only differential equations in t is needed, we could have ignored the factor, i.e, the reciprocal of $B(a,b)$ at first.

Finally, let us consider the multivariate gamma distribution with the density function

$$\begin{aligned}u_n(x;a) &= u(x_1,\ldots,x_n;a_1,\ldots,a_n) \\ &:= \frac{1}{\Gamma(a_1)\cdots\Gamma(a_n)}(x_1)_+^{a_1-1}\cdots(x_n)_+^{a_n-1}e^{-x_1-\cdots-x_n}\end{aligned}$$

with $x = (x_1, \dots, x_n)$ and $a = (a_1, \dots, a_n)$. It is annihilated by

$$x_i \partial_{x_i} + x_i - a_i + 1, \quad a_i E_{a_i} - x_i \qquad (1 \le i \le n)$$

which generate a holonomic $D_n\langle a, E_a\rangle$-ideal. Hence Algorithm 6.22 produces a holonomic ideal of $D_1\langle a, E_a\rangle$ which annihilates the density function

$$v(t; a) := \int_{\mathbb{R}^n} \delta(t - f(x)) u_n(x; a)\, dx$$

for an arbitrary real polynomial $f(x)$ as a random variable. Here we can regard the integrand as an element of $\mathcal{O}(\Omega, \mathcal{SS}'(\mathbb{R}^n_x \times \mathbb{R}_t))$ with

$$\Omega = \{a \in \mathbb{C}^n \mid \operatorname{Re} a_j > 0\ (1 \le j \le n)\}.$$

Example 6.24. Set

$$v(t; a_1, a_2) := \int_{\mathbb{R}^2} \delta(t - x_1 x_2) u_2(x; a_1, a_2)\, dx_1 dx_2.$$

By Algorithm 6.22 we know that it is annihilated by the differential operator

$$t^2 {\partial_t}^2 + (-a_1 - a_2 + 3) t\partial_t - t + a_1 a_2 - a_1 - a_2 + 1$$

whose indicial polynomial at 0 is $(s - a_1 + 1)(s - a_2 + 1)$, as well as by difference-differential operators

$$t\partial_t + a_1(E_{a_1} - 1) + 1, \qquad t\partial_t + a_2(E_{a_2} - 1) + 1.$$

These three operators generate a holonomic ideal of $D_1\langle a_1, a_2, E_{a_1}, E_{a_2}\rangle$.

Example 6.25. Set

$$v(t; a_1, a_2) := \int_{\mathbb{R}^2} \delta(t - x_1^2 - x_2^2) u_1(x; a_1, a_2)\, dx_1 dx_2.$$

It is annihilated by the differential operator

$$
\begin{aligned}
P &= 32t^4{\partial_t}^6 + (-64a_1 - 64a_2 + 480)t^3{\partial_t}^5 \\
&+ \{-32t^3 + (48a_1^2 + (96a_2 - 624)a_1 + 48a_2^2 - 624a_2 + 2040)t^2\}{\partial_t}^4 \\
&+ \{(40a_1 + 40a_2 - 256)t^2 + (-16a_1^3 + (-48a_2 + 264)a_1^2 \\
&\quad + (-48a_2^2 + 528a_2 - 1448)a_1 - 16a_2^3 + 264a_2^2 - 1448a_2 + 2640)t\}{\partial_t}^3 \\
&+ \{10t^2 + (-16a_1^2 + (-32a_2 + 172)a_1 - 16a_2^2 + 172a_2 - 456)t \\
&\quad + 2a_1^4 + (8a_2 - 36)a_1^3 + (12a_2^2 - 108a_2 + 238)a_1^2 \\
&\quad + (8a_2^3 - 108a_2^2 + 476a_2 - 684)a_1 \\
&\quad + 2a_2^4 - 36a_2^3 + 238a_2^2 - 684a_2 + 720\}{\partial_t}^2 \\
&+ \{(-6a_1 - 6a_2 + 30)t + 2a_1^3 + (6a_2 - 26)a_1^2 \\
&\quad + (6a_2^2 - 52a_2 + 108)a_1 + 2a_2^3 - 26a_2^2 + 108a_2 - 144\}\partial_t \\
&- t + a_1^2 - 5a_1 + a_2^2 - 5a_2 + 10.
\end{aligned}
$$

The indicial polynomial of P at 0 is

$$b(s) = s(s-1)(2s-a_1-a_2)(2s-a_1-a_2-1)(2s-a_1-a_2+1)(2s-a_1-a_2+2)$$

up to a constant multiple.

References

[1] Assi, A., Castro-Jiménez, F. J., Granger, J. M., How to calculate the slopes of a $\mathcal{D}$-module, *Compositio Math.* **104** (1996), 107–123.

[2] Bernstein, I. N., Modules over a ring of differential operators: study of the fundamental solutions of equations with constant coefficients, *Funct. Anal. Appl.* **5** (1971), 1–16.

[3] Bernstein, I., N., The analytic continuation of generalized functions with respect to a parameter, *Funct. Anal. Appl.* **6** (1972), 26–40.

[4] Björk, J.-E., "Rings of Differential Operators", North-Holland Publishing Co., Amsterdam–New York, 1979, xvii+374 pp., ISBN: 0-444-85292-1.

[5] Bruns, W., Herzog, J., "Cohen-Macaulay Rings" (revised version), Cambridge University Press, Cambridge, 1998, xiv+453 pp., ISBN: 0-521-56674-6.

[6] Cox, D., Little, J., O'Shea, D., "Ideals, Varieties, and Algorithms: an Introduction to Computational Algebraic Geometry and Commutative Algebra" (second edition), Springer-Verlag, New York, 1997, xii+536 pp., ISBN: 0-387-94680–2.

[7] Decker, W., Greuel, G. M., Pfister, G., Schönemann, H., SINGULAR: a computer algebra system for polynomial computations, http://www.singular.uni-kl.de.

[8] Eisenbud, D., "Commutative Algebra with a View toward Algebraic Geometry", Springer-Verlag, New York, 1995, xvi+785 pp., ISBN: 0-387-94268-8.

[9] Gabber, O., The integrability of the characteristic variety, *Amer. J. Math.* **103** (1981), 445–468.

[10] Gel'fand, I. M., Shilov, G. E., "Generalized Functions. Vol. 1: Properties and Operations", translated from the Russian by E. Saletan. Academic Press, New York–London, 1964, ix+423 pp., ISBN: 0-12-279501-6.

[11] Grayson, R., Stillman, M., Macaulay2, a software system for research in algebraic geometry, http://www.math.uiuc.edu/Macaulay2/.

[12] Hotta, R., "Introduction to Algebra: Groups, Rings and Modules" (in Japanese), Shokabo, Tokyo, 1987, ix+216 pp., ISBN: 4-7853-1402-8.

[13] Kashiwara, M., On the holonomic systems of linear differential equations, II, *Invent. Math.* **49** (1978), 121–135.

[14] Kashiwara, M., "Systems of Microdifferential Equations", Birkhäuser, Boston, MA, 1983, xv+159 pp., ISBN: 0-8176-3138-0.

[15] Kashiwara, M., "D-modules and Microlocal Calculus", translated from the Japanese by M. Saito, American Mathematical Society, Providence, RI, 2003, xvi+254 pp., ISBN: 0-8218-2766-9.

[16] Kashiwara, M., Kawai, T., On the characteristic variety of a holonomic system with regular singularities, *Adv. in Math.* **34** (1979), 163–184.

[17] Maconnell, J. C., Robson, J. C., "Noncommutative Noetherian Rings" (revised edition), American Mathematical Society, Providence, RI, 2001, xii+636 pp., ISBN: 0-8218-2169-5.

[18] Marumo, N., Oaku, T., Takemura, A., Properties of powers of functions satisfying second-order linear differential equations with applications to statistics, *Jpn. J. Ind. Appl. Math.* **32** (2015), 553-572.

[19] Noro, M., An efficient modular algorithm for computing the global b-function, Mathematical software (Beijing, 2002), 147–157, World Sci. Publ., River Edge, NJ, 2002.

[20] Noro, M., Takayama, N., Nakayama, H., Nishiyama, K., Ohara, K, Risa/Asir: a computer algebra system, http://www.math.kobe-u.ac.jp/Asir/asir.html.

[21] Oaku, T., Computation of the characteristic variety and the singular locus of a system of differential equations with polynomial coefficients, *Japan J. Indust. Appl. Math.* **11** (1994), 485–497.

[22] Oaku, T., Algorithms for the b-function and D-modules associated with a polynomial, *J. Pure Appl. Algebra* **117/118** (1997), 495–518.

[23] Oaku, T., "D-modules and Computational Mathematics" (in Japanese), Asakura-Shoten, Tokyo, 2002, iv+196 pp., ISBN: 4-254-11555-5.

[24] Oaku, T., Algorithms for integrals of holonomic functions over domains defined by polynomial inequalities, *J. Symbolic Comput.* **50** (2013), 1–27.

[25] Oaku, T., Localization, local cohomology, and the b-function of a D-module with respect to a polynomial, Adv. Stud. Pure Math. **77** (2018), pp. 353–398.

[26] Oaku, T., Takayama, N., An algorithm for de Rham cohomology groups of the complement of an affine variety, *J. Pure Appl. Algebra* **139** (1999), 201–233.

[27] Oaku, T., Takayama, N., Algorithms for D-modules — restriction, tensor product, localization, and local cohomology groups, *J. Pure Appl. Algebra* **156** (2001), 267–308.

[28] Oaku, T., Takayama, N., Walther, U., A localization algorithm for D-modules, *J. Symbolic Comput.* **29** (2000), 721–728.

[29] Saito, M., Sturmfels, B., Takayama, N., "Gröbner Deformations of Hypergeometric Differential Equations", Springer-Verlag, Berlin, 2000, viii+254 pp., ISBN: 3-540-66065-8.

[30] Sato, M., Theory of hyperfunctions, I, *J. Fac. Sci. Univ. Tokyo, Sect. I* **8** (1959), 139–193.

[31] Sato, M., Kawai, T., Kashiwara M., Microfunctions and pseudo-differential equations, Lecture Notes in Math. Vol. 287, Springer, Berlin, pp. 265–529, 1973.

[32] Schwartz, L., "Théorie des distributions", Hermann, Paris, 1966, xii+420 pp.

[33] Smith, G. G., Irreducible components of characteristic varieties, *J. Pure Appl. Algebra* **165** (2001), 291–306.

[34] Takayama, N., Kan: a system for computation in algebraic analysis, 1991–, http://www.math.kobe-u.ac.jp/KAN.

[35] Walther, U., Algorithmic computation of local cohomology modules and the local cohomological dimension of algebraic varieties, *J. Pure Appl. Algebra* **139** (1999), 303–321.

[36] Wishart, J., Bartlett, M. S., The distribution of second order moment statistics in a normal system, *Math. Proc. Cambridge Philos. Soc.* **28** (1932), 455–459.

Department of Mathematics, Tokyo Woman's Christian University,
Suginami-ku, Tokyo 167-8585, Japan
E-mail address: `oaku@lab.twcu.ac.jp`

Advanced Studies in Pure Mathematics 77, 2018
The 50th Anniversary of Gröbner Bases
pp. 353–398

Localization, local cohomology, and the b-function of a D-module with respect to a polynomial

Toshinori Oaku

Abstract.

Given a D-module M generated by a single element, and a polynomial f, one can construct several D-modules attached to M and f and can define the notion of the (generalized) b-function following M. Kashiwara. These modules are closely related to the localization and the local cohomology of M. We show that the b-function, if it exists, controls these modules and present general algorithms for computing these modules and the b-function if it exists without any further assumptions. We also give some examples of multiplicity computation of such D-modules including a possibly well-known explicit formula for the localization of the polynomial ring by a hyperplane arrangement.

Contents

§1. Introduction

Let K be an algebraically closed field of characteristic zero and $K[x] = K[x_1, \ldots, x_n]$ be the polynomial ring with $x = (x_1, \ldots, x_n)$. Let $D_n = K[x]\langle\partial\rangle = K[x]\langle\partial_1, \ldots, \partial_n\rangle$ be the n-th Weyl algebra, i.e., the ring

Received September 14, 2016.
2010 *Mathematics Subject Classification.* 14F10, 13P10, 13D45, 14B15, 52C35.
Key words and phrases. D-module, Gröbner basis, localization, local cohomology, b-function, hyperplane arrangement, multiplicity.

of differential operators with polynomial coefficients with respect to the variables x, where we denote $\partial = (\partial_1, \ldots, \partial_x)$ with $\partial_i = \partial_{x_i} = \partial/\partial x_i$ being the derivation with respect to x_i. An arbitrary element P of D_n is written in a finite sum

$$P = \sum_{\alpha \in \mathbb{N}^n} a_\alpha(x)\partial^\alpha \quad \text{with} \quad a_\alpha(x) \in K[x],$$

where we denote $\partial^\alpha = \partial_1^{\alpha_1} \cdots \partial_n^{\alpha_n}$ for a multi-index $\alpha = (\alpha_1, \ldots, \alpha_n) \in \mathbb{N}^n$ with $\mathbb{N}$ being the set of non-negative integers. One can define the dimension of a finitely generated left D_n-module M; J. Bernstein [3], [4] proved that the dimension of M is not less than n unless M is the zero module. A finitely generated left D_n-module is called *holonomic* if its dimension is n or else it is the zero module.

Let M be a finitely generated left D_n-module and $f \in K[x]$ be a non-constant polynomial. Then the localization $M[f^{-1}]$ and the local cohomology groups $H^j_{(f)}(M)$ have natural structures of left D_n-module and are holonomic if so is M, as was shown by Kashiwara [13]. More generally, one can construct a left $D_n[s]$-module

$$M(u, f, s) = D_n[s](u \otimes f^s)$$

with an indeterminate s. Suppose that M is generated by u over D_n. Then the (*generalized*) *b-function* for u and f is defined to be the univariate (and monic) polynomial $b_{u,f}(s)$ of the least degree such that

$$b_{u,f}(s)(u \otimes f^s) \in D_n[s](u \otimes f^{s+1})$$

holds. The existence of $b_{u,f}(s)$ was proved by Kashiwara [13] under the assumption that M is holonomic outside of the hypersurface $f = 0$. If M is the polynomial ring $K[x]$ with $u = 1$, then $b_{u,f}(s)$ is nothing but the classical Bernstein-Sato polynomial, or simply the b-function, of f. In the same way as the Bernstein-Sato polynomial controls the localization of the polynomial ring as a D_n-module, the b-function controls the localization $M[f^{-1}]$ or its generalization $D_n(u \otimes f^\lambda)$.

On the other hand, algorithms to compute $M(u, f, s)$ and the b-function if it exists were introduced in [17] under the assumption that M is f-torsion free. These algorithms are based on various Gröbner bases over the ring of differential operators as is presented, e.g., in [23] and [18]. Torrelli [24] studied the b-function $b_{u,f}(s)$ systematically when M is the local cohomology group $H^k_{(f_1,\ldots,f_k)}(K[x])$ under the assumption that $f_1, \ldots, f_k, f$ define a quasi-homogeneous non-isolated singularity, together with the general property of $M(u, f, s)$ under the assumption that M is holonomic without f-torsion.

The purpose of our study on the b-function and $M(u, f, s)$ is twofold: first, we want to clarify how the b-function controls the module $M(u, f, s)$ and the localization $M[f^{-1}]$ as well as the local cohomology $H^1_{(f)}(M)$. This will be performed in Sections 2 and 5. These results should be more or less well-known under some stronger conditions. See, e.g., [24] and Chapter VI of [5], where M is assumed to be f-torsion free, or regular holonomic. The second purpose is to remove the assumption of f-saturatedness from our former algorithms in [17]. For this purpose, we reinterpret the algorithm introduced in [21] for the localization $M[f^{-1}]$ in Section 3. Our algorithms work at least if M is holonomic outside of $f = 0$ without any further assumptions.

In the latter half of this article (Sections 5 and 6), we study the multiplicity (in the sense of Bernstein [3]) and the length of a holonomic D-module, as the most fundamental numerical invariants. This can be also used to prove a relation between $b_{u,f}(s)$ and $M(u, f, \lambda)$. We also give some examples of the multiplicity computation of the localization or the local cohomology. In the last section, we present, together with an elementary proof, a possibly well-known formula on the length and the multiplicity of the localization of the polynomial ring by a polynomial f which defines a hyperplane arrangement. The result is that the length and the multiplicity of $K[x, f^{-1}]$ both coincide with $\pi(1)$, where $\pi(t)$ is what is called the Poincaré polynomial of the hyperplane arrangement.

We use computer algebra system Risa/Asir [16] for computation of Gröbner bases over the ring of differential operators, and in particular, for computation of D-module theoretic integration, which is needed in the localization algorithm.

We would like to thank the organizers of MSJ SI 2015 for the invitation both to the conference and to the proceedings. We would be pleased if we could convince the reader who is interested in D-module theory of the usefulness of Gröbner bases, which are the main theme of MSJ SI 2015, over the ring of differential operators in our case. This work was supported by JSPS Grant-in-Aid for Scientific Research (C) 26400123.

§2. The b-function for a D-module and a polynomial

Let K be an arbitrary field of characteristic zero and $X = K^n$ be the n-dimensional affine space over K. We denote by D_X the n-th Weyl algebra D_n over K. Let M be a left D_X-module and $f \in K[x]$ a non-constant polynomial. We can associate several D_X-modules with M and f by translating the definitions by Kashiwara [13] for analytic D-modules to algebraic setting. First, the localization $M[f^{-1}] := M \otimes_{K[x]} K[x, f^{-1}]$

and the local cohomology groups $H^j_{(f)}(M)$ $(j = 0, 1)$ are defined with M being regarded as a $K[x]$-module; they become again left D_X-modules.

Introducing an indeterminate s, let

$$\mathcal{L}_f := K[x, f^{-1}, s]f^s$$

be the free $K[x, f^{-1}, s]$-module with a free generator f^s. Then $\mathcal{L}_f$ has a natural structure of left $D_X[s]$-module through the action of ∂_{x_i} on $\mathcal{L}_f$ defined by

$$\partial_{x_i}(a(x,s)f^{-k}f^s) = \left(\frac{\partial a(x,s)}{\partial x_i}f^{-k} + (s-k)f_i a(x,s)f^{-k-1}\right)f^s$$

for $j = 1, \ldots, n$ with $f_i := \partial f/\partial x_i$. Sometimes $f^{-k}f^s$ is abbreviated to f^{s-k}.

The tensor product $M \otimes_{K[x]} \mathcal{L}_f$ has a natural structure of left $D_X[s]$-module induced by

$$\partial_{x_i}(u \otimes a(x,s)f^s) = (\partial_{x_i}u) \otimes a(x,s)f^s + u \otimes \partial_{x_i}(a(x,s)f^s) \quad (1 \le i \le n)$$

for $u \in M$ and $a(x,s) \in K[x,s]$. In what follows, we fix an arbitrary nonzero element u of M. Let

$$M(u,f,s) := D_X[s](u \otimes f^s)$$

be the left $D_X[s]$-submodule of $M \otimes_{K[x]} \mathcal{L}_f$ generated by $u \otimes f^s$. In a special case where $M = K[x]$ and $u = 1$, let us denote by

$$\mathcal{N}_f = K[x](1,f,s) = D_X[s]f^s$$

the left $D_X[s]$-submodule of $\mathcal{L}_f$ generated by f^s. Set

$$I(u,f) := \{b(s) \in K[x] \mid b(s)(u \otimes f^s) \in D_n[s](fu \otimes f^s)\}.$$

If $I(u,f) \neq \{0\}$, then the (monic) generator $b_{u,f}(s)$ of $I(u,f)$ is called the (*generalized*) *b-function* for u and f. It was defined by Kashiwara [13] with the following existence theorem.

Theorem 2.1 (Kashiwara [13]). *Let D_X be defined over an algebraically closed field K of characteristic zero. If a left D_X-module M is holonomic on $X_f = \{x \in X \mid f(x) \neq 0\}$, then one has $I(u,f) \neq \{0\}$ for any $u \in M$.*

When $M = K[x]$ and $u = 1$, the b-function $b_{1,f}(s)$ is nothing but what is called the Bernstein-Sato polynomial, or the b-function, associated with f. In fact, Kashiwara proved this theorem for a module

M over the ring of differential operators with analytic coefficients and a complex analytic function f. This corresponds to what is called the local b-function. The coincidence of the local b-functions in the algebraic setting and in the analytic setting is noticed, e.g., as Corollary 8.6 of [17]. It will turn out in what follows that the b-function 'controls' the D-modules associated with M and f.

The b-function can exist even if M is not holonomic on X_f.

Example 2.2. Set $n = 2$, $x_1 = x$, $x_2 = y$, and $P = x\partial_x^2 + \partial_y$. Then $M := D_X/D_X P = D_X u$ with u being the residue class of 1 is not holonomic even outside of $x = 0$ (the dimension of M is three), but has the b-functions $b_{u,x}(s) = (s+1)(s+2)$ and $b_{u,y}(s) = s+1$. In fact, one has

$$(-x\partial_x^2 + 2(s+1)\partial_x - \partial_y)(u \otimes x^{s+1}) = (s+1)(s+2)u \otimes x^s,$$
$$P(u \otimes y^{s+1}) = (s+1)u \otimes y^s$$

in $M \otimes_{K[x,y]} K[x,y,x^{-1}]x^s$ and in $M \otimes_{K[x,y]} K[x,y,y^{-1}]y^s$ respectively.

Definition 2.3. A left D_X-module M is said to be *f-saturated* or *f-torsion free* if the homomorphism $f : M \to M$ is injective. This is equivalent to $H^0_{(f)}(M) = 0$.

An algorithm to determine if there exists the b-function and to compute it if it exists was given in [17] under the assumption that $M = D_X u$ is f-torsion free.

Let us define a D_X-automorphism $t : \mathcal{L}_f \to \mathcal{L}_f$ by

$$t(a(x,s)f^{-k}f^s) = a(x,s+1)f^{-k+1}f^s$$

for $a(x,s) \in K[x,s]$ and $k \in \mathbb{N}$. The inverse t^{-1} is defined by

$$t^{-1}(a(x,s)f^{-k}f^s) = a(x,s-1)f^{-k-1}f^s.$$

It induces a D_X-automorphism

$$t : M \otimes_{K[x]} \mathcal{L}_f \longrightarrow M \otimes_{K[x]} \mathcal{L}_f,$$

which also induces a D_X-endomorphism of $M(u,f,s)$. Note that the actions of t and s on $M(u,f,s)$ satisfies the commutation relation $st = t(s-1)$. It follows that $tM(u,f,s)$ is a left $D_X[s]$-module. It also follows from the definition that $b_{u,f}(s)$ is the minimal polynomial of s acting on the left D_X-module $M(u,f,s)/tM(u,f,s)$ since $P(s)(fu \otimes f^s) = t(P(s-1)(u \otimes f^s))$.

Let $\lambda \in K$ be a constant. Then specializing the parameter s to λ, we obtain left D_X-modules

$$\mathcal{L}_f(\lambda) := \mathcal{L}_f/(s-\lambda)\mathcal{L}_f, \qquad \mathcal{N}_f(\lambda) := \mathcal{N}_f/(s-\lambda)\mathcal{N}_f.$$

Let us denote by f^λ and $f^s|_{s=\lambda}$ the residue class of f^s in $\mathcal{L}_f(\lambda)$, and that of f^s in $\mathcal{N}_f(\lambda)$ respectively. In particular, $\mathcal{L}_f(\lambda) = K[x, f^{-1}]f^\lambda$ is a free $K[x, f^{-1}]$-module generated by f^λ. In the same way, we define a left D_X-module

$$M(u, f, \lambda) = M(u, f, s)/(s-\lambda)M(u, f, s)$$

and denote the residue class of $u \otimes f^s$ in $M(u, f, \lambda)$ by $(u \otimes f^s)|_{s=\lambda}$.

Kashiwara also proved the following fundamental fact, to which we shall give an elementary proof in Section 5.

Theorem 2.4 (Kashiwara [13])**.** *Let K be algebraically closed. If M is holonomic on X_f, then $M(u, f, \lambda)$ is a holonomic D_X-module for any $u \in M$ and $\lambda \in K$.*

Let us define the specialization homomorphism

$$\rho_\lambda : M \otimes_{K[x]} \mathcal{L}_f \longrightarrow M \otimes_{K[x]} \mathcal{L}_f(\lambda)$$

by

$$\rho_\lambda(v \otimes a(x, s)f^{-k}f^s) = v \otimes a(x, \lambda)f^{-k}f^\lambda$$

for $v \in M$, $a(x, s) \in K[x, s]$, and $k \in \mathbb{N}$. Then $\rho_\lambda(P(s)w) = P(\lambda)\rho_\lambda(w)$ holds for any $w \in M \otimes_{K[x]} \mathcal{L}_f$ and $P(s) \in D_X[s]$. Since any element of $(s-\lambda)M(u, f, s)$ is sent by ρ_λ to zero, ρ_λ induces a surjective D_X-homomorphism

$$\tilde{\rho}_\lambda : M(u, f, \lambda) \longrightarrow D_X(u \otimes f^\lambda) \subset M \otimes_{K[x]} \mathcal{L}_f(\lambda),$$

which sends $(u \otimes f^s)|_{s=\lambda}$ to $u \otimes f^\lambda$. It is not injective in general even if $M = K[x]$ and $u = 1$. For example, $\partial_x x^0 = 0$ holds in $\mathcal{L}_x(0)$ but $\partial_x(x^s|_{s=0})$ does not vanish in $\mathcal{N}_x(0)$.

Lemma 2.5. *Let $M = D_X u$ be a left D_X-module generated by u.*

(1) *Every element of $M \otimes_{K[x]} \mathcal{L}_f$ can be expressed as $Q(s)(u \otimes f^{s-k})$ with some $Q(s) \in D_X[s]$ and $k \in \mathbb{N}$.*

(2) *Let λ be an arbitrary element of K. Then every element of $M \otimes_{K[x]} \mathcal{L}_f(\lambda)$ can be expressed as $Q(u \otimes f^{\lambda-k})$ with some $Q \in D_X$ and $k \in \mathbb{N}$.*

Proof. From the identity

$$\partial_{x_i}(v \otimes f^{s-k}) = (\partial_{x_i} v) \otimes f^{s-k} + v \otimes (s-k) f_i f^{s-k-1}$$

for any $v \in M$ and $k \in \mathbb{Z}$, we get

$$(\partial_{x_i} v) \otimes f^{s-k} = (\partial_{x_i} f - (s-k) f_i)(v \otimes f^{s-k-1}).$$

By induction, we can show that for any multi-index $\alpha \in \mathbb{N}^n$ and $k \in \mathbb{Z}$, there exists $Q_\alpha(s) \in D_X[s]$ such that

$$(\partial_x^\alpha v) \otimes f^{s-k} = Q_\alpha(s)(v \otimes f^{s-k-|\alpha|}).$$

This proves the statement (1). The statement (2) can be proved similarly. Q.E.D.

The following proposition should be well-known; see, e.g., Propositions 7.1 and 7.4 of [17]. The case $M = K[x]$ and $f = 1$ was proved by Kashiwara [12].

Proposition 2.6. *Let M be a left D_X-module generated by $u \in M$ and assume that there exists the b-function $b_{u,f}(s)$. Let λ be an element of K and suppose that $b_{u,f}(\lambda - k) \neq 0$ for any positive integer k. Then*

(1) *The image $\rho_\lambda(M(u,f,s)) = D_X(u \otimes f^\lambda)$ coincides with $M \otimes_{K[x]} \mathcal{L}_f(\lambda)$. In other words, $M \otimes_{K[x]} \mathcal{L}_f(\lambda)$ is generated by $u \otimes f^\lambda$ over D_X.*

(2) *$\ker \rho_\lambda \cap M(u,f,s)$ coincides with $(s-\lambda)M(u,f,s)$. Hence $\tilde{\rho}_\lambda : M(u,f,\lambda) \to D_X(u \otimes f^\lambda)$ is an isomorphism of left D_X-modules.*

Proof. (1) In view of Lemma 2.5, we have only to show that $u \otimes f^{\lambda-k}$ belongs to $\rho_\lambda(M(u,f,s))$ for any $k \in \mathbb{N}$. This is obvious for $k = 0$ since $\rho_\lambda(u \otimes f^s) = u \otimes f^\lambda$.

Let us show that $u \otimes f^{\lambda-k}$ belongs to $\rho_\lambda(M(u,f,s))$. Suppose $k \geq 1$. There exists $P(s) \in D_X[s]$ such that $P(s)(u \otimes f^{s+1}) = b_{u,f}(s)(u \otimes f^s)$. Applying t^{-k}, we get

$$P(s-k)(u \otimes f^{s+1-k}) = b_{u,f}(s-k)(u \otimes f^{s-k})$$

in $M \otimes_{K[x]} \mathcal{L}_f$. Proceeding inductively, we see that there exists $\tilde{P}(s) \in D_X[s]$ such that

$$\tilde{P}(s)(u \otimes f^s) = b_{u,f}(s-1) \cdots b_{u,f}(s-k) u \otimes f^{s-k} \tag{1}$$

holds in $M \otimes_{K[x]} \mathcal{L}_f$. The homomorphism ρ_λ gives an identity

$$\tilde{P}(\lambda)(u \otimes f^\lambda) = b_{u,f}(\lambda-1) \cdots b_{u,f}(\lambda-k) u \otimes f^{\lambda-k}$$

in $M \otimes_{K[x]} \mathcal{L}_f(\lambda)$. Since $b_{u,f}(\lambda - j) \neq 0$ for $j = 1, \dots, k$ by the assumption, it follows that

$$u \otimes f^{\lambda-k} = \frac{1}{b_{u,f}(\lambda-1)\cdots b_{u,f}(\lambda-k)}\tilde{P}(\lambda)(u \otimes f^{\lambda}).$$

The right-hand side belongs to $\rho_\lambda(M(u,f,s))$. This completes the proof of (1).

(2) Assume $\rho_\lambda(Q(s)(u \otimes f^s)) = 0$ with $Q(s) \in D_X[s]$. There exist $l \in \mathbb{N}$ and $Q_j \in D_X$ which are zero except finitely many indices j such that

$$Q(s)(u \otimes f^s) = \sum_{j \geq 0} (Q_j u) \otimes (s-\lambda)^j f^{s-l}.$$

By the assumption, $\rho_\lambda(Q(s)(u \otimes f^s)) = (Q_0 u) \otimes f^{\lambda - l}$ vanishes in $M \otimes_{K[x]} \mathcal{L}_f(\lambda)$, which means that $(Q_0 u) \otimes f^{-l}$ vanishes in $M \otimes_{K[x]} K[x, f^{-1}]$. It follows that $(Q_0 u) \otimes 1 = f^l (Q_0 u) \otimes f^{-l} = 0$ in $M \otimes_{K[x]} K[x, f^{-1}]$. Consequently, $(Q_0 u) \otimes f^s$ vanishes in $M \otimes_{K[x]} \mathcal{L}_f$. Thus we have

$$Q(s)(u \otimes f^s) = (s-\lambda)\sum_{j \geq 1} (Q_j u) \otimes (s-\lambda)^{j-1} f^{s-l} = (s-\lambda)Q'(s)(u \otimes f^{s-k})$$

with some $k \in \mathbb{N}$ and $Q'(s) \in D_X[s]$ in view of the proof of Lemma 2.5. By using (1) we obtain

$$b_{u,f}(s-1)\cdots b_{u,f}(s-k)Q(s)(u \otimes f^s) = (s-\lambda)Q'(s)\tilde{P}(s)(u \otimes f^s).$$

Hence $b_{u,f}(\lambda-1)\cdots b_{u,f}(\lambda-k)Q(s)(u \otimes f^s)$ belongs to $(s-\lambda)M(u,f,s)$. This completes the proof of (2). Q.E.D.

The following proposition extends Lemma 1.3 of Walther [27] for the case $M = K[x]$ and $u = 1$ almost verbatim.

Lemma 2.7. *Under the same assumption as in the preceding proposition, assume moreover that $b_{u,f}(\lambda) = 0$. Then one has*

$$D_X(fu \otimes f^\lambda) \subsetneq D_X(u \otimes f^\lambda)$$

in $M \otimes_{K[x]} \mathcal{L}_f(\lambda)$. In particular, $M \otimes_{K[x]} \mathcal{L}_f(\lambda)$ is generated by $u \otimes f^\lambda$, but not by $u \otimes f^{\lambda+1} = fu \otimes f^\lambda$, over D_X.

Proof. There exists $P(s) \in D_X[s]$ such that $P(s)(fu \otimes f^s) = b_{u,f}(s)(u \otimes f^s)$. Assume $D_X(fu \otimes f^\lambda) = D_X(u \otimes f^\lambda)$. Then there exists $A \in D_X$ such that $(1 - Af)(u \otimes f^\lambda) = 0$. By virtue of (2) of the preceding proposition, there exist $Q(s), R(s) \in D_X[s]$ such that

$$1 - Af = Q(s) + (s-\lambda)R(s), \qquad Q(s)(u \otimes f^s) = 0.$$

It follows that

$$\begin{aligned}\frac{b_{u,f}(s)}{s-\lambda}(u\otimes f^s) &= \frac{b_{u,f}(s)}{s-\lambda}Af(u\otimes f^s) + b_{u,f}(s)R(s)(u\otimes f^s)\\ &= \left(\frac{b_{u,f}(s)}{s-\lambda}A + R(s)P(s)\right)(fu\otimes f^s).\end{aligned}$$

This means that $b_{u,f}(s)/(s-\lambda)$ belongs to the ideal $I(u,f)$, which contradicts the definition of $b_{u,f}(s)$. This completes the proof. Q.E.D.

Summing up we obtain

Theorem 2.8. *Let $M = D_X u$ be a left D_X-module generated by $u \in M$ and $f \in K[x]$ be a non-constant polynomial. Assume that there exists the b-function $b_{u,f}(s)$ for u and f. Then the following conditions on $\lambda \in K$ are equivalent:*

(1) $b_{u,f}(\lambda - k) \neq 0$ *for any positive integer* k.

(2) $M \otimes_{K[x]} \mathcal{L}_f(\lambda)$ *is generated by* $u \otimes f^\lambda$ *over* D_X.

Proof. Assume $b_{u,f}(\lambda - k) = 0$ for some positive integer k and let k_0 be the maximum among such k. Then by (1) of Proposition 2.6 and Lemma 2.7, we have

$$\begin{aligned}\mathcal{L}_f(\lambda) &= \mathcal{L}_f(\lambda - k_0) = D_X(u\otimes f^{\lambda-k_0})\\ &\supsetneq D_X(u\otimes f^{\lambda-k_0+1}) \supset D_X(u\otimes f^{\lambda}).\end{aligned}$$

Hence $\mathcal{L}_f(\lambda)$ is not generated by $u\otimes f^\lambda$. Q.E.D.

The converse of the statement (2) of Proposition 2.6 will be given in Theorem 5.9 of Section 5 under the additional assumption that M is holonomic on X_f.

Let us recall local cohomology of D-modules. Let M be a finitely generated left D_X-module, and I be an ideal of $K[x]$. Then the k-th local cohomology group $H_I^k(M)$ supported by I is defined to be the k-th derived functor of the functor

$$M \longmapsto H_I^0(M) = \{u \in M \mid I^k u = 0 \text{ for some } k \in \mathbb{N}\}.$$

They have natural structure of left D_X-module, and they are holonomic if so is M as was proved by Kashiwara [13] in the analytic category.

If I is the principal ideal (f) generated by $f \in K[x]$, then there exists an exact sequence

$$0 \longrightarrow H^0_{(f)}(M) \longrightarrow M \stackrel{\iota}{\longrightarrow} M[f^{-1}] \longrightarrow H^1_{(f)}(M) \longrightarrow 0$$

of left D_X-modules, where ι stands for the natural homomorphism such that $\iota(v) = v \otimes 1$ in $M[f^{-1}] = M \otimes_{K[x]} K[x, f^{-1}]$ for $v \in M$. Hence there is an isomorphism $H^1_{(f)}(M) \cong M[f^{-1}]/\iota(M)$ as left D_X-module.

In general, algorithms to compute $H^k_I(M)$ as left D_X-module were given in [17] for the case I is principal, and in [26] and [20] for general I, under the condition that M is holonomic. See also [25], where related topics such as associated primes and the Weyl closure are also discussed.

Corollary 2.9. *Let $M = D_X u$ be a left D_X-module generated by $u \in M$ and $f \in K[x]$ be a non-constant polynomial. Assume that there exists the b-function $b_{u,f}(s)$ for u and f. Then the following conditions are equivalent:*

(1) *$b_{u,f}(j) \neq 0$ for any integer $j < k$.*
(2) *The localization $M[f^{-1}]$ is generated by $u \otimes f^{-k}$ over D_X.*
(3) *The local cohomology group $H^1_{(f)}(M)$ is generated by the cohomology class $[u \otimes f^{-k}]$ over D_X.*

Proof. The equivalence of (1) and (2) is a special case of Theorem 2.8. In general, if $M[f^{-1}]$ is generated by $u \otimes f^{-k}$, then $H^1_{(f)}(M) = M[f^{-1}]/\iota(M)$ is generated by its residue class. Conversely, assume that $M[f^{-1}]/\iota(M)$ is generated by $[u \otimes f^{-k}]$. Then for any $w \in M[f^{-1}]$, there exist $P, Q \in D_X$ such that

$$w = P(u \otimes f^{-k}) + (Qu) \otimes 1 = (P + Qf^k)(u \otimes f^{-k}).$$

Henece $M[f^{-1}]$ is generated by $u \otimes f^{-k}$. Q.E.D.

§3. Localization algorithm revisited

Let $X = K^n$ be the n-dimensional affine space over K. Let M be a left module over $D_X = D_n$ and $f \in K[x]$ be a non-constant polynomial. Then $X_f := \{x \in X \mid f(x) \neq 0\}$ is an affine open subset of X. Our purpose is to reformulate the algorithm given in [21] for computing the localization $M[f^{-1}] := M \otimes_{K[x]} K[x, f^{-1}]$ as left D_X-module by using local cohomology, hoping to clarify the meaning of the algorithm as well as to make the canonical homomorphism $\iota : M \to M[f^{-1}]$ more explicit.

We assume in what follows, as well as in [21], that M is a holonomic D_X-module, or else it is holonomic on X_f with K being algebraically closed; i.e., $\mathrm{Char}(M) \cap \pi^{-1}(X_f)$ is an n-dimensional algebraic set of $\pi^{-1}(X_f)$, where $\mathrm{Char}(M)$ is the characteristic variety of M, which is an algebraic set of the cotangent bundle $T^*X = \{(x, \xi) \in K^n \times K^n\}$ and $\pi : T^*X \to X$ is the projection (see e.g., 2.4 of [18]).

Introducing a new variable t, set $Y = X \times K \ni (x,t)$ and

$$Z := \{(x,t) \in Y \mid tf(x) = 1\}.$$

Then Z is an affine subset of Y which is isomorphic to X_f. Let

$$B_{Z|Y} := H^1_{(tf(x)-1)}(K[x,t]) = K[x,t,(tf-1)^{-1}]/K[x,t]$$

be the first local cohomology group of $K[x,t]$ with support in Z, which we regard as a left D_Y-module. An arbitrary element of $B_{Z|Y}$ is expressed as

$$\left[\frac{a(x,t)}{(tf(x)-1)^{k+1}}\right] \qquad (k \in \mathbb{N},\ a(x,t) \in K[x,t]),$$

where the bracket denotes the residue class in $B_{Z|Y}$.

Set $f_i = \partial f/\partial x_i$ for $i = 1,\dots,n$ and define

$$\delta^{(k,l)} := \left[\frac{f^{l+1}}{(tf-1)^{k+1}}\right]$$

for $k,l \in \mathbb{Z}$ with $l \geq -1$. Note that $\delta^{(k,l)} = 0$ by the definition if $k < 0$. As left $K[x,t]$-module, $B_{Z|Y}$ is generated by $\delta^{(k,-1)}$ with $k \in \mathbb{N}$.

We have the following identities for $k,l \geq 0$:

$$\begin{aligned}
\partial_t \delta^{(k,l)} &= -(k+1)\left[\frac{f^{l+2}}{(tf-1)^{k+2}}\right] = -(k+1)\delta^{(k+1,l+1)},\\
\partial_{x_i}\delta^{(k,l)} &= (l+1)\left[\frac{f_i f^l}{(tf-1)^{k+1}}\right] - (k+1)\left[\frac{tf_i f^{l+1}}{(tf-1)^{k+2}}\right]\\
&= (l+1)f_i\delta^{(k,l-1)} - (k+1)\left[\frac{f_i(tf-1+1)f^l}{(tf-1)^{k+2}}\right]\\
&= (l+1)f_i\delta^{(k,l-1)} - (k+1)f_i(\delta^{(k,l-1)} + \delta^{(k+1,l-1)})\\
&= (l-k)f_i\delta^{(k,l-1)} - (k+1)f_i\delta^{(k+1,l-1)},\\
t\delta^{(k,l)} &= \left[\frac{tf^{l+1}}{(tf-1)^{k+1}}\right] = \left[\frac{\{(tf-1)+1\}f^l}{(tf-1)^{k+1}}\right] = \delta^{(k-1,l-1)} + \delta^{(k,l-1)}.
\end{aligned}$$

In particular, we have

$$(\partial_t t + k)\delta^{(k,l)} = -(k+1)\delta^{(k+1,l)}, \qquad t\delta^{(0,0)} = \delta^{(0,-1)}.$$

Hence $B_{Z|Y}$ is generated by $\delta^{(0,0)} = [f(tf-1)^{-1}]$ as a left D_Y-module.

Lemma 3.1. *One has $(tf-1)\delta^{(0,0)} = 0$ and $(\partial_{x_i} - f_i\partial_t t^2)\delta^{(0,0)} = 0$ for $i = 1,\dots,n$.*

Proof. The first equality follows immediately from the definition. The second equality follows from

$$\partial_t t^2 \delta^{(0,0)} = \partial_t t \delta^{(0,-1)} = -\delta^{(1,-1)}$$

in view of the formulae above. Q.E.D.

Let us regard $B_{Z|Y}$ as a module over the subring $K[x]$ of D_Y and consider the localization

$$\begin{aligned} B_{Z|Y}[f^{-1}] &:= B_{Z|Y} \otimes_{K[x]} K[x, f^{-1}] \\ &= K[x, t, f^{-1}, (tf-1)^{-1}]/K[x, t, f^{-1}] \end{aligned}$$

with respect to f. Let us denote the residue class in $B_{Z|Y}[f^{-1}]$ by $[\bullet]'$ in order to distinguish it from the residue class in $B_{Z|Y}$ which is denoted $[\bullet]$.

Lemma 3.2. *The natural homomorphism*

$$\iota' : B_{Z|Y} \ni \left[\frac{a(x,t)}{(tf-1)^{k+1}} \right] \longmapsto \left[\frac{a(x,t)}{(tf-1)^{k+1}} \right]' \in B_{Z|Y}[f^{-1}]$$

is an isomorphism of left D_Y-modules.

Proof. Assume $\iota'([a(x,t)(tf-1)^{-k-1}]) = 0$ with $a(x,t) \in K[x,t]$. Then there exists an integer l such that $f^l a(x,t)$ is divisible by $(tf-1)^{k+1}$ in $K[x,t]$. Since f and $tf-1$ are relatively prime, $a(x,t)$ must be divisible by $(1-tf)^{k+1}$. This proves that ι' is injective.

Let us show that ι' is surjective. It suffices to show that

$$[f^{-m}(tf-1)^{-k-1}]' \in \iota'(B_{Z|Y})$$

for any $k, m \in \mathbb{N}$ by induction on $k+m$, which obviously holds for $k = m = 0$. Suppose $k + m \geq 1$. We have

$$\left[\frac{tf}{(tf-1)^{k+1}} \right] = \left[\frac{1 + (tf-1)}{(tf-1)^{k+1}} \right] = \left[\frac{1}{(tf-1)^{k+1}} \right] + \left[\frac{1}{(tf-1)^{k}} \right].$$

It follows that

$$\left[\frac{f^{-m}}{(tf-1)^{k+1}} \right]' = \left[\frac{tf^{1-m}}{(tf-1)^{k+1}} \right]' - \left[\frac{f^{-m}}{(tf-1)^{k}} \right]'.$$

By the induction hypothesis, the right-hand side belongs to the image of ι'. This completes the proof. Q.E.D.

Proposition 3.3. *Let M be a finitely generated left D_X-module. Then the homomorphism*

$$B_{Z|Y} \otimes_{K[x]} M \xrightarrow{\sim} B_{Z|Y} \otimes_{K[x]} M[f^{-1}]$$

of left D_Y-modules, which is induced by the natural homomorphism $\iota : M \to M[f^{-1}]$ is an isomorphism.

Proof. We have

$$\begin{aligned} B_{Z|Y} \otimes_{K[x]} M[f^{-1}] &= B_{Z|Y} \otimes_{K[x]} (K[x, f^{-1}] \otimes_{K[x]} M) \\ &= (B_{Z|Y} \otimes_{K[x]} K[x, f^{-1}]) \otimes_{K[x]} M \\ &= B_{Z|Y}[f^{-1}] \otimes_{K[x]} M. \end{aligned}$$

Hence the isomorphism ι' induces an isomorphism

$$B_{Z|Y} \otimes_{K[x]} M \xrightarrow{\sim} B_{Z|Y}[f^{-1}] \otimes_{K[x]} M = B_{Z|Y} \otimes_{K[x]} M[f^{-1}].$$

Q.E.D.

Proposition 3.4. *Let M be a finitely generated left D_X-module. Then there exists an isomorphism*

$$B_{Z|Y} \otimes_{K[x]} M \xrightarrow{\sim} B_{Z|Y}[f^{-1}] \otimes_{K[x,f^{-1}]} M[f^{-1}]$$

of left D_Y-modules.

Proof. We have

$$\begin{aligned} &B_{Z|Y}[f^{-1}] \otimes_{K[x,f^{-1}]} M[f^{-1}] \\ &= (B_{Z|Y} \otimes_{K[x]} K[x, f^{-1}]) \otimes_{K[x,f^{-1}]} M[f^{-1}] = B_{Z|Y} \otimes_{K[x]} M[f^{-1}]. \end{aligned}$$

This completes the proof combined with Proposition 3.4. Q.E.D.

Let $D_X[f^{-1}] := K[x, f^{-1}] \otimes_{K[x]} D_X$ and $D_Y[f^{-1}] := K[x, f^{-1}] \otimes_{K[x]} D_Y$ be the localization of D_X and D_Y by f, which can be regarded as ring extensions of D_X and of D_Y respectively. Then $B_{Z|Y}[f^{-1}]$ and $B_{Z|Y}[f^{-1}] \otimes_{K[x,f^{-1}]} M[f^{-1}]$ have natural structures of left $D_Y[f^{-1}]$-module.

Definition 3.5. We set $\delta^{(j)} = \iota'(\delta^{(j,j)})$ for $j \in \mathbb{N}$. We denote $\delta = \delta^{(0)}$.

Lemma 3.6. *As an element of the left $D_Y[f^{-1}]$-module $B_{Z|Y}[f^{-1}]$, the annihilator of δ coincides with the left ideal of $D_Y[f^{-1}]$ generated by*

$$t - f^{-1}, \quad \partial_{x_i} - f_i \partial_t t^2 \quad (i = 1, \dots, n).$$

Proof. By Lemma 3.1, these operators annihilate δ. Assume $P \in D_Y[f^{-1}]$ annihilates δ. There exist elements Q_i and R of $D_Y[f^{-1}]$ such that

$$P = Q_0(t - f^{-1}) + \sum_{i=1}^{n} Q_i(\partial_{x_i} - f_i \partial_t t^2) + R$$

and that R belongs to $K[x, f^{-1}, \partial_t]$. Writing R in a finite sum

$$R = \sum_{j=0}^{l} r_j(x) f^{-k} \partial_t^j$$

with $r_j(x) \in K[x]$ and $k, l \in \mathbb{N}$, we have

$$\begin{aligned} 0 = R\delta &= \sum_{j=0}^{l} f^{-k} r_j(x) \partial_t^j \delta = \sum_{j=0}^{l} (-1)^j j! f^{-k} r_j(x) \delta^{(j)} \\ &= f^{-k} \left[\frac{\sum_{j=0}^{l} (-1)^j j! (tf-1)^{l-j} r_j(x)}{(tf-1)^{l+1}} \right]'. \end{aligned}$$

Since ι' is injective, this implies that $r_j(x) = 0$ for any $j \geq 0$, that is, $R = 0$. Q.E.D.

Proposition 3.7. *Any element of* $B_{Z|Y}[f^{-1}] \otimes_{K[x,f^{-1}]} M[f^{-1}]$ *is uniquely expressed as a finite sum* $\sum_{j \geq 0} \delta^{(j)} \otimes v_j$ *with* $v_j \in M[f^{-1}]$.

Proof. By Lemma 3.6, $B_{Z|Y}[f^{-1}]$ is isomorphic to $K[x, f^{-1}, \partial_t]$ as left $K[x, f^{-1}]$-module. Hence $\delta^{(j)} = (-1)^j (1/j!) \partial_t^j \delta$ $(j \in \mathbb{N})$ constitute a free basis of $B_{Z|Y}[f^{-1}]$ over $K[x, f^{-1}]$. This implies the assertion of the proposition. Q.E.D.

Definition 3.8. Set $\vartheta_i := \partial_{x_i} - f_i \partial_t t^2$ for $i = 1, \ldots, n$. Define a ring homomorphism τ from D_X to D_Y by

$$\tau : D_X \ni P(x, \partial_x) \longmapsto P(x, \vartheta_1, \ldots, \vartheta_n) \in D_Y.$$

Since $\vartheta_1, \ldots, \vartheta_n$ commute with each other, and $[\vartheta_i, x_j] = \delta_{ij}$, this substitution is a well-defined ring homomorphism.

Lemma 3.9. *One has*

$$\delta^{(0,0)} \otimes Pv = \tau(P)(\delta^{(0,0)} \otimes v), \qquad \delta \otimes Pv' = \tau(P)(\delta \otimes v')$$

in $B_{Z|Y} \otimes_{K[x]} M$ *and in* $B_{Z|Y}[f^{-1}] \otimes_{K[x,f^{-1}]} M[f^{-1}]$ *respectively for any* $v \in M$, $v' \in M[f^{-1}]$ *and* $P \in D_X$.

Proof. We have only to show the first equality. Since

$$(\partial_{x_i} - f_i\partial_t t^2)\delta^{(0,0)} = 0,$$

we have

$$\begin{aligned}\tau(\partial_{x_i})(\delta^{(0,0)} \otimes v) &= (\partial_{x_i} - f_i\partial_t t^2)\delta^{(0,0)} \otimes v + \delta^{(0,0)} \otimes \partial_{x_i} v \\ &= \delta^{(0,0)} \otimes \partial_{x_i} v.\end{aligned}$$

We can verify that

$$\tau(P)(\delta^{(0,0)} \otimes v) = \delta^{(0,0)} \otimes (Pv)$$

holds by induction on the order of P. Q.E.D.

Proposition 3.10. *Let $v \in M[f^{-1}]$, $P \in D_X$, and $k \in \mathbb{N}$. Then $P(f^{-k}v) = 0$ holds in $M[f^{-1}]$ if and only if $\tau(P)t^k(\delta \otimes v) = 0$ holds in $B_{Z|Y}[f^{-1}] \otimes_{K[x,f^{-1}]} M[f^{-1}]$.*

Proof. Since $t^k f^k \delta = (1 + t^k f^k - 1)\delta = \delta$, we have by Lemma 3.9

$$\begin{aligned}\delta \otimes P(f^{-k}v) &= \tau(P)(\delta \otimes f^{-k}v) \\ &= \tau(P)t^k f^k(\delta \otimes f^{-k}v) = \tau(P)t^k(\delta \otimes v).\end{aligned}$$

This vanishes if and only if $P(f^{-k}v) = 0$ by Proposition 3.7. Q.E.D.

Summing up we obtain

Theorem 3.11. *Let $M = D_X u$ be a left D_X-module generated by u and $I = \mathrm{Ann}_{D_X} u$ the annihilator of u so that $M = D_X/I$. Let $\iota : M \to M[f^{-1}]$ be the canonical homomorphism which sends $u \in M$ to $u \otimes 1$. Let G be a finite set of generators of I, and J be the left ideal of D_Y generated by $\{\tau(P) \mid P \in G\}$ and $tf - 1$. Then*

(1) *J coincides with the annihilator $\mathrm{Ann}_{D_Y}(\delta \otimes \iota(u))$ of $\delta \otimes \iota(u)$ in $B_{Z|Y}[f^{-1}] \otimes_{K[x,f^{-1}]} M[f^{-1}]$.*

(2) *$B_{Z|Y}[f^{-1}] \otimes_{K[x,f^{-1}]} M[f^{-1}]$ is generated by $\delta \otimes \iota(u)$ as a left D_Y-module.*

(3) *As a left D_Y-module, $B_{Z|Y} \otimes_{K[x]} M$ is isomorphic to D_Y/J.*

Proof. (1) It is obvious that J is contained in $\mathrm{Ann}_{D_Y}(\delta \otimes \iota(u))$. Suppose $P(\delta \otimes \iota(u)) = 0$ with $P \in D_Y$. There exist $R \in D_Y[f^{-1}]$ and $a_{\alpha,j}(x) \in K[x, f^{-1}]$ which are zero except finitely many $(\alpha, j) \in \mathbb{N}^n \times \mathbb{N}$

such that

$$\begin{aligned} P = & \sum_{\alpha \in \mathbb{N}^n, j \geq 0} a_{\alpha,j}(x) \partial_t^j (\partial_{x_1} - f_1 \partial_t t^2)^{\alpha_1} \cdots (\partial_{x_n} - f_n \partial_t t^2)^{\alpha_n} \\ & + R(t - f^{-1}) \\ = & \sum_{j \geq 0} \partial_t^j \tau(Q_j) + R(t - f^{-1}) \end{aligned}$$

with $Q_j := \sum_{\alpha \in \mathbb{N}^n} a_{\alpha,j}(x) \partial_x^\alpha \in D_X[f^{-1}]$. Then we have

$$0 = P(\delta \otimes \iota(u)) = \sum_{j \geq 0} \partial_t^j \tau(Q_j)(\delta \otimes \iota(u)) = \sum_{j \geq 0} (-1)^j j! \delta^{(j)} \otimes Q_j \iota(u)$$

and consequently $Q_j \iota(u) = 0$ for each $j \geq 0$ by Proposition 3.7. This implies that $f^l Q_j u = 0$ holds in M, that is, $f^l Q_j$ belongs to I, for some $l \in \mathbb{N}$ independent of j. We may also assume that $f^l R$ belongs to $D_Y f$. Hence $f^l P = \sum_{j \geq 0} \partial_t^j \tau(f^l Q_j) + f^l R(t - f^{-1})$ belongs to J. Since $(1 - t^l f^l)^k P$ belongs to $D_Y(1 - t^l f^l)$, and hence to J, if we take $k \in \mathbb{N}$ sufficiently large, and $t^l f^l P$ belongs to J, we conclude that P itself belongs to J.

(2) By the assumption, Lemma 2.5, and Proposition 3.7, an arbitrary element of $B_{Z|Y}[f^{-1}] \otimes_{K[x,f^{-1}]} M[f^{-1}]$ is expressed as a finite sum

$$\sum_{j \geq 0} \partial_t^j \delta \otimes P_j(u \otimes f^{-k})$$

with $P_j \in D_X$ and $k \in \mathbb{N}$. We get

$$\begin{aligned} \sum_{j \geq 0} \partial_t^j \delta \otimes P_j(u \otimes f^{-k}) &= \sum_{j \geq 0} \partial_t^j \tau(P_j)(\delta \otimes (u \otimes f^{-k})) \\ &= \sum_{j \geq 0} \partial_t^j \tau(P_j) t^k f^k (\delta \otimes (u \otimes f^{-k})) \\ &= \sum_{j \geq 0} \partial_t^j \tau(P_j) t^k (\delta \otimes \iota(u)). \end{aligned}$$

This completes the proof of (2).

(3) follows from (1), (2) and Proposition 3.4. Q.E.D.

Definition 3.12. For the sake of simplicity of the notation, let us set

$$\tilde{M} := B_{Z|Y}[f^{-1}] \otimes_{K[x,f^{-1}]} M[f^{-1}] \cong B_{Z|Y} \otimes_{K[x]} M$$

and define a homomorphism $\varphi : M[f^{-1}] \to \tilde{M}/\partial_t\tilde{M}$ by

$$\varphi(v) = \delta \otimes v \mod \partial_t\tilde{M}.$$

for $v \in M[f^{-1}]$.

Theorem 3.13. *The homomorphism $\varphi : M[f^{-1}] \to \tilde{M}/\partial_t\tilde{M}$ is an isomorphism of left $D_X[f^{-1}]$-modules, and consequently of D_X-modules.*

Proof. By Proposition 3.7 one has direct sum decompositions

$$\tilde{M} = (\delta \otimes M[f^{-1}]) \oplus (\delta^{(1)} \otimes M[f^{-1}]) \oplus (\delta^{(2)} \otimes M[f^{-1}]) \oplus \cdots,$$
$$\partial_t\tilde{M} = (\delta^{(1)} \otimes M[f^{-1}]) \oplus (\delta^{(2)} \otimes M[f^{-1}]) \oplus \cdots$$

as $K[x, f^{-1}]$-modules. Hence φ is an isomorphism of $K[x, f^{-1}]$-modules. For $v \in M[f^{-1}]$ and $P \in D_X[f^{-1}]$, one has

$$P(\delta \otimes v) \equiv \tau(P)(\delta \otimes v) = \delta \otimes Pv \mod \partial_t\tilde{M}$$

since $P - \tau(P)$ belongs to $\partial_t D_Y$. Hence φ is an isomorphism of left $D_X[f^{-1}]$-modules. Q.E.D.

Theorem 3.14. *Assume K is algebraically closed. If M is holonomic on X_f, i.e., if* $\mathrm{Char}(M) \cap \pi^{-1}(X_f)$ *is an n-dimensional algebraic set, then D_Y/J is a holonomic D_Y-module.*

Proof. We may assume $M = D_X/I$. By the definition, we have

$$\begin{aligned}
&\mathrm{Char}(D_Y/J) = \{(x, t; \xi, \tau) \in K^{2n+2} \mid \\
&\quad \sigma(P)(x, \xi_1 - f_1 t^2\tau, \ldots, \xi_n - f_n t^2\tau) \quad (\forall P \in I), \quad tf(x) = 1\} \\
&= \{(x, t; \xi, \tau) \mid (x, \xi_1 - f_1 t^2\tau, \ldots, \xi_n - f_n t^2\tau) \in \mathrm{Char}(M),\ tf(x) = 1\}.
\end{aligned}$$

Hence there is a bijection

$$\begin{aligned}
&(\mathrm{Char}(M) \cap \pi^{-1}(X_f)) \times K \ni (x, \xi, \tau) \\
&\qquad \longmapsto (x, 1/f(x); \xi_1 + f_1 t^2\tau, \ldots, \xi_n + f_n t^2\tau) \in \mathrm{Char}(D_Y/J).
\end{aligned}$$

This implies that $\mathrm{Char}(D_Y/J)$ is of dimension $n + 1$. Q.E.D.

The D_X-module $\tilde{M}/\partial_t\tilde{M}$ is nothing but the integral of the D_Y-module $\tilde{M}$ with respect to t, and $\tilde{M}$ is isomorphic to D_Y/J by Theorem 3.11. Suppose that M is holonomic on X_f. Then $\tilde{M} = D_Y/J$ is a holonomic D_Y-module by the theorem above. Hence $\tilde{M}/\partial_t\tilde{M}$ is also a holonomic D_X-module. In particular, there exists $k_0 \in \mathbb{N}$, or else

$k_0 = -1$, such that $\tilde{M}/\partial_t\tilde{M}$ is generated by residue classes $[t^j\delta \otimes u] = \varphi(u \otimes f^{-j})$ for $0 \le j \le k_0$.

The relations among these generators, i.e., a presentation of the D_X-module $\tilde{M}/\partial_t\tilde{M}$ can be computed by the integration algorithm for D-modules under the assumption that what is called the b-function $b(s)$ with respect to the weight vector $w = (0, ..., 0, 1; 0, ..., 0, -1)$ for $(x_1, \dots, x_n, t; \partial_{x_1}, \dots, \partial_{x_n}, \partial_t)$ exists. The integer k_0 above can be taken as the maximum integer root of $b(s)$. See [19], [23], [18] for details. This condition is fulfilled if $B_{Z|Y} \otimes_{K[x]} M = D_Y/J$ is holonomic, which is the case if M is holonomic on X_f.

In conclusion, we get

Algorithm 3.15 (localization)**.** Input: A set G_0 of generators of a left ideal $I = \mathrm{Ann}_{D_X} u$ of D_X such that $M = D_X u$.
Output: a presentation of $M[f^{-1}]$.

(1) Let J be the left ideal of D_Y generated by $\tau(G_0) \cup \{tf - 1\}$.
(2) Compute the b-function $b(s)$ of D_Y/J with respect to the weight vector $(0, ..., 0, 1; 0, ..., 0, -1)$ by e.g., Algorithm 5.6 of [18]. Quit if $b(s)$ does not exist; the computation fails.
(3) Let k_0 be the maximum integer root of $b(s)$ if any. Then $D_Y/(J + \partial_t D_Y)$, which is isomorphic to $\tilde{M}/\partial_t\tilde{M}$, is generated by the residue classes $[t^j]$ with $0 \le j \le k_0$. If $k_0 < 0$ or $b(s)$ does not have any integer root, then $M[f^{-1}] = 0$; quit.
(4) Compute a set G_1 of generators of the left D_X-submodule

$$N := \left\{ (P_0, P_1, \dots, P_{k_0}) \in (D_X)^{k_0+1} \mid \sum_{j=0}^{k_0} P_j[t^j] = 0 \right\}$$

of $(D_X)^{k_0+1}$ by using the integration algorithm (e.g., Algorithm 5.10 of [18]). Then one has isomorphisms

$$M[f^{-1}] \cong D_Y/(J + \partial_t D_Y) \cong (D_X)^{k_0+1}/N,$$

by which $u \otimes 1, \dots, u \otimes f^{-k_0}$ correspond to the residue classes $[(1, 0, \dots, 0)], \dots, [(0, \dots, 0, 1)]$.

In [21], the homomorphism τ above (denoted ϕ in [21]) is defined with ϑ_i replaced by $\partial_{x_i} - f_i t^2 \partial_t$. This induces the homomorphism $M \to M[f^{-1}]$ which sends u to $u \otimes f^{-2}$ instead of $u \otimes 1$.

The algorithm above and the isomorphisms

$$H^0_{(f)}(M) = \ker\iota \cong \mathrm{Ann}_{D_X}\iota(u)/\mathrm{Ann}_{D_X} u, \quad H^1_{(f)}(M) = M[f^{-1}]/\iota(M)$$

also provide us with algorithms to compute $H^j_{(f)}(M)$ for $j = 0, 1$, which work at least if M is holonomic on X_f. We remark that H. Tsai gave an algorithm for $H^0_{(f)}(M)$ without any assumption on M; see Algorithms 4.3 and 4.5 of [25].

Algorithm 3.16 ($\iota(M)$ and $H^j_{(f)}(M)$)**.** Input: A set G_0 of generators of a left ideal $I = \mathrm{Ann}_{D_X} u$ of D_X such that $M = D_X u$. Output: presentations of $\iota(M)$, $H^1_{(f)}(M)$, and $H^0_{(f)}(M)$ as D_X-modules.

(1) Compute the integer k_0 and a set G_1 of generators of the module N of Algorithm 3.15.

(2) Compute a set G_2 of generators of the left ideal

$$\tilde{I} := \{P \in D_X \mid (P, 0, \ldots, 0) \in N\} = \mathrm{Ann}_{D_X} \iota(u)$$

of D_X by using a Gröbner basis of N with respect to a POT (position-over-term) order (see e.g., Chapter 5 of [9]). Then one has $\iota(M) \cong D_X/\tilde{I}$ with the correspondence $\iota(u) \leftrightarrow \overline{1}$.

(3) Set

$$G_1' := \{(P_1, \ldots, P_{k_0}) \mid (P_0, P_1, \ldots, P_{k_0}) \in G_1 \ (\exists P_0 \in D_X)\}$$

and N' be the left D_X-submodule of $(D_X)^{k_0}$ generated by G_1'. Compute a set G_3 of generators of the left ideal

$$I_1 := \{P \in D_X \mid (0, \ldots, 0, P) \in N'\} = \mathrm{Ann}_{D_X}[f^{-k_0}\iota(u)]$$

of D_X by using a Gröbner basis of N' with respect to a POT order. Then one has $H^1_{(f)}(M) \cong D_X/I_1$ with the correspondence $[f^{-k_0}\iota(u)] \leftrightarrow \overline{1}$.

(4) Set $G_2 = \{P_1, \ldots, P_l\}$. Compute a set of generators of the left D_X-module

$$N_0 := \left\{(Q_1, \ldots, Q_l) \in (D_X)^l \mid \sum_{j=1}^{l} Q_j P_j \in I\right\}$$

through syzygies among $\{P_1, \ldots, P_l\} \cup G_0$. Then one has

$$H^0_{(f)}(M) \cong \tilde{I}/I \cong (D_X)^l/N_0$$

with the correspondence

$$P_1 u \leftrightarrow (1, 0, \ldots, 0) \mod N_0, \quad \ldots, \quad P_l u \leftrightarrow (0, \ldots, 0, 1) \mod N_0.$$

In what follows, we freely use the notation and the terminology introduced in Chapters 2, 3, 5 of [18] concerning weight vectors, Gröbner bases, and b-functions.

Example 3.17. Set $n = 1$ and $x = x_1$. It is easy to see that

$$\mathcal{N}_x = D_X x^s \cong D_X / D_X(x\partial_x - s).$$

Since the b-function of x is $s+1$, $\mathcal{N}_x(\lambda)$ is isomorphic to the submodule $D_X x^\lambda$ of $\mathcal{L}_x(\lambda)$ and hence x-saturated, if $\lambda \notin \mathbb{N}$. So let us consider $M = \mathcal{N}_f(0) = D_X/D_X x\partial_x$. Let u be the residue class of 1 in M. The left ideal J of D_Y defined in Theorem 3.11 is generated by $tx - 1$ and $\partial_x - \partial_t t^2$. A Gröbner basis of J with respect to a monomial order adapted to the weight vector $(1, 0; -1, 0)$ for $(t, x, \partial_t, \partial_x)$ is

$$tx-1, \quad x^2\partial_x - \partial_t, \quad t\partial_t - x\partial_x + 1 = \partial_t t - x\partial_x, \quad t^2\partial_t - \partial_x + 2t = \partial_t t^2 - \partial_x.$$

The b-function of J with respect to the weight vector above (see 5.2 of [18]) is divisible by $s(s-1)$ by virtue of the operator $\partial_t t^2 - \partial_x$. Hence the integration module $\tilde{M}/\partial_t \tilde{M} = D_Y/(J + \partial_t D_Y)$ is generated by the residue classes $[u]$ and $[tu]$, which correspond to $u \otimes 1$ and $u \otimes x^{-1}$ in $M[x^{-1}]$ respectively, over D_X. The fundamental relations among the generators can be read off from the Gröbner basis above as follows (see Algorithm 5.10 of [18]):

$$x[tu] - [u] = 0, \quad \partial_x[u] = 0, \quad x^2\partial_x[tu] + [u] = 0, \quad (x\partial_x + 1)[tu] = 0.$$

We translate these relations to vectors

$$(-1, x), \quad (\partial_x, 0), \quad (1, x^2\partial_x), \quad (0, x\partial_x + 1)$$

in the free module D_X^2. Let N be the left D_X-submodule of D_X^2 generated by these vectors. By using Gröbner bases of N with respect to POT orders we can confirm that $\mathrm{Ann}_{D_X}[u] = D_X\partial_x$ and $\mathrm{Ann}_{D_X}[tu] = D_X(x\partial_x + 1)$. Hence $M[x^{-1}]$ is generated by $u \otimes x^{-1}$ and isomorphic to $D_X/D_X(x\partial_x + 1) \cong K[x, x^{-1}]$ with the correspondence $u \otimes x^{-1} \leftrightarrow \overline{1}$. The image $\iota(M)$ of $\iota : M \to M[x^{-1}]$ is isomorphic to $D_X/D_X\partial_x \cong K[x]$ with the correspondence $u \otimes 1 \leftrightarrow \overline{1}$. Finally we get

$$H^0_{(x)}(M) = \ker \iota = D_X\partial_x u = K[\partial_x]u \cong D_X/D_X x,$$
$$H^1_{(x)}(M) = D_X[tu]/D_X[u] \cong D_X/D_X x.$$

The following is an example of non-holonomic M:

Example 3.18. Set $n = 2$, $x_1 = x$, $x_2 = y$, $P = x\partial_x^2 + \partial_y$, and $M = D_X/D_XP = D_Xu$ as in Example 2.2. Then M is x- and y-saturated. The localizations of M are

$$\begin{aligned} M[x^{-1}] &= D_X(u \otimes x^{-2}) = D_X/D_X(x^2\partial_x{}^2 + 4x\partial_x + x\partial_y + 2), \\ M[y^{-1}] &= D_X(u \otimes y^{-1}) = D_X/D_X(xy\partial_x{}^2 + y\partial_y + 1). \end{aligned}$$

The first local cohomology groups are

$$\begin{aligned} H^1_{(x)}(M) &= D_X[u \otimes x^{-2}] = D_X/(D_Xx^2 + D_Xx\partial_y), \\ H^1_{(y)}(M) &= D_X[u \otimes y^{-1}] = D_X/D_Xy, \end{aligned}$$

both of which are not holonomic.

Example 3.19. Set $n = 2$, $x_1 = x$, $x_2 = y$, and $f = x^3 - y^2$. Let us consider

$$M = D_X/(D_X\partial_x f + D_X\partial_y f) = D_Xu \quad (u = \overline{1}).$$

The characteristic variety of M is

$$\mathrm{Char}(M) = \{(x, y; \xi, \eta) \in K^4 \mid x^3 - y^2 = 0\} \cup \{(x, y; 0, 0) \mid (x, y) \in X\}.$$

Hence M is holonomic on X_f but not on X. The localization $M[f^{-1}]$ is given by

$$M[f^{-1}] = D_X/(D_X(2x\partial_x + 3y\partial_y + 6) + D_X(2y\partial_x + 3x^2\partial_y))$$

with the correspondence $u \otimes 1 \leftrightarrow \overline{1}$ and is holonomic on X. Hence we have $\iota(M) = M[f^{-1}]$ and $H^1_{(f)}(M) = 0$. Algorithm 3.16 also gives a presentation of $H^0_{(f)}(M)$, which is rather complicated. We can verify that its characteristic variety is

$$\mathrm{Char}(H^0_{(f)}(M)) = \{(x, y; \xi, \eta) \in K^4 \mid x^3 - y^2 = 0\}$$

as is expected.

§4. Algorithms for $u \otimes f^s$ and the b-function

The purpose here is to give algorithms to compute $M(u, f, s)$ and $M(u, f, \lambda)$ as well as the b-function $b_{u,f}(s)$ for an arbitrary D_X-module $M = D_Xu$ that is holonomic on X_f, and an arbitrary non-constant polynomial f. Algorithms for these objects were already given in [17] under the additional assumption that M is f-saturated. We remove this assumption by using the localization algorithm.

Set $X = K^n$ and $Y = K^{n+1}$ with coordinates $x = (x_1, \ldots, x_n)$ of X and (x, t) of Y. Let $f = f(x) \in K[x]$ be a non-constant polynomial and let Z be the affine subset $Z = \{(x,t) \mid t = f(x)\}$ of Y. (Note that Z is different from what was defined in the previous section.) We regard the local cohomology group

$$B_{Z|Y} := H^1_Z(K[x,t]) = K[x,t,(t-f)^{-1}]/K[x,t]$$

as a left D_Y-module. For $k \in \mathbb{N}$, let

$$\delta^{(k)}(t - f(x)) = \left[\frac{(-1)^k k!}{(t - f(x))^{k+1}}\right]$$

be the residue class in $B_{Z|X}$ and denote $\delta(t-f) = \delta^{(0)}(t-f)$. Then $\delta(t-f)$ satisfies a holonomic system

$$(t-f)\delta(t-f) = (\partial_{x_i} + f_i\partial_t)\delta(t-f) = 0 \qquad (1 \leq i \leq n)$$

with $f_i = \partial f/\partial x_i$. Hence there exists an isomorphism $B_{Z|Y} \cong D_Y/J_0$ with

$$J_0 := D_Y(t-f) + D_Y(\partial_{x_1} + f_1\partial_t) + \cdots + D_Y(\partial_{x_n} + f_n\partial_t)$$

as left D_Y-module since J_0 is maximal. In particular, $\delta^{(k)}(t-f)$ $(k \in \mathbb{N})$ constitute a free basis of $B_{Z|Y}$ over $K[x]$.

Following Malgrange [15], let us give $\mathcal{L}_f = D_X[s]f^s$ a structure of left D_Y-module by

$$\begin{aligned} t(a(x,s)f^{-k}f^s) &= a(x,s+1)f^{-k+1}f^s, \\ \partial_t(a(x,s)f^{-k}f^s) &= -sa(x,s-1)f^{-k-1}f^s \end{aligned}$$

for $a(x,s) \in K[x,s]$ and $k \in \mathbb{N}$. The actions of t and ∂_t on $\mathcal{L}_f$ satisfy $[\partial_t, t] = 1$, and they commute with x_i, ∂_{x_i}. Hence the definition above extends to the action of D_Y on $\mathcal{L}_f$. In particular, $\partial_t t f^s = -sf^s$ holds, which will play an important role in what follows.

With respect to this action, we can regard $B_{Z|Y}$ as a left D_Y-submodule of $\mathcal{L}_f$ by identifying $\delta^{(k)}(t-f)$ with $(-1)^k s(s-1)\cdots(s-k+1)f^{s-k}$ in $\mathcal{L}_f$. In fact, we have

$$J_0 = \mathrm{Ann}_{D_Y} f^s = \mathrm{Ann}_{D_Y} \delta(t-f)$$

since J_0 annihilates f^s as well as $\delta(t-f)$ and J_0 is a maximal left ideal.

For a left D_X-module $M = D_X u$, let us consider the tensor product $M \otimes_{K[x]} B_{Z|Y}$, which is a left D_Y-module.

Definition 4.1. Set $\vartheta_i' := \partial_{x_i} + f_i\partial_t$ for $i = 1, \dots, n$. Define a ring homomorphism τ' from D_X to D_Y by

$$\tau' : D_X \ni P(x, \partial_x) \longmapsto P(x, \vartheta_1', \dots, \vartheta_n') \in D_Y.$$

Since $\vartheta_1', \dots, \vartheta_n'$ commute with each other, and $[\vartheta_i', x_j] = \delta_{ij}$, this substitution is a well-defined ring homomorphism.

Since $\vartheta_i'\delta(t-f) = 0$ for $1 \le i \le n$,

$$\tau'(P)(v \otimes \delta(t-f)) = Pv \otimes \delta(t-f)$$

holds for any $v \in M$ and $P \in D_X$. Hence we have

$$Pu \otimes \partial_t^k \delta(t-f) = \partial_t^k (Pu \otimes \delta(t-f)) = \partial_t^k \tau'(P)(u \otimes \delta(t-f)).$$

This proves

Lemma 4.2. *If $M = D_X u$, then the left D_Y-module $M \otimes_{K[x]} B_{Z|Y}$ is generated by $u \otimes \delta(t-f)$.*

Theorem 4.3. *Let $M = D_X u$ be a left D_X-module generated by u and $f \in K[x]$. Let G be a finite set of generators of $I := \mathrm{Ann}_{D_X} u$ and let J be the left ideal of D_Y generated by $\{\tau'(P) \mid P \in G\} \cup \{t - f\}$. Then J coincides with the annihilator $\mathrm{Ann}_{D_Y}(u \otimes \delta(t-f))$. Moreover, if M is a holonomic D_X-module, then $M \otimes_{K[x]} B_{Z|Y}$ is a holonomic D_Y-module.*

The first part of this proposition was proved by Walther [26]. The proof is almost the same as the proof of Theorem 3.11. The last assertion can be proved in the same way as Theorem 3.14.

Thus we have an algorithm to compute $M \otimes_{K[x]} B_{Z|Y}$. The inclusion $B_{Z|X} \subset \mathcal{L}_f$ induces a homomorphism

$$\psi : M \otimes_{K[x]} B_{Z|Y} \longrightarrow M \otimes_{K[x]} \mathcal{L}_f$$

of left D_Y-modules. Our main aim is to compute the $D_X[s]$-submodule $M(u, f, s) = D_X[s](u \otimes f^s)$ of $M \otimes_{K[x]} \mathcal{L}_f$, which is the image of the submodule $D_X[s](u \otimes f^s)$ of $M \otimes_{K[x]} B_{Z|Y}$ by ψ. The following lemma was proved in [17] as Proposition 6.13.

Lemma 4.4. *The homomorphism ψ above is injective if and only if M is f-saturated.*

Proof. An arbitrary element of $M \otimes_{K[x]} B_{Z|Y}$ is expressed uniquely as

$$w = \sum_{j=0}^{k} v_j \otimes \delta^{(j)}(t-f)$$

with $k \in \mathbb{N}$ and $v_j \in M$. Then

$$\psi(w) = \sum_{j=0}^{k} (-1)^j v_j \otimes (s(s-1)\cdots(s-j+1)f^{-j}f^s)$$

vanishes if and only if each $v_j \otimes f^{-j}$ vanishes in $M \otimes_{K[x]} K[x, f^{-1}]$, which is equivalent to $v_j \in H^0_{(f)}(M)$. Q.E.D.

Lemma 4.5. *Let $M = D_X u$ be a left D_X-module generated by u and $f \in K[x]$ be a non-constant polynomial. Let $\iota : M \to M[f^{-1}]$ be the canonical homomorphism. Then ι induces isomorphisms*

$$M \otimes_{K[x]} \mathcal{L}_f \xrightarrow{\sim} \iota(M) \otimes_{K[x]} \mathcal{L}_f, \qquad M(u, f, s) \xrightarrow{\sim} \iota(M)(u, f, s)$$

of left $D_X[s]$-modules.

Proof. Since $\mathcal{L}_f$ is isomorphic to $K[x, f^{-1}, s]$ as a $K[x, s]$-module, we have only to show that the natural homomorphism

$$M[f^{-1}] \to \iota(M)[f^{-1}]$$

is an isomorphism, which is obvious by the definition. The second isomorphism follows from the first. Q.E.D.

Summing up we obtain

Algorithm 4.6 ($M(u, f, s)$ and $M(u, f, \lambda)$)**.** Input: A finite set of generators of a left ideal I of D_X so that $M = D_X u = D_X / I$, a non-constant $f \in K[x]$, and $\lambda \in K$.
Output: presentation of $M(u, f, s)$ and of $M(u, f, \lambda)$.

1. (1) Compute $\iota(M) = D_X / \mathrm{Ann}_{D_Y} \iota(u)$ by Algorithm 3.16.
2. (2) Compute $J = \mathrm{Ann}_{D_Y}(\iota(u) \otimes \delta(t - f))$ by using Theorem 4.3.
3. (3) Compute $J \cap D_X[s]$, which is equal to $\mathrm{Ann}_{D_X[s]}(\iota(u) \otimes f^s) = \mathrm{Ann}_{D_X[s]}(u \otimes f^s)$.
4. (4) The substitution $s = \lambda$ for generators of $J \cap D_X[s]$ gives a set of generators of $\mathrm{Ann}_{D_X}(u \otimes f^s)|_{s=\lambda}$.

Proposition 4.7. *Let $M = D_X u$ be a left D_X-module generated by u and $f \in K[x]$ a non-constant polynomial. Then the b-function $b_{u,f}(x)$ exists if and only if there exists a nonzero polynomial $b(s) \in K[s]$ such that*

(2)

$$b(-\partial_t t)(u \otimes \delta(t - f)) \in t D_X[t\partial_t](u \otimes \delta(t - f)) = D_X[t\partial_t] t (u \otimes \delta(t - f)).$$

If M is f-saturated and such $b(s)$ exists, then $b_{u,f}(s)$ is the monic polynomial of the minimum degree among such $b(s)$.

Proof. Assume that (2) holds. Then there exists $P(s) \in D_X[s]$ such that

$$b(-\partial_t t)(u\otimes\delta(t-f)) = P(-\partial_t t)t(u\otimes\delta(t-f)) = P(-\partial_t t)f(u\otimes\delta(t-f)).$$

Applying the homomorphism ψ we get $b(s)(u \otimes f^s) = P(s)(u \otimes f^{s+1})$. Hence $b_{u,f}(s)$ exists and divides $b(s)$.

On the other hand, assume that there exist nonzero $b(s) \in K[s]$ and $P(s) \in D_X[s]$ such that $b(s)(u \otimes f^s) = P(s)(u \otimes f^{s+1})$ holds in $M \otimes_{K[x]} \mathcal{L}_f$. Then as is seen by the proof of Lemma 4.4, there exists $k \in \mathbb{N}$ such that

$$f^k b(-\partial_t t)(u \otimes \delta(t-f)) = f^k P(-\partial_t t)f(u \otimes \delta(t-f)).$$

Since

$$f^k b(-\partial_t t)(u\otimes\delta(t-f)) = b(-\partial_t t)f^k(u\otimes\delta(t-f)) = b(-\partial_t t)t^k(u\otimes\delta(t-f))$$

holds, we get

$$\begin{aligned}\partial_t^k b(-\partial_t t)t^k(u \otimes \delta(t-f))\\ = \partial_t^k f^k P(-\partial_t t)f(u \otimes \delta(t-f)) \in D_X[t\partial_t]t(u \otimes \delta(t-f)).\end{aligned}$$

This completes the proof because there exists $c(s) \in K[x]$ such that $\partial_t^k b(-\partial_t t)t^k = c(-\partial_t t)$. Q.E.D.

Now we obtain an algorithm to determine whether the b-function exists and to compute it if it does:

Algorithm 4.8 ($b_{u,f}(s)$). Input: $M = D_X u = D_X/I$ with a finite set of generators of I and a non-constant $f \in K[x]$.
Output: the b-function $b_{u,f}(s)$ if it exists. 'No' if it does not exist.

(1) Compute $J' := \mathrm{Ann}_{D_Y}(u \otimes \delta(t-f))$ by using Theorem 4.3.
(2) Compute $I' = (J' + D_X[s]f) \cap K[s]$ by elimination. If $I' \neq \{0\}$, then there exists $b_{u,f}(s)$. Otherwise, the b-function does not exist; output 'No' and quit.
(3) Compute a set of generators of $J := \mathrm{Ann}_{D_X[s]}(u \otimes f^s)$ by Algorithm 4.6.
(4) Compute $I(u, f) = (J + D_X[s]f) \cap K[s]$ by elimination. Let $b_{u,f}(s)$ be the monic generator of $I(u, f)$.

Algorithm 4.9 ($D_X(u \otimes f^\lambda)$). Input: $M = D_X u = D_X/I$ with a finite set of generators of I, a non-constant $f \in K[x]$, and $\lambda \in K$.
Output: presentation of $D_X(u \otimes f^\lambda)$, i.e., $\mathrm{Ann}_{D_X}(u \otimes f^\lambda)$ if $b_{u,f}(s)$ exists.

(1) Compute $M(u,f,s)$ and $b_{u,f}(s)$ by Algorithms 4.6 and 4.8. Quit if $b_{u,f}(s)$ does not exist.
(2) Let k_0 be the maximum nonzero integer, if any, such that $b_{u,f}(\lambda - k_0) = 0$. If there is no such k_0, then set $k_0 = 0$.
(3) Compute $I := \mathrm{Ann}_{D_X}(u \otimes f^{\lambda - k_0})$ from $\mathrm{Ann}_{D_X[s]}(u \otimes f^s)$ by substitution $s = \lambda - k_0$.
(4) Compute the left ideal

$$I : f^{k_0} := \{P \in D_X \mid P f^{k_0} \in I\}$$

by an appropriate Gröbner basis. Then one has

$$I : f^{k_0} = \mathrm{Ann}_{D_X}(u \otimes f^{\lambda}).$$

Example 4.10. Set $n = 2$, $x_1 = x$, $x_2 = y$, $P = x\partial_x^2 + \partial_y$, and $M = D_X/D_X P = D_X u$ as in Example 2.2. By Algorithms 4.6 and 4.8 we obtain $b_{u,x}(s) = (s+1)(s+2)$, $b_{u,y}(s) = s+1$, and

$$\begin{aligned} M(u,x,s) &:= D_X[s](u \otimes x^s) \\ &= D_X[s]/D_X[s](x^2{\partial_x}^2 - 2sx\partial_x + x\partial_y + s^2 + s), \\ M(u,y,s) &:= D_X[s](u \otimes y^s) = D_X[s]/D_X[s](xy{\partial_x}^2 + y\partial_y - s). \end{aligned}$$

Example 4.11. Setting $n = 2$ and $x_1 = x$, $x_2 = y$, let us consider $f(x,y) = x^3 - y^2$ and

$$\mathcal{N}_f = D_X[s]f^s, \qquad M := \mathcal{N}_f\Big(\frac{1}{6}\Big) = \mathcal{N}_f\big/\Big(s - \frac{1}{6}\Big)\mathcal{N}_f.$$

We have $M = D_X/I = D_X u$ with $u := f^s|_{s=1/6}$ and

$$I = \mathrm{Ann}_{D_X} u = D_X(2x\partial_x + 3y\partial_y - 1) + D_X(2y\partial_x + 3x^2\partial_y).$$

Then Algorithm 3.15 shows that $M[f^{-1}]$ is generated by $u \otimes f^{-1}$ and is isomorphic to D_X/J with

$$J = \mathrm{Ann}_{D_X} u \otimes f^{-1} = D_X(2x\partial_x + 3y\partial_y + 5) + D_X(2y\partial_x + 3x^2\partial_y).$$

In fact, $M[f^{-1}]$ is isomorphic to $\mathcal{L}_f(1/6) = K[x, f^{-1}]f^{1/6} = D_X f^{-5/6}$, and $\iota(M)$ to the submodule $D_X f^{1/6}$.

By Algorithm 3.16, we have $\iota(M) = D_X/J_0$ with the left ideal J_0 generated by

$$2x\partial_x + 3y\partial_y - 1, \quad 2y\partial_x + 3x^2\partial_y, \quad 8\partial_x^3 + 27y\partial_y^3 + 9\partial_y^2.$$

Note that J_0 is strictly larger than I because of the last generator. Algorithm 3.16 also yields presentations of the local cohomology groups

$$H^0_{(f)}(M) \cong D_X/(D_X x + D_X y) \quad \text{with} \quad (8\partial_x^3 + 27y\partial_y^3 + 9\partial_y^2)u \longleftrightarrow \overline{1},$$
$$H^1_{(f)}(M) \cong D_X/(D_X x + D_X y) \quad \text{with} \quad [u \otimes f^{-1}] \longleftrightarrow \overline{1}.$$

The b-function for u and f is

$$b_{u,f}(s) = (s+1)\Big(s+\frac{4}{3}\Big)\Big(s+\frac{7}{6}\Big) = b_f\Big(s+\frac{1}{6}\Big),$$

where $b_f(s) = (s+1)(s+5/6)(s+7/6)$ is the b-function of f.

Example 4.12. Set $n = 2$ and $x_1 = x$, $x_2 = y$ and consider

$$M = H^1_{(xy)}(K[x,y]) = D_X/(D_X xy + D_X(x\partial_x + 1) + D_Y(y\partial_y + 1)) = D_X u$$

with $u = [(xy)^{-1}]$. Then $M[x^{-1}]$ is generated by $u \otimes 1$, and hence $M[^{-1}] = \iota(M)$ and $H^1_{(x)}(M) = 0$. We have

$$M[x^{-1}] = D_X/(D_X(x\partial_x + 1) + D_X y),$$
$$H^0_{(x)}(M) \cong D_X/(D_X x + D_X \partial_y)$$

with the correspondence $u \otimes 1 \leftrightarrow \overline{1}$ and $yu \leftrightarrow \overline{1}$ respectively. The b-function of $u := [(xy)^{-1}]$ and x is $b_{u,x}(s) = s$. The module $M(u,x,s)$ is

$$M(u,x,s) = D_X[s]/(D_X[s](x\partial_x - s + 1) + D_X[s]y).$$

Example 4.13. Set $n = 3$, $x_1 = x$, $x_2 = y$, $x_3 = z$, and $f = x^3 - y^2z^2$. Let us consider

$$M = H^1_{(xf)}(K[x,y]) = D_X u \qquad (u = [(xf)^{-1}]).$$

The localizations $M[x^{-1}]$ and $M[f^{-1}]$ are given by

$$\begin{aligned} M[x^{-1}] &= D_X(u \otimes 1) \\ &= D_X/(D_X f + D_X(2x\partial_x + 3z\partial_z + 8) + D_X(y\partial_y - z\partial_z), \\ M[f^{-1}] &= D_X(u \otimes 1) \\ &= D_X/(D_X x + D_X(y\partial_y + 2) + D_X(z\partial_z + 2). \end{aligned}$$

In particular, we have $H^1_{(x)}(M) = H^1_{(f)}(M) = 0$. The zeroth cohomology groups are

$$H^0_{(x)}(M) \cong D_X/(D_X x + D_X \partial_y + D_X \partial_z)$$

with the correspondence $fu \leftrightarrow \overline{1}$, and $H^0_{(f)}(M) \cong D_X/I$ with the correspondence $(y\partial_y + 2)u \leftrightarrow \overline{1}$, where I is the left ideal generated by

$$\begin{aligned} &f^2, \quad y\partial_y - z\partial_z, \quad 2x\partial_x + 3z\partial_z + 8, \quad x^3\partial_y - z^3 y\partial_z - 4z^2 y, \\ &f\partial_z - 4zy^2, \quad x^3\partial_y^2 - z^4\partial_z^2 - 6z^3\partial_z - 4z^2, \\ &8z^2 y\partial_x^3 + (27z^2\partial_z^2 + 81z\partial_z + 24)\partial_y. \end{aligned}$$

We also have

$$\begin{aligned} M(u, x, s) &= D_X[s]/(D_X[s]f + D_X[s](2x\partial_x + 3y\partial_y - 2s + 8) \\ &\qquad + D_X[s](y\partial_y - z\partial_z)), \\ M(u, f, s) &= D_X[s]/(D_X[s]x + D_X[s](y\partial_y - 2s - 2) \\ &\qquad + D_X[s](y\partial_y - z\partial_z)). \end{aligned}$$

The corresponding b-functions are

$$b_{u,x}(s) = s^2\Big(s - \frac{3}{2}\Big)^2, \qquad b_{u,f}(s) = s^2\Big(s - \frac{1}{2}\Big)^2.$$

§5. Length and multiplicity of D-modules

W set $X = K^n$ as in the preceding sections. First let us recall basic facts about the length and the multiplicity of a left D_X-module following J. Bernstein ([3],[4]). Let M be a finitely generated left D_X-module. A composition series of M of length k is a sequence

$$M = M_0 \supsetneqq M_1 \supsetneqq \cdots \supsetneqq M_k = 0$$

of left D_n-submodules such that M_i/M_{i-1} is a nonzero simple left D_X-module (i.e. having no proper left D_X-submodule other than 0) for $i = 1, \dots k$. The *length* of M, which we denote by $\operatorname{length} M$, is the least length of composition series (if any) of M. If there is no composition series, the length of M is defined to be infinite. The length is additive in the sense that if

$$0 \longrightarrow N \longrightarrow M \longrightarrow L \longrightarrow 0$$

is an exact sequence of left D_X-modules of finite length, then $\operatorname{length} M = \operatorname{length} N + \operatorname{length} L$ holds.

For each integer k, set

$$F_k(D_X) = \Big\{ \sum_{|\alpha|+|\beta|\le k} a_{\alpha\beta} x^\alpha \partial^\beta \mid a_{\alpha\beta} \in K \Big\}.$$

In particular, we have $F_k(D_X) = 0$ for $k < 0$ and $F_0(D_X) = K$. The filtration $\{F_k(D_X)\}_{k\in\mathbb{Z}}$ is called the Bernstein filtration on D_X.

Let M be a finitely generated left D_X-module. A family $\{F_k(M)\}_{k\in\mathbb{Z}}$ of K-subspaces of M is called a *Bernstein filtration* on M if it satisfies

(1) $F_k(M) \subset F_{k+1}(M) \quad (\forall k \in \mathbb{Z}), \qquad \bigcup_{k\in\mathbb{Z}} F_k(M) = M$,
(2) $F_j(D_X)F_k(M) \subset F_{j+k}(M) \quad (\forall j, k \in \mathbb{Z})$.

Moreover, $\{F_k(M)\}$ is called a *good Bernstein filtration* if there exist $u_i \in F_{k_i}(M)$ $(i = 1, \dots, m)$ such that

$$F_k(M) = F_{k-k_1}(D_X)u_1 + \cdots + F_{k-k_m}(D_X)u_m \qquad (\forall k \in \mathbb{Z}).$$

If $\{F_k(M)\}$ is a good Bernstein filtration, then each $F_k(M)$ is a finite dimensional vector space over K and $F_k(M) = 0$ for $k \ll 0$ (see e.g., 2.3 of [18]).

Let $\{F_k(M)\}$ be a good Bernstein filtration on M. Then there exists a polynomial $p(T) = c_dT^d + c_{d-1}T^{d-1} + \cdots + c_0 \in \mathbb{Q}[T]$ such that

$$\dim_K F_k(M) = p(k) \quad (k \gg 0)$$

and $d!c_d$ is a positive integer. We call $p(T)$ the Hilbert polynomial of M with respect to the filtration $\{F_k(M)\}$. The leading term of $p(T)$ does not depend on the choice of a good Bernstein filtration $\{F_k(M)\}$. The degree d of the Hilbert polynomial $p(T)$ is called the *dimension* of M and denoted $\dim M$. The *multiplicity* of M, denoted $\operatorname{mult} M$ is defined to be the positive integer $d!c_d$. The dimension and the multiplicity are invariants of a finitely generated left D_X-module.

If $M \neq 0$, then the dimension of M is not less than n (Bernstein's inequality). By definition, M is holonomic if $M = 0$ or $\dim M = n$. If M is a holonomic left D_X-module, we have an inequality $\operatorname{length} M \leq \operatorname{mult} M$ and hence M is of finite length in particular. Moreover, the multiplicity is additive for holonomic left D_X-modules.

We can compute the dimension and the multiplicity of a given finitely generated (not necessarily holonomic) D_X-module by using a Gröbner basis with respect to a term order compatible with the Bernstein filtration.

Example 5.1. Let M be the D_X-module with $X = K^2$ defined in Example 4.11. We get exact sequences

$$0 \longrightarrow H^0_{(f)}(M) \longrightarrow M \longrightarrow \iota(M) \longrightarrow 0,$$
$$0 \longrightarrow H^0_{(f)}(M) \longrightarrow M \longrightarrow M[f^{-1}] \longrightarrow H^1_{(f)}(M) \longrightarrow 0$$

with $H^0_{(f)}(M) \cong H^2_{(x,y)}(K[x,y]) \cong H^1_{(f)}(M)$. We have

$$\begin{aligned}&\operatorname{mult} M = \operatorname{mult} M[f^{-1}] = 6, \quad \operatorname{mult} \iota(M) = 5,\\ &\operatorname{mult} H^0_{(f)}(M) = \operatorname{mult} H^1_{(f)}(M) = 1.\end{aligned}$$

The following two propositions are easy and should be well-known.

Proposition 5.2. *Let f be a non-constant polynomial in $x_1, \dots, x_n$. Then the multiplicity of $K[x, f^{-1}]$ is at most $(\deg f + 1)^n$, where $\deg f$ stands for the total degree of f.*

Proof. Let d be the degree of f. Then

$$F_k(K[x, f^{-1}]) := \left\{ \frac{a}{f^{k+1}} \mid a \in K[x_1, \dots, x_n], \deg a \leq (d+1)k \right\}$$

for $k \in \mathbb{Z}$ constitute a (not necessarily good) Bernstein filtration on $K[x, f^{-1}]$, which is finitely generated over D_n. Since

$$\dim_K F_k(K[x, f^{-1}]) = \binom{n + (d+1)k}{n},$$

we have $\dim K[x, f^{-1}] = n$ and $\operatorname{mult} M \leq (d+1)^n$. Q.E.D.

Proposition 5.3. *Let $n = 1$ and $f \in K[x] = K[x_1]$ be non-constant square free. Then one has* $\operatorname{mult} K[x, f^{-1}] = \deg f + 1$.

Proof. Set $M := H^1_{(f)}(K[x])$. Then M is isomorphic to $D_X/D_X f$ since f is square-free. Hence

$$F_k(M) := F_k(D_1)[f^{-1}] \cong F_k(D_1)/F_{k-d}(D_1)f$$

with $d := \deg f$ constitute a good Bernstein filtration on M. Since

$$\begin{aligned}\dim F_k(M) &= \dim F_k(D_1) - \dim F_{k-d}(D_1)\\ &= \binom{k+2}{2} - \binom{k-d+2}{2} = dk - \frac{1}{2}d(d-3)\end{aligned}$$

holds for $k \geq d$, the multiplicity of M is d. Q.E.D.

We shall give two examples in two variables.

Proposition 5.4. *Set $X = K^2$ and write $x_1 = x$, $x_2 = y$. Set $f = x^m + y^l$ with positive integers l, m. Then the multiplicity of $K[x, y, f^{-1}]$ equals* $2\max\{l, m\}$.

Proof. We may assume $m \le l$. Set $M := H^1_{(f)}(K[x,y])$. Since the b-function $b_f(s)$ of f does not have any negative integer ≤ -2 as a root (see e.g., 6.4 of [14]), M is generated by $u := [f^{-1}] \in M$ over D_X. The annihilator $\mathrm{Ann}_{D_X} u$ is generated by

$$f, \quad E := lx\partial_x + my\partial_y + ml, \quad P := ly^{l-1}\partial_x - mx^{m-1}\partial_y$$

(see also 6.4 of [14]). A Gröbner basis of $\mathrm{Ann}_D[f^{-1}]$ with respect to a total-degree reverse lexicographic order $\prec$ such that $x \succ y \succ \xi \succ \eta$ is $G = \{f, E, P\}$, where ξ and η are the commutative variables corresponding to ∂_x and ∂_y respectively. In fact, in case $m < l$ the S-pairs (see Chapter 3 of [18]) are divisible by G:

$$\mathrm{sp}_\prec(f,E) = lx\partial_x f - y^l E = x^m E - my\partial_y f,$$
$$\mathrm{sp}_\prec(f,P) = l\partial_x f - yP = x^{m-1}E, \quad \mathrm{sp}_\prec(E,P) = y^{l-1}E - xP = m\partial_y f.$$

The initial monomials of the Gröbner basis G are $\mathrm{in}_\prec(f) = y^l$, $\mathrm{in}_\prec(E) = x\xi$, $\mathrm{in}_\prec(P) = y^{l-1}\xi$. Hence for $k \ge l$ we obtain

$$\begin{aligned}
&\dim_K F_k(D_X)/(\mathrm{Ann}_{D_X}[f^{-1}] \cap F_k(D_X)) \\
&= \sharp(\{x^i y^j \xi^\mu \eta^\nu \mid i+j+\mu+\nu \le k\} \setminus \langle y^l, x\xi, y^{l-1}\xi\rangle) \\
&= \sharp\{x^i y^j \eta^\nu \mid i+j+\nu \le k,\ j \le l-1\} \\
&\quad + \sharp\{y^j \xi^\mu \eta^\nu \mid j+\mu+\nu \le k,\ j \le l-2,\ \mu \ge 1\} \\
&= \sum_{j=0}^{l-1} \binom{2+k-j}{2} + \sum_{j=0}^{l-2} \binom{2+k-j-1}{2} = \frac{2l-1}{2}k^2 + \cdots .
\end{aligned}$$

On the other hand, in case $m = l$ we have

$$\begin{aligned}
&\mathrm{sp}_\prec(f,E) = l\partial_x f - x^{l-1}E = yP, \\
&\mathrm{sp}_\prec(f,P) = ly^{l-1}\partial_x f - x^l P = y^l P + lx^{l-1}\partial_y f \\
&\mathrm{sp}_\prec(E,P) = y^{l-1}E - xP = l\partial_y f.
\end{aligned}$$

The initial monomials are $\mathrm{in}_\prec(f) = x^l$, $\mathrm{in}_\prec(E) = x\xi$, $\mathrm{in}_\prec(P) = y^{l-1}\xi$. (Note that $y^{l-1}\xi \succ x^{l-1}\eta$ holds.) Hence for $k \geq l$ we obtain

$$\begin{aligned}
&\dim_K F_k(D_X)/(\mathrm{Ann}_{D_X}[f^{-1}] \cap F_k(D_X)) \\
&= \sharp(\{x^i y^j \xi^\mu \eta^\nu \mid i+j+\mu+\nu \leq k\} \setminus \langle x^l, x\xi, y^{l-1}\xi\rangle) \\
&= \sharp\{x^i y^j \eta^\nu \mid i+j+\nu \leq k,\ i \leq l-1\} \\
&\quad + \sharp\{y^j \xi^\mu \eta^\nu \mid j+\mu+\nu \leq k,\ j \leq l-2,\ \mu \geq 1\} \\
&= \sum_{i=0}^{l-1} \binom{2+k-i}{2} + \sum_{j=0}^{l-2} \binom{2+k-j-1}{2} = \frac{2l-1}{2}k^2 + \cdots.
\end{aligned}$$

Hence the multiplicity of M is $2l-1$ in both cases. This proves the assertion. Q.E.D.

Proposition 5.5. *Set $X = K^2$ with $x_1 = x$ and $x_2 = y$. Set $f = x^m + y^l + 1$ with positive integers l, m. Then the multiplicity of $K[x, y, f^{-1}]$ equals $lm + |l-m| + 1$.*

Proof. We may assume $m \leq l$. Set $M := H^1_{(f)}(K[x,y])$. Since the curve $f = 0$ is non-singular, the b-function is $b_f(s) = s+1$. Hence M is generated by $u := [f^{-1}]$. The annihilator $\mathrm{Ann}_{D_X} u$ is generated by f and $P := ly^{l-1}\partial_x - mx^{m-1}\partial_y$ since $f = 0$ is non-singular.

In case $l = m$, $G = \{f, P\}$ is a Gröbner basis of $\mathrm{Ann}_{D_X}[f^{-1}]$ with respect to a total-degree reverse lexicographic order $\prec$ such that $x \succ y \succ \xi \succ \eta$. In fact, we have

$$\mathrm{sp}_\prec(f, P) = ly^{l-1}\partial_x f - x^l P = ly^{l-1}\partial_x f + x^l P.$$

Since $\mathrm{in}_\prec(f) = x^l$ and $\mathrm{in}_\prec(P) = y^{l-1}\xi$, we have for $k \geq 2l$

$$\begin{aligned}
&\dim_K F_k(D_X)/(\mathrm{Ann}_{D_X}[f^{-1}] \cap F_k(D_X)) \\
&= \sharp(\{x^i y^j \xi^\mu \eta^\nu \mid i+j+\mu+\nu \leq N\} \setminus \langle x^l, y^{l-1}\xi\rangle) \\
&= \sharp\{x^i y^j \eta^\nu \mid i+j+\nu \leq k,\ i \leq l-1\} \\
&\quad + \sharp\{x^i y^j \xi^\mu \eta^\nu \mid i+j+\mu+\nu \leq k,\ i \leq l-1,\ 0 \leq j \leq l-2,\ \mu \geq 1\} \\
&= \sum_{i=0}^{l-1} \binom{2+k-i}{2} + \sum_{i=0}^{l-1}\sum_{j=0}^{l-2} \binom{2+k-i-j-1}{2} = \frac{l^2}{2}k^2 + \cdots.
\end{aligned}$$

In case $m < l$, the Gröbner basis of $\mathrm{Ann}_D[f^{-1}]$ with respect to the same order as above is $G = \{f, P, Q\}$ with

$$Q := l(x^m + 1)\partial_x + mx^{m-1}y\partial_y + mlx^{m-1}.$$

In fact, we have

$$\begin{aligned}
\mathrm{sp}_{\prec}(f,P) &= l\partial_x f - yP = Q,\\
\mathrm{sp}_{\prec}(f,Q) &= lx^m\partial_x f - y^l Q = -mx^{m-1}y\partial_y f - yP + x^m Q,\\
\mathrm{sp}_{\prec}(P,Q) &= x^m P - y^{l-1}Q = -mx^{m-1}\partial_y f - P.
\end{aligned}$$

Since $\mathrm{in}_{\prec}(f) = y^l$, $\mathrm{in}_{\prec}(P) = y^{l-1}\xi$, $\mathrm{in}_{\prec}(Q) = x^m\xi$, we have for $k \geq l+m$,

$$\begin{aligned}
&\dim_K F_k(D_X)/(\mathrm{Ann}_{D_X}[f^{-1}] \cap F_k(D_X))\\
&= \sharp(\{x^i y^j \xi^\mu \eta^\nu \mid i+j+\mu+\nu \leq k\} \setminus \langle y^l, y^{l-1}\xi, x^m\xi\rangle)\\
&= \sharp\{x^i y^j \eta^\nu \mid i+j+\nu \leq k,\ i \leq l-1\}\\
&\quad + \sharp\{x^i y^j \xi^\mu \eta^\nu \mid i+j+\mu+\nu \leq k,\ i \leq m-1,\ j \leq l-2,\ \mu \geq 1\}\\
&= \sum_{i=0}^{l-1}\binom{2+k-i}{2} + \sum_{i=0}^{m-1}\sum_{j=0}^{l-2}\binom{2+k-i-j-1}{2}\\
&= \frac{l+m(l-1)}{2}k^2 + \cdots .
\end{aligned}$$

Hence the multiplicity of M is $l + m(l-1) = ml + l - m$. Q.E.D.

Now let us resume the study on $M(u,f,s)$ for a D_X-module $M = D_X u$ and a polynomial f. As in the preceding section, set $Y = X \times K$ and $Z = \{(x,t) \in Y \mid t = f(x)\}$.

Lemma 5.6. *Let $M = D_X u$ be a left D_X-module generated by u. For any $\lambda \in K$, the endomorphism of $M(u,f,s)$ defined by $s-\lambda$ is injective. Hence the sequence*

$$0 \longrightarrow M(u,f,s) \xrightarrow{s-\lambda} M(u,f,s) \longrightarrow M(u,f,\lambda) \longrightarrow 0$$

of left D_X-modules is exact.

Proof. We may assume that M is f-saturated as was seen in the previous section. The homomorphism $\psi : M \otimes_{K[x]} B_{Z|Y} \to M \otimes_{K[x]} \mathcal{L}_f$ is injective by Lemma 4.4.

Hence we have only to show that $s - \lambda = -\partial_t t - \lambda$ is an injective endomorphism of $M \otimes_{K[x]} B_{Z|Y}$. Let

$$v = \sum_{j=0}^{k} v_j \otimes \delta^{(j)}(t-f)$$

be an arbitrary element of $M \otimes_{K[x]} B_{Z|Y}$ with $k \in \mathbb{N}$ and $v_j \in M$. Then one has

$$\begin{aligned}(s-\lambda)v &= -\sum_{j=0}^{k} v_j \otimes (t\partial_t + \lambda + 1)\delta^{(j)}(t-f) \\ &= -\sum_{j=0}^{k} v_j \otimes (f\delta^{(j+1)}(t-f) + (\lambda - j)\delta^{(j)}(t-f)) \\ &= -\lambda v_0 \otimes \delta(t-f) - \sum_{j=1}^{k}(fv_{j-1} + (\lambda - j)v_j) \otimes \delta^{(j)}(t-f) \\ &\quad - fv_k \otimes \delta^{(k+1)}(t-f).\end{aligned}$$

Thus $(s-\lambda)v = 0$ is equivalent to

$$\lambda v_0 = fv_k = fv_{j-1} + (\lambda - j)v_j = 0 \quad (1 \le j \le k),$$

which implies $v_k = v_{k-1} = \cdots = v_0 = 0$ since M is f-saturated. Q.E.D.

Theorem 5.7. *Let $f \in K[x]$ be a non-constant polynomial with K being algebraically closed. Let $M = D_X u$ be a left D_X-module generated by u which is holonomic on $X_f := \{x \in X \mid f(x) \neq 0\}$. Then $M(u, f, \lambda)$ and $M(u, f, s)/tM(u, f, s)$ are holonomic D_X-modules for any $\lambda \in K$.*

Proof. Since $M(u, f, s) = \iota(M)(\iota(u), f, s)$, we may assume M to be a nonzero holonomic D_X-module and f-saturated replacing M by $\iota(M)$. Set

$$N := M \otimes_{K[x]} B_{Z|Y}, \qquad F_k(N) := F_k(D_Y)\delta(t - f(x)) \qquad (k \in \mathbb{Z}).$$

Since N is holonomic, there exists a polynomial $p(k)$ of degree $n+1$ such that $p(k) = \dim_K F_k(N)$ for any sufficiently large k. Let us define a filtration $F_k(D_X[s])$ on the ring $D_X[s]$ by

$$F_k(D_X[s]) = \Big\{ \sum_{\alpha,\beta,j} a_{\alpha,\beta,j} x^\alpha \partial_x^\beta s^j \mid |\alpha| + |\beta| + 2j \le k \Big\}.$$

Set

$$F_k(M(u, f, s)) := F_k(D_X[s])(u \otimes f^s) \quad (k \in \mathbb{Z}).$$

Applying a well-known fact in commutative algebra (see e.g., Theorem 4.4.3 in [7]) to the graded module $\mathrm{gr}(M(u, f, s))$ over the graded ring $\mathrm{gr}(D_X[s])$, we can show that there exist two polynomials $q_1(k)$ and $q_2(k)$ of the same degree d such that

$$\dim_K F_{2k}(M(u, f, s)) = q_1(2k), \quad \dim_K F_{2k+1}(M(u, f, s)) = q_2(2k+1)$$

for any sufficiently large k. We have $d \leq n+1$ since $F_k(M(u,f,s))$ is contained in $\psi(F_k(N))$ and ψ is an injective homomorphism from $M \otimes_{K[x]} B_{Z|Y}$ to $M \otimes_{K[x]} \mathcal{L}_f$.

Set $\mathcal{M} = M(u,f,s)/tM(u,f,s)$ and

$$F_k(\mathcal{M}) = F_k(M(u,f,s))/(tM(u,f,s) \cap F_k(M(u,f,s)).$$

Then $\{F_k(\mathcal{M})\}$ is a Bernstein filtration on the left D_X-module $\mathcal{M}$ (i.e., the action of s being ignored) although we do not know at this stage whether $\mathcal{M}$ is finitely generated over D_X or not.

Since $t : M(u,f,s) \to M(u,f,s)$ is injective, we have

$$\begin{aligned}\dim_K F_k(\mathcal{M}) &= \dim_K F_k(M(u,f,s)) \\ &\qquad - \dim_K(tM(u,f,s) \cap F_k(M(u,f,s)) \\ &\leq \dim_K F_k(M(u,f,s)) - \dim_K t^2 F_{k-2}(M(u,f,s)) \\ &= \dim_K F_k(M(u,f,s)) - \dim_K F_{k-2}(M(u,f,s)) \\ &= \begin{cases} q_1(k) - q_1(k-2) & \text{if } k \gg 0 \text{ is even,} \\ q_2(k) - q_2(k-2) & \text{if } k \gg 0 \text{ is odd.} \end{cases}\end{aligned}$$

Since the degree of $q_i(k) - q_i(k-2)$ $(i = 1,2)$ is $d-1 \leq n$, this inequality implies that an arbitrary finitely generated D_X-submodule of $\mathcal{M}$ is holonomic and its multiplicity is bounded in terms of the leading coefficients of $q_1(k)$ and $q_2(k)$. Hence we conclude that $\mathcal{M}$ itself is holonomic.

We can prove the holonomicity of $M(u,f,\lambda)$ in the same way replacing t^2 by $s - \lambda$. This fact is a special case of Theorem 6.10 in [18]. Q.E.D.

The first statement of the following theorem is given in 6.5 of [14] for the case $M = K[x]$ and $u = 1$.

Theorem 5.8. *Let $M = D_X u$ be a D_X-module generated by u and $f \in K[x]$ be a non-constant polynomial. Assume that the b-function $b_{u,f}(s)$ exists. For $\lambda \in K$ let $\varphi_\lambda : M(u,f,\lambda+1) \to M(u,f,\lambda)$ be the D_X-homomorphism induced by t, which sends $(u \otimes f^s)|_{s=\lambda+1}$ to $(fu \otimes f^s)|_{s=\lambda}$.*

(1) *φ_λ is an isomorphism if and only if $b_{u,f}(\lambda) \neq 0$.*

(2) *Assume that M is holonomic on X_f with K being algebraically closed. Then one has*

$$\begin{aligned}\operatorname{mult} M(u,f,\lambda+k) &= \operatorname{mult} M(u,f,\lambda), \\ \operatorname{length} M(u,f,\lambda+k) &= \operatorname{length} M(u,f,\lambda)\end{aligned}$$

for any $\lambda \in K$ and any integer k. In particular, one has

$$\operatorname{mult} M[f^{-1}] = \operatorname{mult} M(u,f,k), \qquad \operatorname{length} M[f^{-1}] = \operatorname{length} M(u,f,k)$$

for any integer k.

Proof. There exists a commutative diagram

$$
\begin{array}{ccccccccc}
 & & & & & & 0 & & \\
 & & & & & & \downarrow & & \\
 & & 0 & & 0 & & \mathcal{K}_0(\lambda) & & \\
 & & \downarrow & & \downarrow & & \downarrow & & \\
0 & \longrightarrow & M(u,f,s) & \xrightarrow{t} & M(u,f,s) & \longrightarrow & \mathcal{M} & \longrightarrow & 0 \\
 & & \downarrow{\scriptstyle s-\lambda-1} & & \downarrow{\scriptstyle s-\lambda} & & \downarrow{\scriptstyle s-\lambda} & & \\
0 & \longrightarrow & M(u,f,s) & \xrightarrow{t} & M(u,f,s) & \longrightarrow & \mathcal{M} & \longrightarrow & 0 \\
 & & \downarrow & & \downarrow & & \downarrow & & \\
 & & M(u,f,\lambda+1) & \xrightarrow{\varphi_\lambda} & M(u,f,\lambda) & \longrightarrow & \mathcal{K}_1(\lambda) & & \\
 & & \downarrow & & \downarrow & & \downarrow & & \\
 & & 0 & & 0 & & 0 & &
\end{array}
$$

with $\mathcal{M} = M(u, f, s)/tM(u, f, s)$ and some left D_X-modules $\mathcal{K}_0(\lambda)$, $\mathcal{K}_1(\lambda)$, where the three vertical sequences and the upper two horizontal sequences are exact in view of Lemma 5.6. Hence by the snake lemma we obtain an exact sequence

$$(3) \quad 0 \longrightarrow \mathcal{K}_0(\lambda) \longrightarrow M(u, f, \lambda + 1) \xrightarrow{\varphi_\lambda} M(u, f, \lambda) \longrightarrow \mathcal{K}_1(\lambda) \longrightarrow 0$$

of left D_X-modules.

(1) Assume $b_{u,f}(\lambda) \neq 0$. Then there exist $a(s), c(s) \in K[s]$ such that $a(s)(s - \lambda) + c(s)b_{u,f}(s) = 1$. Hence for any $Q(s) \in D_X[s]$,

$$Q(s)(u \otimes f^s) = Q(s)c(s)b_{u,f}(s)(u \otimes f^s) + (s - \lambda)Q(s)a(s)(u \otimes f^s)$$

belongs to $tM(u, f, s) + (s - \lambda)M(u, f, s)$. If $(s - \lambda)Q(s)(u \otimes f^s)$ belongs to $tM(u, f, s)$, then $Q(s)(u \otimes f^s)$ also belongs to $tM(u, f, s)$. Hence $s - \lambda$ is an automorphism of $\mathcal{M}$.

Conversely, assume that $s - \lambda$ is an automorphism of $\mathcal{M}$. Then the minimal polynomial $b_{u,f}(s)$ of s on this module cannot be a multiple of $s - \lambda$. Summing up we have shown that $b_{u,f}(\lambda) \neq 0$ if and only if $\mathcal{K}_0(\lambda) = \mathcal{K}_1(\lambda) = 0$. In view of the exact sequence (3), this is also equivalent to φ_λ being an isomorphism.

(2) We may assume that M is a holonomic D_X-module and that M is f-saturated replacing M by $\iota(M)$. Since $\mathcal{M}$ is holonomic by Theorem 5.7, the length (and the multiplicity) of $\mathcal{K}_0(\lambda)$ and the length (and the multiplicity respectively) of $\mathcal{K}_1(\lambda)$ are the same in view of the rightmost vertical exact sequence. Combined with this fact the exact sequence (3) proves the statement (2). Q.E.D.

This theorem provides us with an algorithm to compute the multiplicity of $M[f^{-1}]$ without any information on $b_{u,f}(s)$; we have only to compute a Gröbner basis, e.g., of $M(u,f,0)$ with respect to a term order compatible with the Bernstein filtration.

Theorem 5.9. *Let $M = D_X u$ be a D_X-module generated by u and $f \in K[x]$ be a non-constant polynomial with K being algebraically closed. Assume that M is holonomic on X_f. Then the homomorphism $\tilde{\rho}_\lambda$: $M(u,f,\lambda) \to D_X(u \otimes f^\lambda)$ is an isomorphism if and only if $b_{u,f}(\lambda - k) \neq 0$ for any positive integer k.*

Proof. If $b_{u,f}(\lambda - k) \neq 0$ for any positive integer k, then $\tilde{\rho}_\lambda$ is an isomorphism by virtue of Proposition 2.6. Now suppose $b_{u,f}(\lambda - k) = 0$ holds for some positive integer k and let k_0 be the maximum among such k. Then Proposition 2.6 and Lemma 2.7 imply that $\tilde{\rho}_{\lambda-k_0}$ is an isomorphism and that $D_X(u \otimes f^{\lambda-k_0+1}) \subsetneq D_X(u \otimes f^{\lambda-k_0})$. Hence by (2) of Theorem 5.8 we have

$$\begin{aligned}\operatorname{length} M(u,f,\lambda) &= \operatorname{length} M(u,f,\lambda-k_0) = \operatorname{length} D_X(u \otimes f^{\lambda-k_0}) \\ &> \operatorname{length} D_X(u \otimes f^{\lambda-k_0+1}) \geq \operatorname{length} D_X(u \otimes f^\lambda).\end{aligned}$$

Thus $\tilde{\rho}_\lambda$ is not an isomorphism. Q.E.D.

Corollary 5.10. *Under the same assumptions as in Theorem 5.9, $M(u,f,\lambda)$ is f-saturated if and only if $b_{u,f}(\lambda - k) \neq 0$ for any positive integer k. In general, $\iota(M(u,f,\lambda))$ is isomorphic to $D_X(u \otimes f^\lambda)$.*

Proof. We may assume M to be f-saturated. First note that $M \otimes_{K[x]} \mathcal{L}_f(\lambda)$ is f-saturated for any $\lambda \in K$ since it is isomorphic to $M[f^{-1}]$ as $K[x]$-module. Hence $M(u,f,\lambda) \cong D_X(u \otimes f^\lambda)$ is also f-saturated under the assumption on $b_{u,f}(s)$.

Now assume $b_{u,f}(\lambda - k) = 0$ for some positive integer k. Then $\tilde{\rho}_\lambda$ is not injective. Thus there exists $P \in D_X$ such that $P((u \otimes f^s)|_{s=\lambda}) \neq 0$ but $P(u \otimes f^\lambda) = 0$. There exist $P_j \in D_X$ and $k, l \in \mathbb{N}$ such that

$$P(u \otimes f^s) = \sum_{j=0}^{k} (P_j u) \otimes (s-\lambda)^j f^{s-l}.$$

Then the equality $P(u \otimes u^\lambda) = 0$ means $P_0 u = 0$. Hence by Lemma 2.5, there exist $Q(s) \in D_X[s]$ and $m \in \mathbb{N}$ such that

$$P(u \otimes f^s) = (s - \lambda)Q(s)(u \otimes f^{s-m}).$$

Take a sufficiently large $l \in \mathbb{N}$ so that $f^l Q(s) f^{-m}$ belongs to $D_X[s]$. Then we have

$$f^l P(u \otimes f^s) = (s - \lambda)(f^l Q(s) f^{-m})(u \otimes f^s),$$

and consequently $f^l P((u \otimes f^s)|_{s=\lambda}) = 0$. Hence $M(u, f, \lambda)$ is not f-saturated. The last statement also follows from this argument. Q.E.D.

Example 5.11. Set $n = 2$ and write $x_1 = x$, $x_2 = y$. Let u be the residue class of 1 in $M = D_X/I$ with I being the left ideal of D_X generated by two operators

$$P_1 = x(1-x)\partial_x{}^2 + y(1-x)\partial_x\partial_y, \quad P_2 = y(1-y)\partial_y{}^2 + x(1-y)\partial_x\partial_y.$$

This is Appell's hypergeometric system F_1 with all parameters equal to zero. The singular locus of M is a line arrangement defined by

$$f := x(x-1)y(y-1)(x-y) = 0.$$

Let $\iota : M \to M[f^{-1}]$ be the canonical homomorphism. Then $M[f^{-1}]$ is generated by $f^{-2}\iota(u)$ and $\iota(M)$ is given by

$$\begin{aligned}\iota(M) &= D_X \iota(u) \\ &= D_X/(D_X\partial_x\partial_y + D_X((1-x)\partial_x{}^2 - \partial_x) + D_X((1-y)\partial_y{}^2 - \partial_y)).\end{aligned}$$

The b-function with respect to u and f is

$$b_{u,f}(s) = (s+1)^3(s+2)^2\left(s+\frac{2}{3}\right)^2\left(s+\frac{4}{3}\right)^2\left(s+\frac{5}{3}\right).$$

As to multiplicities we have

$$\begin{aligned}&\operatorname{mult} M = 10, \quad \operatorname{mult}\iota(M) = 5, \quad \operatorname{mult} M[f^{-1}] = 36, \\ &\operatorname{mult} H^1_{(f)}(M) = 31, \quad \operatorname{mult} H^0_{(f)}(M) = 5.\end{aligned}$$

The characteristic varieties are

$$\begin{aligned}
&\mathrm{Char}(M) = \{(x,y;\xi,\eta) \mid x = y = 0\} \cup \{x = y = 1\} \cup \{x = \eta = 0\}\\
&\quad \cup \{x-1 = \eta = 0\} \cup \{y = \xi = 0\} \cup \{y-1 = \xi = 0\}\\
&\quad \cup \{x-y = \xi+\eta = 0\} \cup \{\xi = \eta = 0\},\\
&\mathrm{Char}(\iota(M)) = \{x-1 = \eta = 0\} \cup \{y-1 = \xi = 0\} \cup \{\xi = \eta = 0\},\\
&\mathrm{Char}(M[f^{-1}]) = \{x = y = 0\} \cup \{x-1 = y = 0\} \cup \{x = y-1 = 0\}\\
&\quad \cup \{x = y = 1\} \cup \{x = \eta = 0\} \cup \{x-1 = \eta = 0\}\\
&\quad \cup \{y = \xi = 0\} \cup \{y-1 = \xi = 0\} \cup \{x-y = \xi+\eta = 0\}\\
&\quad \cup \{\xi = \eta = 0\},\\
&\mathrm{Char}(H^1_{(f)}(M)) = \{x = y = 0\} \cup \{x-1 = y = 0\} \cup \{x = y-1 = 0\}\\
&\quad \cup \{x = y = 1\} \cup \{x = \eta = 0\} \cup \{x-1 = \eta = 0\}\\
&\quad \cup \{y = \xi = 0\} \cup \{y-1 = \xi = 0\} \cup \{x-y = \xi+\eta = 0\},\\
&\mathrm{Char}(H^0_{(f)}(M)) = \{x = y = 0\} \cup \{x = y = 1\} \cup \{x = \eta = 0\}\\
&\quad \cup \{y = \xi = 0\} \cup \{x-y = \xi+\eta = 0\}.
\end{aligned}$$

Example 5.12. Set $X = K^4$ and consider the A-hypergeometric system associated with the matrix $A = \begin{pmatrix} 1 & 1 & 1 & 1 \\ 0 & 1 & 3 & 4 \end{pmatrix}$, which is taken from Example 4.3.9 of [23]. It is the left D_X-module $M_A(b_1,b_2) = D_X/H_A(b_1,b_2)$ with the left ideal $H_A(b_1,b_2)$ of D_X generated by operators

$$\begin{aligned}
&x_1\partial_1 + x_2\partial_2 + x_3\partial_3 + x_4\partial_4 - b_1, \quad x_2\partial_2 + 3x_3\partial_3 + 4x_4\partial_4 - b_2,\\
&\partial_2\partial_4^2 - \partial_3^3, \quad \partial_1\partial_4 - \partial_2\partial_3, \quad \partial_2^2\partial_4 - \partial_1\partial_3^2, \quad \partial_1^2\partial_3 - \partial_2^3
\end{aligned}$$

with parameters $b_1, b_2 \in K$. Computing a Gröbner basis of the left ideal of $D_X[b_1,b_2]$ (i.e., regarding b_1 and b_2 as indeterminates) generated by these operators with respect to a term order $\prec$ such that

$$|\alpha| + |\beta| < |\alpha'| + |\beta'| \quad \Rightarrow \quad b_1^i b_2^j x^\alpha \partial_x^\beta \prec b_1^k b_2^l x^{\alpha'} \partial_x^{\beta'}$$

and that $b_1^i b_2^j \preceq x^\alpha \partial_x^\beta$ for any i, j, α, β, we can verify that the multiplicity of $M_A(b_1,b_2)$ is 16 unless $b_1 = 1$ and $b_2 = 2$, while the multiplicity of $M_A(1,2)$ is 17. A similar phenomenon with respect to the holonomic rank was shown in [23].

On the other hand, the characteristic variety of $M_A(b_1, b_2)$ does not depend on b_1, b_2 and its singular locus is the zero set of

$$g(x) = x_1x_4(256x_1^3x_4^3 + (-192x_1^2x_2x_3 - 27x_2^4)x_4^2 - 6x_1x_2^2x_3^2x_4 - 27x_1^2x_3^4 - 4x_2^3x_3^3).$$

For example, the b-functions of $u := \overline{1} \in M_A(1,2)$ and x_1 is $b_{u,x_1}(s) = s(s+1)(s+2)$, while the b-functions of $v := \overline{1} \in M_A(0,0)$ and x_1 is $b_{v,x_1}(s) = (s+1)^2$. Algorithm 3.16 ensures that $M_A(0,0)$ and $M_A(1,2)$ are x_1- and x_4-saturated. The computation of the localization with respect to g is intractable.

§6. Hyperplane arrangements

Let us prove a formula on the multiplicity and the length of the local cohomology $H^1_{(f)}(K[x])$ or the localization $K[x, f^{-1}]$ of the polynomial ring when f defines a hyperplane arrangement in the affine space $X = K^n$. We set $R = K[x]$ in what follows.

The length of such modules was studied e.g., in [1], [27]. The characteristic cycle of the local cohomology with respect to an arrangement of linear subvarieties was studied in [2]. Although not explicitly stated, Corollary 1.3 of [2] should yield main results of this section. Nero Budur informed the author that Theorem 6.4 below follows from results in Section 1.7 of [8]. We give an elementary direct proof with a hope to make both the statement and the proof more accessible.

Lemma 6.1. *Let $h_0 = h_0(x) \in K[x]$ be a linear polynomial and I be an ideal of $R = K[x]$. Let $R' := R/Rh_0$ be the affine ring associated with the hyperplane $h_0(x) = 0$ and set $I' = (I + Rh_0)/Rh_0$. Then we have*

$$\operatorname{length} H^i_{I+Rh_0}(R) = \operatorname{length} H^{i-1}_{I'}(R'),$$
$$\operatorname{mult} H^i_{I+Rh_0}(R) = \operatorname{mult} H^{i-1}_{I'}(R')$$

for any integer i.

Proof. Since $H^i_{Rh_0}(R) = 0$ for $i \neq 1$, there is an isomorphism

$$H^i_{I+Rh_0}(R) \cong H^{i-1}_{I+Rh_0}(H^1_{Rh_0}(R)).$$

We may assume by an affine coordinate transformation, which preserves the Bernstein filtration, that $h_0(x) = x_n$. Then we may regard $R' = K[x_1, \dots, x_{n-1}]$ and have an isomorphism

$$H^1_{Rh_0}(R) \cong R' \otimes_K H^1_{(x_n)}(K[x_n]),$$

where the tensor product on the right-hand side is a left module over $D_n = D_{n-1} \otimes_K D_1$ with D_1 being the ring of differential operators in the variable x_n.

Let $\{f_1, \dots, f_r\}$ be a set of generators of I. We may assume that $f_1, \dots, f_r$ belong to R'. (We replace f_i with its homomorphic image in R'.) Then for $0 \leq i_1 < \cdots < i_k \leq r$ with $k \in \mathbb{N}$, the localization by $f_{i_1} \cdots f_{i_k}$ yields

$$\begin{aligned}(H^1_{Rh_0}(R))_{f_{i_1}\cdots f_{i_k}} &:= H^1_{Rh_0}(R)[(f_{i_1} \cdots f_{i_k})^{-1}] \\ &= R'_{f_{i_1}\cdots f_{i_k}} \otimes_K H^1_{(x_n)}(K[x_n]).\end{aligned}$$

On the other hand, we have

$$(H^1_{Rh_0}(R))_{x_n} := H^1_{Rh_0}(R)[x_n^{-1}] = R' \otimes_K (H^1_{(x_n)}(K[x_n]))_{x_n} = 0.$$

Hence $H^{i-1}_{I+Rh_0}(H^1_{Rh_0}(R))$ is the $(i-1)$-th cohomology group of the Čech complex

$$\begin{aligned}0 &\longrightarrow R' \otimes_K H^1_{(x_n)}(K[x_n]) \longrightarrow \bigoplus_{1\leq i\leq r} R'_{f_i} \otimes_K H^1_{(x_n)}(K[x_n]) \\ &\longrightarrow \bigoplus_{1\leq i_1<i_2\leq r} R'_{f_{i_1} f_{i_2}} \otimes_K H^1_{(x_n)}(K[x_n]) \longrightarrow \\ &\cdots \longrightarrow R'_{f_1\cdots f_r} \otimes_K H^1_{(x_n)}(K[x_n]) \longrightarrow 0,\end{aligned}$$

which is isomorphic to $H^{i-1}_{I'}(R') \otimes_K H^1_{(x_n)}(K[x_n])$ (see e.g., Theorem 7.13 in [11]). This implies

$$\begin{aligned}H^i_{I+Rh_0}(R) &\cong H^{i-1}_{I'}(R') \otimes_K H^1_{(x_n)}(K[x_n]) \\ &\cong H^{i-1}_{I'}(R') \otimes_K (D_1/D_1 x_n) \\ &\cong (D_n/D_n x_n) \otimes_{D_{n-1}} H^{i-1}_{I'}(R'),\end{aligned}$$

where $D_n/D_n x_n$ is regarded as a (D_n, D_{n-1})-bimodule. The rightmost term is the D-module theoretic direct image of $H^{i-1}_{I'}(R')$ with respect to the inclusion $H_0 := \{x \in X \mid x_n = 0\} \to X$. In view of Kashiwara's equivalence in the category of algebraic D-modules (see e.g., Theorem 7.11 of [6] or Theorem 1.6.1 of [10]), there is a one-to-one correspondence between the D_{n-1}-submodules M of $H^{i-1}_{I'}(R')$ and the D_n-submodules $M \otimes_K H^1_{(x_n)}(K[x_n])$ of $H^i_{I+Rh_0}(R)$. This implies

$$\operatorname{length} H^i_{I+Rh_0}(R) = \operatorname{length} H^{i-1}_{I'}(R').$$

Next, let us show

$$\operatorname{mult} H^i_{I+Rh_0}(R) = \operatorname{mult} H^{i-1}_{I'}(R').$$

Let $\{F_k\}$ be a good Bernstein filtration on $H^{i-1}_{I'}(R')$ and set $m = \operatorname{mult} H^{i-1}_{I'}(R')$. Then there exists a polynomial $p(k)$ and $k_1 \in \mathbb{Z}$ such that

$$\dim_K F_k = p(k) = \frac{m}{(n-1)!}k^{n-1} + (\text{terms with degree} < n-1)$$

holds for $k \geq k_1$. Define a filtration $\{G_k\}$ on $H^{i-1}_{I'}(R') \otimes_K H^1_{(x_n)}(K[x_n])$ by

$$G_k := \sum_{j=0}^{k} F_j \otimes_K (K[x_n^{-1}] + \cdots + K[x_n^{-(k-j)-1}]) = \bigoplus_{j=0}^{k} F_j \otimes_K K[x_n^{-(k-j)-1}],$$

where $K[x_n^{-j}]$ denotes the K-space spanned by the residue class of x_n^{-j}. It is easy to see that $\{G_k\}$ is a good Bernstein filtration. Hence we have

$$\dim_K G_k = \sum_{j=0}^{k} \dim_K F_j = \sum_{j=0}^{k_1-1} \dim_K F_j + \sum_{j=k_1}^{k} p(j).$$

By the assumption, there exists a polynomial $q(k)$ of degree $\leq n-2$ such that

$$p(j) = \frac{m}{(n-1)!} j(j+1)\cdots(j+n-2) + q(j).$$

Since

$$\sum_{j=k_1}^{k} j(j+1)\cdots(j+n-2)$$
$$= \frac{1}{n}\{k(k+1)\cdots(k+n-1) - (k_1-1)k_1\cdots(k_1+n-2)\},$$

we have

$$\dim_K G_k = \frac{m}{n!}k^n + (\text{terms with degree} < n) \qquad (\forall k \geq k_1).$$

Thus we also have $\operatorname{mult} H^i_{I+Rx_n}(R) = m$. This completes the proof. Q.E.D.

Theorem 6.2. *Let $f \in K[x]$ be a multiple of essentially distinct linear polynomials and $\mathcal{A}$ be the hyperplane arrangement in $X = K^n$ defined by f. Let H_0 be an element of $\mathcal{A}$. Set $\mathcal{A}' := \mathcal{A} \setminus \{H_0\}$ and let f' be the product of the defining polynomials of hyperplanes belonging to $\mathcal{A}'$. Let us regard*

$$\mathcal{A}'' := \{H \cap H_0 \mid H \in \mathcal{A}',\ H \cap H_0 \neq \emptyset\}$$

as a hyperplane arrangement in the affine space H_0. Let $R' = R/Rh_0$ be the affine ring of H_0, where h_0 is a polynomial of first degree defining H_0. Let $f'' \in R'$ be the product of the defining polynomials of the elements of $\mathcal{A}''$. (Set $f'' = 1$ if $\mathcal{A}'' = \emptyset$.) Then one has

$$\begin{aligned} \operatorname{length} H^1_{(f)}(R) &= \operatorname{length} H^1_{(f')}(R) + \operatorname{length} H^1_{(f'')}(R') + 1, \\ \operatorname{mult} H^1_{(f)}(R) &= \operatorname{mult} H^1_{(f')}(R) + \operatorname{mult} H^1_{(f'')}(R') + 1. \end{aligned}$$

Proof. By the Mayer-Vietoris exact sequence (see e.g., Theorem 15.1 in [11]), we get an exact sequence

$$0 \longrightarrow H^1_{(f')}(R) \oplus H^1_{(h_0)}(R) \longrightarrow H^1_{(f)}(R) \longrightarrow H^2_{(f')+(h_0)}(R) \longrightarrow 0$$

of holonomic left D_n-modules because $H^1_{(f')+(h_0)}(R) = 0$. Since the length and the multiplicity of $H^1_{(h_0)}(R)$ are both one, it follows that

$$\begin{aligned} \operatorname{length} H^1_{(f)}(R) &= \operatorname{length} H^1_{(f')}(R) + \operatorname{length} H^2_{(f')+(h_0)}(R) + 1, \\ \operatorname{mult} H^1_{(f)}(R) &= \operatorname{mult} H^1_{(f')}(R) + \operatorname{mult} H^2_{(f')+(h_0)}(R) + 1. \end{aligned} \tag{4}$$

Since $(f'') = R'f'' \cong (Rf' + Rh_0)/Rh_0$, Lemma 6.1 implies

$$\begin{aligned} \operatorname{mult} H^2_{(f')+(h_0)}(R) &= \operatorname{mult} H^1_{(f'')}(R'), \\ \operatorname{length} H^2_{(f')+(h_0)}(R) &= \operatorname{length} H^1_{(f'')}(R'). \end{aligned}$$

This completes the proof in view of (4). Q.E.D.

Corollary 6.3. $\operatorname{length} H^1_{(f)}(R) = \operatorname{mult} H^1_{(f)}(R)$.

Proof. This can be easily proved by induction on $\sharp\mathcal{A}$ by using Theorem 6.2. Q.E.D.

The intersection poset $L(\mathcal{A})$ is the set of the non-empty intersections of elements of $\mathcal{A}$ including X. For $Y, Z \in L(\mathcal{A})$, the Möbius function $\mu(Y, Z)$ is defined recursively by

$$\mu(Y,Z) = \begin{cases} -\displaystyle\sum_{Z \subsetneq W \subset Y} \mu(Y,W) & \text{if } Z \subsetneq Y \\ 1 & \text{if } Z = Y \\ 0 & \text{otherwise.} \end{cases}$$

Set $\mu(Y) = \mu(X, Y)$. Then $(-1)^{\operatorname{codim} X}\mu(X)$ is positive (see e.g. Theorem 2.47 of [22]). The Poincaré polynomial of the arrangement $\mathcal{A}$ is defined by

$$\pi(\mathcal{A}, t) = \sum_{Y \in L(\mathcal{A})} \mu(Y)(-t)^{\operatorname{codim} Y}.$$

Theorem 6.4. *Let $\mathcal{A}$ be a hyperplane arrangement in $X = K^n$ defined by a polynomial $f \in R = K[x]$. Then the length of $H^1_{(f)}(R)$ is $\pi(\mathcal{A}, 1) - 1$.*

Proof. Let H_0 be an element of $\mathcal{A}$ defined by a first degree polynomial h_0. Let us prove the equality by induction on $\sharp\mathcal{A}$. Since $H_{(h_0)}(R)$ is a simple left D_n-module and $\pi(\{H_0\}, t) = t + 1$, the equality holds if $\mathcal{A} = \{H_0\}$. Let $\mathcal{A}'$, $\mathcal{A}''$ be as in the proof of Theorem 6.2.

By the induction hypothesis, we have

$$\operatorname{length} H^1_{(f')}(R) = \pi(\mathcal{A}', 1) - 1, \qquad \operatorname{length} H^1_{(f'')}(R') = \pi(\mathcal{A}'', 1) - 1.$$

Hence by Theorem 6.2 we get

$$\begin{aligned}\operatorname{length} H^1_{(f)}(R) &= \operatorname{length} H^1_{(f')}(R) + \operatorname{length} H^1_{(f'')}(R') + 1 \\ &= \pi(\mathcal{A}', 1) + \pi(\mathcal{A}'', 1) - 1.\end{aligned}$$

On the other hand, $\pi(\mathcal{A}, t) = \pi(\mathcal{A}', t) + t\pi(\mathcal{A}'', t)$ holds (see e.g., Theorem 2.56 of [22]). Thus we get

$$\operatorname{length} H^1_{(f)}(R) = \pi(\mathcal{A}', 1) + \pi(\mathcal{A}'', 1) - 1 = \pi(\mathcal{A}, 1) - 1.$$

This completes the proof. Q.E.D.

Corollary 6.5. *Let $\mathcal{A}$ be a hyperplane arrangement in $X = K^n$ defined by a polynomial $f \in R = K[x]$. Then the length of $R[f^{-1}]$ is $\pi(\mathcal{A}, 1)$.*

Actual computation can be done effectively by using the recursive formula of Theorem 6.2. For example, we have $\pi(\mathcal{A}, t) = (2t+1)(3t+1)$ and hence $\pi(\mathcal{A}, 1) = 12$ if $f = xy(x-1)(y-1)(x-y)$ with $X = K^2$. If

$$f = xyz(x+y)(x-y)(x+z)(x-z)(y+z)(y-z)$$

with $X = K^3$, then we have $\pi(\mathcal{A}, t) = (t+1)(3t+1)(5t+1)$ and hence $\pi(\mathcal{A}, 1) = 48$. For these relatively small examples, direct computation of the local cohomology group is also possible.

References

[1] Abebaw, T., Bøgvad, R., Decomposition factors of D-modules on hyperplane configurations in general position, *Proc. Amer. Math. Soc.* **140** (2012), 2699–2711.

[2] Àlvarez Montaner, J., García López, R., Zarzuela Armengou, S., Local cohomology, arrangements of subspaces and monomial ideals, *Adv. Math.* **174** (2003), 35–56.

[3] Bernstein, I. N., Modules over a ring of differential operators: study of the fundamental solutions of equations with constant coefficients, *Funct. Anal. Appl.* **5** (1971), 1–16.

[4] Bernstein, I., N., The analytic continuation of generalized functions with respect to a parameter, *Funct. Anal. Appl.* **6** (1972), 26–40.

[5] Björk, J.-E., "Analytic D-modules and Applications", Kluwer Academic Publishers, Dordrecht–Boston–London, 1993, xi+581 pp., ISBN: 0-7923-2114-6.

[6] Borel, A., et al., "Algebraic D-modules", Academic Press, Boston, 1987, xii+355 pp., ISBN: 0-12-117740-8.

[7] Bruns, W., Herzog, J., "Cohen-Macaulay Rings" (revised version), Cambridge University Press, Cambridge, 1998, xiv+453 pp., ISBN: 0-521-56674-6.

[8] Budur, N., Saito, M., Jumping coefficients and spectrum of a hyperplane arrangement, *Math. Ann.* **347** (2010), 545–579.

[9] Cox, D., Little, J., O'Shea, D., "Using Algebraic Geometry", Springer-Verlag, New York–Berlin–Heidelberg, 1998, xii+499 pp., ISBN: 0-387-98487-9.

[10] Hotta, R., Takeuchi, K., Tanisaki, T., "D-modules, Perverse Sheaves, and Representation Theory", Birkhauser, Boston-Basel-Berlin, 2008, x+407 pp., ISBN: 978-0-8176-4363-8.

[11] Iyengar, S. B., et al., "Twenty-Four Hours of Local Cohomology", American Mathematical Society, Providence, RI, 2007, xi+282 pp., ISBN: 978-0-8218-4126-6.

[12] Kashiwara, M., B-functions and holonomic systems—rationality of roots of b-functions, *Invent. Math.* **38** (1976), 33–53.

[13] Kashiwara, M., On the holonomic systems of linear differential equations, II, *Invent. Math.* **49** (1978), 121–135.

[14] Kashiwara, M., "D-modules and Microlocal Calculus", translated from the Japanese by M. Saito, American Mathematical Society, Providence, RI, 2003, xvi+254 pp., ISBN: 0-8218-2766-9.

[15] Malgrange, B., Le polynôme de Bernstein d'une singularité isolée, Lecture Notes in Math., Vol. 459, Springer, Berlin, pp. 98–119, 1975.

[16] Noro, M., Takayama, N., Nakayama, H., Nishiyama, K., Ohara, K, Risa/Asir: a computer algebra system, http://www.math.kobe-u.ac.jp/Asir/asir.html.

[17] Oaku, T., Algorithms for b-functions, restrictions, and algebraic local cohomology groups of D-modules, *Adv. in Appl. Math.* **19** (1997), 61–105.

[18] Oaku, T., Algorithms for D-modules, integration, and generalized functions with applications to statistics, Adv. Stud. Pure Math. **77** (2018), pp. 253–352.

[19] Oaku, T., Takayama, N., An algorithm for de Rham cohomology groups of the complement of an affine variety, *J. Pure Appl. Algebra* **139** (1999), 201–233.

[20] Oaku, T., Takayama, N., Algorithms for D-modules — restriction, tensor product, localization, and local cohomology groups, *J. Pure Appl. Algebra* **156** (2001), 267–308.

[21] Oaku, T,. Takayama, N., Walther, U., A localization algorithm for D-modules, *J. Symbolic Comput.* **29** (2000), 721–728.

[22] Orlik, P., Terao, H., "Arrangements of Hyperplanes", Springer-Verlag, Berlin-Heidelberg, 1992, xiii+325 pp., ISBN: 3-540-55259-6.

[23] Saito, M., Sturmfels, B., Takayama, N., "Gröbner Deformations of Hypergeometric Differential Equations", Springer-Verlag, Berlin, 2000, viii+254 pp., ISBN: 3-540-66065-8.

[24] Torrelli, T., Polynômes de Bernstein associés à une fonction sur une intersection complète à singularité isolée, *Ann. Inst. Fourier* (*Grenoble*) **52** (2002), 221–244.

[25] Tsai, H., Algorithms for associated primes, Weyl closure, and local cohomology of D-modules, in 'Local cohomology and its applications' edited by G. Lyubeznik, Marcel Dekker, Inc., New York, Basel, 2002, pp. 169–194.

[26] Walther, U., Algorithmic computation of local cohomology modules and the local cohomological dimension of algebraic varieties, *J. Pure Appl. Algebra* **139** (1999), 303–321.

[27] Walther, U., Bernstein-Sato polynomial versus cohomology of the Milnor fiber for generic hyperplane arrangements, *Compos. Math.* **141** (2005), 121–145.

Department of Mathematics, Tokyo Woman's Christian University,
Suginami-ku, Tokyo 167-8585, Japan
E-mail address: oaku@lab.twcu.ac.jp

Advanced Studies in Pure Mathematics 77, 2018
The 50th Anniversary of Gröbner Bases
pp. 399–409

A characterization of Gorenstein planar graphs

Tran Nam Trung

Abstract.

We graph-theoretically classify all Gorenstein planar graphs.

§1. Introduction

Let $R = k[x_1, \dots, x_n]$ be a polynomial ring over a field k. Let Δ be a *simplicial complex* on the vertex set $\{1, \dots, n\}$ and I_Δ the *Stanley-Reisner* ideal of Δ in R. We say that Δ is *Cohen-Macaulay* (resp. *Gorenstein*) over k if so is R/I_Δ. One of the main problems is to characterize the Cohen-Macaulay and the Gorenstein property of Δ from the combinatorial data of Δ (see e.g. [13]). For example, Δ is Gorenstein whenever its geometric realization, denoted by $|\Delta|$, is isomorphic to a *sphere* according to [13, Corollary 5.2]. In general, these properties depend on not only the structure of Δ but also the characteristic of k (see [3, 12]).

In this paper we are concerned only with *flag complexes*. A simplicial complex is a flag complex if all of its minimal non-faces are two element sets. Flag complexes are closely related to graphs.

Let G be a simple graph with the vertex set $V(G) = \{1, \dots, n\}$ and the edge set $E(G)$. We associate to G a quadratic squarefree monomial ideal

$$I(G) = (x_i x_j \mid ij \in E(G)) \subset R,$$

which is called the *edge ideal* of G. An *independent set* in G is a set of vertices no two of which are adjacent to each other. The *independence complex* of G, denoted by $\Delta(G)$, is the set of independent sets of G. It is well-known that $I_{\Delta(G)} = I(G)$, so that $\Delta(G)$ is a flag complex. Conversely, any flag complex is the independence complex of some graph.

We say that G is Cohen-Macaulay (resp. Gorenstein) over k if so is $R/I(G)$; G is *well-covered* if very maximal independent set of G has the

Received March 12, 2016.
Revised February 27, 2017.
2010 *Mathematics Subject Classification.* 13D45, 05C90, 05E40, 05E45.
Key words and phrases. planar graph, edge ideal, Gorenstein ring.

same size, that is $\alpha(G)$, the *independence number* of G. A well-covered graph G is said to be a member of the class W_2 if $G \setminus v$ is well-covered with $\alpha(G \setminus v) = \alpha(G)$ for every vertex v (see [11, 14]); where $G \setminus v$ stands for the induced subgraph of G on the vertex set $V(G) \setminus \{v\}$. At first sight, if G is Gorenstein then G is in W_2 (see [7, Lemma 3.1]). In general we cannot read off the Gorenstein property of a graph just from its structure since this property as usual depends on the characteristic of k (see [8, Proposition 3.1]), so now we focus on some classes of graphs such as (see [4, 6, 8]):

(1) A bipartite graph is Gorenstein if and only if it is consists of disjoint edges.
(2) A chordal graph is Gorenstein if and only if it is consists of disjoint edges.
(3) A triangle-free graph is Gorenstein if and only if it is a member of W_2.

The main theme of this paper is to characterize Gorenstein graphs among *planar graphs*. A graph which can be drawn in the plane in such a way that edges meet only at points corresponding to their common ends is called a planar graph. Because a graph is Gorenstein if and only if every its connected component is Gorenstein by Lemma 2.2, so that it suffices to characterize Gorentein planar connected graphs.

This work is done for triangle-free planar graphs (see [7, 8]). Namely, for $m \geqslant 1$, let G_m be the graph with the vertex set $\{1, 2, \ldots, 3m-1\}$ and the edge set

$$\{\{1,2\}, \{\{3s-1,3s\}, \{3s,3s+1\}, \{3s+1,3s+2\}, \{3s+2,3s-2\}\}_{s=1,\ldots,m-1}, \{\{3t-3,3t\}\}_{t=2,\ldots,m-1}\}.$$

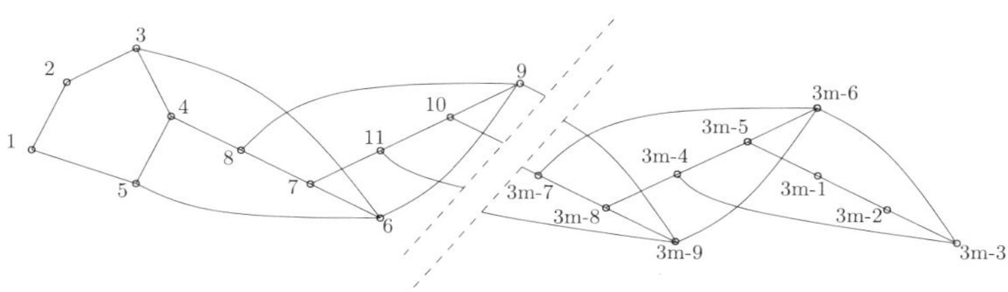

Fig. 1. The graph G_m

Then the family $\mathcal{G} = \{G_m\}_{m\geqslant 1}$ is exactly Gorenstein triangle-free planar connected graphs (see [8, Remark 4.5]). It is worth mentioning that this family can be visualized by a recursive construction due to Pinter [10]:

(1) Begin with $G_1 = K_2$ and $G_2 = C_5$;

(2) Given the graph G_m for $m \geqslant 2$, let x be a vertex of degree 2 in G_m and $N_G(x) = \{y, z\}$. Then construct the graph G_{m+1} with precisely three more points than G_m as follows. Let the three new points be a, b and c. Then join a to three points b, y, z; b to c and c to x (see Figure 2).

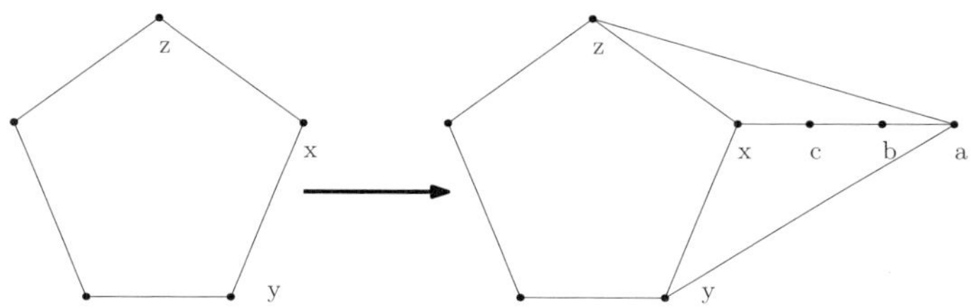

Fig. 2. The construction of G_3 from G_2

Moving away from triangle-free graphs, we get a new Gorenstein planar graph R_3 (the name due to [11]) as in Figure 3. The main result of the paper says that Gorenstein planar connected graphs are exactly the family $\mathcal{G}$ and R_3.

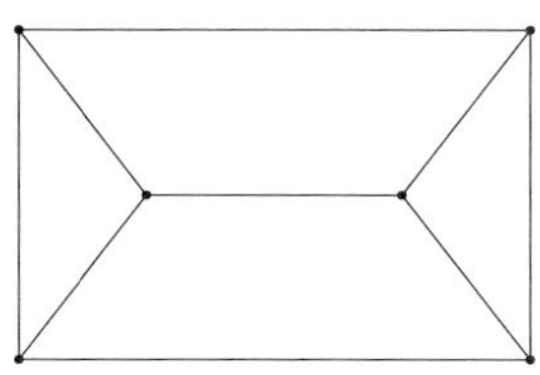

Fig. 3. The graph R_3

Theorem 3.7 *Let G be a planar connected graph without isolated vertices. Then, G is Gorenstein if and only if G is either G_m, where $m = \alpha(G)$, or R_3.*

The paper is organized as follows. In Section 1 we set up some basic terminology for simplicial complexes and graphs. In Section 2 we characterize Gorenstein planar graphs.

§2. Preliminaries

Let Δ be a simplicial complex on the vertex set $V = \{1, \dots, n\}$. Thus, Δ is a collection of subsets of V closed under taking subsets; that is, if $\sigma \in \Delta$ and $\tau \subseteq \sigma$ then $\tau \in \Delta$. Given a field k we now define the Stanley-Reisner ideal of Δ to be the square-free monomial ideal

$$I_\Delta = (x_{j_1} \cdots x_{j_i} \mid j_1 < \cdots < j_i \text{ and } \{j_1, \dots, j_i\} \notin \Delta)$$

in $R = k[x_1, \dots, x_n]$ and the *Stanley-Reisner* ring of Δ to be the quotient ring $k[\Delta] = R/I_\Delta$. We say that Δ is Cohen-Macaulay (resp. Gorenstein) if $k[\Delta]$ is Cohen-Macaulay (resp. Gorenstein). For $F \in \Delta$, we define the dimension of $F \in \Delta$ to be $\dim F = |F| - 1$ and the dimension of Δ to be $\dim \Delta = \max\{\dim F \mid F \in \Delta\}$. The link of F inside Δ is its subcomplex:

$$\mathrm{lk}_\Delta F = \{H \in \Delta \mid H \cup F \in \Delta \text{ and } H \cap F = \emptyset\}.$$

Let f_i be the number of i-dimensional faces of Δ. Then the *reduced Euler characteristic* $\widetilde{\chi}(\Delta)$ of Δ is defined by

$$\widetilde{\chi}(\Delta) := \sum_{i=-1}^{d} (-1)^i f_i,$$

where $d = \dim(\Delta)$.

We call Δ is *pure* if all its facets have the same dimension, and *Eulerian* if it is pure and $\widetilde{\chi}(\mathrm{lk}_\Delta F) = (-1)^{\dim(\mathrm{lk}_\Delta F)}$ for all $F \in \Delta$.

The restriction of Δ to a subset S of V is $\Delta_S := \{F \in \Delta \mid F \subseteq S\}$. The star of a vertex v in Δ is $\mathrm{st}_\Delta(v) := \{F \in \Delta \mid F \cup \{v\} \in \Delta\}$. Let $\mathrm{core}(V) := \{x \in V \mid \mathrm{st}_\Delta(x) \neq \Delta\}$, then the core of Δ is $\mathrm{core}(\Delta) := \Delta_{\mathrm{core}(V)}$. For a vertex v, $\Delta = \mathrm{st}_\Delta(v)$ means that Δ is a cone over v, hence $\Delta = \mathrm{core}(\Delta)$ means that Δ is not a cone.

Let Δ and Γ be simplicial complexes with disjoint vertex sets V and W, respectively. Define the join $\Delta * \Gamma$ to be the simplicial complex on the vertex set $V \cup W$ with faces $F \cup H$, where $F \in \Delta$ and $H \in \Gamma$. It follows that $\Delta = \mathrm{core}(\Delta) * \langle V \setminus \mathrm{core}(V) \rangle$; where for a finite set P, we denote $\langle P \rangle$ to be the simplex over P.

We then have a criterion for determining when a simplicial complex Δ is Gorenstein due to Stanley (see [13, Theorem 5.1 in Chapter II]).

Lemma 2.1. *Δ is Gorenstein if and only if and only if* $\mathrm{core}(\Delta)$ *is an Euler complex which is Cohen-Macaulay.*

Let G be a simple graph. For a subset S of $V(G)$, the neighborhood of S in G is

$$N_G(S) := \{v \in V(G) \setminus S \mid uv \in E(G) \text{ for some } u \in S\},$$

the close neighborhood of S is $N_G[S] := S \cup N_G(S)$. We denote by G_S the induced subgraph of G on the vertex set $V(G) \setminus N_G[S]$. If v is a vertex of G, we write $N_G(v)$ (resp. $N_G[v]$ and G_v) instead of $N_G(\{v\})$ (resp. $N_G[\{v\}]$ and $G_{\{v\}}$). The *degree* of v in G is $\deg_G(v) := |N_G(v)|$. If $\deg_G(v) = 0$, then v is called an *isolated vertex* of G.

For the independence complex of G, we have $\dim(\Delta(G)) = \alpha(G) - 1$. Clearly, G is well-covered if and only if $\Delta(G)$ is pure and $\Delta(G) = \mathrm{core}(\Delta(G))$ if and only if G has no isolated vertices.

Lemma 2.2. *Let G be a simple graph. Then, G is Gorenstein if and only every connected component of G is also Gorenstein.*

Proof. Let $G_1, \ldots, G_s$ be connected components of G. Then,

$$\Delta(G) = \Delta(G_1) * \cdots * \Delta(G_s).$$

Hence, $\Delta(G)$ is Gorenstein if and only if so are $\Delta(G_1), \ldots, \Delta(G_s)$ by [2], and the lemma follows. Q.E.D.

Lemma 2.3. *Let G be a Gorenstein graph without isolated vertices and let S be an independent subset of G. Then,*

(1) *If G_S is a Gorenstein graph without isolated vertices;*
(2) $\alpha(G_S) = \alpha(G) - |S|$

Proof. (1) Since G has no isolated vertices, $\Delta(G) = \mathrm{core}(\Delta(G))$. By Lemma 2.1, $\Delta(G)$ is Eulerian. Note that $\Delta(G_S) = \mathrm{lk}_{\Delta(G)}(S)$, so $\Delta(G_S)$ is Eulerian. In particular, $\widetilde{\chi}(\Delta(G_S)) \neq 0$, so $\Delta(G)$ is not a cone. It follows that G_S has no isolated vertices.

On the other hand, $\Delta(G_S)$ is Cohen-Macaulay by [5, Corollary 8.1.8]. Therefore, $\Delta(G_S)$ is Gorenstein by Lemma 2.1, i.e. G_S is Gorenstein.

(2) Because $\Delta(G)$ is pure, we have

$$\alpha(G_S) = \dim(\Delta(G_S)) + 1 = \dim(\Delta(G)) - |S| + 1 = \alpha(G) - |S|,$$

and the lemma follows. Q.E.D.

In the sequence we also need a criterion on deciding whether a given graph is planar. A *minor* of a graph G is any graph obtainable from G by means of a sequence of vertex and edge deletions and edge contractions. Alternatively, consider a partition $(V_0, V_1, \ldots, V_m)$ of $V(G)$ such that

$G[V_i]$ is connected, $1 \leqslant i \leqslant m$, and let H be the graph obtained from G by deleting V_0 and shrinking each induced subgraph $G[V_i]$, $1 \leqslant i \leqslant m$, to a single vertex. Then any spanning subgraph of H is a minor of G. A minor which is isomorphic to K_5 or $K_{3,3}$ is called a Kuratowski minor. Then, Wagner's theorem (see [1, Theorem 10.3]) says that: *G is planar if and only if G has no Kuratowski minor.*

§3. A Characterization of Planar Gorenstein graphs

In this section we characterize Gorenstein planar graphs. A graph is called *trivial* if it has only one vertex. We always use the symbol G to indicate a simple graph without isolated vertices. We first consider graphs with small independence numbers.

Lemma 3.1. *Let G be a graph with $\alpha(G) = 1$. Then, G is Gorenstein if and only if G is an edge.*

Proof. Since $\alpha(G) = 1$, G is a complete graph. Note that G is not trivial by our assumption, $|V(G)| \geqslant 2$. By [6, Corollary 2.2], we conclude that G is Gorenstein if and only if it is an edge. Q.E.D.

Lemma 3.2. *R_3 is Gorenstein.*

Proof. Since the geometric realization of $\Delta(R_3)$ is isomorphic to the cycle C_6, the lemma now follows from [13, Theorem 5.1 in Chapter II]. Q.E.D.

Lemma 3.3. *Let G be a planar graph with $\alpha(G) = 2$. Then G is Gorenstein if and only if G is two disjoint edges, or a pentagon, or R_3.*

Proof. Let $n := |V(G)|$. For each vertex v of G, G_v is a Gorenstein graph without isolated vertices by Lemma 2.2. Since, $\alpha(G_v) = \alpha(G) - 1 = 1$, G_v is just one edge by Lemma 3.1. It follows that $\deg_G(v) = n - 3$ for any $v \in V(G)$. Let G^c denote the complement of G. Then, $\deg_{G^c}(v) = 2$ for all $v \in V(G^c)$, and hence G^c is an n-cycle. Since $\alpha(G) = 2$, we have $n \geqslant 4$. We now consider the following cases depending on n:

Case 1: $n = 4$. Then, G is two disjoint edges.

Case 2: $n = 5$. Then, G is a pentagon.

Case 3: $n = 6$. Then, G is just R_3.

Case 4: $n \geqslant 7$. Assume that $V(G) = \{1, 2, \dots, n\}$. If we delete all vertices $\{8, \dots, n\}$ from G, and then contract the edge $\{2, 6\}$, then we obtain a graph that has a minor $K_{3,3}$. It follows that G is not planar by Wagner's theorem, a contradiction.

In summary, G is Gorenstein if and only if G is two disjoint edges, or a pentagon, or R_3. Clearly, these graphs are Gorenstein. Q.E.D.

We call a graph G is *locally triangle-free* if G_v is triangle-free for every vertex v of G. The next result plays a key role in characterizing Gorenstein graphs with higher independence numbers (see [9, Theorem 4.1]).

Lemma 3.4. *Let G be a locally triangle-free graph. Then, G is Gorenstein if and only if G is a triangle-free graph in W_2 except for 4 graphs: P_{10}, P_{12}, Q_9 and Q_{12}* (see Figure 4).

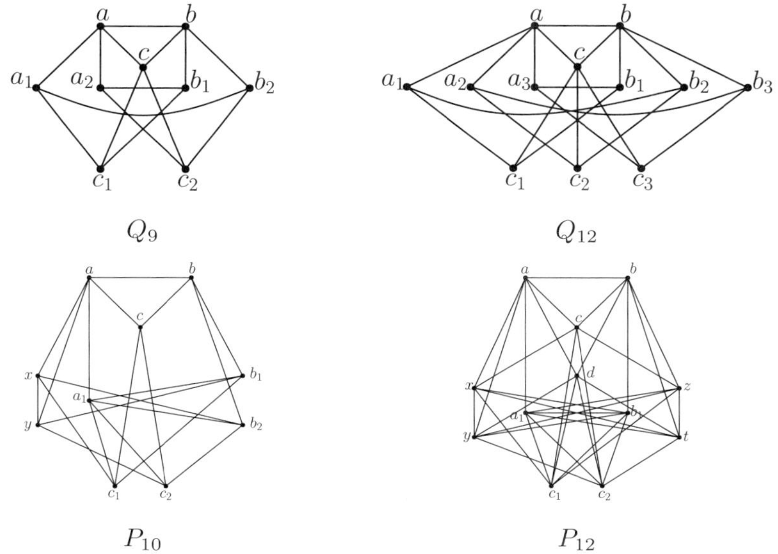

Fig. 4. Four graphs Q_9, Q_{12}, P_{10} and P_{12}

By using this result, we next characterize Gorenstein locally triangle-free planar graph. Namely,

Lemma 3.5. *Let G be a Gorenstein planar connected graph. If G is locally triangle-free, then G is either a triangle-free graph or R_3.*

Proof. If $\alpha(G) \leqslant 2$, the lemma follows from Lemmas 3.1 and 3.3.

Assume that $\alpha(G) \geqslant 3$. By Lemma 3.4, G is a triangle-free graph or one of P_{10}, P_{12}, Q_9 or Q_{12}. Therefore, it remains to verify that P_{10}, P_{12}, Q_9 and Q_{12} are not planar. In order to do that, we use Wagner's theorem.

For the graph Q_9, by shrinking a_1 to c_1, b_2 to c_2, a to a_2 and b to b_1 we see that P_{10} has a minor K_5. Hence, Q_9 is not planar.

For the graph Q_{12}, by shrinking a_1, a_2 and a_3 to a; and b_1, b_2 and b_3 to b we see that Q_{12} has a minor $K_{3,3}$. Hence, Q_{12} is not planar.

For the graph P_{10}, by shrinking a and b to c, b_1 to b_2, x and y to c_1, we see that P_{10} has a minor K_5. Hence, P_{10} is not planar.

Finally, for the graph P_{12}, first deleting two vertices a_1 and b_1; and then shrinking a and b to c, x to y, and z to t, we see that P_{12} has a minor $K_{3,3}$. Hence, P_{12} is not planar, and the lemma follows. Q.E.D.

Lemma 3.6. *If G is a Gorenstein planar connected graph, then it is locally triangle-free.*

Proof. We prove by induction on $\alpha(G)$. If $\alpha(G) \leqslant 2$, the lemma follows from Lemmas 3.1 and 3.3.

If $\alpha(G) \geqslant 3$. Assume on contrary that G is not locally triangle-free, so that G_x would have a triangle for some vertex x of G. We first prove the following claims.

Claim 1: Every connected component of G_v is either a triangle-free non-trivial graph or R_3 for any vertex v of G.

Indeed, by Lemma 2.3 we have G_v is Gorenstein graph with $\alpha(G_v) = \alpha(G) - 1$ which has no isolated vertices. Note that G_v is planar as well. By the induction hypothesis, G_v is locally triangle-free. Thus, every component of G_v is a triangle-free graph or R_3 by Lemmas 2.2 and 3.5, as claimed.

Let $G_1, \ldots, G_m$ be connected components of G_x. By Claim 1, each G_i is either a Gorenstein non-trivial triangle-free graph or R_3. Since G_x has a triangle, we may assume that G_1 is R_3. Let $X := N_G(x)$.

Claim 2: There is a vertex of G_1 which is not adjacent to any vertex in X.

Indeed, assume on contrary that every vertex of G_1 is adjacent to some vertex in X. By deleting $G_2, \ldots, G_m$, and shrinking X into x, we obtain a graph G' as in Figure 5. By shrinking two vertices inside the rectangle, we see that G' has a minor $K_{3,3}$, so G is not planar, and the claim follows.

Because G is connected, there is some vertex in X which is adjacent to some vertex of G_1. Together with By Claim 2, we can partition X into two nonempty subsets $X = Y \cup Z$ such that every vertex in Y is adjacent to some vertex of G_1; but every vertex in Z is not adjacent to any vertex of G_1.

Claim 3: $m = 1$.

Indeed, assume that $m \geqslant 2$. Let v be a vertex of G_i for some $i = 2, \ldots, m$. If v is not adjacent to any vertex in Y, then there would be a connected component of G_v that contains G_1 and the vertex x.

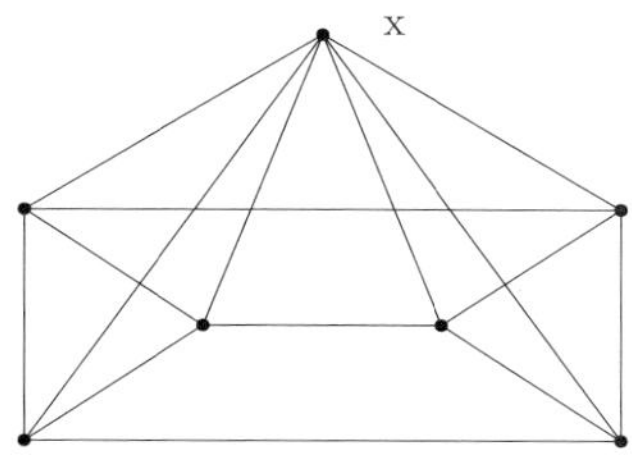

Fig. 5. The graph G'

This is impossible because every connected component of G_v is either a triangle-free graph or R_3 by Claim 1. Hence, every vertex of G_i for every $i = 2, \ldots, m$ is adjacent to all vertices in Y.

By Claim 2, there is a vertex of G_1, say a, which is not adjacent to any vertex in X. Then, G_a has a connected component, say H, that contains $N_G[x]$ and $G_2, \ldots, G_m$. Since G_2 is non-trivial by Claim 1 and every vertex of G_2 is adjacent to all vertices in Y, H must contain triangles, so H is isomorphic to R_3 by Claim 1. Hence, we may assume that $E(H) = \{xy, xz, xw, yz, zu, wu, yv, uv, wv\}$, so that G_2 is just the edge uv. In this case, $Y = \{w\}$ and $Z = \{y, z\}$. But then the graph G_z has a component that contains $G_1 = R_3$ and w. This contradicts Claim 1. Therefore, $m = 1$, as claimed.

We now return to prove the lemma. By Claim 3, we have $G_x = G_1$. Since G_1 is R_3, $\alpha(G) = \alpha(R_3) + 1 = 3$. Assume that $E(G_1) = \{ab, bc, ca, de, ef, fd, ad, be, cf\}$ and a is not adjacent to any vertex in X. Fix any vertex $z \in Z$. Then, the set $S := \{a, z\}$ is an independent set of G, and then G_S is Gorenstein with $\alpha(G_S) = \alpha(G) - |S| = 1$ by Lemma 2.3. Thus, G_S is an edge by Lemma 3.1, and thus it must be the edge ef. Consequently, $N_G[x] = N_G[z]$. Fix any another vertex y of Y. Then, (xyz) is a triangle of G because of $N_G[x] = N_G[z]$.

In the other hand, since $\alpha(G_a) = \alpha(G) - 1 = 2$ and G_a contains the triangle (xyz), G_a must be R_3 by Claim 1. Note also that x, z are vertices of G_a and

$$N_{G_a}[x] = N_G[x] = N_G[z] = N_{G_a}[z].$$

This is impossible since neighborhoods of two distinct vertices of R_3 are different. Therefore, G is locally triangle-free, and the proof is complete. Q.E.D.

We are in position to prove the main result of this paper.

Theorem 3.7. *Let G be a planar connected graph without isolated vertices. Then, G is Gorenstein if and only if G is either G_m, where $m = \alpha(G)$, or R_3.*

Proof. Assume that G is Gorenstein. If $\alpha(G) \leqslant 2$, then the theorem follows from Lemmas 3.1 and 3.3. If $\alpha(G) \geqslant 3$, then G is a triangle-free graph by Lemma 3.6, and then it is a member of $\mathcal{G}$ by [8, Remark 4.5]. Thus, $G = G_m$ for some $m \geqslant 1$. Since $\alpha(G_m) = m$ by [7, Lemma 3.7], we have $m = \alpha(G)$.

The converse implication follows from [8, Remark 4.5] together with Lemma 3.2. Q.E.D.

Acknowledgment

The author is partially supported by the research project:

VAST.HTQT.NHAT.1/16-18

References

[1] J. A. Bondy and U. S. R. Murty, Graph theory, Springer 2008.

[2] R. Fröberg, *A note on the Stanley-Reisner ring of a join and of a suspension*, Manuscripta Math. **60** (1988), no. 1, 89–91.

[3] H.-G. Gräbe, *The Gorenstein property depends on characteristic*, Beiträge Algebra Geom. 17 (1984), 169-174.

[4] J. Herzog and T. Hibi, *Distributive lattices, bipartite graphs and Alexander duality*, J. Algebraic Combin. **22** (2005), no. 3, 289–302.

[5] J. Herzog and T. Hibi, Monomial ideals, Springer, 2010.

[6] J. Herzog, T. Hibi and X. Zheng, *Cohen-Macaulay chordal graphs.* J. Combin. Theory Ser. A, **113** (2006), no. 5, 911–916.

[7] D. T. Hoang, N. C. Minh and T. N. Trung, *Cohen-Macaulay graphs with large girth*, Journal of Algebra and Its Applications, **14** (2015), 16 pages.

[8] D. T. Hoang and T. N. Trung, *A Characterization of Triangle-free Gorenstein graphs and Cohen-Macaulayness of second powers of edge ideals*, J. Algebr. Comb. **43**(2016), no.2, 325–338.

[9] D. T. Hoang and T. N. Trung, *Buchsbaumness of second powers of edge ideals*, arXiv:1606.02815v2.

[10] M. R. Pinter, *A class of planar well-covered graphs with girth four*, J. Graph Theory, **19** (1995), no. 1, 69–81.

[11] M. D. Plummer, *Well-covered graphs: Survey.* Quaestiones Math. **16** (1993), no. 3, 253–287.

[12] G. Reisner, *Cohen-Macaulay quotients of polynomial rings*, Advances in Math. **21** (1976), 30–49.

[13] R. Stanley, Combinatorics and Commutative Algebra, 2. Edition, Birkhäuser, 1996.

[14] J. W. Staples, *On some subclasses of well-covered graphs*, J. Graph Theory **3** (1979), 197–204.

Institute of Mathematics, VAST, 18 Hoang Quoc Viet, Hanoi, Vietnam
E-mail address: tntrung@math.ac.vn

Advanced Studies in Pure Mathematics 77, 2018
The 50th Anniversary of Gröbner Bases
pp. 411–436

Differential relations for the largest root distribution of complex non-central Wishart matrices

Raimundas Vidunas and Akimichi Takemura

Abstract.

A holonomic system for the probability density function of the largest eigenvalue of a non-central complex Wishart distribution with identity covariance matrix is derived. Furthermore a new determinantal formula for the probability density function is derived (for the dimensions $m = 2, 3$) or conjectured.

§1. Introduction

The Wishart distribution is an important higher dimensional generalization of the χ^2-distribution. In many applications the distribution of *roots* (i.e., eigenvalues) of Wishart matrices are needed (see references in Hashiguchi et al. [2]). In this paper we consider complex non-central Wishart matrices, which are important for applications to performance evaluation of wireless communication systems (Siriteanu et al. [7], [8]). The purpose of this paper is to give differential relations for the largest root distribution of complex non-central Wishart matrices based on the result of Kang and Alouini [4].

Suppose we take n random vectors $x_i \in \mathbb{C}^m$, $i = 1, \dots, n$, independently drawn from an m-variate complex Gaussian distribution $\mathcal{CN}(v_i, \Sigma)$, with the mean vector v_i and the covariance matrix Σ. We put those vectors into $n \times m$ matrices X and V. The distribution of the random (symmetric, positive definite) $m \times m$ scatter matrices $S = X^*X$ defines the complex *Wishart distribution* $\mathcal{W}_m(\Sigma, V^*V\Sigma^{-1}, n)$ with degrees of freedom n, covariance matrix Σ and the non-centrality parameter matrix $V^*V\Sigma^{-1}$. We are interested in the distribution of largest root of S.

Received August 31, 2016.
2010 *Mathematics Subject Classification.* 60E05, 13N10, 62H10.
Key words and phrases. complex Wishart distribution, holonomic systems.

In the special case $m = 1$, we have the distribution of the value $|x_1|^2 + \ldots + |x_n|^2$. In the $\mathbb{R}$-valued case, this is the χ^2-distribution. The central χ^2-distribution ($V = 0$) is a special case of the gamma distribution.

The distribution of the largest root of the $\mathbb{R}$-valued central Wishart distribution is known, Muirhead [5]. The probability distribution function for the largest root is expressed in terms of a matrix hypergeometric function:

$$\frac{\Gamma_m\left(\frac{m+1}{2}\right)(\det\Sigma)^{-\frac{n}{2}}}{\Gamma_m\left(\frac{n+m+1}{2}\right)} \exp\!\left(-\frac{x}{2}\operatorname{tr}\Sigma^{-1}\right)\left(\frac{x}{2}\right)^{\frac{nm}{2}} {}_1\mathrm{F}_1\left(\begin{matrix}\frac{m+1}{2}\\ \frac{n+m+1}{2}\end{matrix}\,\middle|\,\frac{x}{2}\Sigma^{-1}\right). \tag{1}$$

Here $\Gamma_m(z) = \pi^{\frac{1}{4}m(m-1)}\prod_{i=1}^m \Gamma\left(z - \frac{i-1}{2}\right)$ is called the multivariate Gamma function and the ${}_1\mathrm{F}_1(M)$ function is defined in terms of symmetric functions (*zonal polynomials*) of the eigenvalues of M; Constantine [1], James [3].

A holonomic system for ${}_1\mathrm{F}_1\left(\begin{smallmatrix}a\\c\end{smallmatrix}\middle| M\right)$ in terms of the eigenvalues λ_i of M was derived by Muirhead [5]. For $i \in \{1, 2, \ldots, m\}$ we have

$$\lambda_i\frac{\partial^2}{\partial\lambda_i^2} + (c - \lambda_i)\frac{\partial}{\partial\lambda_i} + \frac{1}{2}\sum_{j=0, j\neq i}^{m} \frac{\lambda_j}{\lambda_i - \lambda_j}\left(\frac{\partial}{\partial\lambda_i} - \frac{\partial}{\partial\lambda_j}\right) - a. \tag{2}$$

Efficiency of the holonomic gradient method was demonstrated by Hashiguchi et al. [2].

In Section 2.2 we derive differential relations for the density function of the largest root of complex non-central Wishart matrices with the identity covariance matrix $\Sigma = \mathrm{Id}$. Additionally we assume the Gaussian distribution to be *circularly symmetric*; see [6], [11, Complex normal distribution]). A conjectural formula is given in Section 2.3. Later sections are devoted to proofs and additional results for $m \leqslant 3$.

§2. Setting and the contributions

The cumulative distribution function $\mathcal{F}_{n,m}(x)$ for the largest root x in the circularly symmetric case was derived by Kang and Alouini [4]. Let us recall the hypergeometric function

$${}_0\mathrm{F}_1\left(\begin{matrix}z\\ n\end{matrix}\right) := {}_0\mathrm{F}_1\left(\begin{matrix}-\\ n\end{matrix}\,\middle|\, z\right) = \sum_{k=0}^{\infty}\frac{x^k}{(n)_k\,k!}, \tag{3}$$

where $(n)_k = n(n+1)\ldots(n+k-1)$ is the Pochhammer symbol [11]. This function is related to the non-central χ^2-distribution and the modified

Bessel function [9, §9]

$$I_n(z) = \frac{(z/2)^n}{n!}\, {}_0\mathrm{F}_1\binom{z^2/4}{n+1}. \tag{4}$$

We introduce the integral

$$H_n^k(x,y) = \int_0^x t^k\, \mathbf{e}^{-t}\, {}_0\mathrm{F}_1\binom{ty}{n}\, dt \tag{5}$$

related to the Marcum Q-function [9, §9]

$$Q_n(x,y) = \frac{\mathbf{e}^{-x}}{(n-1)!} \int_y^\infty t^{n-1} \mathbf{e}^{-t}\, {}_0\mathrm{F}_1\binom{xt}{n}\, dt. \tag{6}$$

2.1. The distribution functions

Let $\lambda_1, \ldots, \lambda_m$ be the eigenvalues of V^*V. The Kang–Alouini distribution function for the largest root of $\mathcal{W}_m(\mathrm{Id}, V^*V, n)$ is

$$\mathcal{F}_{n,m}(x,\lambda_1,\ldots,\lambda_m) = \frac{\mathbf{e}^{-\lambda_1-\ldots-\lambda_m}}{\{(n-m)!\}^m \prod_{i=1}^m \prod_{j>i}^m (\lambda_i-\lambda_j)} \det\Big(H_{n-m+1}^{n-j}(x,\lambda_i)\, \Big\rangle_{i=1}^m \Big). \tag{7}$$

Here $\rangle_{i=1}^m$ indicates the i-th row of an $m \times m$ matrix, with the column index implicitly taken to be j.

The probability density function is

$$\psi_{n,m}(x,\lambda_1,\ldots,\lambda_m) = \frac{\partial}{\partial x} \mathcal{F}_{n,m}(x,\lambda_1,\ldots,\lambda_m) \tag{8}$$

$$= \frac{\mathbf{e}^{-\lambda_1-\ldots-\lambda_m}}{\{(n-m)!\}^m \prod_{i=1}^m \prod_{j>i}^m (\lambda_i-\lambda_j)} \mathcal{R}_{n,m}(x,\lambda_1,\ldots,\lambda_m), \tag{9}$$

where

$$\mathcal{R}_{n,m}(x,\lambda_1,\ldots,\lambda_m) = \frac{\partial}{\partial x} \det\Big(H_{n-m+1}^{n-j}(x,\lambda_i)\, \Big\rangle_{i=1}^m \Big) \tag{10}$$

$$= \mathbf{e}^{-x} \sum_{k=1}^m {}_0\mathrm{F}_1\binom{x\lambda_k}{n-m+1} \det\begin{pmatrix} H_{n-m+1}^{n-j}(x,\lambda_i)\, \rangle_{i\neq k}^{i\leqslant m} \\ x^{n-j} \qquad \rangle_{i=k} \end{pmatrix}. \tag{11}$$

For example, an expanded expression for $m = 3$ is

$$\mathcal{R}_{n,3}(x,\lambda_1,\lambda_2,\lambda_3) =$$

$$\mathbf{e}^{-x}\,{}_0\mathrm{F}_1\left(\begin{matrix} x\lambda_1 \\ n-2 \end{matrix}\right) \det \begin{pmatrix} x^{n-1} & x^{n-2} & x^{n-3} \\ H_{n-2}^{n-1}(x,\lambda_2) & H_{n-2}^{n-2}(x,\lambda_2) & H_{n-2}^{n-3}(x,\lambda_2) \\ H_{n-2}^{n-1}(x,\lambda_3) & H_{n-2}^{n-2}(x,\lambda_3) & H_{n-2}^{n-3}(x,\lambda_3) \end{pmatrix}$$

$$+\,\mathbf{e}^{-x}\,{}_0\mathrm{F}_1\left(\begin{matrix} x\lambda_2 \\ n-2 \end{matrix}\right) \det \begin{pmatrix} H_{n-2}^{n-1}(x,\lambda_1) & H_{n-2}^{n-2}(x,\lambda_1) & H_{n-2}^{n-3}(x,\lambda_1) \\ x^{n-1} & x^{n-2} & x^{n-3} \\ H_{n-2}^{n-1}(x,\lambda_3) & H_{n-2}^{n-2}(x,\lambda_3) & H_{n-2}^{n-3}(x,\lambda_3) \end{pmatrix}$$

$$+\,x^{n-3}\mathbf{e}^{-x}\,{}_0\mathrm{F}_1\left(\begin{matrix} x\lambda_3 \\ n-2 \end{matrix}\right) \det \begin{pmatrix} H_{n-2}^{n-1}(x,\lambda_1) & H_{n-2}^{n-2}(x,\lambda_1) & H_{n-2}^{n-3}(x,\lambda_1) \\ H_{n-2}^{n-1}(x,\lambda_2) & H_{n-2}^{n-2}(x,\lambda_2) & H_{n-2}^{n-3}(x,\lambda_2) \\ x^2 & x & 1 \end{pmatrix}.$$

2.2. Main results

The main result of this paper is a holonomic system of differential equations for $\mathcal{R}_{n,m}(x,\lambda_1,\ldots,\lambda_m)$, for any dimension m. It is formulated in the following two theorems. The first one introduces a holonomic system with the differentiations $\partial/\partial\lambda_i$ only. Theorem 2.2 allows to introduce or eliminate $\partial/\partial x$.

Recall [10] that a *least common left multiple* (LCLM) of several differential operators $\mathcal{L}_1,\ldots,\mathcal{L}_k$ in the Weyl algebra $\mathbb{C}(x,y)\langle\partial/\partial y\rangle$ is a differential operator $\mathcal{L}^*$ of minimal order such that $\mathcal{L}^*$ is a left multiple of any $\mathcal{L}_j$, $j\in\{1,\ldots,k\}$. An alternative defining property is that $\mathcal{L}^*Y=0$ is a differential equation of minimal order such that all Picard-Vessiot [10] solutions of $\mathcal{L}_jY=0$ are solutions of $\mathcal{L}^*Y=0$.

Theorem 2.1. *Let us define the differential operators*

$$\mathcal{P}_M[y] = y\,\frac{\partial^2}{\partial y^2} + (M+1)\,\frac{\partial}{\partial y} - x, \tag{12}$$

$$\mathcal{Q}_{N,M}[y] = y\,\frac{\partial^3}{\partial y^3} + (M-y+2)\,\frac{\partial^2}{\partial y^2} - (x+N+1)\,\frac{\partial}{\partial y} + x. \tag{13}$$

Let us denote $\mathcal{T}_1[y]=\mathcal{P}_{n-m}[y]$ *and*

$$\mathcal{T}_j[y] = \mathcal{Q}_{n-m+j,n-m}[y] \qquad \text{for } 2\leqslant j\leqslant m. \tag{14}$$

The following operators annihilate $\mathcal{R}_{n,m}(x,\lambda_1,\ldots,\lambda_m)$:

(*i*) *The products* $\mathcal{T}_k[\lambda_1]\cdots\mathcal{T}_k[\lambda_m]$, *for* $k=1,2,\ldots,m$.

(*ii*) *The least common left multiples* $\mathrm{LCLM}(\mathcal{T}_1[\lambda_k],\ldots,\mathcal{T}_m[\lambda_k])$ *with* $k=1,2,\ldots,m$.

Theorem 2.2. *This second order operator annihilates the function* $\mathcal{R}_{n,m}(x, \lambda_1, \ldots, \lambda_m)$:

$$x\frac{\partial}{\partial x} + \sum_{k=1}^{m} \Big(\lambda_k \frac{\partial^2}{\partial \lambda_k^2} + \big(n - m + 1 - \lambda_k\big) \frac{\partial}{\partial \lambda_k} - n \Big) + \frac{m(m-1)}{2} + 1. \tag{15}$$

The theorems are proved in §3.1 and §3.3. To get differential equations for the density function $\psi_{n,m}(x, \lambda_1, \ldots, \lambda_m)$, the presented operators must be modified by the *gauge translations*

$$\frac{\partial}{\partial \lambda_i} \mapsto \frac{\partial}{\partial \lambda_i} + 1 + \sum_{j \neq i} \frac{1}{\lambda_i - \lambda_j}. \tag{16}$$

This is a standard technique to account for the front factor in (9).

By its determinantal form (10), the target function $\mathcal{R}_{n,m}(x, \lambda_1, \ldots, \lambda_m)$ is a non-logarithmic and anti-symmetric function. In particular, it is multiplied by the sign $(-1)^\sigma$ under a permutation σ of the variables $\lambda_1, \ldots, \lambda_m$

Theorem 2.3. (*i*) *The differential operators of Theorem* 2.1 *annihilating* $\mathcal{R}_{n,m}(x, \lambda_1, \ldots, \lambda_m)$ *generate a holonomic system of rank* $2m! \cdot 3^{m-1}$.

(*ii*) *Let* $\mathcal{S}$ *denote the subspace of anti-symmetric solutions in a full solution space* (*of dimension* $2m! \cdot 3^{m-1}$). *Then* $\dim \mathcal{S} = 2 \cdot 3^{m-1}$.

(*iii*) *The subspace of non-logarithmic anti-symmetric solutions has the dimension* 2^{m-1}.

(*iv*) *There exists a holonomic system of rank* $\leqslant 3^m - 1$ *defined over* $\mathbb{Q}(x, \lambda_1, \ldots, \lambda_m)$ *and annihilating* $\mathcal{R}_{n,m}(x, \lambda_1, \ldots, \lambda_m)$.

This theorem is proved in §3.2. Our computations for $m = 2$, $m = 3$ indicate that the lower rank system has markedly more complicated equations and singularities. These computations are presented in §4.

2.3. A conjectural formula

We were led to Theorem 2.1 after computing holonomic systems for $\mathcal{R}_{n,2}(x, \lambda_1, \lambda_2)$ of rank 12 and 8, elimination of $\partial/\partial x$, $\partial/\partial \lambda_2$ and observing a differential operator in $\partial/\partial \lambda_1$ of order 5 with a simple LCLM factorization. Computations for $\mathcal{R}_{n,3}(x, \lambda_1, \lambda_2, \lambda_3)$ led to holonomic systems of rank 108 and 26 cumbersomely, but probing for a differential operator in only $\partial/\partial \lambda_1$ quickly gave one of relatively low order 8 and a remarkable LCLM factorization into operators of order 2 or 3. Theorem 2.1 establishes continuation of this pattern.

The solution space of the holonomic systems in Theorem 2.1 is highly factorizable by specificity of the presented generators. Particular solutions are

$$\det\Big(Y_j(x,\lambda_i) \,\rangle_{i=1}^{m} \Big), \tag{17}$$

where $Y_j(x,y)$ is a solution $\mathcal{T}_j[y]\,Y_j = 0$. The LCLM operator in *(ii)* annihilates the kth row of this matrix, while the product in *(i)* annihilates the kth column. Based on obtained new expressions for $\mathcal{R}_{n,2}(x,\lambda_1,\lambda_2)$, $\mathcal{R}_{n,3}(x,\lambda_1,\lambda_2,\lambda_3)$, solutions of $\mathcal{Q}_{N,M}[y]$ and their recurrences, we conjecture that $\mathcal{R}_{n,m}(x,\lambda_1,\ldots,\lambda_m)$ has a determinantal expression (17). Here is a formulation in the transposed form.

Conjecture 2.4. *Let us define*

$$\begin{aligned} G_{n,2}(x,y) = {} & n\,{}_0\mathrm{F}_1\left(\begin{matrix} xy \\ n \end{matrix}\right) + y\,{}_0\mathrm{F}_1\left(\begin{matrix} xy \\ n+1 \end{matrix}\right) \\ & + (x-y-n+1)\,\mathbf{e}^{y} \int_y^{\infty} \mathbf{e}^{-t}\,{}_0\mathrm{F}_1\left(\begin{matrix} xt \\ n+1 \end{matrix}\right) dt. \end{aligned} \tag{18}$$

For $m \geqslant 2$, we recursively define

$$G_{n,m+1}(x,y) = \left(-y\,\frac{\partial^2}{\partial y^2} - (n-m+1)\,\frac{\partial}{\partial y} + x + m \right) G_{n,m}(x,y). \tag{19}$$

We conjecture that

$$\mathcal{R}_{n,m}(x,\lambda_1,\ldots,\lambda_m) = C(x)\,\det\left(\begin{matrix} {}_0\mathrm{F}_1\left(\begin{smallmatrix} x\lambda_i \\ n-m+1 \end{smallmatrix}\right) \;\rangle_{j=1} \\ G_{n-m+j,j}(x,\lambda_i) \,\rangle_{j=2}^{m} \end{matrix} \right) \tag{20}$$

with

$$C(x) = \frac{(n-m+1)\,x^{mn-\binom{m}{2}-1}\,\mathbf{e}^{-mx}}{\prod_{k=1}^{m}(n-k+1)^k}. \tag{21}$$

Note that ${}_0\mathrm{F}_1\left(\begin{smallmatrix} x\,y \\ n-m+1 \end{smallmatrix}\right)$ is a solution of $\mathcal{P}_{n-m}[y]\,Y = 0$. As we show in §3.4, the function $G_{n,m}(x,y)$ is a solution of $\mathcal{Q}_{n-m+j,n-m}[y] = 0$ for any integers $n \geqslant m > 0$. Recurrence (19) stems from §3.4 as well.

Notably, the integral in (18) is complementary to $H^0_{n+1}(y,x)$.

The conjecture has been fully checked for $m = 2$ and $m = 3$, as described in §4.2 and §4.4. Also, the front factor (21) has been confirmed for $m = 4$. The conjecture happens to be true for $m = 1$ as well. In §4.2 we specifically prove

$$\psi_{n,2}(x,\lambda_1,\lambda_2) = \frac{x^{2n-2}\,\mathbf{e}^{-\lambda_1-\lambda_2-2x}}{n!\,(n-2)!\,(\lambda_1-\lambda_2)} \det\left(\begin{matrix} {}_0\mathrm{F}_1\left(\begin{smallmatrix} x\lambda_1 \\ n-1 \end{smallmatrix}\right) & {}_0\mathrm{F}_1\left(\begin{smallmatrix} x\lambda_2 \\ n-1 \end{smallmatrix}\right) \\ G_{n,2}(x,\lambda_1) & G_{n,2}(x,\lambda_2) \end{matrix} \right). \tag{22}$$

If $\lambda_1 = \lambda_2$, application of l'Hospital's rule leads to differentiating a matrix column. For comparison, expression (8) is

(23)

$$\psi_{n,2}(x,\lambda_1,\lambda_2) = \frac{\mathbf{e}^{-\lambda_1-\lambda_2}}{\{(n-2)!\}^2} \frac{\dfrac{\partial}{\partial x}\det\begin{pmatrix} H_{n-1}^{n-1}(x,\lambda_1) & H_{n-1}^{n-1}(x,\lambda_2) \\ H_{n-1}^{n-2}(x,\lambda_1) & H_{n-1}^{n-2}(x,\lambda_2) \end{pmatrix}}{\lambda_1-\lambda_2}.$$

Not only the differentiation $\partial/\partial x$ is avoided, but the integral in (18) is numerically preferable to the H_{n-1}^{n-1}, H_{n-1}^{n-2} functions.

Significance of the conjectured formula is that it would utilize the factorization structure of the holonomic system in Theorem 2.1. Applying the holonomic gradient method to the entries of the conjectured matrix would be more efficient than employing the same method for the large multi-variate holonomic system.

2.4. Auxiliary integrals

To get the holonomic system, we use recurrences for $H_n^k(x,y)$ in (5) and the generalization

$$H_n^{k,\ell}(x,y) = \int_0^x \mathbf{e}^{-t}\, t^k (x-t)^\ell\, {}_0\mathrm{F}_1\begin{pmatrix} t\,y \\ n \end{pmatrix} dt. \tag{24}$$

Surely, $H_n^{k,0}(x,y) = H_n^k(x,y)$. These differentiations are straightforward:

$$\frac{\partial}{\partial x} H_n^k(x,y) = x^k \mathbf{e}^{-x}\, {}_0\mathrm{F}_1\begin{pmatrix} xy \\ n \end{pmatrix}, \tag{25}$$

$$\frac{\partial}{\partial y} H_n^{k,\ell}(x,y) = \frac{1}{n}\, H_{n+1}^{k+1,\ell}(x,y). \tag{26}$$

Lemma 2.5. *If $k > 0$, then*

$$H_{n-1}^k(x,y) = H_n^k(x,y) + \frac{y}{n(n-1)}\, H_{n+1}^{k+1}(x,y), \tag{27}$$

$$k\, H_n^{k-1}(x,y) = H_n^k(x,y) - \frac{y}{n}\, H_{n+1}^k(x,y) + x^k \mathbf{e}^{-x}\, {}_0\mathrm{F}_1\begin{pmatrix} xy \\ n \end{pmatrix}, \tag{28}$$

$$\begin{aligned} k\, H_{n-1}^{k-1}(x,y) = {} & \frac{n-y-1}{n-1}\, H_n^k(x,y) + \frac{y}{n(n-1)}\, H_{n+1}^{k+1}(x,y) \\ & + x^k \mathbf{e}^{-x}\, {}_0\mathrm{F}_1\begin{pmatrix} xy \\ n-1 \end{pmatrix}. \end{aligned} \tag{29}$$

Proof. The first formula follows from the recurrence

$$ {}_0\mathrm{F}_1\left(\begin{matrix} z \\ n-1 \end{matrix}\right) = {}_0\mathrm{F}_1\left(\begin{matrix} z \\ n \end{matrix}\right) + \frac{z}{n(n-1)}\, {}_0\mathrm{F}_1\left(\begin{matrix} z \\ n+1 \end{matrix}\right) \tag{30} $$

that is equivalent to the hypergeometric equation (with $a = n$)

$$ z\,Y''(z) + a\,Y'(z) - Y(z) = 0 \tag{31} $$

for ${}_0\mathrm{F}_1(z)$. The second formula follows after integration by parts

$$ H_n^k(x,y) = -\int_0^x t^k\, {}_0\mathrm{F}_1\left(\begin{matrix} ty \\ n \end{matrix}\right) d\mathbf{e}^{-t}. \tag{32} $$

The last formula follows after substituting $n \mapsto n-1$ in (28) and eliminating $H_{n-1}^k(x,y)$ using (27). Q.E.D.

Formula (29) is equivalent to the differential equation

$$ \left(y\frac{\partial^2}{\partial y^2} + (n-y)\frac{\partial}{\partial y} - k - 1\right) H_n^k(x,y) = -x^{k+1}\mathbf{e}^{-x}\, {}_0\mathrm{F}_1\left(\begin{matrix} xy \\ n \end{matrix}\right). \tag{33} $$

We can obtain recurrences that shift only the indices k or n, presented in the following lemma. Remarkably, both formulas lose an H-term when $k = n-1$. The simplified formulas are readily applicable to the $j = m$, $j = m-1$ columns in (7).

Lemma 2.6. *If $k > 0$, then*

$$ \begin{aligned} n(n-1)H_{n-1}^k(x,y) &= n\,(y+n-1)H_n^k(x,y) + y\,(k-n+1)H_{n+1}^k(x,y) \\ &\quad - yx^{k+1}\mathbf{e}^{-x}\, {}_0\mathrm{F}_1\left(\begin{matrix} xy \\ n+1 \end{matrix}\right), \end{aligned} \tag{34} $$

$$ \begin{aligned} H_n^{k+1}(x,y) &= (y-n+2k+2)H_n^k(x,y) + k\,(n-k-1)H_n^{k-1}(x,y) \\ &\quad - (n-1)x^k\mathbf{e}^{-x}\, {}_0\mathrm{F}_1\left(\begin{matrix} xy \\ n-1 \end{matrix}\right) + (k-x)x^k\mathbf{e}^{-x}\, {}_0\mathrm{F}_1\left(\begin{matrix} xy \\ n \end{matrix}\right). \end{aligned} \tag{35} $$

Proof. The first formula is obtained by eliminating $H_n^{k-1}(x,y)$, $H_{n-1}^{k-1}(x,y)$ from these 3 equations: (28), the shift $k \mapsto k-1$ of (27), and the shift $n \mapsto n-1$ of (28). For the second formula, we eliminate $H_{n+1}^k(x,y)$, $H_{n-1}^{k-1}(x,y)$, $H_{n+1}^{k+1}(x,y)$ from these 4 equations: (28), (29), the shift $k \mapsto k-1$ of (27), and the shift $k \mapsto k+1$ of (28). Q.E.D.

Lemma 2.7. *The following recurrences with two H-terms hold, for $n>0$:*

$$(n-1)H_{n-1}^{n-1}(x,y) = (y+n-1)H_n^{n-1}(x,y) - \frac{yx^n}{n}\mathbf{e}^{-x}\,{}_0\mathrm{F}_1\left(\begin{matrix} xy \\ n+1 \end{matrix}\right), \tag{36}$$

$$H_n^n(x,y) = (y+n)H_n^{n-1}(x,y) - \frac{yx^n}{n}\mathbf{e}^{-x}\,{}_0\mathrm{F}_1\left(\begin{matrix} xy \\ n+1 \end{matrix}\right) - x^n\mathbf{e}^{-x}\,{}_0\mathrm{F}_1\left(\begin{matrix} xy \\ n \end{matrix}\right), \tag{37}$$

$$H_{n+1}^n(x,y) = n\,H_n^{n-1}(x,y) - x^n\mathbf{e}^{-x}\,{}_0\mathrm{F}_1\left(\begin{matrix} xy \\ n+1 \end{matrix}\right). \tag{38}$$

Proof. The first two formulas constitute the special case $k=n-1$ of the previous lemma. The third formula is obtained by eliminating $H_n^n(x,y)$ from (37) and the shift $n\mapsto n+1$ of (36). Q.E.D.

Formula (38) is comparable to the recurrence for the incomplete gamma function $\gamma(a,x)=\int_0^x t^{a-1}\mathbf{e}^{-t}\,dt$:

$$\gamma(a+1,z) = a\,\gamma(a,z) - x^a\mathbf{e}^{-x}. \tag{39}$$

The presented recurrences can be used to express all matrix entries in (7) in terms of $H_{n-m+1}^{n-m}(x,\lambda_i)$ and two ${}_0\mathrm{F}_1$ functions.

Proposition 2.8. *Any function $H_n^k(x,y)$ with integer $n>1$ and $k\geqslant n-1$ can be expressed as a $\mathbb{Q}(x,y)$-linear combination of*

$$H_N^{N-1}(x,y),\quad x^N\mathbf{e}^{-x}\,{}_0\mathrm{F}_1\left(\begin{matrix} xy \\ N \end{matrix}\right),\quad x^N\mathbf{e}^{-x}\,{}_0\mathrm{F}_1\left(\begin{matrix} xy \\ N+1 \end{matrix}\right)$$

for any $N>1$. The same statement applies to the derivatives of $H_n^k(x,y)$ of any order (with respect to x,y).

Proof. Lemma 2.7 proves the first claim for $k=n-1$ and $k=n$. Lemma 2.6 extends the statement to larger k. Differentiation rules (25)–(26) imply the second claim. Q.E.D.

Recurrence relations $H_n^{k,\ell}(x,y)$ are obtained by a straightforward extension of the results for $H_n^k(x,y)$.

Lemma 2.9. *For $k>0$, $\ell>0$ we have*

$$x\,H_n^{k,\ell}(x,y) = H_n^{k+1,\ell}(x,y) + H_n^{k,\ell+1}(x,y), \tag{40}$$

$$H_{n-1}^{k,\ell}(x,y) = H_n^{k,\ell}(x,y) + \frac{y}{n(n-1)}\,H_{n+1}^{k+1,\ell}(x,y), \tag{41}$$

$$H_n^{k,\ell}(x,y) = k\,H_n^{k-1,\ell}(x,y) - \ell\,H_n^{k,\ell-1}(x,y) + \frac{y}{n}\,H_{n+1}^{k,\ell}(x,y). \tag{42}$$

Proof. The first recurrence is obtained by splitting

$$(x-t)^\ell = x(x-t)^{\ell-1} - t(x-t)^{\ell-1}$$

in the defining integral (24). The other two equations follow similarly as (27)–(28), from the three-term recurrence for the ${}_0\mathrm{F}_1$ function and, respectively, by integration by parts. Q.E.D.

Lemma 2.10. *For $k>0$, $\ell>0$ we have*

$$\begin{aligned}(n-1)H_{n-1}^{k-1,\ell}(x,y) &= H_n^{k,\ell}(x,y) + \ell\, H_n^{k,\ell-1}(x,y) \\ &\quad + (n-k-1)H_n^{k-1,\ell}(x,y),\end{aligned} \tag{43}$$

$$k\,H_k^{k-1,\ell}(x,y) = H_{k+1}^{k,\ell}(x,y) + \ell\, H_{k+1}^{k,\ell-1}(x,y). \tag{44}$$

Proof. First we show this intermediate equation:

$$\begin{aligned}H_{n-1}^{k,\ell+1}(x,y) + \frac{y}{n(n-1)}\, H_{n+1}^{k+2,\ell}(x,y) &= \\ H_n^{k,\ell+1}(x,y) + \frac{xy}{n(n-1)}\, H_{n+1}^{k+1,\ell}(x,y).&\end{aligned} \tag{45}$$

It is annihilated by the relations of Lemma 2.9 as follows. The two terms with denominators are eliminated by (41) and its shift $k\mapsto k+1$. Then elimination of $H_n^{k,\ell}(x,y)$, $H_{n-1}^{k,\ell}(x,y)$ by (40) and its shift $n\mapsto n-1$ leaves no terms.

Now multiply equation (45) by $(n-1)$ and apply the shifts $k\mapsto k-1$, $\ell\mapsto \ell-1$. Then subtract (42) and eliminate the terms with denominators using the shifted version $n\mapsto n+1$, $\ell\mapsto\ell-1$ of (40). The result is (43). The second claimed recurrence is the special case $n=k+1$ of the first one. Q.E.D.

§3. Proofs and analysis

The motivation for this article was potential application of the holonomic gradient method [2] to computation of the probability density function $\psi_{n,m}(x,\lambda_1,\ldots,\lambda_m)$. Our main results are formulated in §2.2 for the function $\mathcal{R}_{n,m}(x,\lambda_1,\ldots,\lambda_m)$ as in (9). The obtained holonomic systems are more complicated than expected, in the simplest cases $m\leqslant 3$ as well. As mentioned at the end of §2.3, application of the holonomic gradient method to the entries of the conjectured matrix in (20) should be more effective than employment of multi-variate holonomic systems.

This section proves the main results presented in §2.2. Additionally, §3.2 discusses the obtained holonomic systems, and §3.4 presents explicit solutions and recurrences relevant to Conjecture 2.4.

3.1. Proof of Theorem 2.1

The product $\mathcal{T}_1[\lambda_1]\dots\mathcal{T}_1[\lambda_m]$ annihilates $\mathcal{R}_{n,m}(x,\lambda_1,\dots,\lambda_m)$ because the operators $\mathcal{T}_1[\lambda_i]=\mathcal{P}_{n,m}[\lambda_i]$ annihilate the front factor ${}_0\mathrm{F}_1\binom{x\lambda_i}{n-m+1}$ of each term in (10). Remarkably, the other operators annihilate each term in (10) as well.

Consider now the action of $\mathcal{T}_m[y]=\mathcal{Q}_{n,m}[y]$. We have

$$\mathcal{T}_m[\lambda_1]\dots\mathcal{T}_m[\lambda_m]\,\mathcal{R}_{n,m}(x,\lambda_1,\dots,\lambda_m)=$$

$$\mathbf{e}^{-x}\sum_{k=1}^{m}\mathcal{T}_m[\lambda_k]{}_0\mathrm{F}_1\binom{x\lambda_k}{n-m+1}\det\begin{pmatrix}\mathcal{T}_m[\lambda_i]H^{n-j}_{n-m+1}(x,\lambda_i)\,\rangle^{i\leqslant m}_{i\neq k}\\ x^{n-j}\qquad\qquad\rangle_{i=k}\end{pmatrix}. \tag{46}$$

We claim that all m determinants are zero, because the matrices have a specific kernel vector

$$\left(\binom{m-1}{j-1}(-x)^j\,\Big\rangle_{j=1}^{m}\right)^T. \tag{47}$$

The scalar product of this vector with the $i=k$ rows

$$\left(x^{n-1},x^{n-2},\dots,x^{n-m}\right)^T \tag{48}$$

equals 0 straightforwardly, since $\sum_{j=1}^m(-1)^j\binom{m-1}{j-1}=0$ as well known. We want to show

$$\sum_{j=1}^{m}\binom{m-1}{j-1}(-x)^j\;\mathcal{T}_m[y]H^{n-j}_{n-m+1}(x,y)=0. \tag{49}$$

By applying the differentiation

$$\mathcal{T}_m[y]=\Big(\frac{\partial}{\partial y}-1\Big)\mathcal{P}_{n,m}[y]-m\,\frac{\partial}{\partial y} \tag{50}$$

and recurrence of Lemma 2.5,

$$\begin{aligned}\mathcal{T}_m[y]H^{n-j}_{n-m+1}(x,y)&=x\,H^{n-j}_{n-m+1}(x,y)-\frac{x+j-1}{n-m+1}\,H^{n-j+1}_{n-m+2}(x,y)\\ &\quad-x^{n-j+2}\mathbf{e}^{-x}{}_0\mathrm{F}_1\binom{x\,y}{n-m+2}.\end{aligned} \tag{51}$$

We ignore the last term, for linear dependency with the $i=k$ row. By permuting the summation and integration, our target is

$$\int_0^x\mathbf{e}^{-t}t^{n-m}(x-t)^{m-1}P(x,t)dt=0 \tag{52}$$

with

$$(53)\quad P(x,t) = (x-t)\,{}_0\mathrm{F}_1\left(\begin{matrix} y\,t \\ n-m+1 \end{matrix}\right) - \frac{t\,(x+m-t)}{n-m+1}\,{}_0\mathrm{F}_1\left(\begin{matrix} y\,t \\ n-m+2 \end{matrix}\right).$$

This integral is equivalent to the recurrence relation

$$(54)\quad (n-m+1)\,H^{n-m,\,m}_{n-m+1}(x,y) = H^{n-m+1,m}_{n-m+2}(x,y) + m\,H^{n-m+1,m-1}_{n-m+2}(x,y)$$

that is equivalent to (43). The claimed relation (49) follows.

Other products $\mathcal{T}_q[\lambda_1]\cdots\mathcal{T}_q[\lambda_m]$ with $2 \leqslant q < m$ annihilate the function $\mathcal{R}_{n,m}(x,\lambda_1,\ldots,\lambda_m)$ similarly, with the kernel vectors

$$(55)\qquad \Big(\underbrace{0,\ldots,0}_{m-q},\ \Big\langle \tbinom{q-1}{j-m+q-1}(-x)^j \Big\rangle_{j=m-q+1}^{m}\Big)^T$$

of the m matrices

$$\begin{pmatrix} \mathcal{T}_q[\lambda_i]H^{n-j}_{n-m+1}(x,\lambda_i) & \rangle^{i\leqslant m}_{i\neq k} \\ x^{n-j} & \rangle_{i=k} \end{pmatrix}$$

in an expression like in (46).

For $\ell = 1,\ldots,m$, the $\mathrm{LCLM}(\mathcal{T}_1[\lambda_\ell],\ldots,\mathcal{T}_m[\lambda_\ell])$ transforms the function $\mathcal{R}_{n,m}(x,\lambda_1,\ldots,\lambda_m)$ to

$$\mathbf{e}^{-x}\sum_{k=1,\neq\ell}^{m} {}_0\mathrm{F}_1\left(\begin{matrix} x\lambda_k \\ n-m+1 \end{matrix}\right) \det\begin{pmatrix} H^{n-j}_{n-m+1}(x,\lambda_i) & \rangle^{i\leqslant m}_{i\neq k,\ell} \\ \mathrm{LCLM}\,H^{n-j}_{n-m+1}(x,\lambda_i) & \rangle_{i=\ell} \\ x^{n-j} & \rangle_{i=k} \end{pmatrix}.$$

The $i=\ell$ row is proportional to the $i=k$ row vector (48), because:

- For $q = 2,\ldots,m$, the operator $\mathcal{T}_q[\lambda_\ell]$ makes the $i=\ell$ row "orthogonal" to (55).
- The LCLM is a left factor of each $\mathcal{T}_q[\lambda_\ell]$, thus preserves the "orthogonality" property.
- The vector (48) is the only vector "orthogonal" to the $m-1$ independent vectors.

Hence the LCLM operators annihilate all m terms of $\mathcal{R}_{n,m}(x,\lambda_1,\ldots,\lambda_m)$.

3.2. Holonomic systems

Here we prove Theorem 2.3. To simplify technical details, we posit that differential Galois theory [10] extends straightforwardly to the considered holonomic systems.

Let $\mathcal{M}$ denote the system of differential operators in Theorem 2.1 annihilating $\mathcal{R}_{n,m}(x,\lambda_1,\ldots,\lambda_m)$. It is a holonomic system, because the

LCLM operators bound the order in each $\partial/\partial\lambda_i$. Since the rank of an LCLM operator equals $3m-1$, a straightforward upper bound for the holonomic rank is $(3m-1)^m$. After a choice of (Picard-Vessiot) solution basis for each $\mathcal{T}_k[y]$, the subsystem of LCLM operators has the following straightforward basis of solutions: $g_{j_1}(\lambda_1)\cdots g_{j_m}(\lambda_m)$, where $g_k(y)$ is a basis solution of $\mathcal{T}_k$. Let $\mathcal{B}$ denote this basis of $(3m-1)^m$ functions.

The solution space of $\mathcal{M}$ is considered inside the span of $\mathcal{B}$. The following $2m!\cdot 3^{m-1}$ functions in $\mathcal{B}$ will be solutions of the differential system $\mathcal{M}$: $g_1(\lambda_{j_1})\cdots g_m(\lambda_{j_m})$, where $g_k(y)$ is a basis solution of $\mathcal{T}_k[y]$, and $(j_1,\ldots,j_m)$ is a permutation of $(1,\ldots,m)$. Any other element of $\mathcal{B}$ is not annihilated by at least one operator in *(i)* of Theorem 2.1, and a linear combination of these elements will not be nullified by the same operator(s). The claim *(i)* of Theorem 2.3 follows.

The solution space of $\mathcal{M}$ splits into a direct sum of $2\cdot 3^{m-1}$ subspaces that are invariant under the permutations of $\lambda_1,\ldots,\lambda_m$. Each of these subspaces gives one independent anti-symmetric solution, and the claim *(ii)* follows.

Each operator $\mathcal{T}_j[y]$ has logarithmic solutions at $y=0$. A broad reason is that appearance of ${}_0\mathrm{F}_1\left({z \atop n}\right)$ functions brings the ill-determined functions ${}_0\mathrm{F}_1\left({z \atop 2-n}\right)$. More precisely, logarithmic solutions appear in a limit $a\to n$ of the general solution

$$C'\,{}_0\mathrm{F}_1\left(\begin{matrix} z \\ a \end{matrix}\right)+C''\,z^{1-a}\,{}_0\mathrm{F}_1\left(\begin{matrix} z \\ 2-a \end{matrix}\right) \tag{56}$$

of the hypergeometric equation (31) with generic $a\in\mathbb{C}$. Analysis of local solutions of $\mathcal{T}_k[y]$ at the singularities $y=0$, $y=\infty$ shows that the space of non-logarithmic solutions of $\mathcal{T}_k[y]$ is one-dimensional for $k=1$ and two-dimensional for $k\geqslant 2$. Explicit instances in §3.4 demonstrate this. Similarly as above, the space of non-logarithmic solutions for $\mathcal{M}$ has the dimension $2^{m-1}m!$, and the space of non-logarithmic anti-symmetric solutions has the dimension 2^{m-1}.

Existence of a holonomic system of rank $\leqslant 3^m-1$ of claim *(iv)* follows from Proposition 2.8. It allows to express $\mathcal{R}_{n,m}(x,\lambda_1,\ldots,\lambda_m)$ and its derivatives as $\mathbb{Q}(x,\lambda_1,\ldots,\lambda_m)$-linear combinations of $f_1(\lambda_1)\cdots f_m(\lambda_m)$, where each

$$f_j(\lambda_j)\in\left\{H_n^{n-1}(x,\lambda_j),\,x^n\mathbf{e}^{-x}{}_0\mathrm{F}_1\left(\begin{matrix} x\lambda_j \\ n \end{matrix}\right),x^n\mathbf{e}^{-x}{}_0\mathrm{F}_1\left(\begin{matrix} x\lambda_j \\ n+1 \end{matrix}\right)\right\}. \tag{57}$$

These functions generate the space of dimension 3^m, but the term

$$H_n^{n-1}(x,\lambda_1)\cdots H_n^{n-1}(x,\lambda_m)$$

does not appear, because $\mathcal{R}_{n,m}(x,\lambda_1,\ldots,\lambda_m)$ is defined after applying $\partial/\partial x$, and further differentiations will not bring this term back. Examples of these linear expressions are given in §4.3.

Remark 3.1. The determinants in (17) linearly generate the space of anti-symmetric solutions in Theorem 2.3 *(ii)*. They form a Grassmanian-like variety in this space. Taking scalar multiplication of the rows and the whole matrix into account, the dimension of this variety equals $(2-1)+(m-1)(3-1)+1=2m$. Similarly, the subvariety of non-logarithmic anti-symmetric determinantal solutions has the dimension $(1-1)+(m-1)\cdot 1+1=m$. For $m=2$, comparison with the dimension count in *(iii)* of Theorem 2.3 implies that a determinantal formula like (22) is inevitable.

Remark 3.2. In the proof of Theorem 2.1 we may start with any $m-1$ independent vectors $(v_1^{(q)},\ldots,v_m^{(q)})$ "orthogonal" to monomial vector $(x,x^2,\ldots,x^m)^T$, and take for $\mathcal{T}_2[y],\ldots,\mathcal{T}_m[y]$ the operators annihilating $\sum_{j=1}^m v_j^{(q)} H_{n-m+1}^{n-j}(x,y)$ up to a term proportional to (48). The alternative operators

$$\mathcal{T}_k[\lambda_1]\cdots\mathcal{T}_k[\lambda_m], \qquad \mathrm{LCLM}(\mathcal{T}_1[\lambda_k],\ldots,\mathcal{T}_m[\lambda_k])$$

would generate a holonomic system annihilating $\mathcal{R}_{n,m}(x,\lambda_1,\ldots,\lambda_m)$ by the same reasons. They would have order 3 as well by Proposition 2.8, but they would be more complicated, with additional singularities. For example, taking $m=3$ and the vector $(1,-x,0)^T$ gives the differential operator

$$\mathcal{Q}_{n,3}[y]+\frac{x}{xy+(n-2)(x+2)}\left(-y\frac{\partial^2}{\partial y^2}+y\frac{\partial}{\partial y}+x+2\right) \tag{58}$$

instead of $\mathcal{Q}_{n-1,2}[y]$ corresponding to $(0,1,-x)^T$. As shown in §4.4, the LCLM operators are apparently the same as in Theorem 2.1. This powerfully demonstrates non-uniqueness of LCLM factorization in non-commutative Weyl algebras [11]. But different products in Theorem 2.1*(i)* lead to different holonomic systems, of the same rank though.

3.3. Proof of Theorem 2.2

We have to prove that

$$x\frac{\partial}{\partial x}+\sum_{k=1}^m\left(\lambda_k\frac{\partial^2}{\partial\lambda_k^2}+(n-m+1-\lambda_k)\frac{\partial}{\partial\lambda_k}\right) \tag{59}$$

multiplies $\mathcal{R}_{n,m}(x,\lambda_1,\ldots,\lambda_m)$ by the constant $mn-\binom{m}{2}-1$. Note that only $\partial/\partial x$ in (59) splits the summation factors in (11) by Leibniz rule, because the determinants do not depend on the respective λ_k in each summand of (11).

The action of (59) on the front factor $\mathbf{e}^{-x}$ in (11) is multiplication by $-x$. This is compensated by the action on the ${}_0\mathrm{F}_1$ factors in (11), because:

- $x\frac{\partial}{\partial x} - \lambda_k \frac{\partial}{\partial \lambda_k}$ nullifies ${}_0\mathrm{F}_1(x\lambda_k)$;
- $\lambda_k \frac{\partial^2}{\partial \lambda_k^2} + (n-m+1)\frac{\partial}{\partial \lambda_k} = \mathcal{P}_{n-m}[\lambda_k] + x.$

The action on the determinants gives

$$(60)\qquad \mathbf{e}^{-x}\sum_{k=1}^{m} {}_0\mathrm{F}_1\begin{pmatrix} x\lambda_k \\ n-m+1 \end{pmatrix} \det\begin{pmatrix} H^{n-j}_{n-m+1}(x,\lambda_i)\ \rangle^{i\leqslant m}_{i\neq k} \\ (n-j)\, x^{n-j} \qquad \rangle_{i=k} \end{pmatrix}$$

$$+\,\mathbf{e}^{-x}\sum_{k,\ell=1,k\neq\ell}^{m}\sum {}_0\mathrm{F}_1\begin{pmatrix} x\lambda_k \\ n-m+1 \end{pmatrix} \det\begin{pmatrix} H^{n-j}_{n-m+1}(x,\lambda_i) & \rangle^{i\leqslant m}_{i\neq k,\ell} \\ (n-j+1)\, H^{n-j}_{n-m+1}(x,\lambda_\ell) & \rangle_{i=\ell} \\ x^{n-j} & \rangle_{i=k} \end{pmatrix},$$

because:

- Applying $\partial/\partial x$ to the rows $i \neq k$ gives linear dependence with the $i = k$ row;
- Formula (33) implies

$$(61)\qquad \Big(\lambda_\ell \frac{\partial^2}{\partial \lambda_\ell^2} + (n-m+1+\lambda_\ell)\,\frac{\partial}{\partial \lambda_\ell}\Big) H^{n-j}_{n-m+1}(x,\lambda_\ell)$$
$$= (n-j+1)\, H^{n-j}_{n-m+1}(x,\lambda_\ell)\ - x^{n-j+1}\mathbf{e}^{-x}\, {}_0\mathrm{F}_1\begin{pmatrix} x\,\lambda_\ell \\ n-m+1 \end{pmatrix};$$

- The new terms with ${}_0\mathrm{F}_1(x\lambda_\ell)$ can be ignored by linear combination with the $i = k$ row.

We split the factors $(n-j+1)$ in the rows $i = \ell$ into $(n-j)$ and $(+1)$. The $(+1)$'s aggregate to multiplication of $\mathcal{R}_{n,m}(x,\lambda_1,\ldots,\lambda_m)$ by $m(m-1)/m$. The modified summation (60) becomes a sum of m special instances $c_j = n - j$ of the following lemma:

Lemma 3.3. *For any $m \times m$ matrix $\big(a_{i,j}\ \rangle^m_{i=1}\big)$ and m scalars $c_1,\ldots,c_m$ we have*

$$\sum_{\ell=1}^{m} \det\begin{pmatrix} a_{i,j} & \rangle^{i\leqslant m}_{i\neq\ell} \\ c_j\, a_{i,j} & \rangle_{i=\ell} \end{pmatrix} = \Big(\sum_{j=1}^{m} c_j\Big) \det\Big(a_{i,j}\ \rangle^m_{i=1}\Big).$$

Proof. Each of the $m!$ expanded terms of $\det\big(a_{i,j}\big)$ gets multiplied by $c_1,\ldots,c_m$ among the $m\cdot m!$ expanded terms on the left-hand side. Q.E.D.

In conclusion, (59) multiplies $\mathcal{R}_{n,m}(x, \lambda_1, \ldots, \lambda_m)$ by

$$m - 1 + \sum_{j=1}^{m} (n - j) = m - 1 + mn - \binom{m+1}{2} = mn - \binom{m}{2} - 1.$$

3.4. Solutions and recurrences

Here we present solutions of the differential equations $\mathcal{Q}_{N,M}[y]Y = 0$ with $N - M \in \{2, 3, 4\}$.

Lemma 3.4. *Let N denote a positive integer, and let*

$$\begin{aligned} L_2 &= y - x + N - 1, \\ L_3 &= L_2^2 + 2x - N + 1, \\ L_4 &= L_2^3 + 3\,(2x - N + 1)\,(L_2 - 1) - N + 1. \end{aligned}$$

- *A solution of $\mathcal{Q}_{N,N-2}[y]\,Y = 0$ is*

$$Y_{N,2}(x, y) = N\,{}_0\mathrm{F}_1\left(\begin{matrix} xy \\ N \end{matrix}\right) + y\,{}_0\mathrm{F}_1\left(\begin{matrix} xy \\ N+1 \end{matrix}\right) + L_2\,\mathbf{e}^y\,H^0_{N+1}(y, x). \tag{62}$$

- *A solution of $\mathcal{Q}_{N,N-3}[y]\,Y = 0$ is*

$$\begin{aligned} Y_{N,3}(x, y) = N\,(L_2 - 2)\,{}_0\mathrm{F}_1\left(\begin{matrix} xy \\ N \end{matrix}\right) + y\,(L_2 - 1)\,{}_0\mathrm{F}_1\left(\begin{matrix} xy \\ N+1 \end{matrix}\right) \\ + L_3\,\mathbf{e}^y\,H^0_{N+1}(y, x). \end{aligned} \tag{63}$$

- *A solution of $\mathcal{Q}_{N,N-4}[y]\,Y = 0$ is*

$$\begin{aligned} Y_{N,4}(x, y) = {} & N\left((L_2 - 2)^2 + 2y + 2x + 2\right){}_0\mathrm{F}_1\left(\begin{matrix} xy \\ N \end{matrix}\right) \\ & + y\left((L_2 - 1)^2 + y + 3x - N + 2\right){}_0\mathrm{F}_1\left(\begin{matrix} xy \\ N+1 \end{matrix}\right) \\ & + L_4\,\mathbf{e}^y\,H^0_{N+1}(y, x). \end{aligned} \tag{64}$$

Proof. The operator $\mathcal{Q}_{N,N-2}[y]$ factors as $\mathcal{K}_2\mathcal{K}_1$ in $\mathbb{R}(x, N, y)\langle \partial/\partial y \rangle$, with

$$\begin{aligned} \mathcal{K}_1 &= \frac{\partial}{\partial y} - 1 - \frac{1}{L_2}, \\ \mathcal{K}_2 &= y\frac{\partial^2}{\partial y^2} + \left(N + \frac{y}{L_2}\right)\frac{\partial}{\partial y} - x - 1 + \frac{y + N}{L_2} - \frac{y}{L_2^2}. \end{aligned}$$

The solution of $\mathcal{K}_1 Y_2 = 0$ is $Y_2(x,y) = L_2 \mathbf{e}^y$. This is a solution of $\mathcal{Q}_{N,N-2}[y]\, Y = \mathcal{K}_2\mathcal{K}_1 Y = 0$ as well. A non-logarithmic solution of $\mathcal{K}_2 Y = 0$ is

$$Y_3(y) = \frac{1}{L_2}\left({}_0\mathrm{F}_1\left(\begin{matrix} xy \\ N \end{matrix}\right) + \frac{y}{N}\,{}_0\mathrm{F}_1\left(\begin{matrix} xy \\ N+1 \end{matrix}\right)\right).$$

Solving $\mathcal{Q}_{N,N-2}[y]Y = 0$ now means solving the non-homogeneous $\mathcal{K}_1 Y = Y_3$. This leads to the integration

$$\int_0^y \frac{\mathbf{e}^{-t}}{(t-x+N-1)^2}\left({}_0\mathrm{F}_1\left(\begin{matrix} xt \\ N \end{matrix}\right) + \frac{t}{N}\,{}_0\mathrm{F}_1\left(\begin{matrix} xt \\ N+1 \end{matrix}\right)\right) dt.$$

After a step of integration by parts (and multiplication by N), we obtain (62).

The other two operators $\mathcal{Q}_{N,N-M}[y]$, $M \in \{3,4\}$ factor similarly $\mathcal{K}_2^{(M)}\mathcal{K}_1^{(M)}$ as $\mathcal{Q}_{N,N-2}[y]$. In the same way, by solving the first order $\mathcal{K}_1^{(M)} Y_2^{(M)} = 0$ and the second order $\mathcal{K}_2^{(M)} Y_3^{(M)} = 0$, we are led to solving the non-homogeneous first order $\mathcal{K}_1^{(M)} Y = Y_3^{(M)}$. The equations $\mathcal{K}_2^{(M)} Y_3^{(M)} = 0$ have $M-1$ apparent singularities defined by $L_M Y = 0$. Maple 18 does not solve them, but looking at (non-logarithmic) power series solutions multiplied by L_M we recognize the solutions

$$Y_3^{(M)} = \frac{1}{L_M}\sum_{j=0}^{M}\binom{M}{j}\frac{y^j}{(N-M+1)_j}\,{}_0\mathrm{F}_1\left(\begin{matrix} xy \\ N-M+j+1 \end{matrix}\right). \tag{65}$$

Similarly as for $M = 2$, the solutions of $\mathcal{K}_1^{(M)} Y = Y_3^{(M)}$ are integrals that can be simplified to (63) or (64) by applying integration by parts a few times. Q.E.D.

A general non-logarithmic solution of $\mathcal{Q}_{N,N-2}[y]\, Y = 0$ is

$$C_1 Y_{N,2}(x,y) + C_2 L_2 \mathbf{e}^y. \tag{66}$$

The function $G_{n,2}(x,y)$ in (18) differs from $Y_{n,2}(x,y)$ by

$$C_2 = -\int_0^\infty \mathbf{e}^{-t}\,{}_0\mathrm{F}_1\left(\begin{matrix} xt \\ n+1 \end{matrix}\right) dt \tag{67}$$

$$= -\,n\,x^{-n}\,\mathbf{e}^x\,\gamma(n,x). \tag{68}$$

The latter expression is obtained by expanding the ${}_0\mathrm{F}_1$-series, and recognizing a ${}_1\mathrm{F}_1$-series for the incomplete gamma function $\gamma(n,x)$ after the definite integration. Similarly, general non-logarithmic solutions of

$\mathcal{Q}_{N,N-3}[y]$, $\mathcal{Q}_{N,N-4}[y]$ are obtained by scalar multiplication (by C_1) and considering the integral $H^0_{N+1}(y,x)$ with an integration constant C_2.

We observe empirically, especially from differentiation relations between L_2, L_3, L_4, that

$$Y_{N,M-1}(x,y) = (M-1)\left(\frac{\partial}{\partial y} - 1\right) Y_{N,M}(x,y) \tag{69}$$

for $M = 3, 4$. This observation indeed generalizes, leading us to lowering and raising operators on non-logarithmic solutions.

Theorem 3.5. *Let $Y_{N,M}(x,y)$ denote a solution of $\mathcal{Q}_{N,N-M}[y]$.*

(*i*) *The differential operator $\dfrac{\partial}{\partial y} - 1$ transforms $Y_{N,M}(x,y)$ to a solution of $\mathcal{Q}_{N,N-M+1}[y]$.*

(*ii*) *The differential operator*

$$y\,\frac{\partial^2}{\partial y^2} + (N-M+1)\,\frac{\partial}{\partial y} - x - M \tag{70}$$

transforms $Y_{N,M}(x,y)$ to a solution of $\mathcal{Q}_{N,N-M-1}[y]$.

(*iii*) *If $Y_{N,M-1}(x,y) = \left(\dfrac{\partial}{\partial y} - 1\right) Y_{N,M}(x,y)$ and $Y_{N,M-2}(x,y) = \left(\dfrac{\partial}{\partial y} - 1\right)^2 Y_{N,M}(x,y)$, then*

$$\begin{aligned}&(y - x + N - 2M + 1)Y_{N,M}(x,y)\\ &+ (2y + N - M + 1)Y_{N,M-1}(x,y) + yY_{N,M-2}(x,y)\end{aligned} \tag{71}$$

is a solution of $\mathcal{Q}_{N,N-M-1}[y]$.

Proof. The first claim follows from the commutation relation

$$\mathcal{Q}_{N,N-M+1}[y]\left(\frac{\partial}{\partial y} - 1\right) = \left(\frac{\partial}{\partial y} - 1\right)\mathcal{Q}_{N,N-M}[y]. \tag{72}$$

The second claim follows by rewriting

$$\mathcal{Q}_{N,N-M-1}[y] = \left(y\,\frac{\partial^2}{\partial y^2} + (N-M+1)\,\frac{\partial}{\partial y} - x - M\right)\left(\frac{\partial}{\partial y} - 1\right) - M.$$

The last claim similarly follows from

$$\begin{aligned}\mathcal{Q}_{N,N-M-1}[y] = {} & y\left(\frac{\partial}{\partial y} - 1\right)^3 + (2y + N - M + 1)\left(\frac{\partial}{\partial y} - 1\right)^2\\ & + (y - x + N - 2M + 1)\left(\frac{\partial}{\partial y} - 1\right) - M.\end{aligned} \tag{73}$$

Q.E.D.

The recurrence (19) in Conjecture 2.4 is a slight modification of the differential operator (70), and the functions $G_{n,m}(x,y)$ differ from the solutions $Y_{n,m}(x,y)$ of this section by the difference (67) and the sign $(-1)^m$.

§4. Holonomic systems for $m \leqslant 3$

The results of this article originated from explicit computations for the matrix dimensions $m = 2$ and $m = 3$. The aim was holonomic systems for $\psi_{n,m}(x, \lambda_1, \ldots, \lambda_m)$, so that the holonomic gradient method [2] could be applied for numeric computation of the probability density function.

4.1. The rank 12 system

With $m = 2$, the holonomic system of Theorem 2.1 has rank 12. It is easy to obtain from standard differential equations for ${}_0\mathrm{F}_1\left({x\lambda_1 \atop n-1}\right)$, ${}_0\mathrm{F}_1\left({x\lambda_2 \atop n-1}\right)$ and the integrals

$$H_{n-1}^{n-1}(x,\lambda_1), \quad H_{n-1}^{n-1}(x,\lambda_2), \quad H_{n-2}^{n-1}(x,\lambda_1), \quad H_{n-2}^{n-1}(x,\lambda_2).$$

This was demonstrated by Christoph Koutschan on his Mathematica package. The singularities of the holonomic system are along

$$x = 0, \quad \lambda_1 = 0, \quad \lambda_2 = 0, \quad \infty\text{-compactifations.} \tag{74}$$

It is generated by these three differential operators of order 2 or 3:

$$\lambda_1 \frac{\partial^2}{\partial \lambda_1^2} + \lambda_2 \frac{\partial^2}{\partial \lambda_2^2} - (\lambda_1 - n + 1)\frac{\partial}{\partial \lambda_1} - (\lambda_2 - n + 1)\frac{\partial}{\partial \lambda_2} + x\frac{\partial}{\partial x} - 2n + 2, \tag{75}$$

$$\frac{\partial^3}{\partial x \partial \lambda_1 \partial \lambda_2} + 2\frac{\partial^2}{\partial \lambda_1 \partial \lambda_2} - \frac{\partial}{\partial \lambda_1} - \frac{\partial}{\partial \lambda_2}, \tag{76}$$

$$\begin{aligned}&\left(\lambda_1\lambda_2\frac{\partial}{\partial \lambda_1} + \lambda_1\lambda_2\frac{\partial}{\partial \lambda_2} + (n-1)(\lambda_1+\lambda_2)\right)\frac{\partial^2}{\partial \lambda_1 \partial \lambda_2} + (n-1)x \\ &+\left(x\frac{\partial}{\partial x} + 2x - n + 2\right)\left(x\frac{\partial}{\partial x} - \lambda_1\frac{\partial}{\partial \lambda_1} - \lambda_2\frac{\partial}{\partial \lambda_2} + x - 2n + 2\right).\end{aligned} \tag{77}$$

The first equation as in Theorem 2.2. For the sake of compactness, the last equation is expressed using non-commutative multiplication (in the last term).

Elimination of $\partial/\partial x$ and $\partial/\partial\lambda_2$ leads to the fifth order operator

$$\begin{aligned} &\lambda_1^2\frac{\partial^5}{\partial\lambda_1^5}+\lambda_1(2n-\lambda_1+2)\frac{\partial^4}{\partial\lambda_1^4}+(n^2+n-2x\lambda_1-2n\lambda_1-3\lambda_1)\frac{\partial^3}{\partial\lambda_1^3}\\ &\qquad -(n^2+2n-2x\lambda_1+2nx)\frac{\partial^2}{\partial\lambda_1^2}+x(2n+x+1)\frac{\partial}{\partial\lambda_1}-x^2. \end{aligned} \tag{78}$$

Consistent with the theory of Gröbner bases, the rank 8 system has elimination equations with the leading monomials $\partial^4/\partial\lambda_1^3\partial\lambda_2+\ldots$ and $\partial^2/\partial\lambda_2^2+\ldots$. The fifth order operator does not involve the variable λ_2 even. It factorizes as the LCLM of $\mathcal{P}_{n-2}[\lambda_1]$ and $\mathcal{Q}_{n,n-2}[\lambda_1]$. Correspondingly, the holonomic system factorizes nicely to a direct sum of two 6-dimensional subspaces, each of those subspaces is a tensor product of a rank 2 system in one variable λ_1 or λ_2, and rank 3 system in the other variable. The factorization corresponds nicely with the terms in the expanded determinantal formula (22), as ${}_0\mathrm{F}_1\left({xy \atop n-1}\right)$ is a solution of $\mathcal{P}_{n-2}[y]$, and $G_{n,2}(x,y)$ is a solution of $\mathcal{Q}_{n,n-2}[y]$.

4.2. Proof of formula (22)

We seek to prove

$$\mathcal{R}_{n,2}(x,\lambda_1,\lambda_2)=\frac{x^{2n-2}\,\mathbf{e}^{-2x}}{n\,(n-1)}\det\begin{pmatrix} {}_0\mathrm{F}_1\left({x\lambda_1 \atop n-1}\right) & {}_0\mathrm{F}_1\left({x\lambda_2 \atop n-1}\right)\\ G_{n,2}(x,\lambda_1) & G_{n,2}(x,\lambda_2)\end{pmatrix}. \tag{79}$$

With the same holonomic system of rank 12 established for both sides, it is enough to compare a few coefficients in the two series expansions in λ_1,λ_2. The subspace of non-logarithmic anti-symmetric solutions is 2-dimensional, hence it is enough to compare 2 pairs of independent coefficients.

After division by $\lambda_1-\lambda_2$ as in (22), a proper general setting is expansion in terms of the symmetric *Schur polynomials* [11] in λ_1,λ_2. The Schur polynomials functions are defined in terms of monomial determinants. Correspondingly, we formulate the following general statement.

Lemma 4.1. *Consider m functions $f_1(y),\ldots,f_m(y)$ defined by the convergent series*

$$f_k(y)=\sum_{j=0}^{\infty}c_j^{(k)}y^j. \tag{80}$$

Then

$$\det\left(\,f_i(\lambda_j)\,\right)_{i=1}^{m}=\sum\cdots\sum_{0\leqslant q_1<q_2<\ldots<q_m}\det\left(c_{q_j}^{(i)}\,\right)_{j=1}^{m}\det\left(\lambda_i^{q_j}\,\right)_{j=1}^{m}. \tag{81}$$

Proof. Intermediate expansions are

$$= \sum_{q_1=0}^{\infty} \cdots \sum_{q_m=0}^{\infty} \det \left(c_{q_j}^{(i)} \right)_{j=1}^{m} \lambda_1^{q_1} \cdots \lambda_m^{q_m} \tag{82}$$

$$= \sum_{q_1=0}^{\infty} \cdots \sum_{q_m=0}^{\infty} c_{q_1}^{(1)} \cdots c_{q_m}^{(m)} \det \left(\lambda_i^{q_j} \right)_{j=1}^{m}. \tag{83}$$

The newest determinants with some $q_i = q_j$ for $i \neq j$ are zero. After collecting the terms with the same sets $\{q_1, \ldots, q_m\}$, we get the result. Q.E.D.

We first apply this lemma to the determinant in (23) with $f_1(y) = H_{n-1}^{n-1}(x,y)$, $f_2(y) = H_{n-1}^{n-2}(x,y)$. Therefore

$$c_j^{(1)} = \frac{\gamma(n+j,x)}{(n-1)_j\, j!}, \qquad c_j^{(2)} = \frac{\gamma(n+j-1,x)}{(n-1)_j\, j!}.$$

We differentiate (81) to get $\mathcal{R}_{n,2}(x,\lambda_1,\lambda_2)$ as

$$\sum_{i=0}^{\infty} \sum_{j=i+1}^{\infty} \frac{\frac{d}{dx} \det \begin{pmatrix} \gamma(n+i,x) & \gamma(n+i-1,x) \\ \gamma(n+j,x) & \gamma(n+j-1,x) \end{pmatrix}}{(n-1)_i (n-1)_j\, i! j!} \det \begin{pmatrix} \lambda_1^i & \lambda_2^i \\ \lambda_1^j & \lambda_2^j \end{pmatrix}$$

$$\begin{aligned} = {} & \frac{d}{dx} \det \begin{pmatrix} \gamma(n,x) & \gamma(n-1,x) \\ \gamma(n+1,x) & \gamma(n,x) \end{pmatrix} \frac{\lambda_2 - \lambda_1}{n-1} \\ & + \frac{d}{dx} \det \begin{pmatrix} \gamma(n,x) & \gamma(n-1,x) \\ \gamma(n+2,x) & \gamma(n+1,x) \end{pmatrix} \frac{\lambda_2^2 - \lambda_1^2}{2n(n-1)} \\ & + \cdots \end{aligned} \tag{84}$$

$$\begin{aligned} = {} & \left(\left(\frac{x^2}{n-1} - 2x + n \right) \gamma(n-1,x) + \frac{x-n}{n-1}\, x^{n-1} \mathbf{e}^{-x} \right) x^{n-2} \mathbf{e}^{-x} (\lambda_1 - \lambda_2) \\ & + \left(\left(\frac{x^3}{n(n-1)} - \frac{x^2}{n} - x + n + 1 \right) \gamma(n-1,x) \right. \\ & \qquad \left. + \frac{(x-n-1)(x+n)}{n(n-1)}\, x^{n-1} \mathbf{e}^{-x} \right) x^{n-2} \mathbf{e}^{-x}\, \frac{\lambda_1^2 - \lambda_2^2}{2} \\ & + \cdots . \end{aligned} \tag{85}$$

Surely, recurrence (39) has been used. Considering the left-hand side of (79), we set $c_j^{(1)} = x^j / \big(j!\,(n-1)_j\big)$. An expansion of (62) with $N = n$ is

$$Y_{n,2}(x,y) = n + ny + \sum_{k=0}^{\infty} \left(n + \sum_{j=0}^{k} \frac{k+1-j}{(n+1)_j}\, x^j \right) \frac{y^{k+2}}{(k+2)!}. \tag{86}$$

We only need the first few terms. To get coefficients of $G_{n,2}(x,y)$, we add the difference $C_2(y-x+n-1)\mathbf{e}^y$ as in (67)–(68), thus adding

$$n\,x^{-n}\,\mathbf{e}^x\,\gamma(n,x)\sum_{k=0}^{\infty}(x-n-k+1)\frac{y^k}{k!}.$$

Therefore

$$\begin{aligned} c_0^{(2)} &= n+n\,(x-n+1)\,x^{-n}\mathbf{e}^x\gamma(n,x),\\ c_1^{(2)} &= n+n\,(x-n)\,x^{-n}\mathbf{e}^x\gamma(n,x),\\ c_2^{(2)} &= \frac{n+1}{2}+\frac{n}{2}\,(x-n-1)\,x^{-n}\mathbf{e}^x\gamma(n,x). \end{aligned}\tag{87}$$

The determinant in (79) expands as

$$\det\begin{pmatrix} 1 & \frac{x}{n-1}\\ c_0^{(2)} & c_1^{(2)}\end{pmatrix}(\lambda_2-\lambda_1)+\det\begin{pmatrix} 1 & \frac{x^2}{2n(n-1)}\\ c_0^{(2)} & c_2^{(2)}\end{pmatrix}(\lambda_2^2-\lambda_1^2).$$

We get the same two terms as in (86).

4.3. The rank 8 system

A rank 8 holonomic system for $\mathcal{R}_{n,2}(x,\lambda_1,\lambda_2)$ is obtained by expressing this function and its derivatives as linear combinations of

$$\begin{aligned} &x^n\,\mathbf{e}^{-x}{}_0\mathrm{F}_1\left(\begin{matrix}x\lambda_2\\ n-1\end{matrix}\right)H_{n-1}^n(x,\lambda_1), && x^n\,\mathbf{e}^{-x}{}_0\mathrm{F}_1\left(\begin{matrix}x\lambda_2\\ n\end{matrix}\right)H_{n-1}^n(x,\lambda_1),\\ &x^n\,\mathbf{e}^{-x}{}_0\mathrm{F}_1\left(\begin{matrix}x\lambda_1\\ n-1\end{matrix}\right)H_{n-1}^n(x,\lambda_2), && x^n\,\mathbf{e}^{-x}{}_0\mathrm{F}_1\left(\begin{matrix}x\lambda_1\\ n\end{matrix}\right)H_{n-1}^n(x,\lambda_2),\\ &x^{2n}\,\mathbf{e}^{-2x}{}_0\mathrm{F}_1\left(\begin{matrix}x\lambda_1\\ n-1\end{matrix}\right){}_0\mathrm{F}_1\left(\begin{matrix}x\lambda_2\\ n-1\end{matrix}\right), && x^{2n}\,\mathbf{e}^{-2x}{}_0\mathrm{F}_1\left(\begin{matrix}x\lambda_1\\ n-1\end{matrix}\right){}_0\mathrm{F}_1\left(\begin{matrix}x\lambda_2\\ n\end{matrix}\right),\\ &x^{2n}\,\mathbf{e}^{-2x}{}_0\mathrm{F}_1\left(\begin{matrix}x\lambda_1\\ n\end{matrix}\right){}_0\mathrm{F}_1\left(\begin{matrix}x\lambda_2\\ n-1\end{matrix}\right), && x^{2n}\,\mathbf{e}^{-2x}{}_0\mathrm{F}_1\left(\begin{matrix}x\lambda_1\\ n\end{matrix}\right){}_0\mathrm{F}_1\left(\begin{matrix}x\lambda_2\\ n\end{matrix}\right). \end{aligned}$$

These expressions follow from Theorem 2.3 *(iv)*. For example, the expression of $\mathcal{R}_{n,2}(x,\lambda_1,\lambda_2)$ has these coefficients, respectively:

$$\frac{\lambda_1-x}{n-1}+1,\ 0,\ -\frac{\lambda_2-x}{n-1}-1,\ 0,\ 0,\ \frac{x}{n-1}-1,\ -\frac{x}{n-1}+1,\ 0.$$

Further, the expression of $(n-1)\,\partial\mathcal{R}_{n,2}/\partial x$ has these coefficients:

$$\begin{aligned} &-1, && \lambda_2\left(\frac{\lambda_1-x}{n-1}+1\right), && 1, && -\lambda_1\left(\frac{\lambda_2-x}{n-1}+1\right),\\ &0, && -\lambda_2+1, && \lambda_1-1, && (\lambda_1-\lambda_2)\left(\frac{x}{n-1}-1\right); \end{aligned}\tag{88}$$

and so on. The rank 8 system has singularities not just along (74), but additionally along $\lambda_1 = \lambda_2$ and the hypersurface $S_n(x,y) = 0$, where

$$S_n(x,y) = (2x-n+1)^2\,(\lambda_1+\lambda_2-2x+2n-2) - \frac{x}{2}\,(\lambda_1-\lambda_2)^2. \tag{89}$$

It contains another second order operator

$$\begin{aligned}
&S_n(x,y)\Big(x\mathcal{D}_x^2+\frac{2\lambda_1\lambda_2}{x}\frac{\partial^2}{\partial\lambda_1\partial\lambda_2}-\mathcal{D}_\lambda(2\mathcal{D}_x+1)+(x-n+1)\mathcal{D}_x+\lambda_1+\lambda_2-1\Big)\\
&+(\lambda_1+\lambda_2-4x+2n-2)\Big((x-n+1)\Big(\mathcal{D}_\lambda\mathcal{D}_x-\frac{2\lambda_1\lambda_2}{x}\frac{\partial^2}{\partial\lambda_1\partial\lambda_2}\Big)\\
&\qquad\qquad+(2x-n)\,(\mathcal{D}_\lambda+(x-n+1)\mathcal{D}_x-1)\Big)\\
&+2(2x-n+1)\Big((\lambda_1+\lambda_2)(\mathcal{D}_\lambda+2x-n+1)-\lambda_1^2\frac{\partial}{\partial\lambda_1^2}-\lambda_2^2\frac{\partial}{\partial\lambda_2^2}-\frac{(\lambda_1+\lambda_2)^2}{2}\Big)\\
&+2\lambda_1\lambda_2\Big(x\mathcal{D}_\lambda\mathcal{D}_x-\frac{\lambda_1+\lambda_2}{2}-2x+n\Big)\\
&+(\lambda_1-\lambda_2)\Big(\lambda_1^2\frac{\partial^2}{\partial\lambda_1^2}-\lambda_2^2\frac{\partial^2}{\partial\lambda_2^2}-\lambda_1^2\frac{\partial}{\partial\lambda_1}+\lambda_2^2\frac{\partial}{\partial\lambda_2}\Big),
\end{aligned} \tag{90}$$

where

$$\mathcal{D}_x = \frac{\partial}{\partial x}+1-\frac{n-2}{x}, \qquad \mathcal{D}_\lambda = \lambda_1\frac{\partial}{\partial\lambda_1}+\lambda_2\frac{\partial}{\partial\lambda_2}. \tag{91}$$

The smaller rank system appears to be more complex. It can be obtained from the rank 12 system by adjoining this 3rd order operator:

$$\begin{aligned}
&\Big(2\lambda_1\frac{\partial}{\partial\lambda_1}+2\lambda_2\frac{\partial}{\partial\lambda_2}-3\lambda_1-3\lambda_2+4n-6\Big)\frac{\partial^2}{\partial\lambda_1\partial\lambda_2}+\lambda_2\frac{\partial}{\partial\lambda_1}+\lambda_1\frac{\partial}{\partial\lambda_2}\\
&-\Big(x\frac{\partial^2}{\partial x^2}+(x-n+3)\frac{\partial}{\partial x}+n\Big)\Big(\frac{\partial}{\partial\lambda_1}+\frac{\partial}{\partial\lambda_2}\Big)+3x\frac{\partial}{\partial x}-2n+6.
\end{aligned} \tag{92}$$

Elimination of $\partial/\partial x$ and $\partial/\partial\lambda_2$ leads to the same fifth order operator (78). The elimination equations with the leading monomials $\partial^4/\partial\lambda_1^2\partial^2\lambda_2+\ldots$ and $\partial^2/\partial\lambda_2^3+\ldots$. Factorization of the solution spaces of operator (78) is harder to follow, as the 6-dimensional subspaces intersect.

4.4. The case $m = 3$

The holonomic system of Theorem 2.1 has rank 108 when $m = 3$. A Gröbner basis computation without $\partial/\partial x$ is fast on Maple 18 (with respect to a total degree ordering in $\partial/\partial\lambda_k$'s, on a 2.8GHz MacBook

Pro of 2014). The lowest order operator in the $\partial/\partial\lambda_k$'s is of order 5. Allowing $\partial/\partial x$, it is equivalent to

$$\frac{\partial^4}{\partial x\partial\lambda_1\partial\lambda_2\partial\lambda_3}+3\frac{\partial^3}{\partial\lambda_1\partial\lambda_2\partial\lambda_3}-\frac{\partial^2}{\partial\lambda_1\partial\lambda_2}-\frac{\partial^2}{\partial\lambda_1\partial\lambda_3}-\frac{\partial^2}{\partial\lambda_2\partial\lambda_3}. \tag{93}$$

This expression is comparable with (76). A Gröbner basis computation with $\partial/\partial x$ leads to rapid increase of memory usage, 4GB in a few minutes.

Replacing $\mathcal{T}_2[y]$ by (58) gives the same LCLM operators, but a different holonomic system of rank 108. A similar Gröbner basis computation without $\partial/\partial x$ takes about 8 minutes. The lowest order operator is

$$\begin{aligned}\sum_{k=1}^{3}\Big(\lambda_k^2\frac{\partial^4}{\partial\lambda_k^4}+(2n-2-\lambda_k)\frac{\partial^3}{\partial\lambda_k^3}+(n^2-3n+2-(x+2n)\lambda_k)\frac{\partial^2}{\partial\lambda_k^2}\\ +(x\lambda_k-(n-2)(x+n+1))\frac{\partial}{\partial\lambda_k}\Big)+(3n-2)x.\end{aligned} \tag{94}$$

Combining both holonomic systems leads to formidable Gröbner basis computation, apparently. Computation of differential operators for the rank $\leqslant 26$ system of Theorem 2.3*(iv)* is barely viable on Singular 4 (given several hours), but further manipulation is hard.

Conjecture 2.4 was checked for $m=4$ by expanding both sides of (20) in the determinants of

$$\begin{pmatrix}1&1&1\\ \lambda_1&\lambda_2&\lambda_3\\ \lambda_1^2&\lambda_2^2&\lambda_3^2\end{pmatrix},\ \begin{pmatrix}1&1&1\\ \lambda_1&\lambda_2&\lambda_3\\ \lambda_1^3&\lambda_2^3&\lambda_3^3\end{pmatrix},\ \begin{pmatrix}1&1&1\\ \lambda_1^2&\lambda_2^2&\lambda_3^2\\ \lambda_1^3&\lambda_2^3&\lambda_3^3\end{pmatrix},\ \begin{pmatrix}\lambda_1&\lambda_2&\lambda_3\\ \lambda_1^2&\lambda_2^2&\lambda_3^2\\ \lambda_1^3&\lambda_2^3&\lambda_3^3\end{pmatrix},$$

etc., and then comparing the coefficients to these four determinants. For example, comparison of the first coefficient by Lemma 4.1 gives

$$\begin{aligned}&\frac{1}{2(n-2)^2(n-1)}\frac{\partial}{\partial x}\det\begin{pmatrix}\gamma(n,x)&\gamma(n-1,x)&\gamma(n-2,x)\\ \gamma(n+1,x)&\gamma(n,x)&\gamma(n-1,x)\\ \gamma(n+2,x)&\gamma(n+1,x)&\gamma(n,x)\end{pmatrix}\\ &=C(x)\det\begin{pmatrix}1&\frac{x}{n-2}&\frac{x^2}{2(n-2)(n-1)}\\ \widetilde{c}_0^{(2)}&\widetilde{c}_1^{(2)}&\widetilde{c}_2^{(2)}\\ c_0^{(3)}&c_1^{(3)}&c_2^{(3)}\end{pmatrix},\end{aligned}$$

where $\widetilde{c}_k^{(2)}$ are the shifted $n \mapsto n-1$ versions of $c_k^{(2)}$ in (87), and $c_k^{(3)}$ are the first coefficients of the expansion of $G_{n,3}(x,y)$:

$$\begin{aligned} c_0^{(3)} &= -nx + n(n-3) - n\left((x-n)^2 + 4x - 3n + 2\right)x^{-n}\mathbf{e}^x\gamma(n,x), \\ c_1^{(3)} &= -nx + n(n-1) - n\left((x-n)^2 + 2x - n\right)x^{-n}\mathbf{e}^x\gamma(n,x), \\ c_2^{(3)} &= \frac{-nx + n(n+1)}{2} - \frac{n}{2}\left((x-n)^2 + n\right)x^{-n}\,\mathbf{e}^x\gamma(n,x). \end{aligned} \tag{95}$$

For an intermediate check, here is a quadratic expression in $A=\gamma(n-1,x)$ and $E = x^{n-1}\mathbf{e}^{-x}$ for the first coefficient:

$$\begin{aligned} \frac{x^{n-3}\mathbf{e}^{-x}}{n-2}\Big(&-\left((x-n+1)^2A + (x-n)E\right)^2 + (n-1)(n-1-4x)A^2 \\ &+ 2(x^2+2x-n^2+n)AE + 2(x+n)E^2\Big). \end{aligned} \tag{96}$$

To set up $C(x)$ in the conjecture, we compared the similar first coefficients for $m = 4$ as well.

Acknowledgement

This work was supported by the JSPS Grant-in-Aid for Scientific Research No. 25220001.

References

[1] A. G. Constantine. Some non-central distribution problems in multivariate analysis. *The Annals of Mathematical Statistics*, 34(4):1270–1285, 1963.

[2] H. Hashiguchi, Y. Numata, N. Takayama, and A. Takemura. The holonomic gradient method for the distribution function of the largest root of a Wishart matrix. *Journal of Multivariate Analysis*, 117:296–312, 2013.

[3] A. T. James. Distributions of matrix variates and latent roots derived from normal samples. *The Annals of Mathematical Statistics*, 35:475–501, 1964.

[4] M. Kang and M.-S. Alouini. Largest eigenvalue of complex Wishart matrices and performance analysis of MIMO MRC systems. *IEEE Journal on Selected Areas in Communications*, 21(3):418–426, 2003.

[5] R. J. Muirhead. Systems of partial differential equations for hypergeometric functions of matrix argument. *The Annals of Mathematical Statistics*, 41(3):991–1001, 1970.

[6] A. J. Paulraj, D. A. Gore, R. U. Nabar, and H. Bolcskei. An overview of MIMO communications — A key to gigabit wireless. *Proc. of the IEEE*, 92(2):198–218, 2004.

[7] C. Siriteanu, A. Takemura, S. Kuriki, D. Richards, and H. Shin. Schur complement based analysis of MIMO zero-forcing for Rician fading. *IEEE Transactions on Wireless Communications*, 14(4):1757–1771, Apr 2015.

[8] C. Siriteanu, A. Takemura, S. Kuriki, H. Shin, and C. Koutschan. MIMO zero-forcing performance evaluation using the holonomic gradient method. *IEEE Transactions on Wireless Communications*, 14(4):2322–2335, Apr 2015.

[9] N. M. Temme. *Special functions: An introduction to the classical functions of mathematical physics*. John Wiley Sons Inc., New York, 1996.

[10] M. van der Put and M. F. Singer. *Galois theory of linear differential equations*. Fundamental Principles of Mathematical Sciences, No 328. Berlin, New York: Springer-Verlag, 2003.

[11] Wikipedia. https://en.wikipedia.org: Complex normal distribution; Noncentral chi-squared distribution; Pochhammer symbol; Schur polynomial; Weyl algebra.

(Raimundas Vidunas) *Department of Pure and Applied Mathematics, Graduate School of Information Science and Technology, Osaka University, 1-5 Yamadaoka, Suita-shi, Osaka 565-0871, Japan*
E-mail address: rvidunas@gmail.com

(Akimichi Takemura) *Faculty of Data Science, Shiga University, 1-1-1 Banba, Hikone-shi, Shiga 522-8522, Japan*
E-mail address: akimichi.takemura@gmail.com

Advanced Studies in Pure Mathematics

A SERIES OF UP-TO-DATE GUIDES OF LASTING INTEREST TO ADVANCED MATHEMATICS

Volume 1 Algebraic Varieties and Analytic Varieties.
Edited by S. Iitaka. February, 1983

Volume 2 Galois Groups and their Representations.
Edited by Y. Ihara. December, 1983

Volume 3 Geometry of Geodesics and Related Topics.
Edited by K. Shiohama. June, 1984

Volume 4 Group Representations and Systems of Differential Equations.
Edited by K. Okamoto. March, 1985

Volume 5 Foliations.
Edited by I. Tamura. February, 1986

Volume 6 Algebraic Groups and Related Topics.
Edited by R. Hotta. March, 1985

Volume 7 Automorphic Forms and Number Theory.
Edited by I. Satake. February, 1986

Volume 8 Complex Analytic Singularities.
Edited by T. Suwa and P. Wagreich. February, 1987

Volume 9 Homotopy Theory and Related Topics.
Edited by H. Toda. February, 1987

Volume 10 Algebraic Geometry, Sendai, 1985.
Edited by T. Oda. July, 1987

Volume 11 Commutative Algebra and Combinatorics.
Edited by M. Nagata and H. Matsumura.
October, 1987

Volume 12 Galois Representations and Arithmetic Algebraic Geometry.
Edited by Y. Ihara. November, 1987

Volume 13 Investigations in Number Theory.
Edited by T. Kubota. March, 1988

Volume 14 Representations of Lie Groups, Kyoto, Hiroshima, 1986.
Edited by K. Okamoto and T. Oshima. February, 1989

Volume 15 Automorphic Forms and Geometry of Arithmetic Varieties.
Edited by K. Hashimoto and Y. Namikawa. July, 1989

Volume 16 Conformal Field Theory and Solvable Lattice Models.
Edited by M. Jimbo, T. Miwa and A. Tsuchiya.
June, 1988

Volume 17 Algebraic Number Theory—in honor of K. Iwasawa.
Edited by J. Coates, R. Greenberg, B. Mazur and I. Satake. August, 1989

Volume 18 I—Recent Topics in Differential and Analytic Geometry.
Edited by T. Ochiai. November, 1990
II—Kähler Metric and Moduli Spaces.
Edited by Edited by T. Ochiai. November, 1990

Volume 19 Integrable Systems in Quantum Field Theory and Statistical Mechanics.
Edited by M. Jimbo, T. Miwa and A. Tsuchiya.
November, 1989

Volume 20 Aspects of Low Dimensional Manifolds.
Edited by Y. Matsumoto and S. Morita.
December, 1992

Volume 21 Zeta Functions in Geometry.
Edited by N. Kurokawa and T. Sunada.
December, 1992

Volume 22 Progress in Differential Geometry.
Edited by K. Shiohama. September, 1993

Volume 23 Spectral and Scattering Theory and Applications.
Edited by K. Yajima. February, 1994

Volume 24 Progress in Algebraic Combinatorics.
Edited by E. Bannai and A. Munemasa. May, 1996

Volume 25 CR-Geometry and Overdetermined Systems.
Edited by T. Akahori, G. Komatsu, K. Miyajima, M. Namba and K. Yamaguchi. July, 1997

Volume 26 Analysis on Homogeneous Spaces and Representation Theory of Lie Groups, Okayama-Kyoto.
Edited by M. Kashiwara, T. Kobayashi, T. Matsuki, K. Nishiyama and T. Oshima. April, 2000

Volume 27 Arrangements – Tokyo 1998.
Edited by M. Falk and H. Terao. August, 2000

Volume 28 Combinatorial Methods in Representation Theory.
Edited by M. Kashiwara, K. Koike, S. Okada, I. Terada, and H.-F. Yamada. December, 2000

Volume 29 Singularities – Sapporo 1998.
Edited by J.-P. Brasselet and T. Suwa.
December, 2000

Volume 30 Class Field Theory – Its Centenary and Prospect.
Edited by K. Miyake. April, 2001

Volume 31 Taniguchi Conference on Mathematics Nara '98.
Edited by M. Maruyama and T. Sunada. May, 2001

Volume 32 Groups and Combinatorics — in memory of Michio Suzuki.
Edited by E. Bannai, H. Suzuki, H. Yamaki and T. Yoshida. October, 2001

Volume 33 Computational Commutative Algebra and Combinatorics.
Edited by T. Hibi. February, 2002

Volume 34 Minimal Surfaces, Geometric Analysis and Symplectic Geometry.
Edited by K. Fukaya, S. Nishikawa and J. Spruck.
June, 2002

Volume 35 Higher Dimensional Birational Geometry.
Edited by S. Mori and Y. Miyaoka. July, 2002

Volume 36 Algebraic Geometry 2000, Azumino.
Edited by S. Usui, M. Green, L. Illusie, K. Kato, E. Looijenga, S. Mukai and S. Saito. November, 2002

Volume 37 Lie Groups, Geometric Structures and Differential Equations — One Hundred Years after Sophus Lie —.
Edited by T. Morimoto, H. Sato and K. Yamaguchi.
December, 2002

Volume 38 Operator Algebras and Applications.
Edited by H. Kosaki. January, 2004

Volume 39 Stochastic Analysis on Large Scale Interacting Systems.
Edited by T. Funaki and H. Osada. March, 2004

Volume 40 Representation Theory of Algebraic Groups and Quantum Groups.
Edited by T. Shoji, M. Kashiwara, N. Kawanaka, G. Lusztig and K. Shinoda. March, 2004

Volume 41 Stochastic Analysis and Related Topics in Kyoto
In honour of Kiyosi Itô.
Edited by H. Kunita, S. Watanabe and Y. Takahashi.
March, 2004

Volume 42 Complex Analysis in Several Variables
– Memorial Conference of Kiyoshi Oka's Centennial Birthday, Kyoto/Nara 2001.
Edited by K. Miyajima, M. Furushima, H. Kazama, A. Kodama, J. Noguchi, T. Ohsawa, H. Tsuji and T. Ueda. June, 2004

Volume 43 Singularity Theory and Its Applications.
Edited by S. Izumiya, G. Ishikawa, H. Tokunaga, I. Shimada and T. Sano. January, 2007

Volume 44 Potential Theory in Matsue.
Edited by H. Aikawa, T. Kumagai, Y. Mizuta and N. Suzuki. August, 2006

Volume 45 Moduli Spaces and Arithmetic Geometry (Kyoto, 2004).
Edited by S. Mukai, Y. Miyaoka, S. Mori, A. Moriwaki and I. Nakamura. January, 2007

Volume 46 Singularities in Geometry and Topology 2004.
Edited by J.-P. Brassele and T. Suwa. February, 2007

Volume 47 Asymptotic Analysis and Singularities.
1—Hyperbolic and dispersive PDEs and fluid mechanics.
2—Elliptic and parabolic PDEs and related problems.
Edited by H. Kozono, T. Ogawa, K. Tanaka, Y. Tsutsumi and E. Yanagida. October, 2007

Volume 48 Finsler Geometry, Sapporo 2005
— In Memory of Makoto Matsumoto.
Edited by S. V. Sabau and H. Shimada.
November, 2007

Volume 49 Probability and Number Theory — Kanazawa 2005.
Edited by S. Akiyama, K. Matsumoto, L. Murata and H. Sugita. December, 2007

Volume 50 Algebraic Geometry in East Asia — Hanoi 2005.
Edited by K. Konno and V. Nguyen-Khac.
February, 2008

Volume 51 Surveys on Geometry and Integrable Systems.
Edited by M. Guest, R. Miyaoka and Y. Ohnita.
November, 2008

Volume 52 Groups of Diffeomorphisms
in honor of Shigeyuki Morita
on the occasion of his 60th birthday.
Edited by R. Penner, D. Kotschick, T. Tsuboi,
N. Kawazumi, T. Kitano and Y. Mitsumatsu.
November, 2008

Volume 53 Advances in Discrete Dynamical Systems.
Edited by S. Elaydi, K. Nishimura, M. Shishikura and
N. Tose. June, 2009

Volume 54 Algebraic Analysis and Around
in honor of Professor Masaki Kashiwara's 60th birthday.
Edited by T. Miwa, A. Matsuo, T. Nakashima and
Y. Saito. February, 2009

Volume 55 Noncommutativity and Singularities
Proceedings of French–Japanese symposia held at IHÉS in 2006.
Edited by J.-P. Bourguignon, M. Kotani, Y. Maeda and
N. Tose. July, 2009

Volume 56 Singularities — Niigata–Toyama 2007.
Edited by J.-P. Brasselet, S. Ishii, T. Suwa and
M. Vaquie. November, 2009

Volume 57 Probabilistic Approach to Geometry.
Edited by M. Kotani, M. Hino and T. Kumagai.
March, 2010

Volume 58 Algebraic and Arithmetic Structures of Moduli Spaces
(Sapporo 2007).
Edited by I. Nakamura and L. Weng. June, 2010

Volume 59 New Developments in Algebraic Geometry, Integrable
Systems and Mirror Symmetry (RIMS, Kyoto, 2008).
Edited by M.-H. Saito, S. Hosono and K. Yoshioka.
August, 2010

Volume 60 Algebraic Geometry in East Asia — Seoul 2008.
Edited by J. Keum, S. Kondō, K. Konno and K. Oguiso.
September, 2010

Volume 61 Exploring New Structures and Natural Constructions
in Mathematical Physics.
Edited by K. Hasegawa, T. Hayashi, S. Hosono and
Y. Yamada. October, 2011

Volume 62 Arrangements of Hyperplanes—Sapporo 2009.
Edited by H. Terao and S. Yuzvinsky. March, 2012

Volume 63 Galois–Teichmüller Theory and Arithmetic Geometry.
Edited by H. Nakamura, F. Pop, L. Schneps and A. Tamagawa. October, 2012

Volume 64 Nonlinear Dynamics in Partial Differential Equations.
Edited by S.-I. Ei, S. Kawashima, M. Kimura and T. Mizumachi. April, 2015

Volume 65 Algebraic Geometry in East Asia — Taipei 2011.
Edited by J. A. Chen, M. Chen, Y. Kawamata and J. Keum. April, 2015

Volume 66 Singularities in Geometry and Topology 2011.
Edited by V. Blanlœil and O. Saeki. May, 2015

Volume 67 Variational Methods for Evolving Objects.
Edited by L. Ambrosio, Y. Giga, P. Rybka and Y. Tonegawa. June, 2015

Volume 68 School on Real and Complex Singularities in São Carlos, 2012.
Edited by O. Saeki, R. N. Araújo dos Santos, T. Nishimura and V. H. Jorge Pérez. July, 2016

Volume 69 Development of Moduli Theory—Kyoto 2013.
Edited by O. Fujino, S. Kondō, A. Moriwaki, M.-H. Saito and K. Yoshioka. July, 2016

Volume 70 Minimal Models and Extremal Rays (Kyoto, 2011).
Edited by J. Kollár, O. Fujino, S. Mukai and N. Nakayama. September, 2016

Volume 71 Schubert Calculus — Osaka 2012.
Edited by H. Naruse, T. Ikeda, M. Masuda and T. Tanisaki. December, 2016

Volume 72 Geometry, Dynamics, and Foliations 2013
In honor of Steven Hurder and Takashi Tsuboi on the occasion of their 60th birthdays.
Edited by T. Asuke, S. Matsumoto and Y. Mitsumatsu. August, 2017

Volume 73 Hyperbolic Geometry and Geometric Group Theory.
Edited by K. Fujiwara, S. Kojima and K. Ohshika. September, 2017

Volume 74 Higher Dimensional Algebraic Geometry
in honour of Professor Yujiro Kawamata's sixtieth birthday.
Edited by K. Oguiso, C. Birkar, S. Ishii and S. Takayama. November, 2017

Volume 75 Algebraic Varieties and Automorphism Groups.
Edited by K. Masuda, T. Kishimoto, H. Kojima,
M. Miyanishi and M. Zaidenberg. December, 2017

Volume 76 Representation Theory, Special Functions and Painlevé
Equations — RIMS 2015.
Edited by H. Konno, H. Sakai, J. Shiraishi, T. Suzuki and
Y. Yamada. June, 2018

Volume 77 The 50th Anniversary of Gröbner Bases.
Edited by T. Hibi. August, 2018

To be continued